## Exponents (See 3.4A and 3.4B)

$(a, b > 0)$

$$a^{x+y} = a^x \cdot a^y$$

$$(a^x)^y = a^{xy}$$

$$(ab)^x = a^x b^x$$

$$a^{-x} = \frac{1}{a^x}, \qquad a^0 = 1$$

## Logarithms (See 3.4C)

$(a, b, y > 0)$

$y = \log_a x$ means $a^y = x$

$$\log_a x_1 x_2 = \log_a x_1 + \log_a x_2$$

$$\log_a \frac{x_1}{x_2} = \log_a x_1 - \log_a x_2$$

$$\log_a x^r = r \log_a x$$

$$\log_a b = \frac{\log_c b}{\log_c a}$$

$$\log_a a = 1$$

$$\ln x = \log_e x \quad (e = 2.71828\ldots)$$

## TRIGONOMETRY

$$\sin \theta = \frac{b}{r} \qquad \csc \theta = \frac{r}{b} = \frac{1}{\sin \theta}$$

$$\cos \theta = \frac{a}{r} \qquad \sec \theta = \frac{r}{a} = \frac{1}{\cos \theta}$$

$$\tan \theta = \frac{b}{a} \qquad \cot \theta = \frac{a}{b} = \frac{1}{\tan \theta}$$

Law of cosines

$$z^2 = x^2 + y^2 - 2xy \cos \theta$$

reduces to the

Pythagorean theorem $z^2 = x^2 + y^2$

when $\theta = \frac{\pi}{2}$

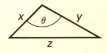
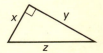

| Degrees | 0° | 30° | 45° | 60° | 90° | 120° | 135° | 150° | 180° | 210° | 225° | 240° | 270° | 300° | 315° | 330° | 360° |
|---------|----|-----|-----|-----|-----|------|------|------|------|------|------|------|------|------|------|------|------|
| Radians | 0 | $\frac{1}{6}\pi$ | $\frac{1}{4}\pi$ | $\frac{1}{3}\pi$ | $\frac{1}{2}\pi$ | $\frac{2}{3}\pi$ | $\frac{3}{4}\pi$ | $\frac{5}{6}\pi$ | $\pi$ | $\frac{7}{6}\pi$ | $\frac{5}{4}\pi$ | $\frac{4}{3}\pi$ | $\frac{3}{2}\pi$ | $\frac{5}{3}\pi$ | $\frac{7}{4}\pi$ | $\frac{11}{6}\pi$ | $2\pi$ |
| $\sin \theta$ | 0 | $\frac{1}{2}$ | $\frac{\sqrt{2}}{2}$ | $\frac{\sqrt{3}}{2}$ | 1 | $\frac{\sqrt{3}}{2}$ | $\frac{\sqrt{2}}{2}$ | $\frac{1}{2}$ | 0 | $\frac{-1}{2}$ | $\frac{-\sqrt{2}}{2}$ | $\frac{-\sqrt{3}}{2}$ | $-1$ | $\frac{-\sqrt{3}}{2}$ | $\frac{-\sqrt{2}}{2}$ | $\frac{-1}{2}$ | 0 |
| $\cos \theta$ | 1 | $\frac{\sqrt{3}}{2}$ | $\frac{\sqrt{2}}{2}$ | $\frac{1}{2}$ | 0 | $\frac{-1}{2}$ | $\frac{-\sqrt{2}}{2}$ | $\frac{-\sqrt{3}}{2}$ | $-1$ | $\frac{-\sqrt{3}}{2}$ | $\frac{-\sqrt{2}}{2}$ | $\frac{-1}{2}$ | 0 | $\frac{1}{2}$ | $\frac{\sqrt{2}}{2}$ | $\frac{\sqrt{3}}{2}$ | 1 |
| $\tan \theta$ | 0 | $\frac{\sqrt{3}}{3}$ | 1 | $\sqrt{3}$ | $\infty$ | $-\sqrt{3}$ | $-1$ | $\frac{-\sqrt{3}}{3}$ | 0 | $\frac{\sqrt{3}}{3}$ | 1 | $\sqrt{3}$ | $\infty$ | $-\sqrt{3}$ | $-1$ | $\frac{-\sqrt{3}}{3}$ | 0 |

## Identities

$$\sin^2 \theta + \cos^2 \theta = 1$$

$$\tan^2 \theta + 1 = \sec^2 \theta$$

$$\cot^2 \theta + 1 = \csc^2 \theta$$

$$\sin 2\theta = 2 \sin \theta \cos \theta$$

$$\cos 2\theta = \cos^2 \theta - \sin^2 \theta$$

$$\sin (\alpha \pm \beta) = \sin \alpha \cos \beta \pm \cos \alpha \sin \beta$$

$$\cos (\alpha \pm \beta) = \cos \alpha \cos \beta \mp \sin \alpha \sin \beta$$

$$\sin^2 \theta = \tfrac{1}{2}(1 - \cos 2\theta)$$

$$\cos^2 \theta = \tfrac{1}{2}(1 + \cos 2\theta)$$

Freshman Calculus

**Robert A. Bonic**

**Edgar DuCasse**
Brooklyn College

**Gabriel Vahan Hajian**
Draper Laboratory, M.I.T.

**Martin M. Lipschutz**
New York University

Freshman CALCULUS

Second
Edition

**D. C. Heath and Company**
Lexington, Massachusetts Toronto London

# Foreword to the First Edition

To follow this book you should start on the left-hand side of each page. Whenever a symbol, word, phrase, formula, or sentence appears in color it means that there is a comment about it that appears directly to the right. After reading the comment move back to the left and continue.

### foreword to the student
### by the fifteen student authors

This book was written so that you might have the easiest and most enjoyable time learning calculus. Often in class you get the impression that the instructor doesn't really understand the difficulties you have in learning the material. The result is a drag for everyone, especially you, and that is who really counts. Our book was written to combat this problem, and we determined its emphasis throughout. If we thought a particular topic was difficult then more space was given to it. When we felt that too much space was being spent on

### comments mainly to the instructor
### by Robert A. Bonic

We assume your students to have a tentative (perhaps only fragile) interest in learning calculus. Thus besides presenting the necessary facts we have attempted to write a book that is interesting and encouraging;

How this book was written by fifteen students, an engineer, and a mathematician is described in the appendix. Basically I determined what was to be written and the engineer added a practical flavor. The students then decided on how it would be written, for they were the real authorities on the problems of learning calculus.

v

some fine point, its importance was minimized. We are not a bunch of junior mathematicians or quiz kids . . . we are people just like you. If you had been at UCSC in the fall of 1968 you might have ended up working on this book too. Our job was simply to be honest enough to say something when we were confused by the presentation, and possibly offer an alternative. We hope that the result of our confusion (and often hysteria) is a book that is near to hassle-free. It is meant to be interesting and maybe even fun. After all, fifteen red-blooded students could not spend a year working on a book without having a good time and having it show in their work. The result is a textbook with a personality, and the whole idea is that calculus doesn't have to be a bore . . . it is really kind of a trippy subject.

The most obvious difference between our book and standard calculus texts is the divided page format. The left side of the page together with material on the right that appears in black is called the text. Here you will find the central ideas, major results, and principal applications of calculus. On the right-hand side of the page, in color print, are the comments. They provide a running commentary about events occurring on the left.

The comments are of various types. Perhaps our most individualized contributions are the signed student comments which definitely should be read. One might answer a question you have, while another may help clarify for you something that was unclear to one of us. Some may be humorous, and at times we may only offer you our sympathy. The remaining comments were written by Professor Bonic, Vahan (Hajian), or other mathematicians. Those with a shaded background are optional and need not be studied on a first reading. They deal with theoretical matters, and may only be of value to you after you are very familiar with the material.

At fourteen places in the book you will find history sections written across the entire page. These will provide you with a better perspective of the subject and an appreciation of some of the men who made it their life's work.

Each section closes with three types of exercises: B (basic), D (drill), and M (miscellaneous). The basic exercises make up a reasonable assignment for the

→ Most of the theory has been postponed until the last two chapters, and in the first eight the approach is intuitive. For example, limits are merely illustrated by examples, the derivative is introduced geometrically, and the definite integral is based on the "intuitively obvious" notion of area. The emphasis is on problem solving and the techniques of using calculus.

→ We learned very early that the student's knowledge of algebra, geometry, and trigonometry needed strengthening. Some of this material is reviewed in the first chapter, and in the next two chapters we spend considerable time computing derivatives. This allows the student time to practice his algebra, and to review the basic properties about the trigonometric, logarithmic, and exponential functions.

→ There is more than one year's worth of material in the book and it can be used in a flexible way. Three different courses are described in the table below. In contrast to the first eight chapters, the last two are exceptionally rigorous. Calculus is redone, but the emphasis now is on its theoretical aspects.

|  | Term equals | Term 1 | Term 2 |
|---|---|---|---|
| Basic course | One quarter 3 hours/week | Chap. 1, 2, 3 | Chap. 4, 5 Sec. 6.1, 6.2, 6.3 |
| Science oriented | One quarter 4 hours/week | Chap. 1, 2, 3, 4 | Chap. 5, 6, 7 Sec. 8.1, 8.2, 8.3 |
| Honors course | One semester 4 hours/week | Basic course | Chap. 9, 10 |

→ It was often difficult to decide whether or not a particular comment should be shaded. Depending on the level of your class, you may wish to shade some more and unshade others. However, to include all the material in the shaded comments would significantly increase the level of the book. In a sense, the nonshaded portion contains the information that "the student is expected to know for the exam."

→ Another reason for including the history sections is to provide evidence for presenting calculus in a more historical fashion. Mathematics is often written in a way that is exactly opposite to its manner of discovery. This tends to make the presentation elegant and logically neat, but it disguises the central ideas, obscures the reasoning, and makes the subject much more difficult to understand.

→ The miscellaneous exercises begin at the level of the basic exercises and then tend to get more difficult. They can be used to supplement the basic assignment. Answers to all the exer-

section. If you are having difficulties with them, then do the drill exercises. They are routine, but solving them will help develop your skill and confidence. Do not go on to the next section until you can do them with ease.

Besides the exercises, the book also contains problems. These should be done as they appear, for they test your understanding of a concept just learned. We have provided space for you to write down your solutions, but do not confine yourself to this. Feel free to write whatever you find useful wherever you find room. In this way the book will transform itself from being ours to being yours. Sample exams appear at the end of each of the first eight chapters. They probably will be similar to the exams you will have in your course and will help you in preparing for them. We advise you to do some of them as a review, but also as a means of testing your working time.

The reference information on the front inside cover and flyleaf is material that you should already know. What appears on the back inside cover and flyleaf are facts that we hope you will know when you finish the book.

If you have any thoughts about our book, or if you find certain points still confusing, we would appreciate it if you would write to Professor Robert Bonic or to one of us in care of him. We hope to improve the book in a second edition and invite you to offer your own signed comments.

We students who have sweated over this book hope you have the most enjoyable time possible learning calculus this year.

*Santa Cruz, California*

cises are given in the back of the book, and solutions to the basic exercises can be found in the companion paperback volume entitled *Exercises and Sample Exams for Freshman Calculus.*

→ These problems are distinct from the examples which are worked out in the book. You may wish to use the examples to illustrate the material, but we suggest that you let your students find their own solutions to the problems.

→ Within each chapter the first exam is quite easy while the fourth is fairly difficult. The answers appear in the back of the book and solutions for some of the exams are given in the exercise manual.

→ We do not think it is necessary for students to memorize any of this reference material. In fact, if they are allowed to use this information during examinations, it will free them to concentrate on the concepts and in reasoning properly.

→ Throughout the writing of this book I have had a great deal of excellent advice, and my acknowledgments will be found in the Appendix. However, I would here like to thank Alan, Beth, Dana, Ed, Estelle, Fred, Greg, Jean, Kent, Lee, Mark, Pat, Bob, Robin, and Steve. Individually, they contributed sparkling ideas, and as a group their influence on the book was profound. They challenged everything I said, and inspired my teaching. This book is different from other texts, and if your students find it enjoyable then they also will be indebted to the student authors.

*Boston, Massachusetts*

# Foreword to the Second Edition

Revising a successful textbook is probably still more of an art than a science. One doesn't want to tamper too much with success, but at the same time there are obviously few books, especially textbooks, which can't stand some improvement. Our approach, therefore, in revising *Freshman Calculus* has been a pragmatic one. We have retained that which seemed to work the first time, we fixed that which needed to be fixed or added to, and we made every effort to eliminate old errors and introduce no new ones. The result of this process is a moderate revision.

The changes in this edition are, to a large extent, the result of suggestions from users of the first edition. The most important changes are the addition of a section on limits and continuity in Chapter 2, the addition of a section on differentials in the last section of Chapter 3, a new treatment of the definite integral in Chapter 5, and the expansion of the exercise sets throughout the book. We have also reorganized and rewritten the softbound student book, now called *Studying Calculus,* that accompanies the text in an effort to make it more useful.

The new section on limits and continuity first defines the limit geometrically in terms of open intervals with centers at a point. The epsilon-delta definition then follows in a natural way. The new paragraphs on differentials include applications to approximations.

In Chapter 5, the opening two sections have been rearranged. The first now treats only antiderivatives, and the second covers the definite integral. The definite integral is now defined as a limit of sums rather than in terms of the antiderivative. A proof of the Fundamental Theorem of Calculus is now included.

Throughout the exercise sets, problems in applied mathematics have been added, primarily to the M problems, which have also been rearranged so that they now appear in ascending order of difficulty. Some additional easy-to-medium level of difficulty problems have been added to the B exercises. The revised student book now includes solutions to selected D problems as well as solutions to all B problems and a few of the more interesting M problems. We anticipate that *Studying Calculus* will now be more valuable to the student in need of assistance outside the classroom as well as to the student who is interested in self-study.

Naturally, this revision would not have been possible without the help of many people. In particular, we would like to acknowledge the contributions of Robert Appleson, Gilbert Sampson, Keyhang Keem, Lawrence Cox, and Inda Lepson, who commented upon their experiences with the first edition. The manuscript was ably reviewed by Dorothy Ryan of Bunker Hill Community College, Robert Nowlan of Southern Connecticut State College, Willard Parker of Kansas State University, Leonard Shapiro of The University of Minnesota, and Leonard Evens of Northwestern.

A special thanks is due Philip Gillett of The University of Wisconsin, Marathon County Center who reviewed the entire manuscript in great detail and whose comments were especially helpful. Also, of course, we'd like to extend a grateful thanks to the staff of D. C. Heath for their support and assistance.

Edgar DuCasse
Martin M. Lipschutz

# Contents

Students enter college with widely varying skills and talents in mathematics. The purpose of this introductory chapter is to review those branches of mathematics that are essential to the study of calculus, which begins in Chapter 2. The student should not deceive himself with overconfidence and should read the chapter carefully to learn the definitions and the notations that will be used throughout the remainder of the book.

# Background Material

1

## *The Need for Calculus*

Calculus, compared to the more elementary areas of mathematics such as introductory algebra, geometry, and trigonometry, may be described as the study of variable rather than fixed quantities. In algebra fixed equations and their resulting solutions are considered. In geometry the figures discussed are fixed triangles, rectangles, and so on, and properties such as similarity or congruence of given figures are investigated. In trigonometry, students learn to solve given triangles for unknown sides or angles. In calculus the emphasis is on changing, dynamic properties of systems as opposed to the fixed characteristics of algebraic or geometric entities.

New branches of mathematics generally develop as a response to the needs of some other area of investigation, and indeed this was the case with calculus. To see the types of problems that can be solved using its techniques, we will look briefly at its historical development.

It is a mistake to think that the methods and concepts of calculus were developed from first principles by one or two men. Although Isaac Newton (1642–1727) and Gottfried Wilhelm von Leibniz (1646–1716) are given most of the credit for the development of calculus, some of their ideas and the ideas of their contemporaries were anticipated by almost two thousand years by the ancient Greeks. However, even though the Greeks were in possession of some of the fundamental concepts of calculus, they never developed a unified theory of the subject like that of Newton, Leibniz, and their contemporaries.

In the sixteenth and seventeenth centuries, astronomy and physics were two of the areas whose needs led to the development of calculus. Among these needs was a method to calculate the paths described by projectiles. This method uses techniques of differentiation, one of the two principal branches of calculus. The differential calculus may be described as that area which treats problems requiring the determination of the rate at which a variable quantity is changing. The other principal branch, the integral calculus, deals with somewhat the reverse situation. Its goal is the determination of functions whose rate of change is known. Integral calculus also treats the problem of finding the lengths of curves and the areas and volumes of figures bounded by curves and surfaces. These problems also confronted the seventeenth-century scientists.

Another class of problems facing these scientists was the determination of maximum and minimum values of functions. Some maxima and minima problems can be solved by employing techniques from elementary algebra or geometry. Many problems, however, can only be solved by the more sophisticated methods that are part of the calculus.

From the list of examples presented above, one might have the impression that calculus is of use only in solving problems that faced mankind several hundred years ago. Actually, the number and variety of applications of this important subject are probably greater today than at any time in the past. Calculus is presently used in the prediction of orbits of artificial satellites, the design of radar systems, the development of sophisticated inertial navigation systems, and the implementation of such advanced tools of modern physics as cyclotrons and linear accelerators. Its applications are not restricted to the physical sciences and engineering. In medicine, biology, oceanography, ecology, or any field of human endeavor dealing with the properties of variable quantities, the techniques of calculus will be useful, if not indispensable to the workers in these fields.

## 1.1 *The Real Numbers*

Calculus begins with analytic geometry, and analytic geometry begins with the real numbers. We start by reviewing their properties and establishing certain notations and conventions to be used from now on. The treatment of the real numbers given here is intuitive.

**A. The Real Line**   We will begin with a few remarks concerning the history of the real numbers.

*The Real Numbers*

By the end of the sixteenth century the real numbers had reached a stage of development that was sufficient for the needs of arithmetic, algebra, and geometry. Then calculus was invented, and with it there came into mathematics a great variety of infinite processes. During the seventeenth and eighteenth centuries the theories and applications growing out of calculus dominated mathematics, and it flourished as never before. The infinite processes were used with marvelous success, but the more complicated the theories became, the more evident it was that the understanding of the processes was very rough. In attempting to comprehend them mathematicians were forced to look at the foundations of mathematics and their description of the real numbers. One of the major accomplishments of the nineteenth century was a thorough reexamination and rebuilding of the foundations. When mathematics emerged into the twentieth century it found itself stamped with a new style: rigor. The real numbers were described by a precise set of axioms, and mathematicians were now exceedingly careful in formulating their concepts, giving definitions, and presenting arguments.

Our presentation of the real numbers will be based on their identification with points on a line as follows: Draw a straight line (Fig. 1) and imagine that it extends indefinitely in both directions.

→ We will not attempt to define the words *line*, *straight*, and *indefinitely*, but will simply rely on the reader's common sense, which will suffice for our purposes. The line should be pictured as being continuous, that is, having no holes or gaps.

**Figure 1**

Now choose a point on the line and label it 0. Choose another point to the right of 0 and label it 1. Based on the unit of length which these points determine, the positions of the *integers*

$$\ldots, -3, -2, -1, 0, 1, 2, 3, \ldots$$

are now fixed on the line (Fig. 2).

**Figure 2**

→ The "..." on the right is to be read as "and so on."
The "..." on the left could be read as "and so it was."
(Steve)

The locations of the *rational numbers,* all numbers of the form $\frac{m}{n}$ where $m$ and $n$ are integers and $n$ is not zero, follow (Fig. 3) from those of the integers.

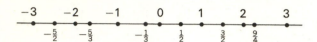

**Figure 3**

The rational numbers, however, do not fill up the line because not all numbers are rational. Those remaining are called *irrational numbers*. We will show below that $\sqrt{2}$ is irrational. The collection of all numbers on the line is called the *real numbers,* and is denoted by $R$.  The line is called the *real line* (Fig. 5).

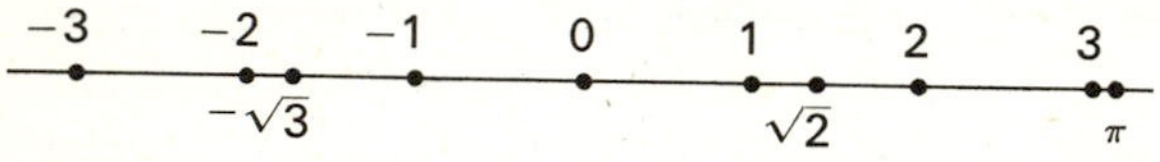

**Figure 5**

We assume that the reader is familiar with the addition and multiplication properties of the real numbers. These are stated in Chapter 9 in the form of axioms. Now we will only remark about division by zero. In the equation

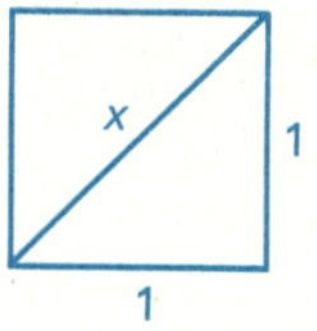

$$ac = bc$$

we may divide out the $c$ to obtain $a = b$ provided $c$ is not zero. An example will show why we cannot allow $c$ to equal zero. When $c = 0$, $a = 7$, and $b = 8$, we have $7 \cdot 0 = 8 \cdot 0$, and dividing out zero yields $7 = 8$.

**THEOREM 1**   $\sqrt{2}$ *is not a rational number.*

(This means that $\sqrt{2}$ cannot be expressed as the ratio of two integers, that is, in the form $\frac{m}{n}$ with $m$ and $n$ integers. The proof shows that if we express $\sqrt{2}$ as the ratio of two integers, a contradiction results.)

 Throughout this book a large period ● will signify the end of a proof. It is to be read as "PERIOD!"

Many early Greeks believed that all numbers were rational, and they became seriously concerned when they realized that the number $x$ representing the length of the diagonal of a unit square could not be described as the ratio of two integers (Fig. 4). Even today fallacious proofs of the rationality of certain nonrational numbers ($\sqrt{2}$ and $\pi$) are produced by certain irrational individuals.

**Figure 4**

In Chapter 9 the real numbers will be described axiomatically. Were we now to discuss the subtleties connected with these foundations, the reader might find his tentative interest in learning calculus extinguished. It is interesting to note that, had our contemporary standards of rigor been imposed upon mathematicians of the sixteenth century, calculus might still belong to the future. Although it is sometimes important to be rigorous, a loose intuitive approach is often more instructive. The presentation in this book will basically be historical and intuitive, but we will often comment on a topic when it will be treated more extensively in the last two chapters. We hope that by the time the reader reaches Chapter 9 he will then be able to appreciate the need to take a more serious look at the foundations of calculus.

**Proof**  Assume that $\qquad \sqrt{2} = \dfrac{m}{n}$

for some integers $m$ and $n$, $n \neq 0$, and that $m$ and $n$ have no common factor. Then $m = \sqrt{2}\, n$, and squaring both sides yields $m^2 = 2n^2$; thus $m^2$ is even. Since the square of an odd number is odd, $m$ must be even. If $n$ were also even, then $m$ and $n$ would both be divisible by 2, contradicting our assumption that they have no common factor. Hence $n$ must be odd; therefore $n^2$ is odd and $2n^2$ is not divisible by 4. But $m^2$ is divisible by 4 because $m$ is even. Thus $m^2 \neq 2n^2$, which is a clear contradiction. Therefore our assumption that $\sqrt{2}$ is rational must be false, and $\sqrt{2}$ is irrational ●

## B. Inequalities, Intervals, and Absolute Value

Given two real numbers $a$ and $b$ we must have either

$$a = b \quad \text{or}$$
$$a < b \quad \text{or}$$
$$a > b.$$

These are read as

"$a$ equals $b$" or
"$a$ is less than $b$" or
→ "$a$ is greater than $b$."

An expression involving $<$, $>$, $\leq$, or $\geq$ is called an → *inequality*. Some of the important properties of inequalities are listed below:

The symbols $\leq$ and $\geq$ stand for "less than or equal to" and "greater than or equal to," respectively. Also, the notation $a < b < c$ means that both $a < b$ and $b < c$ and may be read "$b$ is between $a$ and $c$." The case is similar for $\leq$.

1) $a < b$ implies $-a > -b$
2) $a < b$ and $c > 0$ implies $ac < bc$
3) $ab > 0$ implies either
    $a > 0$ and $b > 0$, or
    $a < 0$ and $b < 0$
4) $ab < 0$ implies either
    $a > 0$ and $b < 0$, or
    $a < 0$ and $b > 0$

*Do not be misled by this list and think that we are now familiar with all the properties of inequalities. For example, we know that $-2 < 2$ and $-\frac{1}{2} < \frac{1}{2}$, however, is it true for any two numbers $a$ and $b$ that if $a < b$ then $\frac{1}{a} < \frac{1}{b}$? Try $a = 2$ and $b = 4$, or $a = -2$ and $b = -4$. If you are interested, try to prove that if $a < b$ then $\frac{1}{a} < \frac{1}{b}$ for $ab < 0$, and $\frac{1}{a} > \frac{1}{b}$ for $ab > 0$.*

*(Vahan)*

The sets to be described below are called *intervals*, and it will be convenient to introduce the notation:

$$\{x \in R: \text{such and such}\},$$

or more simply $\{x: \text{such and such}\}$ to be read as "the set of all real numbers $x$ satisfying such and such." For $a < b$ we denote (Fig. 6):

→ This notation can be used with any set $A$. For example, if $A$ denotes the integers, then

$$\{x \in A: x > 0 \quad \text{and} \quad x \leq 5\} = \{1, 2, 3, 4, 5\}.$$

The Greek letter "$\in$" is read "is an element of."

| | | |
|---|---|---|
| $\{x: a \leq x \leq b\}$ | by | $[a, b]$ |
| $\{x: a < x < b\}$ | by | $(a, b)$ |
| $\{x: a \leq x < b\}$ | by | $[a, b)$ |
| $\{x: a < x \leq b\}$ | by | $(a, b]$ |
| $\{x: a \leq x\}$ | by | $[a, \infty)$ |
| $\{x: a < x\}$ | by | $(a, \infty)$ |
| $\{x: x \leq b\}$ | by | $(-\infty, b]$ |
| $\{x: x < b\}$ | by | $(-\infty, b)$ |
| $\{x: x \text{ is a real number}\}$ | by | $(-\infty, \infty)$ |

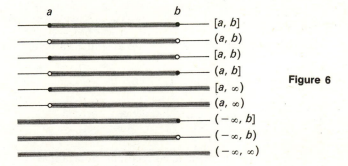

Figure 6

The intervals

$$(a, b), (a, \infty), (-\infty, b), \quad \text{and} \quad (-\infty, \infty)$$

are called *open* intervals, and the intervals

$$[a, b], [a, \infty), (-\infty, b], \quad \text{and} \quad (-\infty, \infty)$$

are called *closed* intervals. Notice that $(-\infty, \infty)$ is both an open and closed interval, whereas $[a, b)$ is neither an open nor closed interval.

*The symbols $\infty$ and $-\infty$ do not represent real numbers.* → *We are using them only to shorten the writing of longer expressions. Their use could be avoided, but experience has shown that their convenience outweighs the confusion that they might engender.*

*(Ed)*

The significant feature of an open interval $I$ is that, given → any point $x \in I$, there are points to the right and left of $x$ in $I$.

*Example 1*  Show that the set of $x$ satisfying both

$$x + 2 \geq 0 \quad \text{and} \quad 2x - 6 < 0$$

is the interval $[-2, 3)$.

*Solution*  The set of $x$ satisfying

$$x + 2 \geq 0 \quad \text{and} \quad 2x - 6 < 0$$

is the same as the set of $x$ satisfying

$$x \geq -2 \quad \text{and} \quad x < 3, \quad \text{or} \quad -2 \leq x < 3,$$

which is the interval $[-2, 3)$. See Figure 7. Observe that the interval $[-2, 3)$ is the intersection of the intervals $[-2, \infty)$ and $(-\infty, 3)$.

**Figure 7**

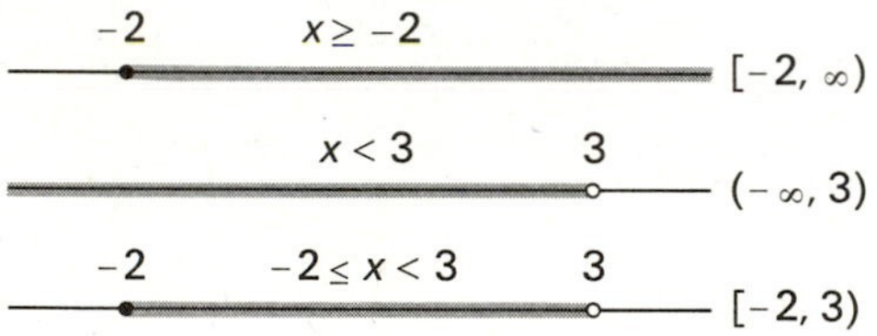

The *absolute value* of a number $a$, denoted by $|a|$, represents its distance from the point 0. Hence

$$|a| = \begin{cases} a & \text{if } a \geq 0 \\ -a & \text{if } a < 0. \end{cases}$$

For example, $|2| = 2$ and $|-2| = -(-2) = 2$.
Some important properties of the absolute value are

1) $|a| = \sqrt{a^2}$
2) $|-a| = |a|$
3) $|ab| = |a|\,|b|$
4) $|a + b| \leq |a| + |b|$

The absolute value is a convenient notation. For example, given any two points $a$ and $b$ on the real line, the

distance between $a$ and $b$ = $|a - b|$

regardless of the order in which they are subtracted. Also certain sets of numbers can be easily described with the absolute value notation. For example, the interval $\{x: -3 < x < 3\}$ can be written as $\{x: |x| < 3\}$. This is evident when $|x|$ is thought of as the

*Example 2*  Find the set of $x$ satisfying the inequality

$$x^2 - x - 2 < 0.$$

*Solution*  The set of $x$ satisfying $x^2 - x - 2 < 0$ is the same as the set of $x$ satisfying $(x - 2)(x + 1) < 0$. Since a product is less than zero if and only if one of the factors is less than zero and the other is greater than zero, $x$ satisfies the given inequality if and only if $x$ satisfies the inequalities

A)  $x - 2 < 0 \quad \text{and} \quad x + 1 > 0$

or the inequalities

B)  $x - 2 > 0 \quad \text{and} \quad x + 1 < 0.$

The set of $x$ satisfying (A) is the same as the set of $x$ satisfying

$$x < 2 \quad \text{and} \quad x > -1,$$

which is the interval $(-1, 2)$. The set of $x$ satisfying (B) is the same as the set of $x$ satisfying

$$x > 2 \quad \text{and} \quad x < -1,$$

which is the empty set. Thus the solution set is the open interval $(-1, 2)$.

→ Proof that $|-a| = |a|$: If $a > 0$, then $-a < 0$ and so $|-a| = -(-a) = a = |a|$. If $a \leq 0$, then $-a \geq 0$, so that $|-a| = -a = |a|$. Thus in all cases $|-a| = |a|$ ●

→ Proof that $|a + b| \leq |a| + |b|$: If $a, b \geq 0$, then $|a + b| = a + b = |a| + |b|$. If $a, b \leq 0$, then $|a + b| = -(a + b) = (-a) + (-b) = |a| + |b|$. If $a \geq 0$ and $b \leq 0$, we have two cases: If $a \geq -b$, then $a + b \geq 0$, so $|a + b| = a + b \leq a = |a| \leq |a| + |b|$. If $a \leq -b$, then $a + b \leq 0$, so $|a + b| = -(a + b) = (-a) + (-b) \leq -b = |b| \leq |a| + |b|$ ●

→ The word *distance* will always be used to refer to a non-negative quantity. Also, whenever the notations $\sqrt{a}$ and $a^{1/2}$ are used, they designate the positive square root.

→ In general for $d > 0$ the statement $|x| < d$ means $-d < x < d$. The statement $|x| > d$ means $x > d$ or $x < -d$.

distance between $x$ and 0. The set $\{x: |x| > 3\}$ is the set of $x$ whose distance from 0 is greater than 3, which is the set $\{x: x > 3 \text{ or } x < -3\}$. More complicated but typical examples are illustrated below.

---

*Example 3*   Show that the set of $x$ satisfying $|x - 2| < 3$ is equal to the interval $(-1, 5)$.

*Solution*   The set of $x$ satisfying

$$|x - 2| < 3$$

is the same as the set of $x$ satisfying

$$-3 < x - 2 < 3$$

or

$$-1 < x < 5,$$

which is the interval $(-1, 5)$, as shown in Figure 8. Observe that this is the set of $x$ whose distance from 2, the midpoint of the interval, is less than 3.

---

*Example 4*   Shade the points on the axis which satisfy $|x - 2| \geq 3$.

*Solution*   The set of $x$ satisfying

$$|x - 2| \geq 3$$

is the same as the set of $x$ satisfying

$$x - 2 \geq 3 \quad \text{or} \quad x - 2 \leq -3,$$

that is,

$$x \geq 5 \quad \text{or} \quad x \leq -1.$$

This is the set of $x$ whose distance from 2 is greater than or equal to 3, as shown in Figure 9.

---

**Figure 8**

**Figure 9**

## Evolution of the Number Concept

The concept of number and the birth of mathematics were coincident. Around 3500 B.C. the Sumerians had elementary hieroglyphs for the rational numbers (except zero). By the time of the Babylonians a highly developed arithmetic had evolved. It was inherited by the Greeks who, in spite of their brilliant insights in geometry, did little to further it. Perhaps their most significant contribution to arithmetic was to show that the length of the diagonal of a square with side of unit length is irrational. The Greeks had thought that all numbers were rational and the existence of irrationals was not warmly received. One school of geometry, the Pythagoreans, outlawed divulgence of this fact under penalty of death.

What might seem strange to us today is the difficulty involved in developing the idea of a symbol for zero. Using "something" to represent "nothing" boggled the mind. As late as the Middle Ages, mystery was associated with the symbol for zero and it even became the icon for several secret societies.

The development of "number" during the Middle Ages will not be dealt with here, except to note the significant contributions of Leonardo Fibonacci of Pisa during the thirteenth century. Fibonacci formalized the idea of negative numbers and used them in the development of algebra. His influence accelerated the use of Arabic numerals over Roman numerals and this greatly simplified mathematical notation.

The advances after the death of Fibonacci eventually led to the permanent establishment, in the sixteenth century, of the modern system of decimal notation and the identification of the real numbers with the points on a line. Shortly thereafter, interest in the nature of numbers virtually ceased, for the degree of sophistication necessary for the invention of analytic geometry and the calculus had already been reached. In the nineteenth century, interest in number theory was reawakened, and the subject underwent an intensive reexamination.

## C. Exercises

**B1.** Locate on the real line the indicated points.

a) 1     c) 2     e) 3     g) $|1 - 2 + 3|$

b) $-1$     d) $|-2|$     f) $-|-3|$     h) $|1 - 2 - 3|$

**B2.** Find the approximate location, on the real line, of the following points.

a) $\sqrt{2}$     c) $\sqrt{4}$     e) $\dfrac{1 + \sqrt{2}}{2}$

b) $\sqrt{3}$     d) $\pi$     f) $\dfrac{1 - \sqrt{2}}{2}$

**B3.** In each of these expressions, insert one of the symbols $<, >, \leq, \geq,$ or $=$ in order to make it a valid statement.

a) 5   7      d) 7   $-5$      g) $-|1|$   1

b) $-|5|$   $-|7|$      e) $-7$   $-5$      h) $(-2)^2$   $2^2$

c) $\frac{1}{5}$   $\frac{1}{7}$      f) 5   $|-7|$      i) $(-2)^3$   $2^3$

**B4.** Match each set described below with the interval which it equals.

a) $\{x: |x| < 4\}$      A) $[2, 4]$

b) $\{x: |x + 1| < 3\}$      B) $[0, 2)$

c) $\{x: |x - 3| \leq 1\}$      C) $(-4, 4)$

d) $\{x: |x| < 4 \text{ and } x > 0\}$      D) $(-4, 2)$

e) $\{x: |x| < 2 \text{ and } x \geq 0\}$      E) $(0, 4)$

**B5.** Describe as an interval the points common to the indicated intervals.

a) $(-2, 2]$ and $(1, 3)$

b) $(-\infty, 4]$ and $(2, \infty)$

c) $[3, 8]$ and $[5, 9]$

d) $(-\infty, 4)$, $(-3, \infty)$, and $(-3, 2]$

e) $(-1, 2)$ and $[0, 2]$

**B6.** Shade the points on the axis that satisfy the given inequalities.

a) $x^2 - 1 \leq 0$

b) $(x - 1)(x + 3) > 0$

**D7.** Locate the following points on the real line.

a) $0, 1, 2, 3, 4, -4, -3, -2, -1$

b) $0^2, 1^2, 2^2, 3^2$

c) $(-0)^3, (-1)^3, (-2)^3, (-3)^3$

d) $1, \frac{1}{2}, \frac{1}{3}, \frac{1}{4}, \frac{1}{5}$

e) $\frac{1}{2}, \frac{2}{3}, \frac{3}{4}, \frac{4}{5}, \frac{5}{6}$

f) $2, 3, \frac{1}{2}, \frac{1}{3}, 4, 5, \frac{1}{4}, \frac{1}{5}$

**D8.** Locate on the real line the points described.

a) The point that is 3 units to the left of $-2$.

b) The point that is 4 units to the right of $-3$.

c) The two points that are at a distance of 3 from the point 4.

d) The point that is equidistant from the points 3 and 7.

e) The negative number at a distance of 5 from the point 3.

**D9.** Each of the open intervals described below can be written in the form

$$\{x: |x - a| < b\}.$$

Find the numbers $a$ and $b$ in each case.

a) $(2, 4)$      e) $(-3, 4)$

b) $(3, 5)$      f) $(-4, -2)$

c) $(3, 7)$      g) $(-4, 0)$

d) $(-3, 3)$      h) $(-2, 10)$

a) $a = 3, b = 1$      e) $a = \frac{1}{2}, b = \frac{7}{2}$

b) $a = 4, b = 1$      f) $a = -3, b = 1$

c) $a = 5, b = 2$      g) $a = -2, b = 2$

d) $a = 0, b = 3$      h) $a = 4, b = 6$

**D10.** Each of the closed intervals described below is of the form $[a, b]$. Find the numbers $a$ and $b$ in each case.

a) $\{x: |x| \leq 3\}$      e) $\{x: |2x - 4| \leq 6\}$

b) $\{x: |-x| \leq 3\}$      f) $\{x: |2x - 3| \leq 5\}$

c) $\{x: |x - 1| \leq 2\}$      g) $\{x: |3x + 2| \leq 4\}$

d) $\{x: |x - 3| \leq 1\}$      h) $\{x: |7x + 5| \leq 11\}$

a) $[-3, 3]$    b) $[-3, 3]$    c) $[-1, 3]$    d) $[2, 4]$

e) $[-1, 5]$    f) $[-1, 4]$    g) $[-2, \frac{2}{3}]$    h) $[-\frac{16}{7}, \frac{6}{7}]$

**D11.** Each of the sets described below can be written in the form

$$\{x: x \in (a, \infty) \text{ or } x \in (-\infty, b)\}.$$

Find the numbers $a$ and $b$.

a) $|x| > 0$      e) $|x + 1| > 1$

b) $|x - 1| > 0$      f) $|x - 1| > 1$

c) $|x - 3| > 2$      g) $|2x - 3| > 5$

d) $|x - 2| > 3$      h) $|3x + 2| > 6$

a) $a = b = 0$      e) $a = 0, b = -2$

b) $a = b = 1$      f) $a = 2, b = 0$

c) $a = 5, b = 1$      g) $a = 4, b = -1$

d) $a = 5, b = -1$      h) $a = \frac{4}{3}, b = -\frac{8}{3}$

**M12.** Find $x$ in each of the cases below.
 a) $|x - 1| = |x - 2|$
 b) $|2x - 1| = |x - 2|$  $(x < 0)$
 c) $|3x - 3| = |x - 4|$  $(x > 0)$
 d) If $3a - 1 < 7$, then $a < x$.
 e) If $4a + 5 < -4$, then $a < x$.

**M13.** Two inequalities $A < B$ and $B < C$ may be written as an inequality of the form $A < B < C$.
 Describe as an interval the points that satisfy the given inequalities.
 a) $-1 < 4x - 9 < 11$
 b) $3x - 2 < 4x + 1 < 2x + 5$
 c) $2 - 2x < 5 - x < 4 - 2x$

**M14.** Describe the points that satisfy the given inequalities.
 a) $x^3 + x < 0$
 b) $(x - 1)(x - 2)(x - 3) \leq 0$

**M15.** The following argument gives a *false proof* that $1 = 2$. Find the error in the reasoning.

Assume
$$x = 1.$$
Then
$$x^2 - x = x^2 - 1.$$
Factoring gives
$$x(x - 1) = (x + 1)(x - 1).$$

Canceling $(x - 1)$ gives
$$x = x + 1.$$
Substituting $x = 1$ we have
$$1 = 1 + 1 \quad \text{or} \quad 1 = 2.$$

**M16.** a) Prove that $2ab \leq a^2 + b^2$ for all real numbers $a$ and $b$.
 (Hint: Start with the inequality $(a - b)^2 \geq 0$.)
 b) Prove that $(a + b)^2 \leq 2a^2 + 2b^2$ for all real numbers $a$ and $b$.
 c) Prove that
$$(a + b + c)^2 \leq 3a^2 + 3b^2 + 3c^2.$$

**M17.** Prove that
$$(a_1 + a_2 + \cdots + a_n)^2 \leq n(a_1^2 + a_2^2 + \cdots + a_n^2)$$
for all real numbers $a_1, a_2, \ldots, a_n$. (Hint: See Exercise M16.)

**M18.** Prove each of the following.
 a) $|a|^2 = a^2$
 b) $|ab| = |a|\,|b|$

**M19.** Prove that $\sqrt{3}$ is not a rational number. (Hint: Assume $\sqrt{3} = \frac{m}{n}$ so $m^2 = 3n^2$ where $m$ and $n$ are integers without a common factor. Then $m$ and $n$ can each be even or odd, and this leads to four cases. Show that each case is impossible. The case where $m$ and $n$ are both odd is the most involved.)

**M20.** Prove that if $m$, $n$, and $r$ are integers and $m^2 = rn^2$, then $r = s^2$ where $s$ is an integer.

---

## 1.2 *The Coordinate Plane*

**Analytic geometry** is the study of geometry from an algebraic point of view. Geometric concepts are replaced by algebraic ones and algebraic ideas are then used to develop the theory. We will begin by replacing the geometric object, a point on the plane, by the algebraic object, a pair of real numbers.

→ The phrase "algebraic geometry" would be a more suggestive title for the subject.

→ This process is called *coordinatization* and is characteristic of the difference between analytic and Euclidean geometry.

**A. Coordinates** Draw two perpendicular real lines, one vertical and the other horizontal, intersecting at

0 (Fig. 1). The horizontal line is called the *x-axis* and

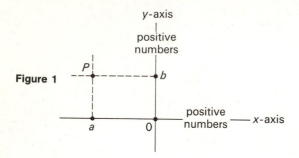

**Figure 1**

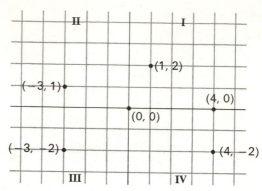

**Figure 2**

the vertical line, the *y-axis*. The positive numbers occur on these lines as shown.

Now let $P$ denote any point in the plane, and draw lines through $P$ parallel to the axes. The lines intersect the axes as shown in Figure 1. Then the pair of real numbers $(a, b)$ is called the *coordinates* of the point $P$. We identify $P$ with this pair $(a, b)$ and write

$$P = (a, b).$$

The number $a$ is called the *x-coordinate* and $b$ is called the *y-coordinate* of the point $P$. The words *abscissa* and *ordinate* are also used.

Recall that the notation $(a, b)$ is also used to denote an open interval. However, the context of the discussion will make it easy to distinguish the two uses of the notation.

(Greg)

Several points and their coordinates are indicated in Figure 2. By convention, the Roman numerals I, II, III, and IV are used in referring to the four subdivisions of the plane known as *quadrants*.

The signs of the coordinates of the points in the respective quadrants are shown in Figure 3.

**Figure 3**

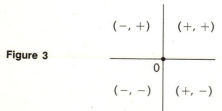

***Conventions*** (*See Figure 4*)  To simplify our figures we will often use the conventions illustrated in Figure 4a rather than those of Figure 4b. They are:

a) Axes are not labeled.

b) Same units on each axis.

c) Only one coordinate used when the point lies on an axis.

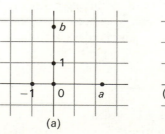

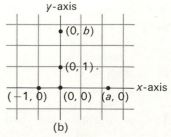

**Figure 4**

*Example 1*　Suppose that $(1, 5)$ and $(6, 1)$ are endpoints of the diagonal of the rectangle shown in Figure 5. Find the values of $(x_1, y_1)$ and $(x_2, y_2)$.

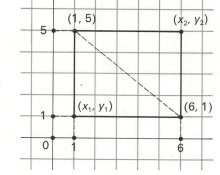

**Figure 5**

*Solution*　Since the lines parallel to the $y$-axis from the points $(x_1, y_1)$ and $(x_2, y_2)$ intersect the $x$-axis at the points 1 and 6 respectively, we see that $x_1 = 1$ and $x_2 = 6$. Also, since the lines parallel to the $x$-axis from these points intersect the $y$-axis at the points 1 and 5, $y_1 = 1$ and $y_2 = 5$. Hence $(x_1, y_1) = (1, 1)$ and $(x_2, y_2) = (6, 5)$.

### B. The Distance Formula

Given two points $(x_1, y_1)$ and $(x_2, y_2)$, we will derive the formula for the distance $d$ between them. Referring to the right triangle in Figure 7, we see that

$$a = |x_2 - x_1| \quad \text{and} \quad b = |y_2 - y_1|.$$

By the Pythagorean Theorem we have

$$d^2 = a^2 + b^2 = |x_2 - x_1|^2 + |y_2 - y_1|^2$$
$$= (x_2 - x_1)^2 + (y_2 - y_1)^2.$$

Now taking square roots of both sides of the equation yields the *distance formula*:

$$(*) \qquad \begin{cases} \text{or} & d = \sqrt{(x_2 - x_1)^2 + (y_2 - y_1)^2} \\ & d = \sqrt{(x_1 - x_2)^2 + (y_1 - y_2)^2} \end{cases}$$

since $(x_1 - x_2)^2 = (x_2 - x_1)^2$ and $(y_1 - y_2)^2 = (y_2 - y_1)^2$.

*Problem 1*　In Figure 6 find, in order, the coordinates of the point $P_1$ and the midpoints $P_2$, $P_3$, and $P_4$.

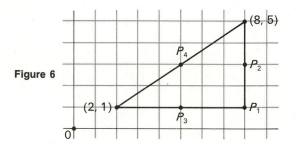

**Figure 6**

*Answer:*　$P_1 = (8, 1)$, $P_2 = (8, 3)$, $P_3 = (5, 1)$ and $P_4 = (5, 3)$

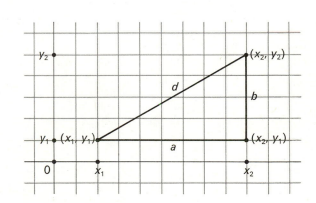

**Figure 7**

→　A simple proof of the Pythagorean Theorem can be constructed as follows:

Draw a right triangle (Fig. 8)

**Figure 8**

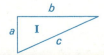

and construct three more congruent to it. Now arrange the four triangles on a square planar surface with sides $a$ and $b$

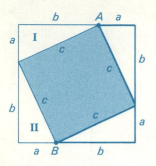

**Figure 9**

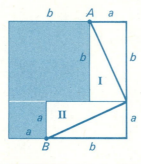

**Figure 10**

as shown in Figure 9. The area of the shaded region is clearly $c^2$.

Now rotate the indicated triangles I and II about $A$ and $B$ respectively to obtain Figure 10.

The shaded region has not changed in area and now clearly equals $a^2 + b^2$.

Hence $c^2 = a^2 + b^2$ ●

*This is only one of the many proofs of the Pythagorean Theorem. In fact, James Garfield, before he became President, was credited with the discovery of one proof of this theorem. He worked it out during an exam when he could not remember the proof in the text.*

*(Robin)*

---

*Example 2* Find the distance between the following pairs of points.
a) $(3, 2)$ and $(0, 6)$
b) $(-3, 5)$ and $(2, -7)$
c) $(5, 1)$ and $(7, 4)$

*Solution* Applying the distance formula (*), we have
a) $\sqrt{(3 - 0)^2 + (2 - 6)^2} = \sqrt{3^2 + 4^2} = \sqrt{25} = 5$
b) $\sqrt{(-3 - 2)^2 + (5 - (-7))^2} = \sqrt{5^2 + 12^2} =$
$\sqrt{169} = 13$
c) $\sqrt{(5 - 7)^2 + (1 - 4)^2} = \sqrt{2^2 + 3^2} = \sqrt{13}$

---

**C. Graphs** An equation in the variables $x$ and $y$ defines a set of points in the plane called the *graph* of the equation. The graph consists of all points $(x, y)$ that satisfy the equation. For example, take the equation

$$x^3 = y.$$

A few of the points in the plane whose coordinates satisfy the equation are $(0, 0)$, $(1, 1)$, $(-1, -1)$, $(\frac{3}{2}, \frac{27}{8})$, and $(-\frac{3}{2}, -\frac{27}{8})$, because $0^3 = 0$, $1^3 = 1$, $(-1)^3 = -1$, $(\frac{3}{2})^3 = \frac{27}{8}$, and $(-\frac{3}{2})^3 = -\frac{27}{8}$. These points are shown in Figure 11. By finding more points that lie on the

*You are probably familiar with the terms* variable *(usually represented by one of the letters x, y, z, u, v, w, r, s, t, and θ) and* constant *(usually represented by one of the letters a, b, c, d, e, A, B, C, D, and K). Remember that a variable may assume more than one value in a given discussion while a constant has a fixed value. For example, in the equation $y = ax + \pi$, the letter a represents a number whose value is fixed and π is the constant that is the ratio of the circumference of a circle to its diameter. The letters x and y are variables and the equation defines a relationship between them.*

*(Vahan)*

graph and then connecting them we obtain Figure 12, which represents the graph of $x^3 = y$. The reader should note that sometimes the graph of an equation contains no points. This occurs when there are no real number pairs that satisfy the equation.

→ *For example, consider the equation $x^2 + y^2 = -1$. For any point $P = (x, y)$ we have $x^2 + y^2 \geq 0$. Hence the coordinates of $P$ cannot satisfy the equation. As an exercise you can check for yourself that the graph of $y^2 + x^2 + 2x + 2 = 0$ contains no points.*

*(Estelle)*

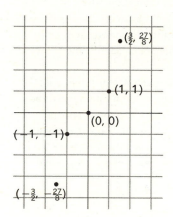

**Figure 11**

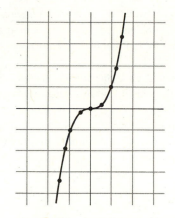

**Figure 12**

In simple cases we can reverse the above process, and, starting with a graph, find an equation that determines it. For example, take a circle of radius 3, centered at the point (2, 1) (Fig. 13). We know that the points on the circumference of a circle are all equidistant from the center. Using the distance formula we see that all points in the circle satisfy the equation

$$(x - 2)^2 + (y - 1)^2 = 3^2.$$

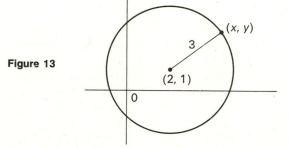

**Figure 13**

More generally, the equation of the circle with center $(h, k)$ and radius $r$ is given by

$$(x - h)^2 + (y - k)^2 = r^2.$$

## The Invention of Analytic Geometry

Analytic geometry, one of the most revolutionary advances in mathematics, was created independently by Pierre de Fermat and René Descartes. Despite this common creation and the fact that they were two of the most brilliant men of the seventeenth century, their life styles were in almost total contrast.

### René Descartes (1596–1650)

... what I have given in the second book on the nature and properties of curved lines, and the method of examining them, is, it seems to me, as far beyond the treatment in the ordinary geometry, as the rhetoric of Cicero is beyond the a, b, c of children.

Descartes was not a modest man, but certainly was a colorful personality. He was a gambler and soldier whose life style vacillated between rogue and scholar. When a drunk once insulted his current girl friend, the short-tempered Descartes drew his sword. As the story goes he would have killed the startled chap, but spared his life because he was too filthy to be slain before the eyes of a beautiful woman. In contrast, by applying the scientific method to natural and metaphysical philosophy, Descartes became "the father of modern philosophy."

Plagued by ill health as a child, he was forced to spend his mornings in bed. Continuing this pleasant habit as an adult, he generally did not rise until past noon. Perhaps his most famous meditation took place in a small sweltering room that was virtually an oven. There Descartes attempted to remove his mind from his body, and questioning all that he knew sought after some fundamental, undeniable truth. From here came his immortal conclusion, "I think, therefore I am."

By 1646, Descartes was famous as a philosopher, mathematician, physiologist, and optician, and ironically this ascendancy brought his ruin. Queen Christina of Sweden, who fancied herself an intellectual, heard of Descartes and obtained him as her personal tutor. A curious woman who was accustomed to little sleep or food and waking before sunrise, she expected the same from those around her. She believed that philosophy was best absorbed at 5:00 A.M. in an unheated room. Although Descartes attempted to oblige, the habits of a lifetime were not easily changed. He lasted until February 15, 1650.

### *Pierre de Fermat (1601–1665)*

Fermat, by profession a lawyer, led a quiet, uneventful life, but his contributions to mathematics were revolutionary. This may seem surprising, since mathematics was only his hobby, taken up in his late twenties. He worked on mathematics because he loved the subject, and this honest motivation no doubt contributed to his success.

Although his letters reveal a lack of modesty, Fermat rarely published, merely jotting his results on the backs of envelopes or in the margins of books. Besides his contributions to geometry, his "principle of least action" laid the foundation for geometric optics. He was a cofounder of probability theory, and years before Newton, anticipated calculus by virtually writing down the definition of the derivative and using it to solve several maximum-minimum problems. His greatest work, however, was in number theory, and he is the father of that branch of mathematics.

Fermat's most famous statement, "I have discovered a truly marvelous demonstration of this theorem which this margin is too narrow to contain," concerns the assertion that there exist no positive integers $a$, $b$, $c$, and $n$ such that

$$a^n + b^n = c^n \quad \text{when} \quad n \geq 3.$$

Although this question can be understood by a schoolboy, the problem is as yet unsolved. Many mathematicians doubt that Fermat's "marvelous demonstration" was valid, but they are still haunted by the fact that he was rarely wrong.

However, as an instance of his fallibility, he conjectured that for all integers $n \geq 0$, the number $2^{(2^n)} + 1$ is prime. This was shown to be false by Euler, who noticed that

$$2^{(2^5)} + 1 = (641)(6700417).$$

It has since been shown that there exists no algebraic formula which generates primes.

Although by 1629 Fermat had obtained results in the new "analytic" geometry, his indifference to publication led to his work not being communicated to others until 1636. As a result, Descartes's work, published in 1637, was not influenced by Fermat. However, centuries earlier,

Nicole Oresme (1323–1382) had devised the method of describing points in the plane by pairs of real numbers and had even found the graphs of certain equations. Following him, Francois Vieta (1540–1603) extended the analytic method and recognized that certain fundamental geometric problems were easily solved by means of coordinate geometry.

One of the reasons that Oresme and Vieta are not often remembered in this context is that they did not realize the full importance of their work. Since they were dominated by the classical Greek geometry, they considered analytic geometry to be but a collection of clever techniques. Fermat and Descartes, on the other hand, had a clear grasp of the significance of their discoveries, but even they treated mathematics as only a hobby and relegated their discoveries in geometry to relative unimportance. They did realize, however, that they now had a powerful tool with which to tackle old problems of geometry and analyze new ones. This realization of the significance of one's discovery is often as important as the discovery itself.

## D. Exercises

**B1.** Locate the following points in the coordinate plane.
a) $(1, 2)$    c) $(1, -2)$    e) $(1, 1)$    g) $(-1, -1)$
b) $(-1, 2)$   d) $(-1, -2)$   f) $(2, 2)$    h) $(-2, -2)$

**B2.** One diagonal of a square has endpoints $(1, 4)$ and $(4, 7)$. Find the coordinates of the endpoints of the other diagonal.

**B3.** Find the distance between the following pairs of points.
a) $(1, 3)$ and $(4, 7)$       c) $(-1, -3)$ and $(4, 7)$
b) $(1, -3)$ and $(4, 7)$      d) $(-1, -3)$ and $(-4, -7)$

**B4.** Show that the triangle with vertices $(1, 1)$, $(3, 5)$ and $(6, 1)$ is isosceles.

**B5.** On the same set of axes sketch, for $-2 \leq x \leq 2$, the graphs of the equations
$$y = x^2, \quad y = -x^2, \quad y = 4x^2, \quad \text{and} \quad y = -4x^2.$$

**B6.** Find the equation that describes the circle of radius 2 and center $(-1, 1)$. Graph the circle.

**B7.** Shade the region in the plane described in each of the following.
a) $\{(x, y): x \geq 0, \ x^2 + y^2 \leq 1\}$
b) $\{(x, y): |x - 1| < 2, \ |y + 2| < 1\}$

**D8.** Find the distance between the following pairs of points.
a) $(2, 3), (3, 2)$        d) $(\frac{1}{2}, 3), (\frac{9}{2}, 6)$
b) $(1, 6), (-3, 9)$       e) $(8, -3), (-4, 2)$
c) $(6, 3), (9, 4)$        f) $(9, -6), (3, 2)$

a) $\sqrt{2}$  b) 5  c) $\sqrt{10}$  d) 5  e) 13  f) 10

**D9.** Find the equation of each of the circles described below.
a) center $(3, 2)$, radius 2
b) center $(3, 2)$, radius $\sqrt{2}$
c) center $(-4, 3)$, radius 3
d) center $(-1, -2)$, radius 1
e) center $(-3, 5)$, diameter 4
f) center $(1, 3)$, area $9\pi$

a) $(x - 3)^2 + (y - 2)^2 = 4$
b) $(x - 3)^2 + (y - 2)^2 = 2$
c) $(x + 4)^2 + (y - 3)^2 = 9$
d) $(x + 1)^2 + (y + 2)^2 = 1$
e) $(x + 3)^2 + (y - 5)^2 = 4$
f) $(x - 1)^2 + (y - 3)^2 = 9$

**D10.** Find the midpoint of each of the segments whose endpoints are
a) $(1, 0), (3, 0)$        d) $(3, 7), (7, 3)$
b) $(0, 5), (0, 9)$        e) $(-5, 8), (-3, -6)$
c) $(3, 3), (7, 7)$        f) $(-12, 4), (4, -8)$

a) $(2, 0)$    b) $(0, 7)$    c) $(5, 5)$    d) $(5, 5)$
e) $(-4, 1)$   f) $(-4, -2)$

**D11.** Verify that each triple of points given below form the vertices of a right triangle. In each case find the length of the hypotenuse.
a) $(0, 0), (1, 1), (2, -2)$       d) $(3, -2), (-2, 3), (0, 4)$
b) $(1, 0), (4, 0), (4, 4)$        e) $(1, 2), (3, 6), (-3, 4)$
c) $(4, 4), (1, 5), (-2, -4)$

a) $\sqrt{10}$  b) 5  c) 10  d) $\sqrt{50}$  e) $\sqrt{40}$

**D12.** Graph each set of equations on the same set of axes.
a) $y = x^2, y = -x^2, x = y^2, x = -y^2$
b) $y = x^2, y = x^2 + 1, y = x^2 - 1$

c) $y = x^2, y = 2x^2, y = \dfrac{x^2}{2}$

d) $y = x, y = x^2, y = x^3, y = x^4$

**D13.** Shade the region in the plane described in each of the following.
a) $\{(x, y): x \geq 0, y \geq 0\}$     e) $\{(x, y): x \geq 0\}$
b) $\{(x, y): x \geq 0, y \leq 0\}$     f) $\{(x, y): y \leq 0\}$
c) $\{(x, y): x \leq 0, y \geq 0\}$     g) $\{(x, y): x \geq y\}$
d) $\{(x, y): xy \geq 0\}$     h) $\{(x, y): 0 \leq x \leq y\}$

**M14.** One vertex of a square is at the origin and the diagonals of the square intersect at the point $(2, 3)$. What is the area of the square?

**M15.** One vertex of an equilateral triangle is at the origin, the second is on the graph of $y = x^2$, and the third lies on the positive $x$-axis. What is the area of this triangle?

**M16.** Two vertices of an equilateral triangle are $(2, -2)$ and $(-2, 2)$, and the third vertex lies in the first quadrant. Find the coordinates of the third vertex.

**M17.** One diagonal of a square has endpoints $(1, 2)$ and $(7, 10)$. Find the coordinates of the endpoints of the other diagonal.

**M18.** Three vertices of a parallelogram in the first quadrant are $(0, 0)$, $(a, 0)$ and $(b, c)$, where $a > b$. Find the coordinates of the fourth vertex and the lengths of the diagonals.

**M19.** Find the coordinates of the point which is the reflection of the point $(a, b)$ with respect to the following:
a) the $x$-axis     c) the point $(0, 0)$
b) the $y$-axis     d) the graph of the equation $y = x$

**M20.** Determine whether or not the points $(-1, -1)$, $(1, 3)$ and $(79, 159)$ lie on a line.

**M21.** The midpoints of the sides of a triangle are $(1, 4)$, $(3, 6)$, and $(4, 3)$. Find the coordinates of the vertices of the triangle.

**M22.** A circle of radius 1 is inside a square whose side has length 2. Show that the area of the largest circle that can be inscribed between the circle and the square is

$$\pi(17 - 12\sqrt{2}).$$

**M23.** One diagonal of a square has endpoints at $(-2, 6)$ and $(1, -3)$. Find the coordinates of the endpoints of the other diagonal.

**M24.** A circle is inscribed in an equilateral triangle of side $a$. Show that the radius of the circle is $\dfrac{a}{\sqrt{3}}$.

## 1.3   *Lines and Their Equations*

Like a point, a line can be identified with an algebraic object, its equation. We will take as undefined the notion of a line.

→   One definition of a line is "a collection of points in a straight row extending from heaven to hell." A precise definition of a line depends on the axiom system used.

**A. Slope**   Let $(x_1, y_1)$ and $(x_2, y_2)$ be distinct points on a nonvertical line $L$ (Fig. 1). We define the slope of $L$ as

$$m = \frac{y_2 - y_1}{x_2 - x_1}.$$

Notice that the slope also equals

$$\frac{-(y_2 - y_1)}{-(x_2 - x_1)} = \frac{y_1 - y_2}{x_1 - x_2}.$$

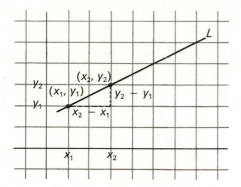

**Figure 1**

Hence the slope does not depend on the order of the points. The slope is merely the *rise* ($y_2 - y_1$), or change in the $y$ direction, over the *run* ($x_2 - x_1$), or change in the $x$ direction.

Some important properties of slope are described below and shown in Figure 3.

1) A horizontal line has slope 0.

2) The slope is positive if the line goes upward as it moves from left to right.

3) The slope is negative if the line goes downward as it moves from left to right.

4) The slope of a vertical line is said to be *undefined* because, according to our definition, the slope would have the form

$$\frac{y_2 - y_1}{x_2 - x_1} = \frac{y_2 - y_1}{0}.$$

**B. Finding the Equation of a Given Line**   The equation of a line is an algebraic expression describing a relationship between the $x$ and $y$ coordinates of the points in the plane which are on that line. Given a

The slope of $L$ can be computed using any two points on $L$. For example, let $(x_3, y_3)$ and $(x_4, y_4)$ be two more points on $L$ (Fig. 2). Then, since corresponding sides of triangles $S$ and $T$ are parallel, the triangles are similar and their sides are proportional. Hence

$$\frac{y_2 - y_1}{x_2 - x_1} = \frac{y_4 - y_3}{x_4 - x_3}.$$

Make sure, however, to subtract the $x$ and $y$ coordinates in the same order, for

$$\frac{y_2 - y_1}{x_2 - x_1} \neq \frac{y_2 - y_1}{x_1 - x_2}.$$

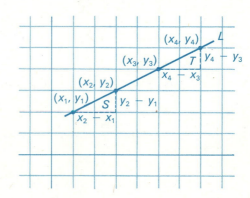

**Figure 2**

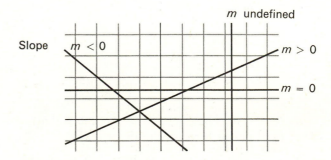

**Figure 3**

A line has a single equation, but the form in which the equation appears can vary. For example, the equation $2x + y = 1$ can be written as $y = 1 - 2x$, $y + 2x - 1 = 0$, $x = \frac{1-y}{2}$, $4x + 2y = 2$, $-6x - 3y + 3 = 0$, etc.

line $L$ we will now show how to derive its equation in two ways.

**THEOREM 1**  *Suppose $L$ is a line passing through the point $(a, b)$ and having slope $m$. Then the equation of $L$ is*

$$y - b = m(x - a).$$

➡ *To see what this theorem says look at Example 1 where the letters have been replaced by numbers.*

*(Kent)*

*Proof*  Suppose $(x, y)$ is any other point on $L$ (Fig. 4). Then the number $\frac{y-b}{x-a}$ is equal to the slope of $L$. Hence $\frac{y-b}{x-a} = m$ or $y - b = m(x - a)$  ●

➡ *This manner of writing the equation is known as the point-slope form.*

**COROLLARY**  *Suppose the line $L$ passes through the two points $(x_1, y_1)$ and $(x_2, y_2)$ and $x_2 \neq x_1$. Then its equation is*

$$y - y_1 = \left(\frac{y_2 - y_1}{x_2 - x_1}\right)(x - x_1).$$

➡ *Don't memorize these formulas, learn to derive them.*

*(Bob)*

*Proof*  The line $L$ passes through the point $(x_1, y_1)$ and its slope is $\dfrac{y_2 - y_1}{x_2 - x_1}$. By replacing $m$ by $\dfrac{y_2 - y_1}{x_2 - x_1}$, $a$ by $x_1$, and $b$ by $y_1$ in the theorem, we obtain the desired result  ●

➡ The astute reader may notice that in this corollary the point $(x_1, y_1)$ seems to play a special role. However, the equation

$$y - y_2 = \left(\frac{y_2 - y_1}{x_2 - x_1}\right)(x - x_2)$$

is merely the one given in Theorem 1, written in a different form. To see this, add $y_2 - y_1$ to each side of the above equation. Then

$$y - y_2 + (y_2 - y_1) = \left(\frac{y_2 - y_1}{x_2 - x_1}\right)(x - x_2) + (y_2 - y_1),$$

so

$$y - y_1 = \left(\frac{y_2 - y_1}{x_2 - x_1}\right)(x - x_2) + \left(\frac{y_2 - y_1}{x_2 - x_1}\right)(x_2 - x_1)$$

$$= \left(\frac{y_2 - y_1}{x_2 - x_1}\right)(x - x_2 + x_2 - x_1)$$

$$= \left(\frac{y_2 - y_1}{x_2 - x_1}\right)(x - x_1).$$

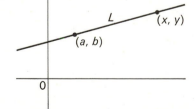

**Figure 4**

**Example 1**  Find the equation of the line $L$ that has slope 2 and passes through the point $(2, 1)$.

*Solution*  Let $(x, y)$ denote another point on the line $L$ as shown in Figure 5. We must then have

$$\frac{y - 1}{x - 2} = 2, \quad \text{so}$$

$$y - 1 = 2x - 4 \quad \text{or}$$

$$y = 2x - 3.$$

**Figure 5**

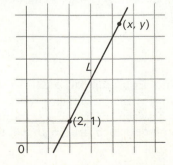

*Example 2*  Find the equation of the line $L$ that passes through the points $(2, 1)$ and $(4, 7)$.

*Solution*  The slope of the line $L$ must be

$$m = \frac{7 - 1}{4 - 2} = \frac{6}{2} = 3.$$

If $(x, y)$ denotes another point on $L$ we then have

$$\frac{y - 1}{x - 2} = 3, \quad \text{which gives} \quad y = 3x - 5.$$

Up to now we have discussed only cases where the slope is defined. Thus we need an expression for describing a vertical line. We see (Fig. 6) that the equation for such a line passing through $(a, b)$ is simply $x = a$.

We now know how to determine the equation of a given line. The equation has the form

$$y - b = m(x - a) \quad \text{(nonvertical)}$$
$$\text{or} \qquad x = a \qquad \text{(vertical)}.$$

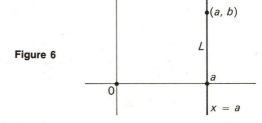

**Figure 6**

Both of these are special cases of

$$Ax + By + C = 0.$$

→  Assuming that $A$ and $B$ are not both zero, this is the most general linear equation in two variables, $x$ and $y$. The equation $y - b = m(x - a)$ reduces to this form with $A = m$, $B = -1$, and $C = b - ma$; $x = a$ can be written in this form with $A = 1$, $B = 0$, and $C = -a$.

**C. Graphing the Line of a Given Equation**  Now starting with an equation we will graph the corresponding line. Suppose we are given

$$\frac{x}{a} + \frac{y}{b} = 1.$$

(This is called the *intercept form* of a line.) Then the points $(a, 0)$ and $(0, b)$ satisfy the equation. The location of these two points then determines the line (Fig. 7).

→  Equations of the form $x = A$, $y = B$, or $y = mx$ cannot be written in the intercept form. However, this is not a serious drawback since the graphs of these equations are easy to find.

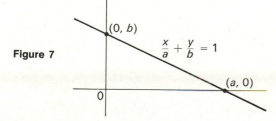

**Figure 7**

*Example 3* Graph $2x - 3y = 6$.

*Solution* Write the equation in the form

$$\frac{2x}{6} - \frac{3y}{6} = 1$$

or

$$\frac{x}{3} + \frac{y}{-2} = 1.$$

Hence the graph is as shown in Figure 8.

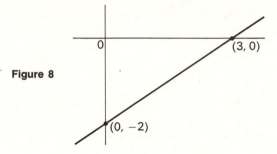

**Figure 8**

Now consider an equation of a line $L$ having the form $y = mx + b$. The point $(0, b)$ clearly satisfies the equation and, rewriting it as $\frac{y-b}{x} = m$, we see that $m$ is the slope of $L$. Hence the graph is as shown in Figure 9.

→ *The equation $y = mx + b$ is known as the* slope-intercept *form of a line. Personally, I found it the most convenient form because from it the graph could easily be found. You don't really need to know all these forms, any one will suffice.*

*(Greg)*

*Example 4* Graph the equation $y = -2x + 3$.

*Solution* From the equation we can read off the facts that the slope of the line is equal to $-2$ and the point $(0, 3)$ lies on the line. The graph must therefore be as shown in Figure 10.

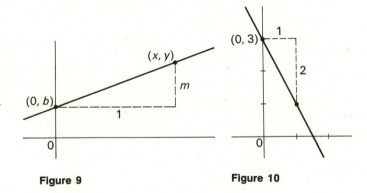

**Figure 9**　　　　**Figure 10**

**D. Perpendicular and Parallel Lines** The geometric property of two lines being perpendicular can be expressed algebraically in terms of their slopes. This is done more precisely in the theorem below:

*THEOREM 2* *Given two lines M and N with slopes m and n respectively, consider the following two statements:*

*A) The lines M and N are perpendicular, and*
*B) mn = −1.*

*Then A is equivalent to B.*

→ Theorem 2 is our first example in which the equivalence of two statements *A* and *B* is shown. This requires two proofs: that *A* implies *B*, and that *B* implies *A*. This is often written "*A* if and only if *B*" or "*A* iff *B*." It is also often said that *A* is a *necessary and sufficient condition for B*.

*Proof of "the lines are perpendicular implies mn = −1."* Since the lines are perpendicular, one of them, say $M$, has positive slope. Let $(a, b)$ denote the point of intersection of the lines.
Construct the triangle $T$ with base 1 and height $m$. Then the triangle $S$ (Fig. 11) having base $m$ is con-

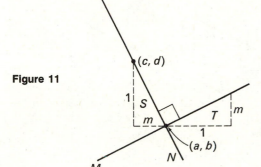

**Figure 11**

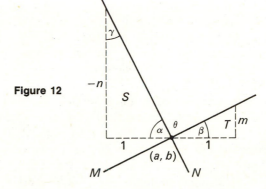

**Figure 12**

gruent to triangle $T$ because $M$ and $N$ are perpendicular. Therefore triangle $S$ must have height 1. Hence the slope $n$ of $N$ is $n = \frac{d-b}{c-a} = \frac{1}{-m}$, so $mn = -1$ ⬤

*Proof of "$mn = -1$ implies the lines are perpendicular."* Since $mn = -1$, one of the lines, say $M$, has positive slope, and the other, negative slope. Construct triangles $S$ and $T$ (Fig. 12), each having base 1.

From the definition of slope it follows that the heights of $S$ and $T$ are $-n$ and $m$, respectively. Since $mn = -1$, the height of $S$, $-n$, is equal to $\frac{1}{m}$, and it follows that the triangles are similar. Therefore the corresponding angles are equal; in particular, $\beta = \gamma$. Since $S$ is a right triangle, $\alpha + \gamma = 90°$, so $\alpha + \beta = 90°$. Hence $\theta = 90°$ and the lines are perpendicular ⬤

⟶ The reader should realize that a statement may be true but its converse, false. For example, the statement "if two lines are perpendicular, then one of them must have a nonnegative slope" is true. However, its converse, "if two lines are given and one of them has a nonnegative slope, then the lines are perpendicular," is false.

---

*Example 5* Find the equation of the line $L$ that passes through the point $(1, 2)$ and is perpendicular to the line $2x - 3y = 12$.

*Solution* We first rewrite the given equation in the form

$$y = \tfrac{2}{3}x - 4.$$

We now see that the slope of this line is $\frac{2}{3}$. Hence the slope of $L$ is $\frac{-3}{2}$. Since $L$ passes through $(1, 2)$ we see by Theorem 1 that

$$y - 2 = \frac{-3}{2}(x - 1) \quad \text{or} \quad 3x + 2y = 7.$$

---

*Problem 1* Find the equation of the line through $(5, 6)$ perpendicular to the line determined by $(1, 2)$ and $(3, 4)$.

*Answer:* $x + y = 11$

The two lines $M$ and $N$ in Figure 13 are parallel if and only if angle $\alpha$ equals angle $\beta$. But $\alpha = \beta$ if and only if the triangles $S$ and $T$ are congruent.

Since the triangles $S$ and $T$ are congruent if and only if $m = n$ (Fig. 13), we have the following result:

**THEOREM 3**  *Two nonvertical lines are parallel if and only if their slopes are equal.*

**E. Intersection of Lines**  Two nonparallel lines intersect at a unique point. When the equations of the lines are given, the point of intersection $(a, b)$ must satisfy both equations. It is the only point with this property; hence it can be found by solving the equations simultaneously.

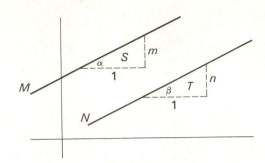

**Figure 13**

---

*Example 6*  Find the equation of the line $L$ through $(-1, 3)$ parallel to $2x - 4y = 5$.

*Solution*  We write the given equation as

$$y = \tfrac{1}{2}x - \tfrac{5}{4}.$$

Thus the slope of this line is $\tfrac{1}{2}$. Since $L$ is parallel to it, its slope is also $\tfrac{1}{2}$. It follows that the equation of $L$ is

$$y - 3 = \tfrac{1}{2}(x + 1) \quad \text{or} \quad x - 2y + 7 = 0.$$

---

In the most general case, the lines have equations of the form

$$a_1 x + b_1 y + c_1 = 0$$

and

$$a_2 x + b_2 y + c_2 = 0.$$

Solving these simultaneously we obtain

$$x = \frac{b_1 c_2 - b_2 c_1}{a_1 b_2 - a_2 b_1} \quad \text{and} \quad y = \frac{a_2 c_1 - a_1 c_2}{a_1 b_2 - a_2 b_1}.$$

This is valid whenever $a_1 b_2 - a_2 b_1 \neq 0$.

When $a_1 b_2 - a_2 b_1 = 0$, it can be verified that the lines are vertical or that the slopes of the two lines are equal; hence the lines are parallel. Obviously there is not a unique point of intersection. Thus the above formula breaks down, exactly as it should, in the case of parallel lines.

---

*Example 7*  Find the point of intersection of the lines whose equations are

$$2x + 3y = 4 \quad \text{and} \quad x + 2y = 5.$$

*Solution*  Solve for $x$ in the second equation to obtain

$$x = 5 - 2y.$$

Putting this expression for $x$ into the first equation gives

$$2(5 - 2y) + 3y = 4 \quad \text{or} \quad 10 - y = 4.$$

Hence $y = 6$, and now letting $y = 6$ in either of the original equations we find that $x = -7$. Hence the point of intersection is $(-7, 6)$.

---

**F. Comparison of Two Methods**  We will compare the techniques of analytic and Euclidean geometry by giving different proofs of the same theorem. The analytic proof is simple and straightforward, and the real power of the analytic method lies in the fact that it can be applied to other theorems in geometry with equal ease and success.

**THEOREM 4**  *The altitudes of a triangle are concurrent (meet in the same point).*

*Analytic Proof*  Locate the triangle with one vertex at the origin, one on the positive $x$-axis, and the third above the $x$-axis. Label the points and lines as indicated in Figure 14.

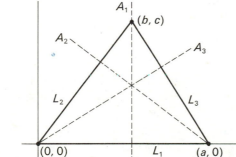

**Figure 14**

The equations of the lines $L_1$, $L_2$, and $L_3$ are

$$y = 0,$$

$$y = \frac{c}{b}x,$$

and

$$y = \frac{c}{b - a}(x - a)$$

respectively. Let $A_1$, $A_2$, $A_3$ denote the lines containing the altitudes from the vertices $(b, c)$ $(a, 0)$ and $(0, 0)$ respectively. By Theorem 2, we know that

slope $A_1$ is undefined,

$$\text{slope } A_2 = -\frac{b}{c},$$

and

$$\text{slope } A_3 = \frac{(a - b)}{c}.$$

The Euclidean proof given below is based on the proof in *Euclid's Elements* and requires thought and ingenuity. In general, Euclidean proofs are much more difficult to discover than analytic ones, but their discovery brings a good deal more satisfaction. Unfortunately the charm of the old ways must give way to the power of the new methods.

*Euclidean Proof*

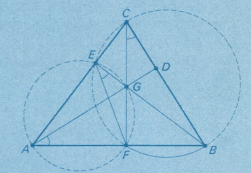

**Figure 15**

Consider the acute triangle $ABC$ where $\overline{CF}$ and $\overline{BE}$ are altitudes from $C$ and $B$ respectively. Let $G$ denote their point of intersection and extend $\overline{AG}$ to meet $\overline{BC}$ at $D$. We will show that $\overline{AD} \perp \overline{BC}$.

Since $AEG$ is a right angle, $\overline{AG}$ is a diameter of the circle containing $A$, $E$, and $G$, for every diameter of a circle subtends an inscribed right angle. Similarly $\overline{AG}$ is a diameter for the circle containing $A$, $F$, and $G$, so these two circles are the same, giving that $A$, $E$, $G$, and $F$ lie on the same circle. Since the chord $\overline{FG}$ of that circle subtends both $\angle FEG$ and $\angle FAG$ we have

$$\angle FAG = \angle FEG.$$

In a like manner, $\overline{BC}$ is a diameter for both the circle containing $C$, $E$, and $B$ and the circle containing $C$, $F$, and $B$. Hence $C$, $E$, $F$, and $B$ lie on the same circle. Since the chord $\overline{FB}$ of that circle subtends both $\angle FEG$ and $\angle FCB$, we have

$$\angle FEG = \angle FCB.$$

Hence

$$\angle BAD = \angle FAG = \angle FEG = \angle FCB.$$

Thus

$$\angle BAD = \angle FCB$$

and, since $\angle CBF = \angle ABD$, it follows that

$$\triangle FCB \text{ is similar to } \triangle DAB.$$

Therefore

$$\angle ADB = \angle CFB = 90°$$

and $\overline{AD}$ is perpendicular to $\overline{CB}$.

The case of an obtuse triangle can be reduced to the above case, and we omit the argument ●

Hence the equations of the lines are

$$x = b, \qquad y = -\frac{b}{c}(x - a),$$

and

$$y = \frac{(a - b)}{c}x.$$

Solving the equations simultaneously for the point of intersection of $A_2$ and $A_3$, we obtain

$$-\frac{b}{c}(x - a) = \frac{(a - b)}{c}x.$$

So $x = b$, $y = \frac{ab - b^2}{c}$, and the point of intersection is $\left(b, \frac{ab - b^2}{c}\right)$. Since the line $A_1$, with equation $x = b$, passes through this point, the altitudes are con-current ●

## The Rise and Fall of Coordinates

The introduction of coordinates into geometry was a significant achievement since it led to the uniform and simple proofs of analytic geometry. By comparison, the methods of Euclid were cumbersome and each theorem often required its own individual argument. Analytic geometry and coordinates, having paved the way for calculus, dominated geometry for 250 years. Then, early in the twentieth century, it was found that the coordinates themselves were now a stumbling block to further progress. One of the main accomplishments of this century has been the virtual removal of coordinates from geometry. Most modern textbooks today treat the subject in a coordinate-free manner. Thus, in a curious way, geometry has returned to Euclid.

## G. Exercises

**B1.** Find the slope of the line passing through the indicated pairs of points.
a) $(1, 3)$ and $(5, 3)$     c) $(3, 6)$ and $(4, 4)$
b) $(-1, 2)$ and $(3, 4)$     d) $(2, 3)$ and $(2, 8)$

**B2.** Using the notion of slope, show that the three given points lie on the same line.
a) $(1, 2)$, $(-3, 4)$, and $(5, 0)$
b) $(0, 1)$, $(2, -1)$, and $(1, 0)$

**B3.** Two perpendicular lines pass through the point $(2, -7)$, and one of them has slope equal to $-3$. Find the equations of these lines.

**B4.** A line passes through the points $(2, 2)$ and $(-1, -4)$. Find its equation, the point where the line intersects the $x$-axis, and the point where the line intersects the $y$-axis.

**B5.** On the same set of axes, graph the lines whose equations are
$$x + y = 1, \; x + y = 2, \; x - y = 1, \; \text{and} \; x - y = 2.$$

**B6.** Find the vertices of the triangle determined by the three lines
$$3x - y = 5, \quad x + y = 3, \quad \text{and} \quad x - y = -1.$$

**D7.** Find the equation of the line passing through the indicated point and having the indicated slope.
a) through $(2, 3)$, slope 4
b) through $(3, 4)$, slope 2
c) through $(-3, 1)$, slope 6
d) through $(1, 7)$, slope 0
e) through $(-3, -5)$, slope undefined
f) through $(-3, 3)$, slope 6
g) through $(0, 0)$, slope 0
h) through $(1, 3)$, slope $\frac{1}{2}$

a) $4x - y - 5 = 0$     e) $x = -3$
b) $2x - y - 2 = 0$     f) $6x - y + 21 = 0$
c) $6x - y + 19 = 0$     g) $y = 0$
d) $y = 7$     h) $x - 2y + 5 = 0$

**D8.** Find the slope of the line passing through the indicated points.
   a) $(1, 2), (2, 1)$     e) $(3, 6), (4, 6)$
   b) $(1, 2), (-1, -2)$     f) $(3, 6), (3, 4)$
   c) $(1, -2), (-1, 2)$     g) $(-3, -2), (-4, 8)$
   d) $(3, 4), (6, -5)$     h) $(1, 6), (7, 1)$

   a) $-1$    b) $2$    c) $-2$    d) $-3$    e) $0$
   f) undefined    g) $-10$    h) $-\frac{5}{6}$

**D9.** Find the point of intersection of each of the following pairs of lines.
   a) $x = 3, y = 7$
   b) $x + y = 0, x - y = 2$
   c) $x = 3, 3x + 7y = 23$
   d) $2x - y = 8, y = 2$
   e) $x - 5y = 6, 2x + 3y = 25$
   f) $2x - y = 6, 3x - 2y = 2$
   g) $4x + 3y = -2, 3x + 2y = 3$
   h) $3x + 4y = 4, 4x + 5y = 3$

   a) $(3, 7)$    b) $(1, -1)$    c) $(3, 2)$    d) $(5, 2)$
   e) $(11, 1)$    f) $(10, 14)$    g) $(13, -18)$    h) $(-8, 7)$

**D10.** Find the equation of the line
   a) through $(1, 2)$ and perpendicular to $x = 3$
   b) through $(7, 11)$ and perpendicular to $y = 4$
   c) through $(3, -1)$ and perpendicular to $2x + 4y = 5$
   d) through $(2, 3)$ and perpendicular to $3x - y = 6$

   a) $y = 2$    b) $x = 7$    c) $2x - y - 7 = 0$
   d) $x + 3y - 11 = 0$

**D11.** Graph each set of lines on its own pair of axes.
   a) $y = x, y = 2x, y = 3x, y = 4x$
   b) $x + y = 1, x - y = 1, -x + y = 1, -x - y = 1$
   c) $y = x, y = x + 1, y = x - 1, y = x + 2, y = x - 2$
   d) $2x + 3y = 4, 3x + 4y = 2, 4x + 2y = 3$

**D12.** Find the equation of the line that is tangent to the circle $x^2 + y^2 = 25$ at each of the following points.
   a) $(3, 4)$      e) $(0, 5)$
   b) $(-3, 4)$     f) $(0, -5)$
   c) $(3, -4)$     g) $(5, 0)$
   d) $(-3, -4)$    h) $(\frac{5}{2}\sqrt{2}, \frac{5}{2}\sqrt{2})$

   a) $3x + 4y - 25 = 0$     e) $y = 5$
   b) $3x - 4y + 25 = 0$     f) $y = -5$
   c) $3x - 4y - 25 = 0$     g) $x = 5$
   d) $3x + 4y + 25 = 0$     h) $x + y = 5\sqrt{2}$

**M13.** Prove that the diagonals of a trapezoid and the line joining the midpoints of the parallel sides are concurrent.

**M14.** Prove by the method of analytic geometry that if a right triangle is inscribed in a circle, then the hypotenuse is a diameter of the circle.

**M15.** Show that the lines drawn from the vertices of a triangle to the midpoints of the opposite sides are concurrent. (Hint: Locate the triangle in a convenient way.)

**M16.** Prove that the diagonals of a square are perpendicular to each other.

**M17.** Show that the two lines

$$ax + by + c = 0$$
$$ax + by + d = 0$$

either coincide or are parallel.

**M18.** Show that the two lines

$$ax + by + c = 0$$
$$bx - ay + d = 0$$

are perpendicular to each other.

**M19.** Suppose lines having slopes $m_1$ and $m_2$ ($m_1 \neq m_2$) intersect at the point $(a, b)$. Find the equations of two lines through the point $(a, b)$, each of which bisects an angle determined by the original two lines. (Hint: Use the fact that $m = \tan \theta$ where $m$ is the slope of a line and $\theta$ is the angle the line makes with the $x$-axis and use the identity

$$\tan (\alpha + \beta) = \frac{\tan \alpha + \tan \beta}{1 - \tan \alpha \tan \beta}.$$

**M20.** Prove that the angle bisectors of a triangle are concurrent. (Hint: Use the formula derived in Exercise M19.)

**M21.** Prove that the distance between a line $L$ whose equation is $ax + by + c = 0$ and a point $(x_0, y_0)$ is

$$\left| \frac{ax_0 + by_0 + c}{\sqrt{a^2 + b^2}} \right|.$$

(Hint: Use the following procedure.
   a) Find the equation of the line $M$ through $(x_0, y_0)$ that is perpendicular to $L$.
   b) Find the point of intersection $(x_1, y_1)$ of the lines $M$ and $L$.
   c) Compute the distance between $(x_0, y_0)$ and $(x_1, y_1)$.)

**M22.** Find the lengths of the altitudes of the triangle with vertices $(1, 0)$, $(3, 2)$, and $(4, 6)$. (Hint: Use M21.)

**M23.** Take a triangle with vertices $A$, $B$, and $C$. Let $D$ denote the midpoint of the segment $\overline{AB}$ and $E$ denote the midpoint of the segment $\overline{BC}$. Prove that
a) $\overline{DE}$ is parallel to $\overline{AC}$ and
b) the length of $\overline{AC}$ is twice the length of $\overline{DE}$.

**M24.** Prove that the diagonals of a rhombus are perpendicular to each other.

**M25.** Find the condition that must exist in order that the angle formed by the line through $(2, 3)$ and $(x, y)$ and the line through $(2, 3)$ and $(5, -6)$ be $45°$.

## 1.4  *Functions*

The notion of a function is central to calculus and we will be concerned with it for the remainder of this book. In this section we give the definitions, establish notations, and derive some general but simple rules.

**A. Definitions**  A *function* is a rule that assigns to each real number in a given set, called the *domain,* another unique real number.

The rule describing a function can be expressed in a number of ways, but the most common is a formula. For example,

$$y = 3x + 4 \quad (x \in R)$$

and

$$V = \tfrac{4}{3}\pi r^3 \qquad (r > 0)$$

describe functions. In the first case the domain consists of all the real numbers and in the second case, the positive numbers. The letters $x$ and $r$, which represent the numbers in the domain, are called *independent variables.* The letters $y$ and $V$, which represent the numbers that are assigned by the function, are called *dependent variables.*

***Convention***  If a function is given by a formula and its domain is not otherwise specified, then we will mean that the domain consists of all numbers for which the formula makes sense. For example, the domain of $y = 2x - \frac{1}{x}$ is the nonzero real numbers, while that of $y = \sqrt{x}$ consists of the nonnegative reals. Often we must do some algebra in order to find the domain. As an illustration, consider:

→   *Most of us had no difficulty understanding what a function is. However, the various notations for function were often confusing. It is important that you understand this section or you will have difficulty with what follows.*

(Mark)

→   Recall that "$x \in R$" simply means that $x$ is an element of $R$, which in turn means $x$ is a real number.

These formulas can be expressed in words, but it is rather cumbersome to do so. Certainly "$y = 3x + 4$" is more convenient than "multiply the number by three and then add four." Not every function, however, can be expressed by a simple formula, as the following examples show:

a) The domain is $R$ and the rule is

$$y = \begin{cases} 1 \text{ if } x \text{ is rational} \\ 0 \text{ if } x \text{ is irrational.} \end{cases}$$

b) The domain is the natural numbers and the rule assigns to each positive integer $r$ the integer representing the number of primes less than $r$. (For many years mathematicians looked for a "simple" formula to describe this function, but it was eventually shown that the search was in vain.)

*Example 1* Find the domain of

$$y = \frac{3x}{2x^2 - x - 6}.$$

*Solution* We factor the denominator. Since

$$2x^2 - x - 6 = (2x + 3)(x - 2),$$

it follows that the denominator is zero when $x = -\frac{3}{2}$ or $x = 2$. Hence the domain is

$$\{x \in R : x \neq -\tfrac{3}{2}, x \neq 2\}.$$

The next two examples illustrate formulas that do not describe functions.

*Example 2* Why does the formula $y = \sqrt{-x^2 - 4}$ *not* define a real valued function?

*Solution* This is because $-x^2 - 4$ is negative for all real $x$, and the square root of a negative number is not real.

*Example 3* Why does the formula $y = \pm\sqrt{x^2 + 4}$ *not* define a function, but the formulas $y = \sqrt{x^2 + 4}$ and $y = -\sqrt{x^2 + 4}$ are functions?

*Solution* The first formula does not represent a function because it does not assign a unique number to each $x$, as is required in the definition of a function. However, the formula

$$y = \sqrt{x^2 + 4}$$

does give a function, as does the formula

$$y = -\sqrt{x^2 + 4},$$

which describes a different function.

**B. Notations and Terminology** The letters used when a function is described by a formula are not essential. For example, the formulas

$$V = \frac{4\pi r^3}{3} \quad (r > 0)$$

and

We could say a great deal more about functions. However, since the notion of function was still vague even 100 years after the invention of calculus, it is not essential that we go into great detail here. We will merely pause to make a few remarks.

Two functions $f$ and $g$ are said to be equal if their domains are the same and $f(x) = g(x)$ for all $x$ in this domain. For example, the functions

$$f(x) = x \quad (x \in R)$$

and

$$g(x) = \frac{x^2}{x} \quad (x \neq 0)$$

are not equal because their domains are different.

In the text we defined what, more exactly, is known as a real valued function of one real variable, but functions can be defined from any set to any other set. Hence there are the notions of complex valued functions and functions of many variables. The general and modern definition of function will be given in Chapter 9, and some of the above topics will be discussed more fully there.

$$y = \frac{4\pi x^3}{3} \quad (x > 0)$$

define the same function. The first formula is to be preferred when we interpret $V$ and $r$ as the volume and radius of a sphere. When no physical or geometrical interpretation is required we will follow tradition by adopting letters $x$ and $y$.

A single letter, such as $f$, $g$, $h$, $F$, $G$, $u$, or $v$, will often be used to denote a function. Given a function $f$, the number that $f$ assigns to a number $a$ in its domain is denoted by $f(a)$ and is called the *value* of $f$ at the point $a$. For example, if $f$ is to denote the function given by $y = x^2 + 1$ then $f(1) = 2$ is the value of $f$ at 1, $f(a) = a^2 + 1$ is the value of $f$ at $a$, and so on.

Expressions like $y = f(x)$, $u = u(t)$, or $V = h(r)$ will be used to denote functions in general and are very convenient.

Given a function $f$, the set of all values

$$\{f(a): a \text{ in the domain of } f\}$$

is called the *image* or *range* of the function $f$. In this book the image will always be assumed to be a subset of the real numbers.

When writing $y = f(x)$ and calling it a function we are actually abusing our notation a bit. Were we to be exceptionally fussy, we would call $f$ the function and $f(x)$ the value of the function at $x$. This would prove very inconvenient because then we could not write expressions as common as $y = x^2$ or $y = \cos x$ and call them functions.

For example, if

a) $f(x) = \dfrac{1}{x^2}$ $(x \neq 0)$, then the domain of $f$ is $\{x: x \neq 0\}$, and the image of $f$ is $(0, \infty)$; and if

b) $g(x) = 1 + x^2$, then the domain of $g$ is $R$, and the image of $g$ is $[1, \infty)$.

**C. The Graph of a Function**  Given a function $f$ with domain $D$, the *graph* of the function consists of all points in the plane of the form

$$(a, f(a)) \quad (a \in D).$$

For example, the graph of the function

$$f(x) = x^2 - 2 \quad (x \in R)$$

is given in Figure 1.

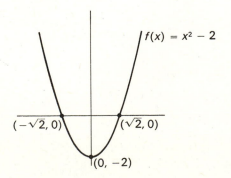

**Figure 1**

When drawing graphs of functions we will adopt the conventions illustrated in Figure 2.

a) The graph of the function $a(x) = x$   $(x \in R)$ extends infinitely in both directions. (We will not use arrowheads to denote this.)

b) The graph of $b(x) = x$   $(0 \le x \le 1)$ has solid endpoints to show that these points are included.

c) The graph of $c(x) = x$   $(0 < x < 1)$ illustrates that the endpoints are not included.

d) The graph of

$$d(x) = \begin{cases} 1 & (-1 \le x \le 0) \\ x & (0 < x < 1) \\ 2 & (1 \le x < 2) \\ \frac{1}{2} & (x = 2) \\ x - 1 & (x > 2) \end{cases}$$

summarizes the first three examples and illustrates the possible variations.

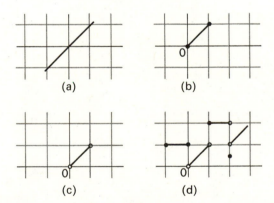

(a)        (b)

(c)        (d)

**Figure 2**

It is important to notice what distinguishes the graph of a function from the graph of an equation that does not represent a function. From the definition of a function we recall the phrase ". . . a rule that assigns to each real number in . . . the domain another *unique* real number." Hence, for each point $a$ in the domain of a function $f$ there can be only one point $(a, f(a))$ on the graph. Geometrically this means that the vertical line $x = a$ intersects the graph in exactly one point. For example, in Figure 3 the curves in (a) and (c) are the graphs of functions, but the curves shown in (b) and (d) are not the graphs of functions.

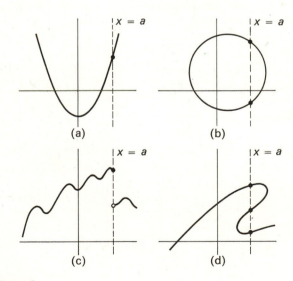

(a)        (b)

(c)        (d)

**Figure 3**

*It is often very useful to think of a function geometrically by considering its graph. A formula that defines a function contains all the information about the function, but from the graph we can immediately see the vital features of the function.*

(Vahan)

**D. Operations with Functions** As with numbers, we may add, subtract, multiply, and divide functions. Given the functions $f$ and $g$ and a real number $c$ we define the functions:

$$(f + g)(x) = f(x) + g(x)$$
$$(f - g)(x) = f(x) - g(x)$$
$$(fg)(x) = f(x) \cdot g(x)$$
$$\left(\frac{f}{g}\right)(x) = \frac{f(x)}{g(x)}$$
$$(cf)(x) = cf(x)$$

For example, if $f(x) = 2x + 3$ and $g(x) = x^2 + 6x - 8$, then

$$(f + g)(x) = x^2 + 8x - 5$$
$$(fg)(x) = (2x + 3)(x^2 + 6x - 8)$$
$$= 2x^3 + 15x^2 + 2x - 24$$

and

$$\left(\frac{f}{g}\right)(x) = \frac{2x + 3}{x^2 + 6x - 8}.$$

Thus we add, multiply, subtract, or divide functions by merely adding, multiplying, subtracting, or dividing their corresponding values. This can be seen pictorially by considering the graphs of the functions.

For example, if $f$ and $g$ have graphs as shown in Figure 4 then the graph of $f + g$ is found by plotting all points of the form $(x, f(x) + g(x))$, as in Figure 5. The graph of $f - g$ is found in a similar fashion (Fig. 6). It is more difficult to find the graphs of $fg$ and $\frac{f}{g}$ because geometrical constructions for products and quotients are usually complicated. In the following sections, we will first compute $f + g$, $f - g$, $fg$, and $\frac{f}{g}$ from the formulas given for $f$ and $g$ and then directly find the desired graphs.

A few words should be said about the domains of these functions. Denote the domains of $f$ and $g$ by $D$ and $E$ respectively. Then the functions $f + g$, $f - g$, $fg$, $f/g$ are only defined for points in both $D$ and $E$. The domain of the quotient $f/g$ must be restricted even further since it is not defined when $g(x) = 0$. For example, if

$$f(x) = \frac{1}{x} \ (x \neq 0) \quad \text{and} \quad g(x) = x^2 - 1 \ (x \in R)$$

then

$$(f + g)(x) = \frac{1}{x} + x^2 - 1 \quad \text{has domain } x \neq 0,$$

while

$$\left(\frac{f}{g}\right)(x) = \frac{1}{x(x^2 - 1)} \quad \text{has domain}$$

$$\{x \in R : x \neq 0, x \neq 1, x \neq -1\}.$$

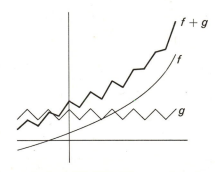

**Figure 5**

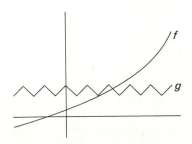

**Figure 4**

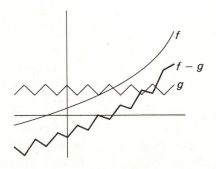

**Figure 6**

## E. Exercises

**B1.** Find the domain of each of the following functions.

a) $y = \dfrac{2x + 1}{x^2 + x - 2}$

b) $y = \sqrt{x^2 - 4}$

c) $y = \dfrac{1}{\sqrt{x + 4}}$

**B2.** Which of the following is a function? Explain.

a) $f(x) = \begin{cases} |x| & \text{for} \quad x < 1 \\ x & \text{for} \quad x > 0 \end{cases}$

b) $g(x) = \begin{cases} x^2 + 1 & \text{for} \quad x < 1 \\ 2x + 1 & \text{for} \quad x > 0 \end{cases}$

**B3.** Graph each of the following functions.

a) $f(x) = x + |x|$

b) $g(x) = |x^3 - 1|$

c) $h(x) = |x(x - 2)|$

**B4.** For $f(x) = 3 - 2x$ write $g(x)$ in terms of $x$ where

a) $g(x) = f(2x)$

b) $g(x) = f(1 - x)$

c) $g(x) = (f(x))^2$

d) $g(x) = f(f(x))$

**B5.** Explain the difference between the functions defined by

$$f(x) = \frac{x^2 - 1}{x + 1} \quad \text{and} \quad g(x) = x - 1.$$

**B6.** Let $f(x) = x^2 + 2$, $g(x) = x$, and $h(x) = x - 1$. Find the equations defining $f + g$, $gh$, and $g - h$ and then graph them.

**B7.** Let $f(x) = x^2 + x - 1$. Find $f(a + b)$, $f(a^2 - 2)$, and $f(a/2b)$.

**D8.** Find the domain of each of the following functions.

a) $y = \sqrt{x - 3}$      e) $y = \sqrt{x^2 - 1}$

b) $y = \sqrt{4 - x}$      f) $y = \sqrt{x^2 - 2x + 1}$

c) $y = \sqrt{x^2}$         g) $y = \sqrt{4 - (x - 1)^2}$

d) $y = \sqrt{4 - x^2}$    h) $y = \sqrt{x - 3} + \sqrt{5 - x}$

a) $x \geq 3$      b) $x \leq 4$      c) all $x$

d) $-2 \leq x \leq 2$      e) $x \leq -1$ or $x \geq 1$      f) all $x$

g) $-1 \leq x \leq 3$      h) $3 \leq x \leq 5$

**D9.** Find the domain of each of the following functions.

a) $y = \dfrac{1}{x}$        b) $y = \dfrac{1}{x - 2}$

c) $y = \dfrac{1}{3 - x}$        f) $y = \dfrac{x}{(x - 2)(x + 3)}$

d) $y = \dfrac{1}{x^2 - 1}$      g) $y = \dfrac{1}{x^2 - x}$

e) $y = \dfrac{3x^2}{(x - 1)^2}$     h) $y = \dfrac{1}{x} + \dfrac{2}{x - 3} + \sqrt{x - 2}$

a) $x \neq 0$      b) $x \neq 2$      c) $x \neq 3$      d) $x \neq \pm 1$

e) $x \neq 1$      f) $x \neq 2, x \neq -3$      g) $x \neq 0, x \neq 1$

h) all $x \geq 2$ except $x = 3$

**D10.** Given $f(x) = 2 - x^2$ find

a) $f(3)$        e) $f(2x)$

b) $f(-7)$       f) $f(1 - x)$

c) $f(f(3))$      g) $f(x^2)$

d) $f(-x)$       h) $(f(x))^2$

a) $-7$      b) $-47$      c) $-47$      d) $2 - x^2$

e) $2 - 4x^2$      f) $1 + 2x - x^2$      g) $2 - x^4$

h) $4 - 4x^2 + x^4$

**D11.** Express the following as functions.

a) The area $A$ of a square in terms of its side $x$.

b) The perimeter $P$ of a square in terms of its side $x$.

c) The volume $V$ of a cube in terms of its side $x$.

d) The area $A$ of an equilateral triangle in terms of its side $x$.

e) The volume $V$ of a cube in terms of its surface area $S$.

f) The diameter $D$ of a circle in terms of its area $A$.

a) $A = x^2$      c) $V = x^3$      e) $V = \left(\dfrac{S}{6}\right)^{3/2}$

b) $P = 4x$      d) $A = \dfrac{x^2\sqrt{3}}{4}$      f) $D = \sqrt{\dfrac{4A}{\pi}}$

**D12.** Graph each of the following functions.

a) $a(x) = \begin{cases} -3x & x \leq 0 \\ 2x & x > 0 \end{cases}$

b) $b(x) = \begin{cases} x + 1 & x \leq 3 \\ 2x - 2 & x > 3 \end{cases}$

c) $c(x) = \begin{cases} 0 & x < 0 \\ x^2 & 0 \leq x < 1 \\ x + 1 & 1 \leq x \end{cases}$

d) $d(x) = \begin{cases} x & x \leq 0 \\ x + 1 & 0 < x < 2 \\ -x + 2 & 2 \leq x \end{cases}$

**D13.** Let $f(x) = x^3 - 1$. Find $f(3a + 2b)$, $f(-a)$, $f(a^2b)$ and $f(1/b^2)$.

**M14.** Suppose a function $f$ satisfies $f(1) = 1$ and $f(x) = f(x - 1) + 2$ for all $x$. Find
  a) $f(3)$
  b) $f(10)$
  c) $f(x) - f(x - 4)$

**M15.** Determine conditions on $a$, $b$, and $c$ so that the function $f(x) = ax^2 + bx + c$ satisfies $f(x) = x^2 f(\frac{1}{x})$ for all $x \neq 0$.

**M16.** Suppose $f(x) = ax + b$ and $g(x) = cx + d$.
  a) Find $f(g(x))$.
  b) Find $g(f(x))$.
  c) When is $f(g(x)) = g(f(x))$ for all $x$?
  d) If $a = 2$, $b = 1$, and $f(g(x)) = x + 2$, find $c$ and $d$.

**M17.** Which functions of the form $f(x) = ax + b$ can be written in the form
  a) $g(g(x))$ where $g(x) = cx + d$,
  b) $(g(x))^2$ where $g(x) = cx + d$?

**M18.** A function $f$ is said to be *even* if $f(x) = f(-x)$ for all $x$ in its domain and *odd* if $-f(x) = f(-x)$.
  a) Given any function $g$, show that the function $g(x) + g(-x)$ is even and that the function $g(x) - g(-x)$ is odd.
  b) Show that any function $g$ can be written in the form $g = h + k$ where $h$ is an even function and $k$ is an odd function.
  c) Show that the decomposition given in (b) is unique. That is, if $g = h_1 + k_1$ where $h_1$ is even and $k_1$ is odd, then $h = h_1$ and $k = k_1$.

**M19.** Show that if $f$ is an odd function for which $f(0)$ is defined, then $f(0) = 0$.

**M20.** a) If $f$ is an even function, show that $f^{[2]}$, defined by $f^{[2]}(x) = f(f(x))$, is also even. What can you say about $f^{[n]}$, where $n$ is any positive integer?
  b) If $f$ is an odd function, show that $f^{[2]}$ is also odd. What can you say about $f^{[n]}$?

**M21.** If $f(x) = ax + b$, $a \neq 0$, find a function $g$ such that $g(f(x)) = f(g(x)) = x$.

**M22.** Suppose a function $f$ satisfies $f(0) = 1$ and $f(n) = nf(n - 1)$ for all positive integers $n$. Find $f(1)$, $f(2)$, $f(3)$, and $f(6)$. Can you identify this familiar function?

---

<table>
<tr><td>1.5</td><td>Some Special Functions<br>and Equations</td></tr>
</table>

In this section we will consider several important functions and equations that will occur frequently in the sequel. Most of this space will be devoted to the trigonometric functions and conic sections.

→    The general definition of function was motivated by the much earlier study of a wide variety of special functions. Even today they still occupy a central position in mathematics.

**A. The Constant Functions**   The function defined by

$$f(x) = C \quad (x \in R),$$

where $C$ is some real number, is called the *constant function C* and is usually denoted simply by $C$. In particular, if $C = 1$, $f(x) = 1 (x \in R)$ and its graph is as shown (Fig. 1).

There are many constant functions since $C$ can be any real number. The graph of a constant function is a horizontal line (Fig. 2a), and not simply a point (Fig. 2b).

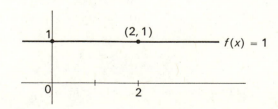

**Figure 1**

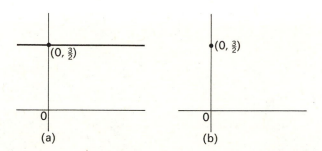

**Figure 2**

### B. Radian Measure

Draw a circle of radius 1 with center $E$, and let $P$ denote a point on its circumference (Fig. 3). Now, given a number $x \geq 0$, move counterclockwise about the circle starting at $P$ until $x$ units of length have been traversed. Letting $Q$ denote the point reached, the angle $PEQ$ is then said to have measure $x$ radians.

Angles can also have a negative measure. Indeed, if $y < 0$ we can start at $P$ and move along the circle (Fig. 4) in a clockwise direction until we arrive $-y$ units later at a point $Q$. Angle $PEQ$ is then said to be $y$ radians.

We also say that angle $PEQ$ (Fig. 3) has measure $x + 2\pi$ radians or, more generally, $x + 2\pi n$ radians ($n$ an integer). This is because each time we traverse the circle the length changes by $2\pi$, but the angle does not change.

The reader is probably more familiar with degrees than radians. However, in this book we will use only radian measure, and a conversion table is provided on the front cover. For a given angle, the relation between its radian measure $x$ and its degree measure $\theta$ is given by the formula

$$x = \frac{\pi}{180}\theta.$$

Two simple, important formulas concerning radian measure are given in the following theorem:

→ *It is customary to refer to this angle as simply x rather than saying that it has measure x radians. This usage will be followed in our book.*

　　　　　　　　　　　　　　(Steve)

→ *This is similar to the fact that an angle of $\theta$ degrees is also an angle of $\theta + 360$ degrees or, more generally, $\theta + 360n$ degrees.*

　　　　　　　　　　　　　　(Ed)

→ By proportions we have (Fig. 5) $\frac{360}{2\pi} = \frac{\theta}{x}$, so

$$x = \frac{\pi}{180}\theta.$$

**Figure 3**　　　**Figure 4**

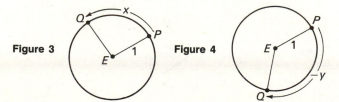

**Figure 5**

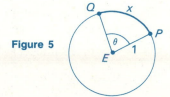

**THEOREM 1** *Draw a circle* (Fig. 6) *of radius r with x denoting the indicated angle. Letting A be the area of the sector and L the length of the arc, we have:*

a) $L = rx$

b) $A = \dfrac{r^2 x}{2}$

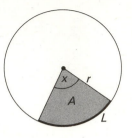

**Figure 6**

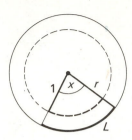

**Figure 7**

*Proof* a) Draw a concentric circle of radius 1 (Fig. 7). Since the angle is $x$ the length of the dotted arc is also $x$, and by proportions we have

$$\frac{x}{L} = \frac{1}{r}$$

or

$$L = rx.$$

b) From Figure 6

$$\frac{A}{\text{area of circle}} = \frac{L}{2\pi r} = \frac{rx}{2\pi r} = \frac{x}{2\pi}.$$

Since the area of the circle is $\pi r^2$,

$$\frac{A}{\pi r^2} = \frac{x}{2\pi}$$

or

$$A = \frac{x \pi r^2}{2\pi} = \frac{r^2 x}{2} \quad \bullet$$

> Recall that the area of a circle of radius $r$ is $\pi r^2$. A derivation of this fact based on the techniques of calculus is given in Example 6 (5.5B).

**C. The Trigonometric Functions** Given a real number $x$, take an angle of $x$ radians, measured in counterclockwise direction if $x \geq 0$ and clockwise direction if $x < 0$ (Fig. 8). Now choose any point $(a, b) \neq (0, 0)$ on the half-line $H$. The *trigonometric functions* are defined by the formulas:

> *In part B of this section we assigned a real number to a given angle. The reader may notice that when we say, "Given a real number x, take an angle of x radians," the reverse process is required.*
>
> (Estelle)

$$\sin x = \frac{b}{c} \qquad \csc x = \frac{c}{b}$$

$$\cos x = \frac{a}{c} \qquad \sec x = \frac{c}{a}$$

$$\tan x = \frac{b}{a} \qquad \cot x = \frac{a}{b}$$

> These definitions do not depend on the choice of the point $(a, b)$ because, by proportions (Fig. 8), if $(a', b') \neq (0, 0)$ is another point on $H$ we have
>
> $$\frac{b}{c} = \frac{b'}{c'}, \frac{a}{c} = \frac{a'}{c'}, \text{ etc.}$$

It is important to notice the periodic nature of the trigonometric functions. For example,

$$\sin x = \sin(x + 2\pi n) \quad (n \text{ an integer}).$$

The main trigonometric identities we will use in this book are:

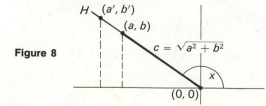

**Figure 8**

a) $\sin^2 x + \cos^2 x = 1$

b) $1 + \tan^2 x = \sec^2 x$

c) $1 + \cot^2 x = \csc^2 x$

d) $\sin (x \pm y) = \sin x \cos y \pm \cos x \sin y$

e) $\cos (x \pm y) = \cos x \cos y \mp \sin x \sin y$

f) $\cos^2 x = \dfrac{1 + \cos 2x}{2}$

g) $\sin^2 x = \dfrac{1 - \cos 2x}{2}$

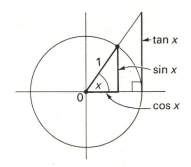

**Figure 9**

The first three identities are easy to verify. For example,

$$\sin^2 x + \cos^2 x = \left(\frac{b}{c}\right)^2 + \left(\frac{a}{c}\right)^2 = \frac{b^2 + a^2}{c^2} = \frac{c^2}{c^2} = 1.$$

However, proofs of identities (d) and (e) are more difficult, and the reader is referred to a trigonometry text. (A sophisticated proof will be given in 10.3C.) Identities (f) and (g) follow from (a) and (e). For example, letting $x = y$ in (e) and taking the plus case we have

$$\cos 2x = \cos^2 x - \sin^2 x.$$

Using (a) this becomes

$$\cos 2x = \cos^2 x - (1 - \cos^2 x) = 2 \cos^2 x - 1.$$

Solving for $\cos^2 x$ gives

$$\cos^2 x = \frac{1 + \cos 2x}{2},$$

which is (f).

Our main interest will be with the sine, cosine, and tangent functions. Their graphs can easily be constructed with the aid of Figure 9.

For example, the graph of $y = \sin x$ $(0 \leq x \leq 2\pi)$ is shown by the heavy line in Figure 10. Then, using the periodicity of the sine function, the graph is extended as shown. The graphs of cosine (Fig. 11) and tangent (Fig. 12) are found in a similar manner. Notice that the domains of sine and cosine are $R$, while their images are $[-1, 1]$. For tangent, however, the domain is

$$\left\{ x \in R : x \neq \pm \frac{\pi}{2}, \pm \frac{3\pi}{2}, \pm \frac{5\pi}{2}, \ldots \right\}$$

and the image is $R$.

This is because the properties of the remaining three functions are easily derived from those of the first three.

Figure 9 refers to the interval $[0, \frac{\pi}{2}]$ but, if one is careful with signs, he may use it for all $x \in R$.

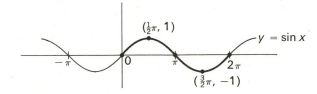

**Figure 10**

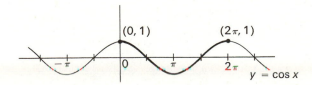

**Figure 11**

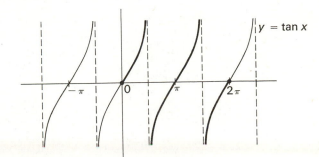

**Figure 12**

**D. The Ellipse, Parabola, and Hyperbola** Recall (1.2C) that the circle centered at $(h, k)$ and having radius $r$ has the equation

$$(x - h)^2 + (y - k)^2 = r^2 \quad \text{(Fig. 13).}$$

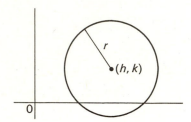

**Figure 13**

These curves, collectively called conic sections, were extensively studied by the Greek geometers, and derive their name from the fact that they occur as plane sections of a cone (Fig. 14).

Descartes was the first to prove that every conic section has an equation of the form

$$Ax^2 + Bxy + Cy^2 + Dx + Ey + F = 0$$

where $A$, $B$, $C$, $D$, $E$, and $F$ are constants. The result is important because with this discovery conic sections were removed from the domain of geometry and were placed into the eager hands of the algebraists. The techniques from algebra brought fresh ideas to geometry, and each subject was enriched.

The circle is a special case of the more general *ellipse*. An equation of the form

$$\frac{(x - h)^2}{a^2} + \frac{(y - k)^2}{b^2} = 1$$

describes an ellipse with *center* $(h, k)$ and it is clear that when $a = b$ the equation reduces to that of a circle. In order to graph the ellipse, it will be easier to consider first the case $h = k = 0$. Hence we have

$$\frac{x^2}{a^2} + \frac{y^2}{b^2} = 1,$$

(of course, in this equation and the previous one, the constants $a$ and $b$ must be nonzero) and clearly the four points $(a, 0)$, $(-a, 0)$, $(0, b)$, $(0, -b)$ (Fig. 15) satisfy the equation. Now, solving for $y$ we have

$$y = \pm \frac{b}{a} \sqrt{a^2 - x^2}.$$

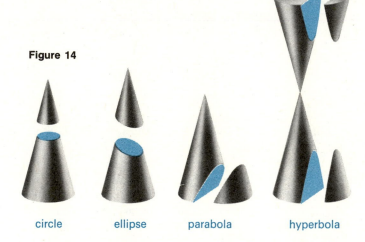

**Figure 14**

circle          ellipse          parabola          hyperbola

In fact we may assume that $a$ and $b$ are both positive, since $(-a)^2 = a^2$.

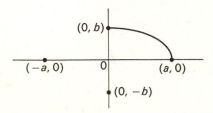

**Figure 15**

For $x, y > 0$ this gives

$$y = +\frac{b}{a} \sqrt{a^2 - x^2}.$$

This equation shows that as $x$ varies from 0 to $a$, $y$ varies from $b$ to 0, and the portion of the graph in the first quadrant is as shown (Fig. 15). The remainder of the graph is obtained by observing that if the point $(x, y)$ lies on it, then so do the points $(-x, y)$, $(-x, -y)$, and $(x, -y)$ because $x$ and $y$ are both squared. Hence we obtain Figure 16. As in the case of the circle, if the equation has the form

$$\frac{(x - h)^2}{a^2} + \frac{(y - k)^2}{b^2} = 1,$$

then the graph is obtained by translating the graph in Figure 16 to have center $(h, k)$ (Fig. 17).

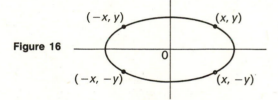

Figure 16

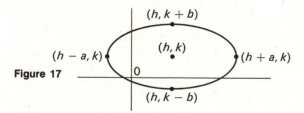

Figure 17

The graphs, easily found, of $y = px^2$ and $x = py^2$ ($p$ a nonzero constant) are compared in Figure 18. As in the case of an ellipse, if $x$ and $y$ are replaced by $x - h$ and $y - k$ the graph is merely translated so that the vertex is at the point $(h, k)$. For example, the graphs of

$$y = x^2$$

and

$$(y + 1) = (x - 3)^2$$

are illustrated in Figure 19. All curves represented in Figures 18 and 19 are known as *parabolas*.

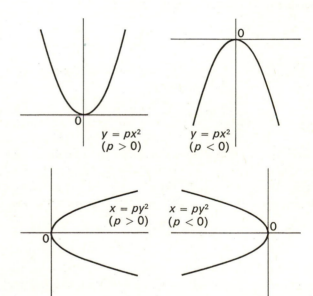

$y = px^2$ ($p > 0$)    $y = px^2$ ($p < 0$)

$x = py^2$ ($p > 0$)    $x = py^2$ ($p < 0$)

Figure 18

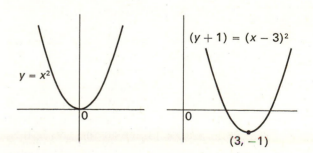

$y = x^2$

$(y + 1) = (x - 3)^2$

$(3, -1)$

Figure 19

Finally, the most common forms for the equation of a *hyperbola* are

a) $y = \dfrac{p}{x}$   ($p$ a nonzero constant)     (Fig. 20)

b) $\dfrac{x^2}{a^2} - \dfrac{y^2}{b^2} = 1$                          (Fig. 21)

c) $\dfrac{y^2}{b^2} - \dfrac{x^2}{a^2} = 1$                          (Fig. 22)

⟶    Here also the constants $a$ and $b$ must be nonzero and may be assumed to be positive.

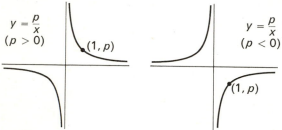

**Figure 20**

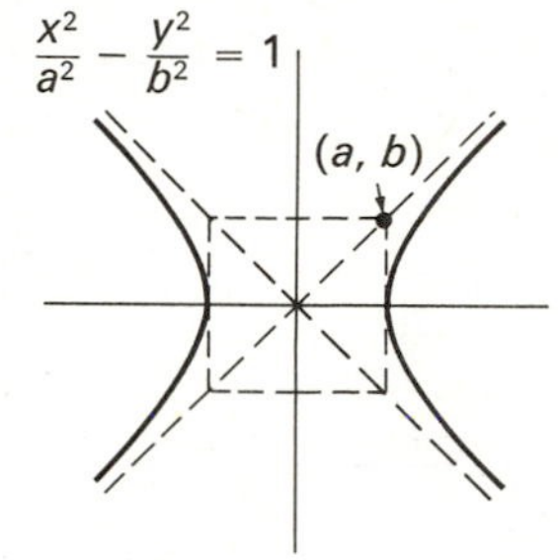

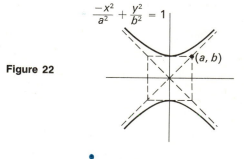

**Figure 21**

**Figure 22**

The graph of (a) for $p > 0$ is obtained by noting that if $(x, y)$ lies on the graph, then $x$ and $y$ have the same sign, so the graph falls in quadrants I and III. Observe that as $x$ increases $y$ must decrease. The case for $p < 0$ is treated similarly. More work must be done to find the graphs of the equations in (b) and (c), and we will only consider the special case of

$$x^2 - y^2 = 1$$

to illustrate the method. Notice that if a point $(x, y)$ satisfies this equation, then we must have $|x| \geq 1$, for if $-1 < x < 1$ then $y^2 = x^2 - 1$ would be negative and this is impossible. Hence the graph is excluded from the shaded region shown (Fig. 23). Restricting to $x, y \geq 0$ and solving for $y$ we obtain

$$y = \sqrt{x^2 - 1}.$$

Clearly $y = 0$ when $x = 1$, and $y$ increases as $x$ does; hence the general shape of the graph in the first quadrant is shown by the heavy curve. Since $y = \sqrt{x^2 - 1} < x$ the curve must lie below the graph of the line $y = x$. In fact the graph gets nearer and nearer to the line $y = x$ as $x$ increases. Finally, as with the ellipse (Fig. 16), if $(x, y)$ lies on the graph, then so do the points $(-x, y)$, $(x, -y)$, and $(-x, -y)$ so the graph is as shown in Figure 23.

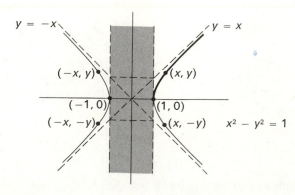

**Figure 23**

Only a few cases of ellipses, parabolas, and hyperbolas have been shown in the figures above. Any curve obtained by rotating one of these is also called respectively an ellipse, a parabola, or a hyperbola.

→ *Fantastically detailed studies of conic sections have been done by certain devoted individuals. A classic example is George Salmon's* A Treatise on Conic Sections *(New York: Chelsea Publishing Co., 1954).*

(Steve)

## E. Exercises

**B1.** Express the following angles in radians.
   a) A right angle
   b) A straight angle
   c) The sum of the interior angles of a triangle
   d) The sum of the interior angles of a quadrilateral (a four-sided figure)
   e) The angle $POQ$ where $P = (1, 1)$, $O = (0, 0)$, and $Q = (0, 1)$
   f) The angle $POQ$ where $P = (1, \sqrt{3})$, $O = (0, 0)$, and $Q = (0, 1)$

**B2.** Use the identities

$$\sin^2 x + \cos^2 x = 1 \quad \text{and} \quad \cos 2x = \cos^2 x - \sin^2 x$$

to establish the following identities.

   a) $\cos^2 x = \dfrac{1 + \cos 2x}{2}$     b) $\sin^2 x = \dfrac{1 - \cos 2x}{2}$

**B3.** Graph the functions

$$f(x) = \sin x, \quad g(x) = \sin 2x, \quad h(x) = \sin \tfrac{x}{2}.$$

**B4.** Graph and identify the following conic sections.
   a) $y = 4x^2 - 8$
   b) $4(y - 3)^2 = 9(x - 2)^2 - 36$
   c) $4x^2 + y^2 = 16$

**B5.** Find the two points of intersection of

$$2x^2 + y^2 = 8 \quad \text{and} \quad y = x^2.$$

**B6.** Show that for any point $P$ on the graph of

$$\frac{x^2}{25} + \frac{y^2}{16} = 1$$

the distance from $P$ to $(3, 0)$ plus the distance from $P$ to $(-3, 0)$ is always the same.

**D7.** Evaluate each of the following quantities.
   a) $\sin \frac{\pi}{6}$     e) $\cos \frac{-\pi}{3}$     i) $\sin \frac{\pi}{2} \tan \frac{\pi}{4}$
   b) $\sin \frac{\pi}{4}$     f) $\cos \frac{2\pi}{3}$     j) $\sin \frac{\pi}{6} \tan \frac{\pi}{6}$
   c) $\sin \frac{2\pi}{3}$     g) $\tan 3\pi$     k) $\sin^2 \frac{\pi}{6} \cos^2 \frac{\pi}{6}$
   d) $\sin 8\pi$     h) $\tan \frac{\pi}{4}$     l) $\sin^2 \frac{\pi}{7} + \cos^2 \frac{\pi}{7}$

   a) $\frac{1}{2}$     b) $\frac{\sqrt{2}}{2}$     c) $\frac{\sqrt{3}}{2}$     d) $0$     e) $\frac{1}{2}$

   f) $\frac{-1}{2}$     g) $0$     h) $1$     i) $1$     j) $\frac{\sqrt{3}}{6}$

   k) $\frac{3}{16}$     l) $1$

**D8.** Graph the following functions.
   a) $y = \cos 2x$          e) $y = \tan \frac{x}{2}$
   b) $y = 2 \cos x$          f) $y = 1 + \sin x$
   c) $y = \cos (x - \frac{\pi}{2})$     g) $y = 1 - \sin x$
   d) $y = \cos (x - \frac{\pi}{3})$     h) $y = \cos x - \sin x$

**D9.** On the same set of axes graph the following.
   a) $\dfrac{x^2}{9} + \dfrac{y^2}{16} = 1$     c) $\dfrac{x^2}{9} + \dfrac{y^2}{9} = 1$
   b) $\dfrac{x^2}{16} + \dfrac{y^2}{9} = 1$     d) $\dfrac{x^2}{16} + \dfrac{y^2}{16} = 1$

**D10.** Graph each of the following equations.
   a) $(y - 1)^2 = (x - 3)^2$
   b) $(y - 1)^2 + (x - 3)^2 = 4$
   c) $x^2 - 4y^2 = 4$
   d) $4x^2 + y^2 = 100$
   e) $y = (x - 1)^2 + 4$
   f) $(y - 1)^2 = (x - 1)^2 + 4$
   g) $(y - 1)^2 + (x - 1)^2 = 4$
   h) $2x^2 + 3y^2 = 4$

**D11.** Find where the graphs of the following pairs of equations intersect.
   a) $x^2 + y^2 = 25$, $(x - 7)^2 + y^2 = 4$
   b) $x^2 + y^2 = 2$, $x = y^2$
   c) $x^2 + y^2 = 4$, $y + 2 = x^2$
   d) $2x^2 + y^2 = 17$, $3x^2 - y^2 = 3$

   a) $(5, 0)$     b) $(1, 1)$, $(1, -1)$
   c) $(0, -2)$, $(\sqrt{3}, 1)$, $(-\sqrt{3}, 1)$
   d) $(2, 3)$, $(2, -3)$, $(-2, 3)$, $(-2, -3)$

**D12.** Find the $x$-coordinate of the points where the graph of the function $f(x) = \sin x$ intersects each of the following curves.
   a) $g(x) = 0$     c) $g(x) = \cos x$
   b) $g(x) = 1$     d) $g(x) = \tan x$

a) $n\pi$    b) $\frac{\pi}{2} + 2n\pi$    c) $\frac{\pi}{4} + n\pi$    d) $n\pi$
(Here $n$ can take any integer value.)

**M13.** Find where the graph of $f(x) = \sin x$ intersects the graph of
a) $g(x) = \sin 2x$    b) $g(x) = \cos 2x$

**M14.** Given the hyperbola $\dfrac{x^2}{9} - \dfrac{y^2}{4} = 1$ find all lines through the origin that do not intersect its graph.

**M15.** Find the equation of the line through the point $(3, 1)$ that intersects the ellipse $x^2 + 2y^2 = 11$ at only one point.

**M16.** There are two lines that pass through $(-3, 4)$ each of which intersects the parabola $y^2 = 16x$ at only one point. Find them.

**M17.** It can be shown that the graph of $y = ax^2 + bx + c$ is a parabola if $a \neq 0$. Find $a$, $b$, and $c$ for the parabola which passes through the points $(1, 3)$, $(2, 4)$ and $(3, 7)$.

**M18.** Find the values of $a$, $b$, and $c$ if the ellipse $4x^2 + y^2 + ax + by + c = 0$ is to be tangent to the $x$-axis at the origin and to pass through the point $(-1, 2)$.

**M19.** Show that the equation

$$\frac{x^2}{7 - a} + \frac{y^2}{4 - a} = 1$$

represents an ellipse if $a < 4$ and a hyperbola if $4 < a < 7$.

**M20.** Every parabola is a set of points each of which is equidistant from a fixed line and a fixed point. Prove this for the parabola $y = (x - a)^2 + b$.

**M21.** Given the ellipse $\dfrac{x^2}{a^2} + \dfrac{y^2}{b^2} = 1$ with $a > b$ and a point $P$ on the ellipse, show that the distance from $P$ to $(\sqrt{a^2 - b^2},\ 0)$ plus the distance from $P$ to $(-\sqrt{a^2 - b^2},\ 0)$ is a constant.

**M22.** Given the hyperbola $\dfrac{x^2}{a^2} - \dfrac{y^2}{b^2} = 1$ and a point $P$ on the hyperbola, show that the absolute value of the difference of the distance from $P$ to $(\sqrt{a^2 + b^2}, 0)$ and the distance from $P$ to $(-\sqrt{a^2 + b^2}, 0)$ is constant.

**M23.** (The parabolic mirror) Consider the parabola $y^2 = 4x$.
a) For each point $P$ on the parabola find the equation of the line that intersects the parabola only at $P$.
b) Let $F = (1, 0)$ and draw a segment from $F$ to a point $P$ on the parabola and then a half line to the right from $P$ parallel to the $x$-axis. Show that the line that bisects the angle formed by the segment and the half line is perpendicular to the line described in (a).

---

## 1.6  *Miscellaneous Topics*

In this section we will discuss completing the square, polar coordinates, and parametric equations.

*→*  *We found it helpful to review these topics immediately before we needed them.*

*(Jean)*

**A. Completing the Square**  The device of completing the square is both useful and simple.
We will start with an expression of the form

$$u^2 + au.$$

Now we add and subtract $\dfrac{a^2}{4}$ to obtain

$$u^2 + au = u^2 + au + \frac{a^2}{4} - \frac{a^2}{4}$$

*→*  *This will be applied below as an aid to graphing conic sections, and will also be used when we study techniques of integration.*
*The phrase "completing the square" refers to writing the original expression as a square plus a constant, which is often much easier to deal with.*

*(Lee)*

$$u^2 + au = \left(u + \frac{a}{2}\right)^2 - \frac{a^2}{4}.$$

Hence we have the formula

$$u^2 + au = \left(u + \frac{a}{2}\right)^2 - \frac{a^2}{4}.$$

An example of how this is used as an aid to graphing is given in:

---

*Example 1*   Find the graph of

$$y = x^2 - 4x + 1$$

by completing the square for $x^2 - 4x$.

*Solution*   Noticing that $a = -4$, we write

$$\begin{aligned}
(x^2 - 4x) + 1 &= (x^2 - 4x + 4 - 4) + 1 \\
&= (x^2 - 4x + 4) - 3 \\
&= (x - 2)^2 - 3.
\end{aligned}$$

Hence our original equation has the form

$$(y + 3) = (x - 2)^2.$$

This is the equation of a parabola whose graph is as shown in Figure 1.

---

*Example 2*   Describe the graph of the equation

$$4x^2 - 16x + 9y^2 + 18y = 11.$$

*Solution*   We write the given equation in the form

$$4(x^2 - 4x) + 9(y^2 + 2y) = 11.$$

Completing the squares gives

$$4(x^2 - 4x + 4 - 4) + 9(y^2 + 2y + 1 - 1) = 11$$

or

$$4(x^2 - 4x + 4) + 9(y^2 + 2y + 1) = 11 + 16 + 9.$$

This is the same as

$$4(x - 2)^2 + 9(y + 1)^2 = 36$$

or

$$\frac{(x - 2)^2}{9} + \frac{(y + 1)^2}{4} = 1,$$

---

→ In each of the equations below we have completed the square, but the reader should work these examples himself.

a) $x^2 - 8x = (x - 4)^2 - 16$

b) $x^2 + 6x = (x + 3)^2 - 9$

c) $y^2 + y = (y + \frac{1}{2})^2 - \frac{1}{4}$

d) $y^2 - 4y + 8 = (y - 2)^2 + 4$

e) $3x^2 + 6x = 3(x + 1)^2 - 3$

f) $x^2 + 2x + y^2 - 6y = (x + 1)^2 + (y - 3)^2 - 10$

g) $2y^2 + 5y + 7 = 2(y + \frac{5}{4})^2 + \frac{31}{8}$

h) $x^2 + 2y^2 - 4x + y - 12 = (x - 2)^2 + 2(y + \frac{1}{4})^2 - \frac{129}{8}$

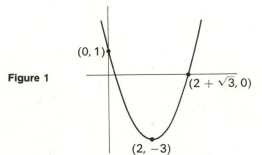

**Figure 1**

(0, 1)

$(2 + \sqrt{3}, 0)$

(2, −3)

which is an ellipse with center at $(2, -1)$.

---

**B. Polar Coordinates** The most common method of locating a point $P$ in the plane is by means of the rectangular coordinates $(x, y)$ used until now (Fig. 2). However, in certain cases, other methods are more convenient. One of the most important of these is *polar coordinates* where $(r, \theta)$ is used to locate the point $P$ (Fig. 3).

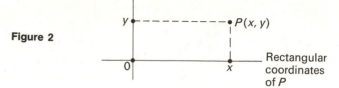

Figure 2

Rectangular coordinates of $P$

The term "polar" derives from the fact that in many physical problems in which circular symmetry exists, the origin is treated as a pole, that is, the center of the symmetry. Rectangular coordinates are also called *Cartesian coordinates* after René Descartes.

Figure 3

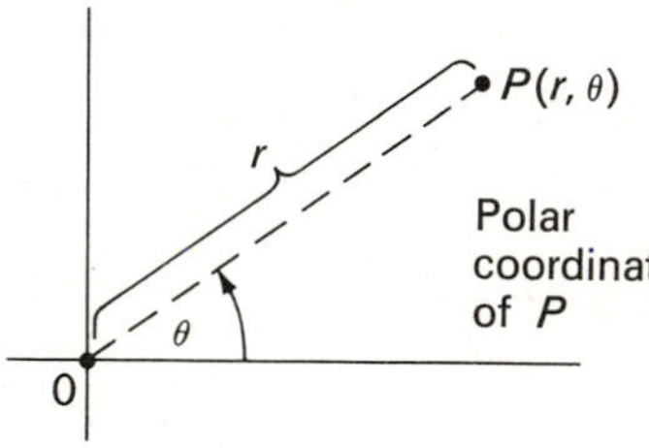

Polar coordinates of $P$

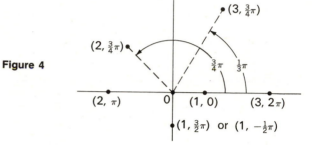

Figure 4

Here $r$ is the distance of the point $P$ from the origin $O$, and $\theta$ the angle from the positive $x$-axis to the line determined by the points $O$ and $P$ (Fig. 4).

Given a point $P$, its rectangular coordinates $(x, y)$ can easily be expressed (Fig. 6) in terms of the polar coordinates $(r, \theta)$ by the formulas

$$x = r \cos \theta \quad \text{and} \quad y = r \sin \theta.$$

Recall that since distance (1.2B) is always nonnegative, it follows that $r \geq 0$.

*In some texts $r$ is a point on a real line through the points $P$ and $O$ and may assume negative values. In this case the points with polar coordinates $(-1, \frac{1}{4}\pi)$ and $(-1, \frac{1}{2}\pi)$ as compared to those with polar coordinates $(1, \frac{1}{4}\pi)$ and $(1, \frac{1}{2}\pi)$ are plotted as shown in Figure 5.*

(Pat)

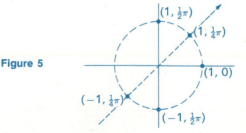

Figure 5

Figure 6

These equations can be reversed to obtain $r$ and $\theta$ in terms of $x$ and $y$ by

$$r^2 = x^2 + y^2 \quad \text{and} \quad \tan \theta = \frac{y}{x}.$$

Each of the coordinate systems, rectangular and polar, has its own advantages. For example, consider the

Because $x^2 + y^2 = (r \cos \theta)^2 + (r \sin \theta)^2$
$$= r^2 \cos^2 \theta + r^2 \sin^2 \theta = r^2,$$
$\sqrt{x^2 + y^2} = r$. Also $\dfrac{y}{x} = \dfrac{r \sin \theta}{r \cos \theta} = \tan \theta.$

Since the representation of an angle is not unique, it follows that the polar coordinates of a point are not unique. For example, $(2, \frac{\pi}{3}) = (2, \frac{7\pi}{3}) = (2, -\frac{5\pi}{3})$. For this reason we cannot solve for $\theta$, but only $\tan \theta$, in terms of $x$ and $y$.

circle of radius 4 centered at 0, and the vertical line through the point $(2, 0)$. In rectangular coordinates the equations are

$$x^2 + y^2 = 16 \text{ (circle)} \quad \text{and} \quad x = 2 \text{ (line)},$$

while in polar coordinates they are

$$r = 4 \text{ (circle)} \quad \text{and} \quad r \cos \theta = 2 \text{ (line)}$$

(Fig. 7). Notice that the equation of the circle is simpler in polar coordinates while that of the line is simpler in rectangular coordinates.

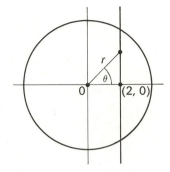

**Figure 7**

---

*Example 3*   Convert

$$x^2 + (y - 3)^2 = 9$$

into an equation in polar coordinates.

*Solution*   We use the relations $y = r \sin \theta$ and $x = r \cos \theta$. Substituting these in the given equation, we have

$$r^2 \cos^2 \theta + (r \sin \theta - 3)^2 = 9$$

or $\qquad r^2 \cos^2 \theta + r^2 \sin^2 \theta - 6r \sin \theta = 0,$

so $\qquad r^2 = 6r \sin \theta \quad \text{and} \quad r = 6 \sin \theta.$

---

*Example 4*   Convert $r^2 + r^2 \cos^2 \theta = 4$ into an equation in rectangular coordinates.

*Solution*   We first replace $\cos \theta$ by $\dfrac{x}{r}$ giving

$$r^2 + r^2 \frac{x^2}{r^2} = 4 \quad \text{or} \quad r^2 + x^2 = 4.$$

Since $r^2 = x^2 + y^2$, we have

$$x^2 + y^2 + x^2 = 4 \quad \text{or} \quad 2x^2 + y^2 = 4.$$

---

*Example 5*   Graph the equation

$$r = 2 \cos^2 \theta.$$

*Solution*   We will plot some points and use the table.

| When $\theta$ goes from | 0 to $\frac{\pi}{2}$ | $\frac{\pi}{2}$ to $\pi$ | $\pi$ to $\frac{3\pi}{2}$ | $\frac{3\pi}{2}$ to $2\pi$ |
|---|---|---|---|---|
| $\cos \theta$ goes from | 1 to 0 | 0 to $-1$ | $-1$ to 0 | 0 to 1 |
| $r = 2 \cos^2 \theta$ goes from | 2 to 0 | 0 to 2 | 2 to 0 | 0 to 2 |

The points $(2, 0)$, $(1, \frac{\pi}{4})$, and $(0, \frac{\pi}{2})$ lie on the graph, and using the table we see that the graph in the first quadrant is as shown by the heavy line in Figure 8.

*Problem 2*   Graph the equation $r = 1 - \sin \theta$.

*Add segments of the line $r = \dfrac{1}{\cos \theta - \sin \theta}$ to the graph in Problem 2 and fill in initials. If my comment does not make sense to you, then you probably made a mistake in your solution.*

(Mark)

---

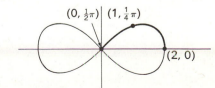

$(0, \frac{1}{2}\pi)$   $(1, \frac{1}{4}\pi)$      $(2, 0)$

**Figure 8**

The graph in the remaining quadrants is found similarly.

Recall that in Cartesian coordinates a vertical line intersects the graph in at most a single point iff the graph is that of a function. With r as a function of θ, however, this test is not valid in polar coordinates. Example 5 is a good illustration, for it is the graph of a function, yet many vertical lines intersect it twice.

(Steve)

### C. Parametric Equations  Equations of the form

$$x = f(t) \quad \text{and} \quad y = g(t)$$

are said to be *parametric equations* for the variables $x$ and $y$ in terms of the *parameter t*. Each value of the parameter $t$ specifies values for $x$ and $y$, given by the equations, and determines a point $(x, y)$ in the plane. The set of points obtained for different values of $t$ then defines a curve. As an illustration:

Such equations are often written in the form $x = x(t)$ and $y = y(t)$ to simplify notation. The parameter can be denoted by any letter, but $t$ is the most popular because it often represents time.

*Example 6*  Find the curve determined by the equations

$$x = t^2 \quad \text{and} \quad y = t + 1.$$

*Solution*  We start by plotting several points.

In higher dimensions (three or more) the definition of a curve in terms of a parameter is very convenient. Our use of parameters will be limited to some examples dealing with physical problems.

| When $t =$ | 0 | 1 | $-1$ | 2 | $-2$ | 3 | $-3$ | $\frac{1}{2}$ | $-\frac{1}{2}$ |
|---|---|---|---|---|---|---|---|---|---|
| $x =$ | 0 | 1 | 1 | 4 | 4 | 9 | 9 | $\frac{1}{4}$ | $\frac{1}{4}$ |
| $y =$ | 1 | 2 | 0 | 3 | $-1$ | 4 | $-2$ | $\frac{3}{2}$ | $\frac{1}{2}$ |

Hence the points $(0, 1)$, $(1, 2)$, $(1, 0)$, $(4, 3)$, $(4, -1)$, $(9, 4)$, $(9, -2)$, $(\frac{1}{4}, \frac{3}{2})$, and $(\frac{1}{4}, \frac{1}{2})$ lie on the curve (Fig. 9). Now, connecting the points we obtain the illustrated curve.

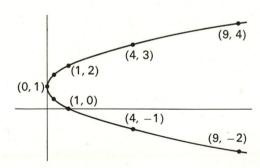

**Figure 9**

A simpler method of finding the graph, given parametric equations, may be used when we can solve for the parameter in one equation and substitute that value in the other equation. As an illustration, in Example 6 we have

$$x = t^2 \quad \text{and} \quad y = t + 1.$$

Solving for $t$ in the second equation gives $t = y - 1$, and putting this in the first equation gives

$$x = (y - 1)^2,$$

which, as we know, is the equation of a parabola.

Sometimes it is either very difficult or impossible to solve for one of the variables in terms of the other, and in this case one must rely on the slow but reliable method of plotting points. However, we can sometimes simplify this process by carefully observing the behavior of $x$ and $y$ as $t$ varies. An instance of this is given in Example 8.

Another simple method is shown in the following example.

---

*Example 7*  Find the curve determined by the equations

$$x = a \cos t \quad \text{and} \quad y = a \sin t.$$

*Solution*

$$\frac{x}{a} = \cos t \quad \text{and} \quad \frac{y}{a} = \sin t,$$

so

$$\left(\frac{x}{a}\right)^2 + \left(\frac{y}{a}\right)^2 = \cos^2 t + \sin^2 t = 1.$$

Hence $x^2 + y^2 = a^2$, and the graph of this equation is a circle centered at $(0,0)$ and having radius $a$ (Fig. 10).

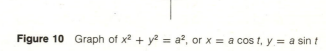

**Figure 10**  Graph of $x^2 + y^2 = a^2$, or $x = a \cos t$, $y = a \sin t$

---

*Example 8*  Find the graph of the equations

$$x = t - \sin t, \quad y = 1 - \cos t.$$

*Solution*  We observe carefully the behavior of the variables with respect to the parameter $t$. Notice the following values and how the graph varies between them. Corresponding to $t = 0$, $\pi$, and $2\pi$ we have $(0, 0)$, $(\pi, 2)$, and $(2\pi, 0)$. Between $t = 0$ and $\pi$, $y$ increases from 0 to 2 and then returns to 0 as $t$ goes from $\pi$ to $2\pi$ (Fig. 11). Since $(t + 2\pi - \sin t, 1 - \cos t)$ lies on the graph whenever $(t - \sin t, 1 - \cos t)$ does, the graph is periodic.

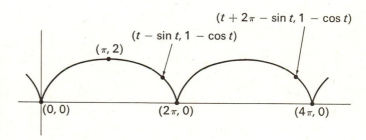

**Figure 11**

---

### D. Exercises

**B1.** Find the graphs of the following equations.
  a) $x^2 + 4y^2 - 2x - 16y + 13 = 0$
  b) $y^2 - 2x - 6y = -4$
  c) $9x^2 - 4y^2 - 36x - 24y = 36$

**B2.** Convert the following expressions from rectangular to polar form.
  a) $x^2 + y^2 - 4x = 0$
  b) $y^2 = \dfrac{x^3}{2 - x}$
  c) $(x^2 + y^2)^{3/2} = 2xy$

**B3.** Convert the following expressions from polar to rectangular form and then find the graph.
  a) $\sin \theta = \dfrac{6}{r}$
  b) $\sin \theta = \dfrac{r}{6}$
  c) $r^2 \sin 2\theta = 8$

**B4.** Given a circle of radius 2 centered at the origin, describe it
  a) using Cartesian coordinates,
  b) using polar coordinates,
  c) using parametric equations.

**B5.** Convert the following pairs of parametric equations to a single equation in rectangular coordinates.
a) $x = 3 \cos t, y = 3 \sin t$
b) $x = t + 1, y = t^2 + 2t + 1$

**B6.** Graph the polar equations

$$r = 4 + 4 \cos \theta \quad \text{and} \quad \frac{3}{1 - \cos \theta}$$

and find the four points of intersection in polar form.

**B7.** Show that the parametric equations $x = 2t + 3$ and $y = t^2 - 4$ define a parabola, and then graph this parabola.

**D8.** Graph the curve defined parametrically by the equations

$$x = \frac{t^2}{1 + t^2}, \quad y = \frac{t^3}{1 + t^2}.$$

**D9.** Complete the square and then graph the following parabolas.
a) $y = x^2 + 2x$               e) $y = 2x^2 + 4x$
b) $y = x^2 - 4x + 1$           f) $y = 2x^2 - 8x + 12$
c) $y = x^2 + 4x + 6$           g) $2y = x^2 + 14x + 11$
d) $y = 2x - x^2$               h) $y = 6 + 4x - 2x^2$

a) $y + 1 = (x + 1)^2$          e) $y + 2 = 2(x + 1)^2$
b) $y + 3 = (x - 2)^2$          f) $y - 4 = 2(x - 2)^2$
c) $y - 2 = (x + 2)^2$          g) $2(y + 19) = (x + 7)^2$
d) $y - 1 = -(x - 1)^2$         h) $y - 8 = -2(x - 1)^2$

**D10.** Convert each of the following equations to an equation in rectangular coordinates and identify its graph.

a) $r = \dfrac{1}{3 \sin \theta - 4 \cos \theta}$       d) $r = 8 \sin \theta$

b) $r = 8$                        e) $r \cos \theta = 4$
c) $r = 4 \cos \theta$            f) $\theta = \frac{\pi}{4}$

a) $3y - 4x = 1$                  d) $x^2 + (y - 4)^2 = 16$
b) $x^2 + y^2 = 64$               e) $x = 4$
c) $(x - 2)^2 + y^2 = 4$          f) $y = x$ where $x \geq 0$

**D11.** Graph the following polar equations.
a) $r = 1 + \cos \theta$         e) $r = 1 + \sin \theta$
b) $r = 2 + \cos \theta$         f) $r = 1 + 2 \sin \theta$
c) $r = 1 + 2 \cos \theta$       g) $r = 2 + \sin \theta$
d) $r = 1 - \cos \theta$         h) $r = \sin \theta + \cos \theta$

**D12.** Graph the following polar equations.
a) $r = \sin 2\theta$            c) $r = \sin 3\theta$
b) $r = \cos 2\theta$            d) $r = \sin 4\theta$

**D13.** Convert the following pairs of parametric equations to a single equation in rectangular coordinates.
a) $x = 4t, y = 2t + 1$
b) $x = t - 5, y = 3t^2 + 4$
c) $x^2 = t + 1, y = 3t + 1$
d) $x = 3 \tan t, y = 3 \sec t$
e) $x = \cos t - \sin t, y = \cos t + \sin t$
f) $x = t^2 + 1, y = t^3 - 4$

a) $x - 2y + 2 = 0$              d) $-x^2 + y^2 = 9$
b) $y - 4 = 3(x + 5)^2$          e) $x^2 + y^2 = 2$
c) $y + 2 = 3x^2$                f) $x - 1 = (y + 4)^{2/3}$

**M14.** Graph the curves represented parametrically by
a) $x = t, y = \cos t$
b) $x = \sin t, y = t$
c) $x = 1 + \sin t, y = t - \cos t$
d) $x = t + \cos t, y = 1 - \sin t$

**M15.** a) Using Cartesian coordinates graph the following:

$$x = 2, \qquad y = 2, \qquad x = y,$$

and

$$x + y = 2.$$

b) Using polar coordinates graph the following:

$$r = 2, \qquad \theta = 2, \qquad r = \theta,$$

and

$$r + \theta = 2.$$

**M16.** Show that $x = \sin^2 t, y = \cos^2 t$ represents the part of the line $x + y = 1$ in the first quadrant.

**M17.** Show that the equations

$$x = a + k \cos \theta$$
$$y = b + k \sin \theta \quad (k \neq 0)$$

form parametric equations for a circle. Find the center and radius of this circle.

**M18.** Compare the graphs of

$$r^2 + \theta^2 = 1 \quad \text{(polar coordinates)}$$

and

$$x^2 + y^2 = 1 \quad \text{(Cartesian coordinates)}.$$

**M19.** Given the points $(r_1, \theta_1)$ and $(r_2, \theta_2)$ in polar coordinates, derive a formula for the distance between them.

**M20.** Consider the curve given parametrically by $x = t$, $y = t^2$ ($t \in R$) and compare it with the curve that is parametrically given by $x = \sin t, y = \sin^2 t$ ($t \in R$).

**M21.** Find three distinct pairs of parametric equations each of which describes the line $2x + y = 4$.

**M22.** A wheel of radius 1 is to be rolled along the $x$-axis. At time $t = 0$ the center of the wheel is at $(0, 1)$ and a black dot is placed on the wheel at $(0, 0)$. Assuming the wheel is moved to the right at a constant rate, show that the position $(x, y)$ of the black dot at time $t$ is given by

$$x = t - \sin t, \quad y = 1 - \cos t.$$

**M23.** (Old Baldy.) On a *single* sheet of paper sketch the graphs of the following equations.

a) $9x^2 + 4y^2 + 8y = 32$
b) $x^2 + y^2 + 8x - 8y + 28 = 0$
c) $16x^2 + 320x + y^2 + 1584 = 0$
d) $16x^2 + y^2 + 16y + 48 = 0$
e) $4x^2 + 4y^2 - 32x - 32y + 127 = 0$
f) $16x^2 - 320x + y^2 + 1584 = 0$
g) $x^2 + y^2 - 8x - 8y + 28 = 0$
h) $196x^2 + 81y^2 = 15{,}876$
i) $4x^2 + 4y^2 + 32x - 32y + 127 = 0$

## 1.7 Sample Exams

### Sample Exam 1 (45–60 minutes)

**1.** Find the equation of the line passing through the point $(2, 3)$ and perpendicular to the line $-2y + x + 7 = 0$.

**2.** Draw the graph of the function defined by

$$f(x) = \begin{cases} 2 & (x \le 0) \\ 3 & (0 < x < 2\pi) \\ 1 & (x = 2\pi) \\ \sin x & (x > 2\pi) \end{cases}$$

**3.** What point on the line $y = \frac{x}{2} + 5$ is at a distance of $3\sqrt{5}$ from the point $(-1, 12)$?

**4.** Find the graph of the equation $2x^2 + y^2 = 4x$.

**5.** Find the graph of the equation $r = \dfrac{2}{1 + \cos \theta}$ where $r$ and $\theta$ are the polar coordinates of a point.

### Sample Exam 2 (45–60 minutes)

**1.** Find the intervals determined by the points $x$ satisfying
a) $|x - 1| \le 5$     b) $|x - 1| < 5$
and give a sketch of each.

**2.** Determine the point of intersection of the lines

$$2x + 3y = 1 \quad \text{and} \quad y = 6x - 3.$$

**3.** Given $f(x) = x^2 - x$ find
a) $f(0)$     b) $f(2x)$     c) $f(2x) - 2f(x)$

**4.** Plot the points $(\sin \frac{5\pi}{2}, \cos \frac{5\pi}{2})$ and $(\tan \pi, \cos (\frac{\pi}{3} + \pi))$ and find the distance between them.

**5.** For what values of $x$ are the following equations undefined?

a) $f(x) = \dfrac{2 \cos x}{\sin x}$     b) $g(x) = \sqrt{x^2 - 3}$.

**6.** Show that the parametric equations $x = t^2 + 6t + 3$, $y = t + 1$ define a parabola, and then graph the parabola.

### Sample Exam 3 (45–60 minutes)

**1.** Convert $r(\sin \theta + 2 \cos \theta) = 1 + r^2 \cos^2 \theta$ to rectangular form and then determine the graph of the resulting equation.

**2.** Find the points common to the following pairs of intervals.
a) $(-1, 3)$, $\{x : |x - 1| \le 2\}$
b) $\{x : |x - 2| < 3\}$, $\{x : |x - 1| \le 4\}$

**3.** Complete the square in the following equations, and determine the type of conic section each represents.
a) $x^2 - 2y^2 - 2x + 4y + 7 = 0$
b) $x^2 - 10x - y - 26 = 0$
c) $4x^2 + 4y^2 - 8x - 4y + 1 = 0$

**4.** Graph the curve given parametrically by $x = t$, $y = 2 \sin 2t$.

**5.** Find the area of the triangle determined by the $x$-axis and the lines

$$2y - 3x - 6 = 0 \quad \text{and} \quad 3x + y - 12 = 0.$$

**6.** A line $L$ has equation $2x + 3y = 8$. Find the value of $b$ so that the line $x + by = 10$
   a) intersects $L$ at $(1, 2)$
   b) is perpendicular to $L$
   c) is parallel to $L$

**7.** Given $f(x) = 2x + 1$
   a) Find two values of $a$ satisfying $f(a^2) = (f(a))^2$.
   b) Show that no $b$ satisfies $f(2b) = 2f(b)$.

*Sample Exam 4 (45–60 minutes)*

**1.** Suppose $f(x) = \begin{cases} 3 & (x \le 1) \\ 4 - x & (x > 1) \end{cases}$

and

$$g(x) = \begin{cases} x + 2 & (x \le 0) \\ 2x + 2 & (x > 0) \end{cases}.$$

Describe the function $h(x) = f(g(x))$ and find its graph.

**2.** Where do the perpendicular bisectors of the sides of the triangle formed by the points $(0, 8)$, $(-2, 2)$, and $(4, 6)$ meet? (It suffices to find where any two of these lines intersect.)

**3.** Find the circle with center $(1, 2)$ which is tangent to the line $x + y = 7$.

**4.** Find the graph of the curve that is defined parametrically by the equations

$$x = \frac{1 - t^2}{1 + t^2}, \quad y = \frac{2t}{1 + t^2}.$$

(Warning: It is important to miss a point.)

**5.** Graph the following polar equations over the indicated regions.
   a) $r = \sin 3\theta \quad (0 \le \theta \le \frac{\pi}{6})$
   b) $r = \sin \frac{\theta}{3} \quad (0 \le \theta \le \frac{3\pi}{2})$.

In this chapter we introduce the derivative. By means of the derivative we can find the tangent to a curve at a point. The definition of the derivative depends upon the concept of a limit, the fundamental concept of calculus. The systematic use of this concept distinguishes calculus from all previous mathematics.

We will determine the derivatives of the polynomial and trigonometric functions and establish the first basic rules for the calculation of derivatives. The material presented in this chapter and the next one will enable the reader to differentiate the most common functions. In Chapter 4, he will then learn how these techniques are applied to many different types of problems.

# The Derivative

49

## 2.1   *The Tangent to a Curve at a Point*

Differentiation is fundamentally related to the intuitive idea of a tangent to a curve. To make this notion precise we will have to introduce a "limiting process." →

*If you just rely on what you already know about tangents and what you think a limit is, it might be easier to understand what's going on.*

(Dana)

**A. Motivation**   Let $P$ be a point on the circle $C$ with center $A$ (Fig. 1). The tangent $T$ to the circle at the point $P$ is the line through $P$ that intersects the circle only at the point $P$. As we know from Euclidean geometry, $T$ can also be described as the line through $P$ that is perpendicular to the segment $AP$.

So far, this is all very easy, but now consider the curve $D$ (Fig. 2) and the point $P$ as indicated. There is no line through $P$ that touches $D$ only at $P$. Nevertheless, it is possible to define a tangent to $D$ at $P$, but to do so requires the use of a new idea.

**B. A New Method of Defining Tangents**   We will illustrate this new idea by first applying it to the case of a circle. Let $C$ denote the circle, $P$ be a point on the circumference, and let $T$ be the tangent through $P$ (Fig. 3).

Choose another point $Q$ on the circle, with $Q \neq P$, and let $L_{PQ}$ denote the line passing through the points $P$ and $Q$. Consider what happens to the line $L_{PQ}$ as the point $Q$ moves toward $P$. →

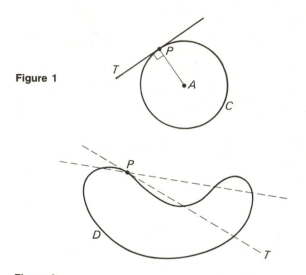

Figure 1

Figure 2

The idea of finding tangents based on a "limiting notion" was known to the Frenchmen Blaise Pascal and Pierre Fermat, but neither recognized its full significance. Leibniz, one of the co-inventors of calculus, used Pascal's work, and was at a loss to understand why Pascal did not see the importance of developing it.

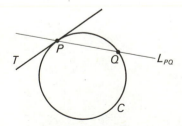

Figure 3

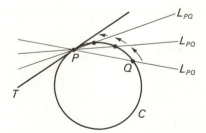

Figure 4

As indicated in Figure 4, $L_{PQ}$ clearly approaches the tangent $T$. This is the basic idea. The tangent $T$ at the point $P$ is the limit of the lines $L_{PQ}$ as $Q$ approaches $P$. →

Notice that the same phenomenon occurs if $Q$ is to the left of $P$ (Fig. 5).

Now let us return to the curve $D$. Let $P$ denote a point on $D$, $Q$ another point on $D$, and $L_{PQ}$ the line determined by $P$ and $Q$ (Fig. 6).

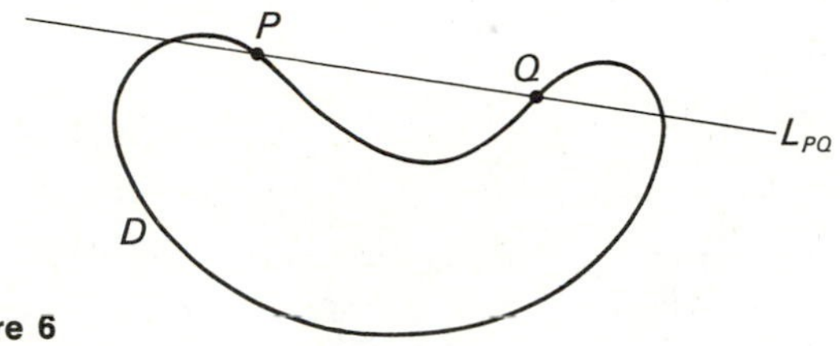


**Figure 6**

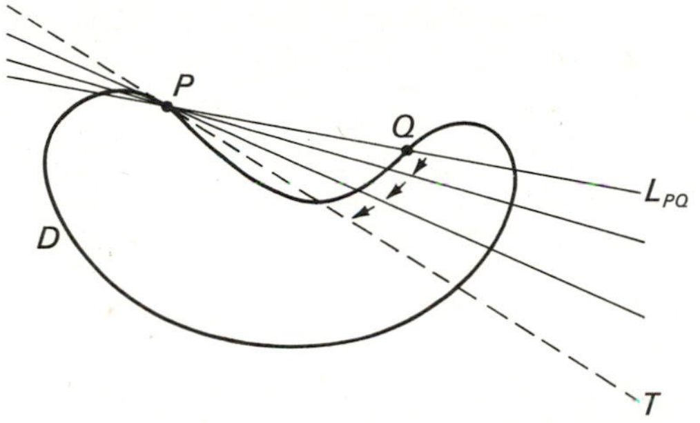


**Figure 7** As $Q$ approaches $P$, the line $L_{PQ}$ approaches the limiting position $T$ as indicated by the dotted line.

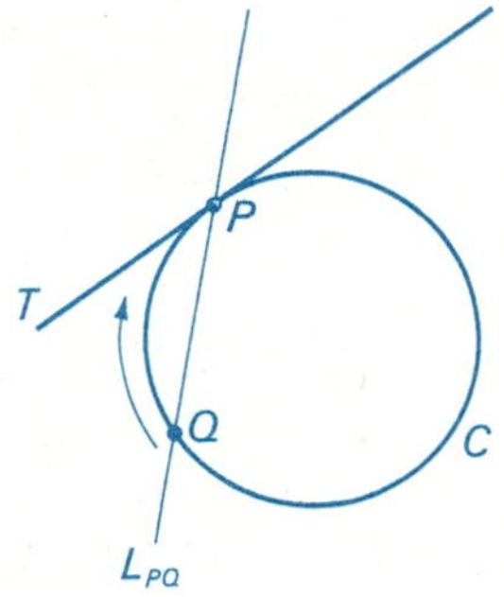


**Figure 5**

Then, if the lines $L_{PQ}$ approach a unique line $T$ as $Q$ approaches $P$, this limiting line is called the *tangent to D at P* (Fig. 7). More simply, we then say that a tangent to $D$ at $P$ exists. It is important to note, however, that we only say a tangent exists when the lines $L_{PQ}$ approach a unique limiting line $T$ as $Q$ approaches $P$ from either side.

This type of limiting notion is fundamental to calculus and will occur frequently in many forms.

→ An example in which the tangent does not exist is shown in Figure 8.

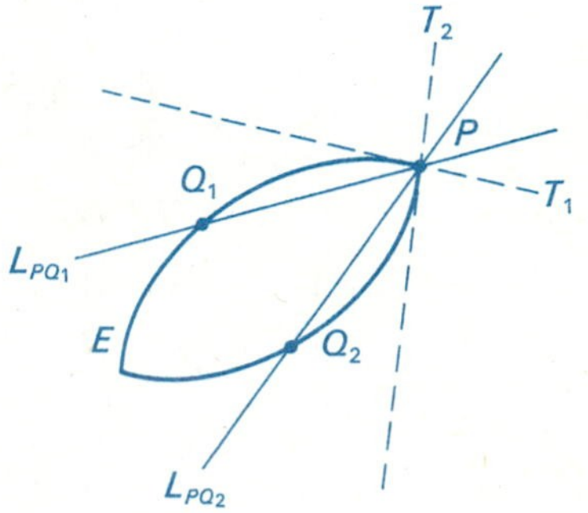


**Figure 8** The lines $L_{PQ_1}$ approach $T_1$ while the lines $L_{PQ_2}$ approach $T_2$. Since $T_1 \neq T_2$, no tangent to the circle $E$ at $P$ exists.

**C. The Derivative of a Function at a Point** We will now apply these ideas to a special class of curves, namely those which are graphs of functions.

Let $f$ be a function whose graph is as shown in Figure 9. Choose a point $x$ and let $P$ denote the point $(x, f(x))$ on the graph. Next let $h$ be a nonzero number and let $Q$ denote the point $(x + h, f(x + h))$. We then have

$$\text{slope of } L_{PQ} = \frac{f(x+h) - f(x)}{h}.$$

Now let the point $Q$ approach $P$. This is clearly the same as letting $h$ approach zero. Then, if the numbers

$$\frac{f(x + h) - f(x)}{h}$$

approach a unique number, that number is called the *derivative of f at x*. We will denote it by $f'(x)$.

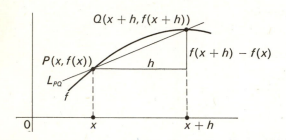

**Figure 9**

To simplify our writing we will use an arrow "$\rightarrow$" in place of the word "approach." Hence, if the derivative of $f$ at $x$ exists we have

$$\frac{f(x + h) - f(x)}{h} \rightarrow f'(x) \quad \text{as} \quad h \rightarrow 0.$$

This is also written in the following way:

$$\lim_{\substack{h \rightarrow 0 \\ h \neq 0}} \frac{f(x + h) - f(x)}{h} = f'(x).$$

The tangent line to the graph of $f$ at $P$ is now defined as the line through $(x, f(x))$ that has slope equal to $f'(x)$ (Fig. 10). Briefly, as $h \rightarrow 0$,

$$\text{slope of } L_{PQ} = \frac{f(x + h) - f(x)}{h}$$

approaches

$$f'(x) = \text{slope of tangent at } P.$$

## D. Equations of the Tangent and Normal Lines

Given a function $f$, suppose we know its derivative $f'(a)$ at $a$. We may compute the equation of the tangent line $T$ through the point $(a, f(a))$. We see (Fig. 11) that

$$\frac{y - f(a)}{x - a} = f'(a) \quad \text{or}$$

$$y = f(a) + f'(a)(x - a)$$

Note that if $f$ is defined in an open interval about the point $x$, as in Figure 9, then $\frac{f(x + h) - f(x)}{h}$ must approach a unique number as $h$ approaches zero independent of $h$ being positive or negative; that is, $Q$ may approach $P$ from either side. On the other hand, suppose there is an open interval about $x$ such that $f$ is defined at $x$ and only on one side of $x$ in the interval, for example, as would be the case if the domain of $f$ were a closed interval $[a, b]$, $a < b$, and $x$ were one of the endpoints. Then only positive or negative values of $h$ need be considered, depending upon which side of $x$ the function is defined on. For our purposes, the derivative shall only be defined at $x$ if one of the above cases occurs.

Throughout the first five sections of this chapter this intuitive notion of limit will be sufficient. We will return to this concept in Section 2.6, where it will be studied rigorously and in greater detail. This information is essential for the development of more advanced branches of analysis and takes considerable time to master. The reader is advised to be patient. He may look ahead to see what is coming, but most of his time should be spent in understanding the fundamental ideas.

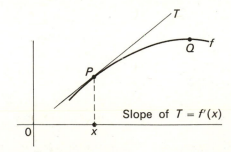

**Figure 10**

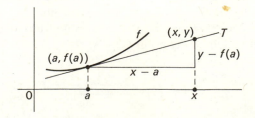

**Figure 11**

which is the equation of the tangent line.

The *normal* line $N$ (Fig. 12) at the point $(a, f(a))$ is defined to be the line through $(a, f(a))$ perpendicular to the tangent $T$ through $(a, f(a))$.

---

**Example 1** Find the equation of the tangent to the graph of $f(x) = x^2 - 2x + 3$ at the point $(-1, 6)$.

*Solution* We first compute $f'(x)$. We have

$$\frac{f(x + h) - f(x)}{h}$$

$$= \frac{((x + h)^2 - 2(x + h) + 3) - (x^2 - 2x + 3)}{h}$$

$$= \frac{x^2 + 2xh + h^2 - 2x - 2h + 3 - x^2 + 2x - 3}{h}$$

$$= \frac{2xh + h^2 - 2h}{h}$$

$$= 2x + h - 2.$$

Now $2x + h - 2 \to 2x - 2$ as $h \to 0$, so

$$f'(x) = 2x - 2.$$

Hence the slope of the tangent at $(-1, 6)$ equals

$$f'(-1) = 2(-1) - 2 = -4.$$

The equation of the tangent line is then given by

$$\frac{y - 6}{x - (-1)} = -4 \quad \text{or} \quad y + 4x = 2.$$

---

Sometimes we will have occasion to use the phrase *the slope of a curve at a point*. By this we will mean the slope of the tangent line at that point. Also, two curves will be said to be *perpendicular* at a point of intersection if their respective tangent lines are perpendicular (Fig. 13).

**E. Terminology** We will list a few definitions pertaining to what we have been discussing.

Suppose we are given a function $f$. Then:

a) The *number $f'(x)$* is called the *derivative of $f$ at* $x$. →

b) The *function $f'$* that assigns to each number $x$ the number $f'(x)$ is called the *derivative of $f$*.

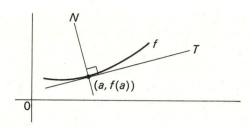

**Figure 12**

---

**Problem 1** Using the formula derived in Example 1 for the derivative of $f(x) = x^2 - 2x + 3$, find the equation of the normal line at the point $(2, 3)$.

*Answer:* $x + 2y = 8$

---

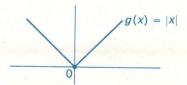

**Figure 13** Curves $C$ and $D$ are perpendicular at $P$, but not at $Q$.

---

Recall that this number need not exist. For example if $g(x) = |x|$ (Fig. 14), then $g'(0)$ does not exist. Indeed

$$\frac{g(h) - g(0)}{h} = \frac{|h| - 0}{h} = \begin{cases} 1 & \text{if } h > 0. \\ -1 & \text{if } h < 0. \end{cases}$$

Hence, there can be no number $L$ satisfying

$$\frac{g(h) - g(0)}{h} \to L \quad \text{as} \quad h \to 0.$$

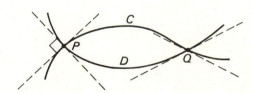

**Figure 14**

c) The *process* of going from the function $f$ to the function $f'$ is called *differentiation*.

d) The *reverse process* of going from the function $f'$ to the function $f$ is called *integration*.

e) The *subject* that consists of the study of differentiation and integration is *calculus*.

→ In this book we will study calculus, but will also learn how its techniques can be applied to solve numerous problems that arise in both the physical and social sciences.

## F. Exercises

In the problems below find the derivatives, when asked, directly from the definition. More exactly, to find the derivative of $f$, you should find the limit of $\dfrac{f(x + h) - f(x)}{h}$ as $h$ approaches zero. Do *not* use formulas from the following sections.

**B1.** Find the derivatives of the following functions.

   a) $f(x) = 2x^2 - x$     b) $g(x) = \dfrac{1}{x}$

**B2.** Find the equation of the tangent line at the point indicated.
   a) $y = x^2 - 2x$           $(2, 0)$
   b) $y = x^2 + 4x + 2$       $(-2, -2)$

**B3.** Find where the derivative of each of the following functions is equal to zero.

   a) $y = x^2 + 2x$     b) $y = \dfrac{x^2}{2} - \dfrac{1}{x}$

**B4.** Find the point where the slope of $y = x^2 + 2x - 1$ is equal to 2.

**B5.** What is the equation of the normal to the graph of $f(x) = x^3 + 1$ at the point $(2, 9)$?

**B6.** Show that the derivative of $f(x) = x|x|$ is
   a)      0   when   $x = 0$,
   b)    $2x$   when   $x > 0$,
   c)   $-2x$   when   $x < 0$.

**D7.** Find the derivatives of the following functions.
   a) $y = 2x$          e) $y = 3x^2 + 4x$
   b) $y = 8x - 6$      f) $y = -3 + 6x - x^2$
   c) $y = 4x^2$        g) $y = (x + 2)^2$
   d) $y = 5x^2 + 7$    h) $y = (x - 3)(x + 4)$

   a) 2    b) 8    c) $8x$    d) $10x$    e) $6x + 4$
   f) $6 - 2x$    g) $2x + 4$    h) $2x + 1$

**D8.** Given $f(x) = 2x^2 + 4x - 4$ find the equations of the lines described below.
   a) The tangent line at $(0, -4)$
   b) The tangent line at $(1, 2)$
   c) The tangent line at $(-1, -6)$
   d) The normal line at $(1, 2)$
   e) The normal line at $(-1, -6)$

   a) $4x - y = 4$     b) $8x - y = 6$     c) $y = -6$
   d) $x + 8y = 17$     e) $x = -1$

**D9.** Find where the derivative of each of the following functions is equal to zero.
   a) $y = 2x^2$                 c) $y = 4x^2 - 7x + 95$
   b) $y = x^2 - 2x$             d) $y = \dfrac{3}{x}$

   a) $x = 0$     b) $x = 1$     c) $x = \tfrac{7}{8}$     d) no $x$

**M10.** Show that no two tangents of $y = x^2$ are parallel.

**M11.** Find the equation of the line that passes through the point $(2, 0)$ and is perpendicular to the tangent line to $f(x) = 2x^2 - 4x$ at that point.

**M12.** Show that the graphs of $f(x) = 3x^2$ and $g(x) = 2x^3 + 1$ have the same tangent lines at the point $(1, 3)$.

**M13.** What is the equation of the line having slope 4 that is tangent to the graph of $f(x) = 3x^2 + 1$?

**M14.** Show that the graphs of

$$f(x) = \frac{x^2}{2} \quad \text{and} \quad g(x) = \frac{(x - 2)^2}{2}$$

are perpendicular at their point of intersection.

**M15.** Prove that the tangent line to

$$f(x) = \frac{1}{x} \quad \text{at} \quad \left(a, \frac{1}{a}\right)(a \neq 0)$$

intersects the graph of $f$ only at $(a, \tfrac{1}{a})$.

**M16.** Show that the tangent line to $g(x) = \frac{1}{x^2}$ at $\left(a, \frac{1}{a^2}\right)$ $(a \neq 0)$ always intersects the graph of $g$ at exactly two points.

**M17.** Given $f(x) = \sqrt{x}$, prove that $f'(x) = \frac{1}{2\sqrt{x}}$.

$\left(\text{Hint: Use the identity } \sqrt{a} - \sqrt{b} = \frac{a - b}{\sqrt{a} + \sqrt{b}}.\right)$

**M18.** Given $f(x) = \sqrt[3]{x}$, show that $f'(x) = \frac{1}{3\sqrt[3]{x^2}}$.

$\left(\text{Hint: Use the identity}\right.$

$$\sqrt[3]{a} - \sqrt[3]{b} = \frac{a - b}{(\sqrt[3]{a})^2 + \sqrt[3]{a}\sqrt[3]{b} + (\sqrt[3]{b})^2}.\bigg)$$

**M19.** Show that if $f(x)$ is differentiable for all $x$ and periodic with period $p$—that is $f(x + p) = f(x)$ for all $x$, $p = $ constant—then $f'(x)$ is periodic with period $p$.

**M20.** Where is the function $f(x) = \frac{1 + |x|}{1 - |x|}$ not differentiable?

**M21.** Show that a function $f$ satisfying $|f(x)| \leq x^2$ for all $x$ must be differentiable at $x = 0$.

**M22.** Suppose $f(0) = 0$ and $|f(x)| \geq |x|^{1/2}$ for all $x \neq 0$. Show that $f$ is not differentiable at $x = 0$.

**M23.** Suppose that $f$ is differentiable at all points and $f(x) = f(-x)$. Show that $f'(x) = -f'(-x)$.

**M24.** Prove that if $f$ is differentiable at $x$ and $f(x) \neq 0$ then $g = \frac{1}{f}$ is differentiable at $x$ and $g'(x) = \frac{-f'(x)}{f^2(x)}$.

**M25.** Suppose $f(x) = 0$ if $x$ is irrational, $f(0) = 0$, and $f(x) = \frac{1}{q^2}$ if $x = \frac{p}{q}$ where $q > 0$ and $p$ and $q$ are integers with no common factor. Show that $f$ is differentiable at $x = 0$.

---

## 2.2   *The Derivative of $x^n$*

We will prove (Theorem 4 of 2.2C) that the derivative of $x^n$ is $nx^{n-1}$. Since the method of proof is very important, we will first illustrate it with some special cases.

⟶   In this discussion $n$ denotes a nonnegative integer, but later it will be shown that $n$ can be any real number. (See the Corollary (2.5C), Theorem 1 (3.2B), and Theorem 2 (3.4D).)

**A. The Method** In what follows we will repeatedly encounter expressions $E(h)$ involving a variable $h$. We will need to investigate their behavior as $h$ approaches some number $b$. If $E(h)$ approaches a number $L$, then $L$ is called the *limit* of $E(h)$ as $h$ approaches $b$. We write

⟶   Neither Newton nor Leibniz, the creators of calculus, had a precise notion of limit. Newton thought in very physical terms while Leibniz had only a vague mathematical conception of limit. Our notion of limit, which is at least as clear as that of Newton and Leibniz, will carry us through Section 2.5.

$$E(h) \to L \quad \text{as} \quad h \to b$$

or, more succinctly,

$$\lim_{h \to b} E(h) = L.$$

For example, if $A$ and $B$ are constants, then

$$A + Bh^n \to A \quad \text{as} \quad h \to 0.$$

⟶   This is true because as $h \to 0$, $h^n \to 0$ and $Bh^n \to 0$, whereas $A$ remains constant. Thus $A + Bh^n \to A + B \cdot 0 = A$ as $h \to 0$.

In this section, every proof about derivatives begins

with a function $f$ which is to be differentiated. We will compute

$$\frac{f(x + h) - f(x)}{h}$$

and then find its limit as $h \to 0$. In this analysis we use the particular limit mentioned above.

**B. Special Cases**  Since the graph of a constant function $f(x) = K$ is a horizontal line, its derivative is zero. Almost as easy to see is:

**THEOREM 1**   *If $f(x) = x$ (Fig. 1), then $f'(x) = 1$.*

*Proof*  Since

$$\frac{f(x + h) - f(x)}{h} = \frac{x + h - x}{h} = \frac{h}{h} = 1,$$

the limit of the Newton Quotient equals 1  ●

**THEOREM 2**   *If $f(x) = x^2$ (Fig. 2), then $f'(x) = 2x$.*

*Proof*  In this case

$$\frac{f(x + h) - f(x)}{h} = \frac{(x + h)^2 - x^2}{h}$$

$$= \frac{x^2 + 2xh + h^2 - x^2}{h}$$

$$= \frac{2xh + h^2}{h}$$

$$= 2x + h.$$

Since $2x + h \to 2x$ as $h \to 0$, we have

$$f'(x) = \lim_{h \to 0} \frac{f(x + h) - f(x)}{h} = 2x  ●$$

**THEOREM 3**   *If $f(x) = x^3$ (Fig. 3), then $f'(x) = 3x^2$.*

*Proof*  In this case

$$\frac{f(x + h) - f(x)}{h} = \frac{(x + h)^3 - x^3}{h}$$

$$= \frac{x^3 + 3hx^2 + 3h^2x + h^3 - x^3}{h}$$

$$= \frac{3hx^2 + 3h^2x + h^3}{h}$$

→ The expression $\dfrac{f(x + h) - f(x)}{h}$ will be used again and again. It is called the *Newton Quotient* and we will refer to it as such.

This can be seen algebraically because the Newton Quotient

$$\frac{f(x + h) - f(x)}{h} = \frac{K - K}{h} = \frac{0}{h}$$

is equal to zero, so its limit, $f'(x)$, must equal zero.

→ This result is obvious from the graph of $f(x) = x$ because it is merely a line with slope 1; hence the tangent line at each point has slope 1.

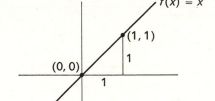

**Figure 1**

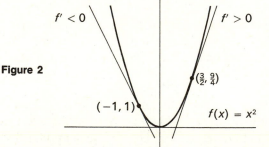

**Figure 2**

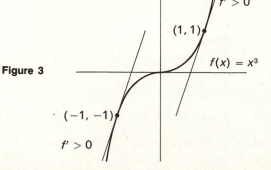

**Figure 3**

$$= 3x^2 + 3hx + h^2.$$

Since $3x^2 + 3hx + h^2 \to 3x^2$ as $h \to 0$, we have

$$f'(x) = \lim_{h \to 0} \frac{f(x + h) - f(x)}{h} = 3x^2 \quad \bullet$$

## C. The General Case

**THEOREM 4**　If $f(x) = x^n$ (*n a positive integer), then*

$$f'(x) = nx^{n-1}.$$

*Proof*　The Newton Quotient is

$$\frac{f(x + h) - f(x)}{h} = \frac{(x + h)^n - x^n}{h}$$

$$= \frac{x^n + nx^{n-1}h + \dfrac{n(n - 1)}{2}x^{n-2}h^2 + \cdots + h^n - x^n}{h}$$

$$= \frac{nx^{n-1}h + \dfrac{n(n - 1)}{2}x^{n-2}h^2 + \cdots + h^n}{h}$$

$$= nx^{n-1} + \frac{n(n - 1)}{2}x^{n-2}h + \cdots + h^{n-1}.$$

Since all terms but the first approach 0 as $h \to 0$, we have

$$f'(x) = \lim_{h \to 0} \frac{f(x + h) - f(x)}{h} = nx^{n-1} \quad \bullet$$

Theorem 4 is also valid when $n = 0$, as is obvious from the graph of $f(x) = x^0 = 1$. Moreover, the theorem will later be proven (Theorem 2 of 3.4D) for all real numbers, not just positive integers.

---

**Example 1**　Given the function $f(x) = x^4$, find the equation of the tangent line through the point (2, 16).

*Solution*　Since $f'(x) = 4x^3$,

$$f'(2) = 4 \cdot 2^3 = 32$$

is the slope of the tangent line through (2, 16). Its equation is therefore

$$y - 16 = 32(x - 2) \quad \text{or}$$
$$32x - y - 48 = 0.$$

*The reader may wonder why he is spending all this time deriving rules to differentiate functions. One reason, of course, is to increase the number of tools to perform engineering tasks. Such a task might be to draw an accurate graph of f(x) = x³ (Fig. 3).*

*The fact that we know f'(x) = 3x² and that the derivative f'(c) equals the slope of x³ at x = c helps us in drawing an accurate curve. After plotting the points (−1, −1), (0, 0), and (1, 1) we check the slope of the curve and see that for x = −1 and x = 1 the slope of x³ is f'(−1) = f'(1) = 3 while at x = 0 the slope is zero. This means that at x = 0 the graph of x³ becomes horizontal, a fact overlooked by many people not familiar with calculus.*

*(Vahan)*

We are using here the Binomial Theorem (first proven by Newton) which states that

$$(x + h)^n = x^n + nx^{n-1}h + \frac{n(n - 1)}{2}x^{n-2}h^2 + \cdots$$

$$+ \frac{n!}{(n - k)!k!}x^{n-k}h^k + \cdots + h^n.$$

The notation *r*! (read "*r* factorial") means

$$r \cdot (r - 1) \cdot (r - 2) \cdots 3 \cdot 2 \cdot 1$$

where *r* is a positive integer.

When $n = 0$, the formula reads $f'(x) = 0 \cdot x^{0-1} = \frac{0}{x} = 0$ if $x \neq 0$, but makes no sense when $x = 0$.

---

**Problem 1**　Find the equation of the normal to the graph of $f(x) = x^5$ at the point (−1, −1).

*Answer: $x + 5y + 6 = 0$*

## D. Exercises

**B1.** Find the derivatives of the following functions at the indicated points.
a) $f(x) = x^3$ at $x = 2$ and $x = -2$
b) $f(x) = x^2$ at $x = 2$ and $x = -2$
c) $f(x) = x^3$ at $x = 1$ and $x = 3$

**B2.** Find the equations of the tangent lines to $f(x) = x^6$ at the points
a) $(0, 0)$, b) $(1, 1)$, c) $(-1, 1)$, d) $(2, 64)$.

**B3.** Sketch the graph of $f(x) = x^4$ from $x = 0$ to $x = 1$. What do the values of $f'(0), f'(\frac{1}{4}), f'(\frac{1}{2}), f'(\frac{3}{4})$, and $f'(1)$ tell you about the way the slope is changing? Is this evident from your graph?

**B4.** At what point is the line $12x - y = 16$ tangent to the graph of $y = x^3$?

**B5.** Find the derivative of each of the following functions.
a) $f(x) = x^{3n}$ b) $f(x) = x^{m+n}$

**D6.** Match the expression on the left with its equal expression on the right.
a) $(a + b)^3 b$      A) $a^4 + 4a^3b + 6a^2b^2 + 4ab^3 + b^4$
b) $a^2(a + b)^2$      B) $a^4 + 2a^3b + a^2b^2$
c) $(a + b)^4$      C) $a^4 + 2a^2b^2 + b^4$
d) $(a + b)^2 + a^4 + b^4$      D) $a^3b + 3ab^3 + 3a^2b^2 + b^4$
e) $(a^2 + b^2)^2$      E) $a^4 + a^2 + 2ab + b^2 + b^4$

a) D    b) B    c) A    d) E    e) C

**D7.** Find the derivative of the given function at the indicated points.
a) $f(x) = x^3$ at $x = 2, 3, 4, 5$
b) $f(x) = x^4$ at $x = -1, -2, -3$
c) $f(x) = x^5$ at $x = -1, 0, 1$
d) $f(x) = x$ at $x = 65, 129, 384, 948$

a) $12, 27, 48, 75$    b) $-4, -32, -108$    c) $5, 0, 5$
d) $1, 1, 1, 1$

**M8.** Find the area of the triangle formed by the line $x = 2$, the tangent line to $f(x) = x^2$ at $(1, 1)$, and the tangent line to $f(x) = x^4$ at $(1, 1)$.

**M9.** The line $x + 7y = 8$ is perpendicular to the graph of $y = x^n$ at the point $(1, 1)$. Find $n$.

**M10.** For each integer $n$ the graph of $y = x^{2n}$ has a tangent with slope equal to $-2n$. Find the equations of these lines.

**M11.** Given $f(x) = x^n$ ($n$ a positive integer), find all points $x$ for which $f'(x) = f(x)$.

**M12.** Using the definition of the derivative, prove that if $f(x) = x^n + x^m$, $n$ and $m$ positive integers, then $f'(x) = nx^{n-1} + mx^{m-1}$.

**M13.** Consider the trapezoid formed by the lines $x = 0$, $y = 0, y = 1$, and the tangent to $y = x^n$ at the point $(1, 1)$. Let $A_n$ denote its area and prove that $A_n \to 1$ as $n \to \infty$.

**M14.** Compute the derivative of $f(x) = |x|^n$ for each integer $n > 1$.

**M15.** Suppose $n, m > 1$ are integers with $m \neq n$. Let
$$f(x) = \begin{cases} x^n & \text{for } x < 0 \text{ or } x > 1 \\ x^m & \text{for } 0 \leq x \leq 1. \end{cases}$$
Show that $f$ is differentiable at $x = 0$ but not at $x = 1$.

**M16.** Given $f(x) = x^4$ compute
a) $f'(x)$      c) $f(f'(x))$
b) $f'(f(x))$      d) $f'(f'(x))$

**M17.** Given
$$f(x) = \begin{cases} x^2 & (-\infty < x \leq 1) \\ ax + b & (x \geq 1) \end{cases}$$
find conditions on $a$ and $b$ so that $f$ is differentiable at $x = 1$.

## 2.3  *Polynomials and Notation*

In this section we will prove a general theorem about derivatives, apply it to polynomials, and introduce several notations for the derivative.

**A. A General Theorem**  We know how to differentiate the functions

$$f(x) = x^4 \quad \text{and} \quad g(x) = x^6,$$

but not the function

$$h(x) = 3f(x) + 7g(x) = 3x^4 + 7x^6.$$

Its derivative is found using the following:

**THEOREM 1**  *If f and g are functions and A and B are constants, then the derivative of the function* $p(x) = Af(x) + Bg(x)$ *is*

$$p'(x) = Af'(x) + Bg'(x).$$

*Proof*  We have

$$\frac{p(x+h) - p(x)}{h}$$

$$= \frac{(Af + Bg)(x+h) - (Af + Bg)(x)}{h}$$

$$= \frac{Af(x+h) + Bg(x+h) - Af(x) - Bg(x)}{h}$$

$$= A\left(\frac{f(x+h) - f(x)}{h}\right) + B\left(\frac{g(x+h) - g(x)}{h}\right),$$

and as $h \to 0$ this becomes

$$Af'(x) + Bg'(x) \quad \bullet$$

Two special cases of Theorem 1 are

$$(f + g)'(x) = f'(x) + g'(x)$$

and

$$(Af)'(x) = Af'(x) \quad (A \text{ constant}).$$

Also, letting $g = h + k$, we have

$$(f + h + k)'(x) = f'(x) + (h + k)'(x)$$
$$= f'(x) + h'(x) + k'(x).$$

*If you guessed that the derivative of $h(x) = 3x^4 + 7x^6$ is $h'(x) = 3 \cdot 4x^3 + 7 \cdot 6x^5 = 12x^3 + 42x^5$, you're right.*

(Bob)

Of course we must assume that the functions $f$ and $g$ are differentiable at $x$, otherwise the whole discussion is irrelevant. We actually show that if $f$ and $g$ are differentiable at $x$, then $Af + Bg$ is differentiable at $x$, and its derivative is as indicated.

*If you have difficulty following this proof, read 1.4D again to make sure that you understand it.*

(Beth)

The first follows by letting $A = B = 1$, and the second by letting $B = 0$.

*It is not obvious to some students that $\frac{f}{K}$ ($K$ constant) has the form $Af$ with $A = \frac{1}{K}$. For example,*

$$\frac{x^2}{5} = \frac{1}{5}x^2 \quad \text{so} \quad \left(\frac{x^2}{5}\right)' = \frac{1}{5}(x^2)' = \frac{2x}{5}.$$

(Lee)

More generally, if $f_1, f_2, \ldots, f_n$ are functions, then

$$(f_1 + \cdots + f_n)'(x) = f_1'(x) + \cdots + f_n'(x).$$

*Example 1*   Find the derivative of $h(x) = 3x^4 + 7x^6$.

*Solution*   Using Theorem 1 we now find that if $h(x) = 3x^4 + 7x^6$, then

$$h'(x) = 3 \cdot 4x^3 + 7 \cdot 6x^5$$
$$= 12x^3 + 42x^5.$$

*Problem 1*   Suppose $f(x) = -2x^2 + 7x^4 - \frac{x}{2} + 5$. Find $f'(0)$, $f'(1)$, $f'(\frac{1}{2})$, and $f'(-1)$.

*Answer:*   $-\frac{1}{2}, \frac{47}{2}, 1, -\frac{49}{2}$

---

**B. Polynomials**   A *polynomial* is any function $f$ that can be written in the form

$$f(x) = a_0 + a_1 x + a_2 x^2 + \cdots + a_n x^n$$

where $n$ is a nonnegative integer and the *coefficients* $a_0, a_1, \ldots, a_n$ are real numbers.

A polynomial always consists of the sum of a finite number of terms, and the *degree* of a polynomial is the largest exponent appearing in a term with a nonzero coefficient. For example, the degree of

$$p(x) = 3 + 2x^2 + 7x^4$$

is four. Notice that a polynomial of degree 0 is simply a constant function $f(x) = a_0 \neq 0$.

From Theorem 4 (2.2C) and Theorem 1 (2.3A) it follows that if

$$p(x) = a_0 + a_1 x + a_2 x^2 + \cdots + a_n x^n,$$

then

$$p'(x) = a_1 + 2a_2 x + \cdots + na_n x^{n-1}.$$

For example, $f(x) = 3x^2 - x + \sqrt{2}$ is a polynomial because it can be written in this form. The function $g(x) = (1 + x^2)^3$ is also a polynomial since

$$g(x) = (1 + x^2)^3 = 1 + 3x^2 + 3x^4 + x^6.$$

The function $h(x) = 4 + 2x + 7/x^2$ is not a polynomial because it cannot be written in the required form.

(Fred)

The degree of $g(x) = 2x^3 - x^4 + 3x^5 - 0x^6$ is 5, not 6.

When the polynomial is not written in this form the derivative is harder to compute. For example, to differentiate $g(x) = (1 + x^2)^3$ we multiply it out to obtain

$$g(x) = 1 + 3x^2 + 3x^4 + x^6, \quad \text{so}$$
$$g'(x) = 6x + 12x^3 + 6x^5.$$

Moreover, $h(x) = (1 + x^2)^{30}$ would be much more difficult to differentiate in this manner. In Chapter 3 we will learn a rule that will eliminate this difficulty.

---

**C. Notations for the Derivative**   The notation $f'$ (or $y'$) denotes the derivative of the function $y = f(x)$. When we wish to evaluate the derivative at a point $a$ we write $f'(a)$ or $y'(a)$.

There are many other notations for the derivative. Leibniz introduced

$$\frac{dy}{dx} \quad \text{and} \quad \frac{df}{dx}$$

(read "the derivative of $y$ (or $f$) with respect to $x$") to denote the derivative.

These notations were introduced by the precocious Joseph Louis Lagrange (1736–1813), a professor of mathematics at the age of nineteen.

Leibniz notation proved to be mathematically useful and suggestive of new results. To some extent it accounted for the success of the German school (of Leibniz) as contrasted to the British school (of Newton). The advantages of Leibniz notation, which will soon be clear to the student, were never appreciated by printers, who objected to it because it required two lines to print.

*Typists do not appreciate Leibniz notation either.*

(Randi)

The notations we will use to represent the derivative are

$$y' = f' = \frac{dy}{dx} = \frac{df}{dx}$$

and for the derivative at $a$

$$y'(a) = f'(a) = \frac{dy}{dx}\Big|_a = \frac{dy}{dx}\Big|_{x=a}$$
$$= \frac{df}{dx}\Big|_a = \frac{df}{dx}\Big|_{x=a}.$$

When letters other than $x$, $y$, and $f$ are used, we must change the notation accordingly. For example, if $u = g(s)$ then $u'$, $g'$, $\frac{du}{ds}$, and $\frac{dg}{ds}$ denote the derivative.

It is important that the reader understand the above notations, and the quickest way is to do exercises. We have worked out some examples below and refer you to Problem 2 for more drill.

→ The notations *Dy, Df, Ty, Tf* are also used to denote the derivative, but we will not need them in this book.

*In physics many functions depend on time, and a "dot" is used to denote the derivative with respect to time. For example, if x = g(t) then ẋ or ġ denotes the derivative, and ġ(a) denotes the derivative at a.*

(Alan)

## Example 2

a) Given $u = 3t^2 - 7t + 5$, $\frac{du}{dt} = 6t - 7$.

b) When $f(u) = 2u^3 - 7u + 5$, $\frac{df}{du} = 6u^2 - 7$

so $\frac{df}{du}\Big|_{u=3} = 47$

c) $\frac{d}{dx}(8x^3 + 6x + 2) = 24x^2 + 6$

d) If $x = -2y^8 - 6y^7 + 5y^6$,

$\frac{dx}{dy} = -16y^7 - 42y^6 + 30y^5$

e) If $y = x$, $\frac{dy}{dx} = 1$

f) $\frac{d}{dx}(x^2 + x + 1) = 2x + 1$

g) $\frac{d}{dx}(x) = 1$

h) $\frac{ds}{ds} = 1$

## Problem 2

a) $u = 3x^2 - 7x$. Find $\frac{du}{dx}$.

b) $f(x) = \frac{x^4}{4} + \frac{x^3}{3} + \frac{x^2}{2} + x + 1$. Find $\frac{df}{dx}$.

c) $\alpha = -\beta^2 + \frac{\beta^4}{3} - \frac{7}{6}\beta^5$. Find $\frac{d\alpha}{d\beta}$.

d) $\frac{d}{dx}(5x^2 - 7x^4 + 11x) = ?$

e) $\frac{d}{dv}(3v^2 - 11v) = ?$

f) If $y = 5x^7 - 32x$, find $y'$.

g) Find $\frac{dx}{dt}\Big|_3$ given $x = 4t^2 - 7t$.

h) Find $\frac{dy}{dx}\Big|_{x=2}$ given $y = 7x^2$.

i) Given $g(y) = -2y^5 + 7y$, find $g'(2)$.

j) Let $x = 3y^2 - \frac{2}{3}y^3$. Find $\frac{dx}{dy}$.

k) $\frac{d}{dx}(3x^2 - 7x)\Big|_{x=7} = ?$

l) $\frac{d}{dr}(-2r + 5r^3 + 2r^8)\Big|_1 = ?$

## D. Exercises

**B1.** Find the derivatives of the following polynomials at the indicated points.
 a)  $y = 3x^2 + 7x - 8$,   $x = 3$
 b) $f(x) = 7x^4 - 4x^7$,   $x = -1$
 c)   $u = t^3 - t^2 + t - 1$,   $t = 2$

**B2.** Find the derivatives of the following polynomials at the indicated points.
 a)   $y = (2x + 1)^2$,   $x = 2$
 b) $p(x) = (x^2 - x + 3)^2$,   $x = 1$
 c) $g(x) = (3x + 1)^3$,   $x = -2$

**B3.** Find the equations of the tangent lines at the indicated points.
 a) $y = 7x^2 + 5x - 8$  at  $(-1, -6)$
 b) $y = (3x + 1)^2$      at  $(0, 1)$

**B4.** Given $x = 2t^3 - 12t^2 + 18t + 11$, find the values of $t$ where $\frac{dx}{dt} = 0$.

**B5.** Given $\dfrac{3y}{5x^3 - 7x + 12} = \dfrac{4}{3}$,  find $\dfrac{dy}{dx}\Big|_{x=0}$

**B6.** Suppose that $f(x) = ax^2 + bx + c$ and $f(0) = 4$, $f'(1) = 2$, and $f'(2) = 1$. Find $a, b,$ and $c$.

**B7.** Find the derivative of each of the following polynomials.
 a) $f(x) = (x^m + x^n)^3$    b) $y = (x^m - x^n + x^{mn})^2$

**D8.** Find $\frac{dy}{dx}$ given
 a) $y = 8x^2 + 7x - 9$
 b) $y = 16x^4 - 4x + 20$
 c) $y = \dfrac{x^4}{4} + \dfrac{x^3}{3} + \dfrac{x^2}{2}$
 d) $y = ax^3 + bx^2 + cx + d$
 e) $y = x^2(x^2 - x)$
 f) $y = (x + 1)^4$
 g) $y = x(x - 1)(x + 1)$
 h) $y = 16x^4 + 8x^8 + 2x^{32} + x^{64}$

 a) $16x + 7$    b) $64x^3 - 4$    c) $x^3 + x^2 + x$
 d) $3ax^2 + 2bx + c$    e) $4x^3 - 3x^2$
 f) $4x^3 + 12x^2 + 12x + 4$    g) $3x^2 - 1$
 h) $64(x^3 + x^7 + x^{31} + x^{63})$

**D9.** Evaluate the indicated quantities.
 a) $\dfrac{d}{dx}(3x^2 - 9x + 8)\Big|_{x=1}$   b) $\dfrac{d}{dx}(6x - 9x^3 - x^4)\Big|_{x=0}$

c) $\dfrac{d}{dt}(t^2 + 2at + a^2)\Big|_{t=0}$   e) $\dfrac{d}{dr}(r^3 - 3r)\Big|_{r=3}$

d) $\dfrac{d}{du}(u(u + 1))\Big|_{u=1}$   f) $\dfrac{d}{ds}((s - 3)^3)\Big|_{s=2}$

g) $\dfrac{d}{dx}(a^2 + b^2 + x^2)\Big|_{x=1}$

h) $\dfrac{d}{dx}(x + 1)(x + 2)(x + 3)\Big|_{x=4}$

 a) $-3$    b) $6$    c) $2a$    d) $3$    e) $24$    f) $3$
 g) $2$    h) $107$

**D10.** Find the values of $x$ where $f'(x) = 0$ given
 a) $f(x) = x^2 - 2x + 6$
 b) $f(x) = 3x^2 - 12x + 7$
 c) $f(x) = 7x^2 + 7x + 7$
 d) $f(x) = (2x - 3)(3x - 2)$
 e) $f(x) = 2x^3 - 27x^2 + 120x + 30$
 f) $f(x) = x^3 + 9x^2 + 16$
 g) $f(x) = (3x - 4)^3$
 h) $f(x) = 3x^5 - 25x^3 + 60x + 40$

 a) $x = 1$    b) $x = 2$    c) $x = \dfrac{-1}{2}$    d) $x = \frac{13}{12}$
 e) $x = 4, 5$    f) $x = 0, -6$    g) $x = \frac{4}{3}$
 h) $x = \pm 1, \pm 2$

**M11.** Find a function whose derivative is $2x - 3$. Find another such function. How do the two differ?

**M12.** Find the equations of two lines, each of slope 5, that are tangent to $f(x) = x^3 + 2x$ at some point.

**M13.** Given $f(x) = 2x^3 + 3x^2 - 7x + 4$, find the point where the slope of its *derivative* is equal to zero.

**M14.** Show in each case that the indicated line cannot be tangent to the corresponding curve.
 a) $x + y = 1$,   $y = x^3$
 b) $y = 2x + 5$,   $y = x^3 + 3x$
 c) $y = 2x + 1$,   $y = x^2 + 1$

**M15.** Find the second degree polynomial that has slope 1 at $x = 2$, slope 2 at $x = 1$, and passes through the point $(1, 2)$.

**M16.** Suppose the polynomial $y = x^2 + bx + c$ has a tangent at $(0, 1)$ whose slope is 2. Find $b$ and $c$.

**M17.** a) Show that the polynomial $p(x) = Ax^2 + Bx + C$ is never 0 if and only if $B^2 < 4AC$.

b) Determine conditions on $a, b, c,$ and $d$ which insure that the derivative of $f(x) = ax^3 + bx^2 + cx + d$ is never 0. (Hint: Use (a).)

**M18.** Suppose $y = ax^2 - bx + c$ satisfies the following conditions.

i) $y'(b) = 0$

ii) $y'(0) = 1$

iii) The tangent to the graph at $(0, c)$ intersects the $x$-axis at $(\frac{b}{2}, 0)$.

Find $a, b,$ and $c$.

**M19.** a) Show that the graph of $y = x^2 + bx + c$ does not have two parallel tangents.

b) Show that for any nonhorizontal tangent to

$$y = x^3 + bx^2 + cx + d$$

there is a distinct parallel tangent.

**M20.** What is the result of differentiating the polynomial $a_0 + a_1x + \cdots + a_nx^n$, $n$ times? What is the result of differentiating it $n + 1$ times?

---

## 2.4 The Derivatives of Sine and Cosine

In this section we will find the derivatives of the sine and cosine functions. The proofs are straightforward, but they involve two nontrivial limits which will be evaluated in a lemma on the comment side of the page.

**A. Two Theorems**

**THEOREM 1** $\dfrac{d}{dx}\sin x = \cos x.$

*Proof* Let $f(x) = \sin x$. Then

$$\frac{f(x + h) - f(x)}{h} = \frac{\sin(x + h) - \sin x}{h}$$

and, using the identity

$$\sin(x + h) = \sin x \cos h + \sin h \cos x,$$

the Newton Quotient becomes

$$\frac{\sin x \cos h + \sin h \cos x - \sin x}{h}$$

$$= \sin x \left(\frac{\cos h - 1}{h}\right) + \frac{\sin h}{h} \cos x.$$

Since $\frac{\cos h - 1}{h} \to 0$ and $\frac{\sin h}{h} \to 1$ as $h \to 0$, we have

$$\frac{f(x + h) - f(x)}{h} \to \sin x \cdot 0 + 1 \cdot \cos x = \cos x,$$

→ The reader may have the impression from his secondary school education that trigonometry is mainly used to find heights of trees and widths of rivers. It is certainly true that the subject has practical uses of this sort, but the importance of the trigonometric functions goes far beyond these elementary applications.

*Lemma*

a) $\displaystyle\lim_{h \to 0} \frac{\sin h}{h} = 1,$ b) $\displaystyle\lim_{h \to 0} \frac{\cos h - 1}{h} = 0.$

*Proof of (a)* Consider Figure 1, which shows a sector with angle $h (0 < h < \frac{\pi}{2})$ of a circle of radius 1. (See Figure 9 (1.5C).)

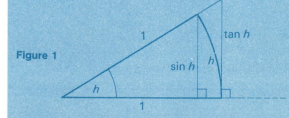

**Figure 1**

→ The length of the arc is $h$ and clearly $\sin h < h$ or $\frac{\sin h}{h} < 1$. Since the area of the sector is $\frac{1^2 h}{2} = \frac{h}{2}$ (front

or

$$f'(x) = \frac{d}{dx} \sin x = \cos x \quad \bullet$$

**THEOREM 2** $\dfrac{d}{dx} \cos x = -\sin x.$

*Proof* Let $g(x) = \cos x$. Then

$$\frac{g(x + h) - g(x)}{h} = \frac{\cos (x + h) - \cos x}{h}$$

and, using the identity

$$\cos (x + h) = \cos x \cos h - \sin x \sin h,$$

the Newton Quotient becomes

$$\frac{\cos x \cos h - \sin x \sin h - \cos x}{h}$$

$$= \left(\frac{\cos h - 1}{h}\right) \cos x - \frac{\sin h}{h} \sin x.$$

Since $\frac{\cos h - 1}{h} \to 0$ and $\frac{\sin h}{h} \to 1$ as $h \to 0$, we have

$$\frac{g(x + h) - g(x)}{h} \to 0 \cdot \cos x - 1 \cdot \sin x = -\sin x,$$

or

$$g'(x) = \frac{d}{dx} \cos x = -\sin x \quad \bullet$$

cover) and the area of the large triangle is $\frac{\tan h}{2}$, we have $\frac{h}{2} < \frac{\tan h}{2}$ or $\cos h < \frac{\sin h}{h}$. Combining this with the above inequality gives

$$\cos h < \frac{\sin h}{h} < 1.$$

Since $\cos h \to 1$ as $h \to 0$ (obvious from its graph), and $\frac{\sin h}{h}$ is squeezed between $\cos h$ and 1, $\frac{\sin h}{h}$ must also approach 1. Since $\cos (-h) = \cos h$ and

$$\frac{\sin (-h)}{-h} = \frac{-\sin h}{-h} = \frac{\sin h}{h},$$

this proof is valid in an open interval about $h = 0$, which is all that is required.

*Proof of (b)* We have

$$\frac{\cos h - 1}{h} = \frac{\cos h - 1}{h} \cdot \left(\frac{\cos h + 1}{\cos h + 1}\right) = \frac{\cos^2 h - 1}{h(\cos h + 1)}$$

$$= \frac{-\sin^2 h}{h(\cos h + 1)} = \frac{\sin h}{h} \cdot (-\sin h) \cdot \frac{1}{\cos h + 1}.$$

As $h \to 0$,

$$\frac{\sin h}{h} \to 1, -\sin h \to 0, \text{ and } \frac{1}{\cos h + 1} \to \frac{1}{2},$$

so

$$\frac{\cos h - 1}{h} \to 1 \cdot 0 \cdot \frac{1}{2} = 0 \quad \bullet$$

## The Importance of Sine and Cosine

Curiously, it was a man who scorned mathematical rigor who showed mathematicians the importance of the sine and cosine functions. In 1822 J. B. J. Fourier of France claimed that every function can be expressed as an infinite sum of sines and cosines. He offered no rigorous proof, and laced his arguments with vague physical and geometrical reasoning. Moreover, he drew from this result a large number of conclusions which had important physical applications. Since Fourier's assertions were nebulous, they were difficult to either prove or disprove. But the outgrowth of Fourier's wild assertions is now the subject of Fourier Series, which occupies a central position in mathematics and mathematical physics.

**B. An Aid to Memory** The following discussion may enable the reader to remember the results of Theorems 1 and 2 more easily.

Restrict the graphs of $\sin x$ and $\cos x$ to the interval $[0, \pi]$ (Fig. 2). Notice from the graph of $\sin x$ that its

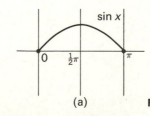

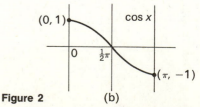

(a)     **Figure 2**     (b)

slope is positive and decreasing in $(0, \frac{\pi}{2})$. The values of $\cos x$, its derivative, behave likewise. Similarly, the slope of $\sin x$ in $[\frac{\pi}{2}, \pi)$ decreases from 0 and so do the values of $\cos x$. From this the reader should remember that

$$\frac{d}{dx} \sin x = \cos x \quad (\text{not } -\cos x).$$

**Problem 1**　Carry out the same analysis for $\cos x$ and its derivative, $-\sin x$, using Figure 3.

With very accurate graphs of the sine and cosine functions, the student could draw several tangents to $\sin x$, measure their slopes and compare them to the values of $\cos x$. The same can be done for $\cos x$ and $-\sin x$. These numbers should be close. If they are not, the drawing is bad.

*It is very interesting, Mr. Bonic, that you can say this with such assurance! After all you are a mathematician, and only prove things about numbers, functions, etc. How do you know about drawings, which involve (presumably) the physical properties of pen, ink, paper, etc.? What if someone made a very careful drawing and the numbers weren't approximately equal? Would that prove math is wrong?—Would it prove the world is wrong?*

(Michael Spivak)

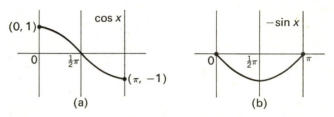

**Figure 3**

## C. Exercises

**B1.** If $f(x) = \cos x$, evaluate $f'(0)$, $f'(\frac{\pi}{4})$, and $f'(\frac{\pi}{2})$.

**B2.** Given $f(x) = \sin x$, compute $f'(0)$, $f'(\frac{\pi}{2})$, $f'(3\pi)$, and $f'(-\frac{3}{4}\pi)$.

**B3.** Find the equation of the tangent line to $y = \sin x$ at the point $(\frac{1}{6}\pi, \frac{1}{2})$.

**B4.** Find the equation of the normal line to $y = \cos x$ at the point $(\frac{1}{3}\pi, \frac{1}{2})$.

**B5.** Find all points where the derivative of the sine function is 0.

**B6.** a) Given $f(x) = 2x + \sin x$, show that $f'(x)$ is never zero.

　　b) Given $g(x) = 3x + x^3 - 3 \sin x$, show that $g'(x) = 0$ at exactly one point.

**D7.** Given $f(x) = \sin x$, find
　　a) $f'(\frac{3}{4}\pi)$　　e) $f'(\frac{16}{3}\pi)$
　　b) $f'(\frac{5}{4}\pi)$　　f) $f'(\sin 0)$
　　c) $f'(\frac{3}{2}\pi)$　　g) $f'(-\frac{3}{2}\pi)$
　　d) $f'(\frac{7}{4}\pi)$　　h) $f'(-\frac{1}{6}\pi)$

　　a) $\dfrac{-1}{\sqrt{2}}$　　b) $\dfrac{-1}{\sqrt{2}}$　　c) $0$　　d) $\dfrac{1}{\sqrt{2}}$　　e) $\dfrac{-1}{2}$

　　f) $1$　　g) $\dfrac{-1}{\sqrt{2}}$　　h) $\dfrac{\sqrt{3}}{2}$

**D8.** Given $g(x) = \cos x$ find
　　a) $g'(88\pi)$　　　　e) $g'(15\pi)$
　　b) $g'(\frac{2001\pi}{2})$　　f) $g'(\frac{17\pi}{6})$
　　c) $g'(\frac{-\pi}{2})$　　　g) $g'(\cos \frac{\pi}{2})$
　　d) $g'(\frac{-\pi}{6})$　　　h) $g'(\sin 8\pi)$

　　a) $0$　　b) $-1$　　c) $1$　　d) $\frac{1}{2}$　　e) $0$　　f) $\dfrac{-1}{2}$

　　g) $0$　　h) $0$

**M9.** Where does the graph of $y = x + \sin x$ have 0 slope?

**M10.** Find all points $x$ where the graph of $\sqrt{2} \sin x$ is perpendicular to the graph of $\sqrt{2} \cos x$.

**M11.** Find the equation of the normal to the graph of $y = 3 \sin x - x^2$ at the point $(\pi, -\pi^2)$.

**M12.** Find the derivative of each of the following functions.
　　a) $\sin(-x)$　　　　d) $\sin(x + 2\pi)$
　　b) $\cos(-x)$　　　　e) $\cos(\frac{\pi}{2} - x)$
　　c) $\sin^2 x + \cos^2 x$

**M13.** Given $f(x) = a \sin x + b \cos x$, $a$, $b$ constant, find $f'(x)$ and denote this function by $g(x)$. Show that $g'(x) + f(x) = 0$ for all $x$.

**M14.** Use the definition of the derivative to prove the following.

a) $\dfrac{d}{dx} \csc x = -\csc x \cot x$

b) $\dfrac{d}{dx} \sec x = \sec x \tan x$

**M15.** a) Use the definition of the derivative to show that $f(x) = \sin^2 x$ satisfies $f'(0) = 0$.
b) Suppose $0 \leq g(x) \leq \sin^2 x$ for all $x$. Show that $g$ is differentiable at 0 and $g'(0) = 0$.

**M16.** Use the definition of the derivative to show that

$$\left. \frac{d}{dx} \sin^n x \right|_{x=0} = 0$$

for any positive integer $n > 1$.

**M17.** Use the definition of the derivative to show that

$$\left. \frac{d}{dx} \cos^n x \right|_{x=0} = 0$$

for any positive integer $n$.

**M18.** Show that the tangent to $f(x) = \sin x$ at the point $(a, \sin a)$ passes through the origin if and only if

$$a = \tan a.$$

---

## 2.5 The Product, Reciprocal, and Quotient Rules

In this section we will learn how to differentiate functions such as

$$x^3 \cos x, \quad \frac{1}{\cos x}, \quad \text{and} \quad \frac{x^3}{\cos x}$$

→ These functions are products, reciprocals, and quotients of functions we already know how to differentiate.

using the three important differentiation rules given below.

**A. A Lemma**  We will need the following result in each of the next two theorems.

*LEMMA*  *Suppose $g$ is differentiable at $x$. Then $g(x + h) \to g(x)$ as $h \to 0$.*

*Proof*  Since $g'(x)$ exists, we have

$$\frac{g(x + h) - g(x)}{h} \to g'(x) \quad \text{as} \quad h \to 0$$

so

$$\frac{g(x + h) - g(x)}{h} - g'(x) = R \to 0 \quad \text{as} \quad h \to 0.$$

Multiplying this equation by $h$ and transposing gives

$$g(x + h) = g(x) + hg'(x) + hR.$$

As $h \to 0$, $hg'(x) \to 0$ and $hR \to 0$ so

$$g(x + h) = g(x) + hg'(x) + hR \to$$

$$g(x) + 0 + 0 = g(x) \quad \bullet$$

## B. The Product Rule

**THEOREM 1**   *If $f = uv$ then $f' = uv' + vu'$*

*Proof*   The Newton Quotient is

$$\frac{f(x + h) - f(x)}{h} = \frac{(uv)(x + h) - (uv)(x)}{h}$$

$$= \frac{u(x + h)v(x + h) - u(x)v(x)}{h}$$

$$= \frac{\begin{aligned}u(x + h)v(x + h) - u(x)v(x + h) \\ + u(x)v(x + h) - u(x)v(x)\end{aligned}}{h}$$

$$= \left[\frac{u(x + h) - u(x)}{h}\right]v(x + h)$$

$$+ u(x)\left[\frac{v(x + h) - v(x)}{h}\right].$$

We know that as $h \to 0$,

$$\frac{u(x + h) - u(x)}{h} \to u'(x), \quad \frac{v(x + h) - v(x)}{h} \to v'(x),$$

and, by the lemma, $v(x + h) \to v(x)$. Therefore,

$$f'(x) = \lim_{h \to 0}\frac{f(x + h) - f(x)}{h} = u'(x)v(x) + u(x)v'(x)$$

$$= u(x)v'(x) + v(x)u'(x) \quad \bullet$$

In Leibniz notation the product rule has the form

$$\frac{d}{dx}(uv) = u\frac{dv}{dx} + v\frac{du}{dx}.$$

---

**Example 1**   Find the derivative of $f(x) = x^3 \cos x$.

*Solution*   Let $u(x) = x^3$ and $v(x) = \cos x$. Using the Product Rule we have

$$f'(x) = u(x)v'(x) + v(x)u'(x)$$

$$= (x^3)(\cos x)' + (\cos x)(x^3)'$$

$$= -x^3 \sin x + 3x^2 \cos x.$$

---

$\longrightarrow$   To be more precise we should state this theorem as follows:

Suppose $f = uv$, and $u$ and $v$ are differentiable at $x$. Then $f$ is differentiable at $x$, and its derivative at $x$ is given by

$$f'(x) = u(x)v'(x) + v(x)u'(x).$$

$\longrightarrow$   *Here we used the trick $a - b = a - y + y - b$, letting $y = u(x)v(x + h)$.*

(Jean)

*It's not just a "trick." The point is, if we hope to find $f'(x)$ in terms of $u'(x)$ and $v'(x)$, we'd better be able to write $\frac{f(x + h) - f(x)}{h}$ in terms of $\frac{u(x + h) - u(x)}{h}$ and $\frac{v(x + h) - v(x)}{h}$. This suggests that we start writing*

$$\frac{u(x + h)v(x + h) - u(x)v(x)}{h} = \frac{u(x + h) - u(x)}{h}v(x + h) + \cdots$$

*and it works out just right.*

(Michael Spivak)

*Call it what you will, Mr. Spivak. I can understand it, but I never would have thought of it. Besides, I'd rather start getting used to the idea of a trick now. In integration there will be lots of them . . . much less obvious than this one.*

(Robin)

$\longrightarrow$   It is interesting to note that both Leibniz and Newton had considerable difficulty in finding the correct formula for the derivative of a product. At one time Leibniz thought that it was $(uv)' = u'v'$.

*Example 2*  Find $\frac{d}{dx}(x^2 \sin x \cos x)$.

*Solution*  Using the product rule we have

$$\frac{d}{dx}(x^2 \sin x \cos x)$$

$$= x^2 \frac{d}{dx}(\sin x \cos x) + \sin x \cos x \frac{d}{dx} x^2.$$

Using the rule again we have

$$\frac{d}{dx}(x^2 \sin x \cos x)$$

$$= x^2 \left[ \sin x \frac{d}{dx} \cos x + \cos x \frac{d}{dx} \sin x \right]$$
$$+ 2x \sin x \cos x$$
$$= x^2 [-\sin^2 x + \cos^2 x] + 2x \sin x \cos x$$
$$= 2x \sin x \cos x + x^2 \cos^2 x - x^2 \sin^2 x.$$

## C. The Reciprocal Rule

**THEOREM 2**  *If* $f = \frac{1}{v}$, *then* $f' = \frac{-v'}{v^2}$.

*Proof*  The Newton Quotient is

$$\frac{f(x+h) - f(x)}{h} = \frac{\dfrac{1}{v(x+h)} - \dfrac{1}{v(x)}}{h}$$

$$= \frac{v(x) - v(x+h)}{hv(x+h)v(x)}$$

$$= -\left[ \frac{v(x+h) - v(x)}{h} \right] \cdot$$
$$\frac{1}{v(x+h)v(x)}.$$

As $h \to 0$,

$$\frac{v(x+h) - v(x)}{h} \to v'(x);$$

and by the lemma,

$$v(x+h) \to v(x).$$

Hence

$$f'(x) = \lim_{h \to 0} \frac{f(x+h) - f(x)}{h}$$

→  More precisely, let $v$ be differentiable at $x$ and let $f(x) = \frac{1}{v(x)}$. Then, if $v(x) \neq 0$, $f$ is differentiable at $x$ and

$$f'(x) = \frac{-v'(x)}{v^2(x)}.$$

*Example 3*  Find the derivative of $f(x) = \sec x$.

*Solution*  We have $f(x) = \sec x = \frac{1}{\cos x}$. Letting $v(x) = \cos x$ then gives

$$f(x) = \frac{1}{v(x)}.$$

Using the Reciprocal Rule we have

$$f'(x) = \frac{-v'(x)}{v^2(x)} = \frac{-(\cos x)'}{\cos^2 x}$$

$$= \frac{\sin x}{\cos^2 x} = \sec x \tan x.$$

$$= \frac{-v'(x)}{v^2(x)} \quad \bullet$$

In Leibniz notation the rule is

$$\frac{d}{dx}\left(\frac{1}{v}\right) = \frac{-\dfrac{dv}{dx}}{v^2}.$$

**COROLLARY** *If* $y = x^n$ $(n = -1, -2, \ldots)$, *then*

$$\frac{dy}{dx} = nx^{n-1}.$$

*Proof* Since $-n$ is a positive integer, the derivative of $x^{-n}$ is $-nx^{-n-1}$ (Theorem 4, 2.2.C). Hence, by Theorem 2,

$$\frac{dy}{dx} = \frac{d}{dx}(x^n) = \frac{d}{dx}\left(\frac{1}{x^{-n}}\right)$$

$$= \frac{-(-nx^{-n-1})}{(x^{-n})^2} = nx^{-n-1+2n}$$

$$= nx^{n-1} \quad \bullet$$

$\longrightarrow$ Our formula $\frac{d}{dx}x^k = kx^{k-1}$ is now valid for all integers $k$.

---

**Example 4** Find $\frac{dx}{dt}$ given $x = -3t^{-3} + \frac{1}{\sin t}$.

*Solution*

$$\frac{dx}{dt} = \frac{d}{dt}\left(-3t^{-3} + \frac{1}{\sin t}\right) = \frac{d}{dt}(-3t^{-3}) + \frac{d}{dt}\frac{1}{\sin t}$$

$$= 9t^{-4} + \frac{-\dfrac{d}{dt}\sin t}{\sin^2 t} = 9t^{-4} - \frac{\cos t}{\sin^2 t}$$

$$= 9t^{-4} - \frac{1}{\sin t}\cdot\frac{\cos t}{\sin t} = 9t^{-4} - \csc t \cot t$$

---

## D. The Quotient Rule

**THEOREM 3** *If* $f = \dfrac{u}{v}$, *then* $f' = \dfrac{vu' - uv'}{v^2}$.

*Proof* Since $f = \frac{u}{v} = u(\frac{1}{v})$, we can use the Product Rule to obtain

$$f' = u'\left(\frac{1}{v}\right) + u\left(\frac{1}{v}\right)'.$$

Then, by the Reciprocal Rule, we have

$$f' = u'\left(\frac{1}{v}\right) + u\left(\frac{-v'}{v^2}\right)$$

$$= \frac{vu' - uv'}{v^2} \quad \bullet$$

In Leibniz notation the quotient rule is

$$\frac{d}{dx}\left(\frac{u}{v}\right) = \frac{v\dfrac{du}{dx} - u\dfrac{dv}{dx}}{v^2}$$

---

**Example 5** Find the derivative of $f(x) = \tan x$.

*Solution* Since $\tan x = \dfrac{\sin x}{\cos x}$, we apply the Quotient Rule letting

$$u(x) = \sin x$$

and

$$v(x) = \cos x.$$

$$f'(x) = \frac{(\cos x)(\sin x)' - (\sin x)(\cos x)'}{\cos^2 x}$$

$$= \frac{(\cos x)(\cos x) - (\sin x)(-\sin x)}{\cos^2 x}$$

$$= \frac{\cos^2 x + \sin^2 x}{\cos^2 x}$$

$$= \frac{1}{\cos^2 x}$$

$$= \sec^2 x$$

---

**Example 6** Find $\dfrac{d}{dx}\left(\dfrac{x^2 + a^2}{x^2 - a^2}\right)$, $a =$ constant.

*Solution*

$$\frac{d}{dx}\left(\frac{x^2 + a^2}{x^2 - a^2}\right)$$

$$= \frac{(x^2 - a^2)\dfrac{d}{dx}(x^2 + a^2) - (x^2 + a^2)\dfrac{d}{dx}(x^2 - a^2)}{(x^2 - a^2)^2}$$

$$= \frac{(x^2 - a^2)2x - (x^2 + a^2)2x}{(x^2 - a^2)^2}$$

$$= \frac{-4a^2x}{(x^2 - a^2)^2}$$

Observe that the derivative of $a^2$ is 0, since $a$ and hence $a^2$ are constants.

---

### Warnings

a) When $y = \frac{u}{C}$ where $C$ is a constant, the Quotient Rule applies but is not necessary. With the Quotient Rule we have

$$y' = \frac{Cu' - uC'}{C^2} = \frac{Cu'}{C^2} = \frac{u'}{C},$$

but $\frac{1}{C}$ is a constant and $(\frac{1}{C}u)' = \frac{1}{C}u'$.

b) Do not commit Leibniz's error $((uv)' = u'v')$ or the analogous one of thinking $(\frac{u}{v})' = \frac{u'}{v'}$.

## E. Derivatives of Trigonometric Functions

The evaluations of $\frac{d}{dx}(\cot x)$ and $\frac{d}{dx}(\csc x)$ are similar to those of $\frac{d}{dx}(\tan x)$ (Example 5) and $\frac{d}{dx}(\sec x)$ (Example 3). We summarize the results for trigonometric functions below:

$$\frac{d}{dx}(\sin x) = \cos x$$

$$\frac{d}{dx}(\cos x) = -\sin x$$

$$\frac{d}{dx}(\tan x) = \sec^2 x$$

$$\frac{d}{dx}(\csc x) = -\csc x \cot x$$

$$\frac{d}{dx}(\sec x) = \sec x \tan x$$

$$\frac{d}{dx}(\cot x) = -\csc^2 x.$$

→ The derivative of $\csc x$ is $-\csc x \cot x$. This can be seen by letting $v = \sin x$ and applying the Reciprocal Rule. Thus

$$\frac{d}{dx}(\csc x) = \frac{-\frac{d}{dx}\sin x}{\sin^2 x}$$

$$= \frac{-\cos x}{\sin^2 x} = \frac{-1}{\sin x} \cdot \frac{\cos x}{\sin x}$$

$$= -\csc x \cot x.$$

The derivative of $\cot x = \frac{\cos x}{\sin x}$ is $-\csc^2 x$. This can be seen by using the Quotient Rule, and letting $u(x) = \cos x$ and $v(x) = \sin x$. Hence

$$\frac{d}{dx}(\cot x) = \frac{\sin x(-\sin x) - (\cos x)\cos x}{\sin^2 x}$$

$$= \frac{-(\sin^2 x + \cos^2 x)}{\sin^2 x} = \frac{-1}{\sin^2 x} = -\csc^2 x.$$

→ *It is a good idea at this time to review what you know about differentiation by looking at the material on the inside back cover. This will help you get into the habit of using the reference material presented. In addition, notice that besides the definition of the derivative $f'(x) = \lim_{h \to 0} \frac{f(x + h) - f(x)}{h}$, you are already familiar with the first ten differentiation rules. Referring to this table should help in solving problems.*

*(Vahan)*

## F. Exercises

**B1.** Differentiate the following:

a) $y = \dfrac{x^3}{\cos x}$     b) $y = 3x + x^3 \tan x$

**B2.** Compute

a) $\left. \dfrac{d}{dx} \dfrac{(x - 1)(x + 2)}{x - 3} \right|_{x=4}$     b) $\left. \dfrac{d}{dx}\left( \dfrac{\sin x}{1 + x^2} \right) \right|_{x=0}$

**B3.** Find where $\frac{dy}{dx} = 0$ given

a) $y = \dfrac{x}{3 - x^2}$     b) $y = \dfrac{x^2}{x^2 + 2x + 3}$

**B4.** Find the normal to the curve $y = \dfrac{x^2}{\cos x \sin x}$ at the point $(\frac{1}{4}\pi, \frac{1}{8}\pi^2)$.

**B5.** Use the Product Rule to show that $\frac{d}{dx} u^3 = 3u^2 \frac{du}{dx}$ where $u$ is a function of $x$.

**B6.** Apply the result from Exercise B5 to find the derivative of

a) $y = (2x + 3)^3$

b) $y = \cos^3 x$     c) $y = \dfrac{1}{(x^2 + 2x + 3)^3}$

**B7.** Find $\frac{dy}{dx}$ given

a) $y = x \sin^2 x$     b) $y = x^2 \cos^3 x$

**D8.** Find $y'$ given

a) $y = x^4 \cos x$     d) $y = \sec^2 x$
b) $y = x^5 \sin x$     e) $y = \tan x \cos x$
c) $y = \sin x \cos x$     f) $y = \sin^2 x + 2 \cos^2 x$

a) $4x^3 \cos x - x^4 \sin x$     b) $5x^4 \sin x + x^5 \cos x$
c) $\cos^2 x - \sin^2 x$     d) $2 \sec^2 x \tan x$     e) $\cos x$
f) $-2 \sin x \cos x$

**D9.** Find $f'(x)$ given

a) $f(x) = \frac{2}{x} + \frac{x}{2}$
b) $f(x) = (1 + 2x)^{-1}$
c) $f(x) = (2 + x)^{-1}$
d) $f(x) = (1 + \sin x)^{-1}$
e) $f(x) = (\sin x \cos x \tan x)^{-1}$
f) $f(x) = (1 + \tan^2 x)^{-1}$

a) $\dfrac{-2}{x^2} + \dfrac{1}{2}$    b) $\dfrac{-2}{(1 + 2x)^2}$

c) $\dfrac{-1}{(2 + x)^2}$    d) $\dfrac{-\cos x}{(1 + \sin x)^2}$

e) $-2 \csc^2 x \cot x$    f) $-2 \sin x \cos x$

**D10.** Find $\dfrac{dy}{dx}$ given

a) $y = \dfrac{5x}{3 - x}$      d) $y = \dfrac{1 + \tan x}{\sec x}$

b) $y = \dfrac{3 - x}{5x}$      e) $y = \dfrac{x^2 - 4}{x^2 + 4}$

c) $y = \dfrac{\cos x}{1 + \sin x}$      f) $y = \dfrac{\cos x - \sin x}{\cos x + \sin x}$

a) $\dfrac{15}{(3 - x)^2}$      b) $\dfrac{-3}{5x^2}$

c) $\dfrac{-1}{1 + \sin x}$      d) $\cos x - \sin x$

e) $\dfrac{16x}{(x^2 + 4)^2}$      f) $\dfrac{-2}{(\cos x + \sin x)^2}$

**M11.** Find where the derivative of $f(x) = \sin^2 x + \cos x$ is equal to zero.

**M12.** For what values of $x$ is $f(x) = \sec x$ equal to its derivative?

**M13.** Given $g(x) = \tan x$, show that no $x$ satisfies $g(x) = g'(x)$.

**M14.** Given $f(x) = \sin^2 x$, find where $f'(x) > 0$.

**M15.** Use the identities

$$\sin 2x = 2 \sin x \cos x$$

and

$$\cos 2x = \cos^2 x - \sin^2 x$$

to find the derivatives of
a) $\sin 2x$    b) $\cos 2x$    c) $\tan 2x$

**M16.** Compute the derivative of

$$f(x) = \frac{(x^2 - 1) \sin x \cot x}{(x + 1) \cos^2 x}$$

and simplify your answer.

**M17.** Use the Product Rule to find the derivatives of the functions given below and then check your answer by multiplying the factors and differentiating directly.
a) $(x + 1)(x - 1)$
b) $(x^2 + 1)(x^2 - x)$
c) $x^{10}(1 + x + x^2)$

**M18.** Prove or give a counterexample to the following assertion: the derivative of a product is the product of the derivatives.

**M19.** Find the derivative of each of the following functions. See problem M17, Section 2.1F.
a) $\sqrt{x}(x^2 - 3x + 9)$
b) $\dfrac{\sqrt{x}}{x - 3}$

**M20.** Prove that the derivative of a rational function (quotient $\frac{P(x)}{Q(x)}$ of polynomials $P(x)$ and $Q(x)$, $Q(x)$ not identically zero), is again a rational function.

**M21.** Use mathematical induction to show that if $f_1, f_2, \ldots, f_n$ are differentiable functions then

$$(f_1 f_2 \cdots f_n)' = f_1' f_2 \cdots f_n$$
$$+ f_1 f_2' f_3 \cdots f_n + \cdots + f_1 f_2 \cdots f_{n-1} f_n'.$$

**M22.** The expression $f/g/h$ is ambiguous because it can mean either $\frac{f}{g}$ divided by $h$ or $f$ divided by $\frac{g}{h}$. Use the Quotient Rule to compute the derivatives of $f/(g/h)$ and $(f/g)/h$.

---

## 2.6   Limits and Continuity

In this section we will consider the concept of a limit in detail and obtain some of its important properties. The notion of continuity, which is intimately related to limits, will also be considered.

**A. Definition of a Limit of a Function**   Consider the function $f$ described in Figure 1(a). Observe that if $x$ is a point close to $a$ then the value $f(x)$ is a point close to the number $L$, the value of $f$ at $a$. In particular, let $I$ be an open interval, which we imagine to be *as small as we wish*, with center at $L$. Since the graph of $f$ is rather smooth around the point $(a, L)$, it is intuitively clear that we will be able to find some open interval $J$ with center at $a$ so that if $x$ is *any* point in $J$, then $f(x)$ is in $I$. Thus the values $f(x)$ are *arbitrarily close* to $L$ for *all* $x$ close *enough* to $a$.

It is the above property that the function $f$ has near the point $a$ that determines the existence of a limit. Namely, we define a function $f(x)$ to have a limit $L$ as $x$ approaches $a$, written

$$f(x) \to L \quad \text{as} \quad x \to a \quad \text{or} \quad \lim_{x \to a} f(x) = L$$

if for every open interval $I$ (however small) with center at $L$, there is an open interval $J$ with center at $a$, such that the values $f(x)$ are in $I$ for all $x \neq a$ in $J$ and in the domain of $f$.

Note in the definition the phrase "for all $x \neq a$ in $J$." Thus $f(a)$ itself need not lie in the interval $I$. See Figure 1(b). It is the behavior of the function $f$ near the point $a$, but excluding $a$ itself, that determines whether or not there is a limit as $x \to a$.

In particular, $f$ need not be defined at $a$ at all. To be precise, all that is required is that there be points in the domain of $f$ arbitrarily close to the point $a$. Thus for example, the point $a$ could be an endpoint of the domain of $f$, as shown in Figure 1(c).

Note finally that the length of the interval $J$ will, in general, depend upon the length of the interval $I$, and for smaller intervals $I$ smaller intervals $J$ will have to be chosen. But as long as one such interval $J$ can be found for each interval $I$, the function has a limit.

As an example, consider the function $f(x) = x + 1$ shown in Figure 2. Here

$$\lim_{x \to 2} f(x) = \lim_{x \to 2} (x + 1) = 3.$$

Because, let $I$ be any open interval with center at 3 and with endpoints say $a < b$. Take $J$ to be the open interval $(a - 1, b - 1)$. Then $J$ has center at 2 and for all $x$ in $J$ we have

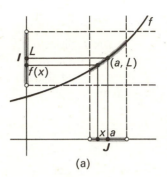

(a)

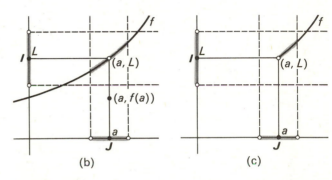

(b)                (c)

**Figure 1**

→    The Newton Quotient $\dfrac{f(x + h) - f(x)}{h}$, considered as a function of $h$, is not defined at $h = 0$. Thus in one of our most important applications of limits the function is not defined where the limit is being considered.

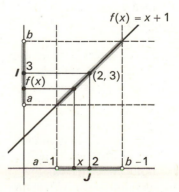

**Figure 2**

$$a - 1 < x < b - 1 \quad \text{or} \quad a < x + 1 < b;$$

that is, $f(x) = x + 1$ is in $I$. Since we have shown that for any open interval $I$ with center at 3 there is an open interval $J$ with center at 2 such that $f(x)$ is in $I$ for all $x$ in $J$, and hence for all $x \neq 2$ in $J$, it follows that

$$\lim_{x \to 2} f(x) = \lim_{x \to 2} (x + 1) = 3.$$

As another example, consider the function

$$H(x) = \begin{cases} 2 \text{ for } x \geq 3 \\ 1 \text{ for } x < 3 \end{cases}$$

shown in Figure 3(a). This function does not have a limit as $x \to 3$. Certainly the number 2 is not a limit. Because, choose an open interval $I$ with center at 2 small enough so that it does not contain the number 1. Since *every* open interval $J$ with center at 3 will contain some points $x < 3$ for which $H(x) = 1$, there is no $J$ such that the values $H(x)$ are in $I$ for *all* $x \neq 3$ in $J$. A similar argument shows that the number 1 cannot be a limit as $x \to 3$. Finally, any other number $L$ can, in fact, be separated from both 2 and 1 by a small enough interval $I$. Thus there is no limit as $x \to 3$. On the other hand, $H(x)$ does have a limit as $x$ approaches any other number. For example, $H(x) \to 2$ as $x \to 4$. Because, let $I$ be any open interval with center at 2, see Figure 3(b). If $J$ is chosen to be an open interval with center at 4 which excludes the point $x = 3$, then for all $x$ in $J$ we have $H(x) = 2$, which is in $I$. Thus $H(x) \to 2$ as $x \to 4$.

For many applications it is easier to have the definition of a limit in a somewhat different form, called the $\varepsilon, \delta$-definition. Namely, let $I$ be an open interval with center at $L$, as shown in Figure 4. If we denote half the length of the interval by $\varepsilon$, then the point $y$ is in $I$ if and only if its distance from $L$ is less than $\varepsilon$, that is if and only if $y$ satisfies the inequality

$$|y - L| < \varepsilon.$$

Note that since any other open interval with center at $L$ will differ from $I$ only by its length, the number $\varepsilon$ uniquely determines the interval $I$.

Similarly, an open interval $J$ with center at the point $a$ is uniquely determined by a positive number $\delta$, rep-

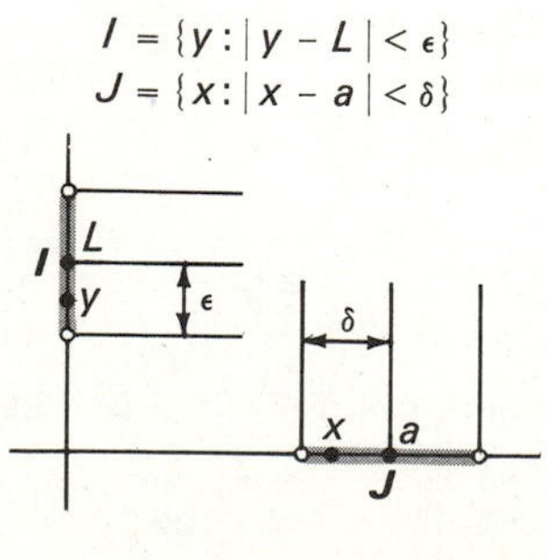

→ The reader should not be led to believe that this function is rather artificial and used mainly for instruction. The fact is that functions consisting of a single step like this one constantly occur in applied mathematics.

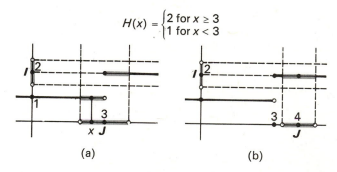

$$H(x) = \begin{cases} 2 \text{ for } x \geq 3 \\ 1 \text{ for } x < 3 \end{cases}$$

(a)          (b)

**Figure 3**

$$I = \{y : |y - L| < \varepsilon\}$$
$$J = \{x : |x - a| < \delta\}$$

**Figure 4**

resenting half its length; and a number $x$ is in $J$ if and only if it satisfies the inequality

$$|x - a| < \delta.$$

It follows from this that

$$\lim_{x \to a} f(x) = L$$

if and only if for every number $\varepsilon > 0$, there is a number $\delta > 0$ such that

$$|f(x) - L| < \varepsilon \qquad (f(x) \text{ is in } I)$$

for all $x \neq a$ in the domain of $f$ and satisfying

$$|x - a| < \delta \qquad (x \text{ is in } J).$$

This is the $\varepsilon,\delta$-definition of a limit.

---

**Example 1**  Use the $\varepsilon,\delta$-definition of a limit to prove that

$$\lim_{x \to 2} (3x + 6) = 12.$$

*Solution*  Given an arbitrary $\varepsilon > 0$, we want to determine a $\delta > 0$ such that for all $x \neq 2$ satisfying $|x - 2| < \delta$, the values $f(x)$ will satisfy

$$|f(x) - L| = |(3x + 6) - 12| < \varepsilon.$$

Now this inequality is the same as

$$|3(x - 2)| < \varepsilon \quad \text{or} \quad 3|x - 2| < \varepsilon \quad \text{or} \quad |x - 2| < \tfrac{\varepsilon}{3}.$$

Thus if we take $\delta = \tfrac{\varepsilon}{3}$, then for all $x$ satisfying $|x - 2| < \delta$, we have

$$|f(x) - L| = |(3x + 6) - 12| = 3|x - 2| < 3\tfrac{\varepsilon}{3} = \varepsilon.$$

---

## B. Some Theorems on Limits

**THEOREM 1**  *If $f(x) \to L$ as $x \to a$ and $g(x) \to M$ as $x \to a$, then*

   a) $(f(x) \pm g(x)) \to (L \pm M)$ *as* $x \to a$
   b) $f(x)g(x) \to LM$ *as* $x \to a$.
   c) *If* $M \neq 0$ *then* $f(x)/g(x) \to L/M$ *as* $x \to a$.

That is, the limit of a sum, difference, product, and quotient is the sum, difference, product, and quotient of the limits, provided one does not divide by zero.

*Proof that* $(f(x) \pm g(x)) \to (L \pm M)$:  Suppose $\varepsilon > 0$ is given. Since $f(x) \to L$ and $g(x) \to M$ as $x \to a$, there is a $\delta_1$ and a $\delta_2$ such that for $x \neq a$ and $|x - a| < \delta_1$ we have $|f(x) - L| < \frac{\varepsilon}{2}$ and for $x \neq a$ and $|x - a| < \delta_2$ we have $|g(x) - M| < \frac{\varepsilon}{2}$. Now take $\delta = \min(\delta_1, \delta_2)$. Then for $x \neq a$ and $|x - a| < \delta$, we have $|x - a| < \delta_1$ and $|x - a| < \delta_2$ and

$$\begin{aligned}
|(f(x) \pm g(x)) - (L \pm M)| &= |(f(x) - L) \pm (g(x) - M)| \\
&\leq |f(x) - L| + |g(x) - M| \\
&< \tfrac{\varepsilon}{2} + \tfrac{\varepsilon}{2} = \varepsilon
\end{aligned}$$

where we used the inequality $|a \pm b| \leq |a| + |b|$ ●

*Proof that* $f(x)g(x) \to LM$:  We assume that neither $L$ nor $M$ equals zero. If $L$ or $M$ equals zero, the proof in fact simplifies and is left for the reader as an exercise. Now, suppose $\varepsilon > 0$ is given. Since $f(x) \to L$ and $g(x) \to M$ as $x \to a$, there is a $\delta_1$ and a $\delta_2$ such that for $x \neq a$ and $|x - a| < \delta_1$ we have $|f(x) - L| < \min(\varepsilon/3|M|, \sqrt{\varepsilon/3})$ and for $x \neq a$ and $|x - a| < \delta_2$ we have $|g(x) - M| < \min(\varepsilon/3|L|, \sqrt{\varepsilon/3})$. Now take $\delta = \min(\delta_1, \delta_2)$. Then for $x \neq a$ and $|x - a| < \delta$ we have

$$\begin{aligned}
|f(x)g(x) - LM| &= |(f(x) - L)(g(x) - M) + M(f(x) - L) \\
&\quad + L(g(x) - M)| \\
&\leq |f(x) - L|\,|g(x) - M| + |M|\,|f(x) - L| \\
&\quad + |L|\,|g(x) - M| \\
&< \sqrt{\varepsilon/3}\,\sqrt{\varepsilon/3} + |M|(\varepsilon/3|M|) + |L|(\varepsilon/3|L|) \\
&= \varepsilon \quad ●
\end{aligned}$$

**THEOREM 2** *If* $p(x) = a_0 + a_1 x + \cdots + a_n x^n$ *is a polynomial in* x, *then for all* a,

$$\lim_{x \to a} p(x) = p(a).$$

*Proof* We leave as exercises for the reader to show that for a constant function $C$, $\lim_{x \to a} C = C$ for all $a$, and that $\lim_{x \to a} x = a$ for all $a$. It then follows from Theorem 1 that

$$\lim_{x \to a} p(x) = \lim_{x \to a} (a_0 + a_1 x + \cdots + a_n x^n)$$

$$= \lim_{x \to a} a_0 + \lim_{x \to a} a_1 \cdot \lim_{x \to a} x + \cdots$$

$$+ \lim_{x \to a} a_n \cdot \overbrace{\lim_{x \to a} x \cdots \lim_{x \to a} x}^{n \text{ times}}$$

$$= a_0 + a_1 a + \cdots + a_n a^n = p(a) \quad \bullet$$

---

*Example 2* Evaluate $\displaystyle\lim_{x \to 2} \frac{(x^2 - 1)(x^3 + 1)}{(x^3 + x + 1)}$.

*Solution*

$$\lim_{x \to 2} \frac{(x^2 - 1)(x^3 + 1)}{(x^3 + x + 1)} = \frac{\displaystyle\lim_{x \to 2}(x^2 - 1)\lim_{x \to 2}(x^3 + 1)}{\displaystyle\lim_{x \to 2}(x^3 + x + 1)}$$

$$= \frac{(3)(9)}{(11)} = \frac{27}{11}$$

---

**THEOREM 3** *If* f *is differentiable at a point* a, *then*

$$\lim_{x \to a} f(x) = f(a).$$

*Proof* In the lemma in the previous section it was shown that if $f$ is differentiable at $a$ then

$$\lim_{h \to 0} f(a + h) = f(a).$$

If we let $x = a + h$, then $h \to 0$ if and only if $x \to a$. Thus if $f$ is differentiable at $a$, then

$$\lim_{x \to a} f(x) = f(a) \quad \bullet$$

**C. The Indeterminate Form** $\frac{0}{0}$ We now want to consider the limit of a quotient when the limit of the denominator is zero. Here two cases are distinguished, depending upon the limit of the numerator. If the limit of the numerator is also zero, that is, if $f(x) \to 0$ as $x \to a$ and $g(x) \to 0$ as $x \to a$, then the function $\frac{f(x)}{g(x)}$ may or may not have a limit as $x \to a$ and only further calculations can determine the limit if there is one. In this case we say that the function $\frac{f(x)}{g(x)}$ has the indeterminate form $\frac{0}{0}$ at $x = a$.

An important example of an indeterminate form is the function $\frac{\sin x}{x}$ at $x = 0$. Recall that in Section 2.4 it was shown that this function has a limit equal to 1 as $x \to 0$.

---

**Example 3**   Evaluate $\lim\limits_{x \to 1} \left( \dfrac{x^2 - 2x + 1}{x - 1} \right)$.

*Solution*

$$\lim_{x \to 1} \left( \frac{x^2 - 2x + 1}{x - 1} \right) = \left( \frac{0}{0} \right) = \lim_{x \to 1} \frac{(x-1)(x-1)}{(x-1)}$$
$$= \lim_{x \to 1} (x - 1) = 0$$

---

**Example 4**   Evaluate $\lim\limits_{x \to 0} \frac{\tan x}{x}$.

*Solution*

$$\lim_{x \to 0} \left( \frac{\tan x}{x} \right) = \left( \frac{0}{0} \right) = \lim_{x \to 0} \left( \frac{1}{\cos x} \right)\left( \frac{\sin x}{x} \right)$$
$$= \lim_{x \to 0} \left( \frac{1}{\cos x} \right) \lim_{x \to 0} \left( \frac{\sin x}{x} \right)$$
$$= (1)(1) = 1$$

where we used the fact that $\lim\limits_{x \to 0} \frac{\sin x}{x} = 1$.

When the limit of the denominator of a quotient is zero and the limit of the numerator is not zero—that is, if $f(x) \to L \neq 0$ as $x \to a$ and $g(x) \to 0$ as $x \to a$—then the function $\frac{f(x)}{g(x)}$ does not have a limit as $x \to a$. Instead the values of the function become arbitrarily large in magnitude near the point $a$ as, for example, the function $h(x) = \frac{1}{x-2}$ near $x = 2$, shown in Figure 5. In this case we say that the quotient approaches infinity as $x$ approaches $a$ and write

Note that if a function $f$ is differentiable at $x$, then the Newton Quotient $\dfrac{f(x + h) - f(x)}{h}$, as a function of $h$, is always an indeterminate form at $h = 0$, since $f(x + h) - f(x) \to 0$ as $h \to 0$ (lemma in previous section) and the denominator $h \to 0$ as $h \to 0$. In the simplest cases we were able to calculate the limit by dividing out the quantity $h$. In the case of $f(x) = \sin x$ we had to first determine the limit of the indeterminate form $\sin h / h$ at $h = 0$, which we recall was no easy task.

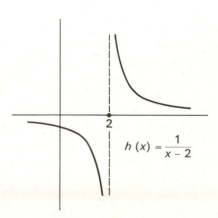

$$h(x) = \frac{1}{x - 2}$$

**Figure 5**

$$\frac{f(x)}{g(x)} \to \infty \text{ as } x \to a \quad \text{or} \quad \lim_{x \to a} \frac{f(x)}{g(x)} = \infty.$$

This is not to say that the quotient has a limit in the sense in which we have defined a limit. A more detailed discussion of this case is considered in Chapter 4.

**D. Continuity**   A function $f$ is said to be continuous at a point $a$ if (1) $a$ is in the domain of $f$, (2) $f(x)$ has a limit $L$ as $x \to a$, and (3) the limit $L = f(a)$. In short, $f$ is continuous at $a$ if

$$\lim_{x \to a} f(x) = f(a).$$

A function is said to be *continuous* on a set $D$ if it is continuous at each point in $D$.

As an immediate consequence of Theorem 1, it follows that if the functions $f$ and $g$ are continuous at a point $a$, then the sum $f \pm g$, the product $fg$, and, if $g(a) \neq 0$, the quotient $f/g$ are continuous at $a$. If $g(a) = 0$ then the quotient $f/g$ is not defined at $a$ and hence it is not continuous at $a$ or, as we say, is *discontinuous* at $a$.

As a consequence of Theorem 2 it follows that polynomials $p(x)$ are continuous for all real $x$.

Finally, it follows from Theorem 3 that if a function $f$ is differentiable at a point $a$, then $f$ is continuous at $a$. The converse of this is not true. For example, the function $f(x) = |x|$ is continuous at $x = 0$ but it is not differentiable there.

→ *Intuitively, a continuous function is one whose graph you can draw without lifting your pencil off the paper.*

*(Vahan)*

---

*Example 5*   Determine where the function

$$f(x) = \frac{x - 1}{x^2 - 1}$$

is continuous and describe the nature of its discontinuities.

*Solution*   The function $f(x)$ is a quotient of two continuous functions (polynomials) and hence it is continuous except where the denominator $x^2 - 1 = 0$, or $x = 1$ and $x = -1$. At $x = 1$ we have

$$\lim_{x \to 1}\left(\frac{x-1}{x^2-1}\right) = \left(\frac{0}{0}\right) = \lim_{x \to 1}\frac{(x-1)}{(x-1)(x+1)}$$

$$= \lim_{x \to 1}\left(\frac{1}{x+1}\right) = \frac{1}{2}.$$

Here there is a limit as $x \to 1$, so $f$ is discontinuous simply because it is not defined at 1, as shown in Figure 6. At $x = -1$ we have

$$\lim_{x \to -1}\left(\frac{x-1}{x^2-1}\right) = \left(\frac{-2}{0}\right) = \infty.$$

Here no limit exists. The function becomes arbitrarily large in magnitude near this point.

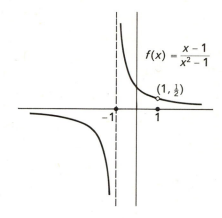

$f(x) = \dfrac{x-1}{x^2-1}$

$(1, \frac{1}{2})$

**Figure 6**

A function which is continuous on a closed interval has some important properties, one of which is given by the Intermediate Value Theorem, which we state below but prove in Chapter 9.

*THEOREM 4*   (Intermediate Value Theorem) *Let f be a continuous function on a closed interval [a, b] and let $y_0$ be a number between $f(a)$ and $f(b)$; that is,*

$$f(a) < y_0 < f(b) \quad \text{or} \quad f(b) < y_0 < f(a).$$

*Then there is at least one point $x_0$ in the open interval (a, b) where*

$$f(x_0) = y_0.$$

See Figure 7.

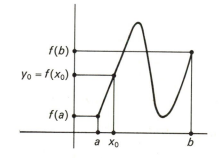

**Figure 7**

**Example 6**   Given $f(x) = x^2 + 2x + 1$, show that there is a number $x_0$ in the open interval $(-2, 2)$ for which

$$f(x_0) = 4.$$

*Solution*   The function $f$ is a polynomial and hence is continuous on the interval $[-2, 2]$. Since $f(-2) = 1$ and $f(2) = 9$ and 4 satisfies $1 < 4 < 9$, there is at least one point $x_0$ in $(-2, 2)$ such that $f(x_0) = 4$. We can find $x_0$ by solving the equation

$$f(x) = x^2 + 2x + 1 = 4$$

for $x$, obtaining $x = -3$ and $x = 1$. Here $x = 1$ is the one in the interval $(-2, 2)$.

## E. Exercises

**B1.** Sketch the following functions and determine from the graphs where each function is continuous and where it is discontinuous.

a) $f(x) = 2x + 1$     c) $f(x) = \begin{cases} x \text{ for } x \geq 2 \\ 1 \text{ for } x < 2 \end{cases}$

b) $f(x) = \dfrac{1}{x}$          d) $f(x) = \dfrac{x^2 - 4}{x - 2}$

**B2.** Evaluate the following limits.

a) $\lim\limits_{x \to 2} \left( \dfrac{x^3 - 27}{x - 3} \right)$     c) $\lim\limits_{h \to 0} \left( \dfrac{(x + h)^3 - x^3}{h} \right)$

b) $\lim\limits_{x \to -4} \left( \dfrac{x^3 + 64}{x^2 - 16} \right)$     d) $\lim\limits_{x \to x_0} \dfrac{(1/x) - (1/x_0)}{x - x_0}$

**B3.** Find the following limits.

a) $\lim\limits_{x \to 0} \left( \dfrac{\sin 2x}{x} \right)$     c) $\lim\limits_{h \to 0} \left( \dfrac{\sin^2 h}{h + h \cos h} \right)$

b) $\lim\limits_{h \to 0} \left( \dfrac{3h}{\tan 5h} \right)$

**B4.** Show that the equation

$$x^3 - 1.8x^2 - 4.2x + 6.9 = 0$$

has a solution in the interval $(2, 3)$.

**D5.** Evaluate the following limits.

a) $\lim\limits_{x \to 2} (x^2 + 2x)$     c) $\lim\limits_{x \to 0} \left( \dfrac{1}{\cos x} \right)$

b) $\lim\limits_{x \to 3} \left( \dfrac{1}{x^3} \right)$          d) $\lim\limits_{x \to x_0} (ax^3 + bx^2 + cx + d)$

a) 8     b) $\frac{1}{27}$     c) 1     d) $ax_0{}^3 + bx_0{}^2 + cx_0 + d$

**D6.** Find the following limits.

a) $\lim\limits_{x \to -2} \left( \dfrac{x^2 - 4}{x + 2} \right)$          c) $\lim\limits_{x \to 0} \left( \dfrac{x}{\sin x} \right)$

b) $\lim\limits_{x \to 0} \left( \dfrac{(3 + x)^3 - 27}{x} \right)$          d) $\lim\limits_{x \to 0} \left( \dfrac{x}{\tan x} \right)$

a) $-4$     b) 27     c) 1     d) 1

**M7.** Evaluate the following limits.

a) $\lim\limits_{x \to 0} \dfrac{\sin 4x}{\sin 3x}$     b) $\lim\limits_{x \to \pi/2} \dfrac{\cos x}{2x - \pi}$

**M8.** Use the $\varepsilon, \delta$-definition of a limit to establish the following.

a) $\lim\limits_{x \to 3} (4x + 3) = 15$     c) $\lim\limits_{x \to 3} (x^2 - 6x) = -9$

b) $\lim\limits_{x \to 2} \left( \dfrac{x^2 - 4}{x - 2} \right) = 4$

**M9.** Prove that if $f$ is a continuous function defined on an interval $[a, b]$, then the image of $f$ is $[f(a), f(b)]$ provided that $f$ is an increasing function; that is, $f(x_1) \leq f(x_2)$ for all $x_1 < x_2$ in the interval $[a, b]$.

**M10.** Prove that the limit of a function, if it exists, is unique.

**M11.** Prove that if $\lim\limits_{x \to a} g(x) = M$ then there is a $\delta > 0$ such that $|g(x)| \geq \frac{|M|}{2}$ for all $x \neq a$ satisfying $|x - a| < \delta$. (Hint: $|M| = |M - g(x) + g(x)| \leq |g(x) - M| + |g(x)|$.)

**M12.** Use the above result to prove that if

$$\lim\limits_{x \to a} g(x) = M \neq 0 \text{ then } \lim\limits_{x \to a} \dfrac{1}{g(x)} = \dfrac{1}{M}.$$

**M13.** Assuming that there are points in the domain of $f$ arbitrarily close to $a$, prove that $f$ is continuous at $a$ if and only if given an $\varepsilon > 0$ there exists a $\delta > 0$ such that $|f(x) - f(a)| < \varepsilon$ whenever $x$ is in the domain of $f$ and $|x - a| < \delta$.

**M14.** Show that $f(x) = |x|$ is continuous at $x = 0$. Show also that this function is not differentiable at $x = 0$. Thus continuity at a point does not imply differentiability at that point.

## 2.7 *Sample Exams*

### *Sample Exam 1 (45–60 minutes)*

1. Find

   a) $\dfrac{d}{dx}\left(3x^2 - \dfrac{3}{x^2} + x\sqrt{2}\right)$

   b) $\dfrac{d}{d\theta}(\sin\theta + \cos\theta + \tan\theta)$

2. Evaluate $\dfrac{d}{dt}(4t^3 - 6t + 3)\big|_{t=2}$.

3. Compute $f'(4)$ where $f(x) = 3x^2 + 5x$ by starting with the Newton Quotient $\dfrac{f(4+h) - f(4)}{h}$ and then taking the limit as $h \to 0$. (No credit will be given if you use the differentiation formula.)

4. Given $y = 2x^3 + 3x^2 - 72x + 13$ find the numbers $x$ where $\dfrac{dy}{dx} = 0$.

5. Find the equation of the tangent line to the graph of $y = 3x - \dfrac{1}{x}$ at the point $(1, 2)$.

6. Use the $\varepsilon,\delta$-definition of a limit to show that $\lim_{x\to 4}(2x - 5) = 3$.

### *Sample Exam 2 (45–60 minutes)*

1. Find $\dfrac{dy}{dx}$ given $y = x^2 \sec x$.

2. What is the area of the triangle formed by the $x$-axis, the $y$-axis and the tangent line to $y = -2x^2 + 6$ at the point $(3, -12)$?

3. Given $f(x) = \dfrac{x^4}{4} - \dfrac{1}{2x^2}$ find the equation of the line that intersects the $x$-axis at $(f'(1), 0)$ and the $y$-axis at $(0, f'(-1))$.

4. Find the points $(x, y)$ where the slope of the tangent to the graph of $y = x^3 + x$ is equal to 13.

5. Find $\dfrac{d}{dx}(2x + 5)^3$ and write your answer in the form $a_0 + a_1 x + a_2 x^2$. (Do not use any formulas except those from Chapter 2.)

6. Sketch the graph of $f(x) = \dfrac{1}{x-1}$ and determine where it is continuous and where it is discontinuous.

### *Sample Exam 3 (45–60 minutes)*

1. Suppose $f(x) = ax^2 + bx + c$, and $f(0) = 5$, $f'(0) = 3$, and $f'(2) = -7$. Find the values of the constants $a$, $b$, and $c$.

2. Where does the normal to the graph of $y = 3x^2 - \dfrac{4}{x}$ at the point $(1, -1)$ intersect the $y$-axis?

3. At what numbers $x$ is the derivative of

   $$y = x^3 + 6x^2 + 45x + 12$$

   equal to 36?

4. Given the function

   $$f(x) = \begin{cases} 1 & (-\infty < x \le -1) \\ |x| & (-1 < x < 1) \\ x & (1 \le x < \infty) \end{cases}$$

   find the points $(x, y)$ where the derivative does not exist.

5. Compute

   a) $\dfrac{d}{dx}\left(\dfrac{x \sin x}{\cos x}\right)$

   b) $\dfrac{d}{dt}(t^3 + \csc^2 t)$

6. Find

   a) $\lim_{h\to 0}(3h^2 + 2(h - 1)^2 + 4)$

   b) $\lim_{h\to 0}\dfrac{(2 + h)^4 - 16}{h}$

### *Sample Exam 4 (45–60 minutes)*

1. At what numbers $x$ is the derivative of $f(x) = x^2 + 3x + 4$ equal to the derivative of $g(x) = x^3 + x^2 + 78$?

2. Given $y = (37x^4 + 9)(-63x^6 + 87)$, find $\dfrac{dy}{dx}\big|_{x=0}$.

3. Find the equation of two lines tangent to the graph of $y = x^2$ that pass through the point $(2, 0)$.

4. The graph of $y = x^2 + ax + b$ passes through the point $(1, 1)$ and its derivative at $x = 0$ is equal to 3. Find the values of $a$ and $b$.

**5.** Find the two values for $x$ at which the tangent to $f(x) = \frac{2x}{x+1}$ is parallel to the line $-2x + y + 8 = 0$.

**6.** Show that the tangent to the graph of $y = 3x^2 - 7x + 8$ at the point $(2, 6)$ is perpendicular to the tangent to the graph of $y = \frac{2x^2}{5} - x$ at the point $\left(1, -\frac{3}{5}\right)$.

## Sample Exam 5 (45–60 minutes)

**1.** Find the equations of two lines tangent to the parabola

$$y = x^2 - 2x + 4$$

that pass through the origin.

**2. a)** Show that the function

$$f(x) = \begin{cases} -x^2 & (-1 < x < 0) \\ x^2 & (0 \le x < 1) \end{cases}$$

is differentiable at all points in the interval $(-1, 1)$.

**b)** On separate pairs of axes sketch the graphs of $f$ and $f'$ over the interval $(-1, 1)$.

**c)** At what point is the function $f'$ not differentiable? Explain.

**3.** Find the indicated limits.

**a)** $\lim\limits_{h \to 0} \dfrac{\sin 3h}{h}$

**b)** $\lim\limits_{h \to 0} 4h \cot 4h$

**c)** $\lim\limits_{x \to 0} \dfrac{\sqrt{4 + x} - 2}{x}$

**4.** Suppose that $p(x) = a_0 + a_1 x + a_2 x^2 + a_3 x^3$ and

$$\left(\frac{d}{dx} p\right)^2 = \frac{d}{dx} (p^2).$$

Prove that $a_1 = a_2 = a_3 = 0$.

**5.** Show that the curves

$$2y + x^2 = 2 \quad \text{and} \quad x^2 - 4y = 2$$

have perpendicular tangents at their points of intersection.

## Sample Exam 6 (Chapters 1 and 2, approximately 2 hours)

**1.** Find the point on the line $y = x + 1$ that is equidistant from the points $(2, 0)$ and $(4, 3)$.

**2.** Given the circle with equation

$$x^2 + y^2 + 4x - 8y - 5 = 0,$$

find the equation of the tangent line at the point $(1, 0)$. (Hint: Use the fact that the tangent is perpendicular to the line through $(1, 0)$ and the center of the circle.)

**3.** Evaluate the following expressions.

**a)** $\dfrac{d}{dx}\left(2x^2 + \dfrac{x^2}{2} + \dfrac{3}{x^3} + \dfrac{1}{3x^3}\right)$

**b)** $\dfrac{d}{dx} (\csc x \cot x)$

**c)** $\dfrac{d}{dt} ((t - 1)(t + 1)(t - 2)(t + 2))$

**4.** Given $f(x) = \dfrac{x^2 + 3x + 4}{x + 3}$, find the set of points where

**a)** $f' = 0$

**b)** $f' > 0$

**5.** Show that the derivatives of the following functions are never zero.

**a)** $f(x) = \dfrac{x + 2}{x^2 - 5}$

**b)** $g(x) = 3x + 2 \cos x$

**c)** $h(x) = |x - 5|$

**6.** What are the coordinates of the four vertices of the smallest rectangle that contains the ellipse whose equation is

$$x^2 + 4y^2 - 6x - 16y + 21 = 0?$$

**7.** Sketch the graph of $r = 2 \cos 2\theta$.

**8.** Explain why the function given by

$$f(x) = \begin{cases} 0 & (-\infty < x < 0) \\ \tan x & (0 \le x < \infty) \end{cases}$$

is not differentiable at $x = 0$.

**9.** Find the indicated limits.

**a)** $\lim\limits_{x \to 0} \dfrac{\sin^2 x}{x + x \cos x}$

**b)** $\lim\limits_{h \to -2} \dfrac{2h^2 + 3h - 2}{h + 2}$

**10.** Given the parabola $y = a + bx + cx^2$, find $a$, $b$, and $c$, assuming that the point $(1, 2)$ satisfies the equation,

$$\left.\frac{dy}{dx}\right|_{x=0} = 3, \quad \text{and} \quad \left.\frac{dy}{dx}\right|_{x=4} = 19.$$

In this chapter we will discuss the very important differentiation principle known as the chain rule. The technique of implicit differentiation is then developed and applied to inverse functions. In particular we study the inverse sine and inverse tangent functions and show how the exponential and logarithm functions are related. We conclude with a section on differentials and higher derivatives. This chapter completes our list of important functions, their derivatives, and the basic differentiation techniques.

# More About Differentiation

## 3.1    *The Chain Rule*

As noted in 2.3B the function $h(x) = (1 + x^2)^{30}$ is not easily differentiated using the rules of Chapter 2. The chain rule broadens the class of functions we can differentiate and, in particular, allows us to quickly differentiate the above function.

**A. Composite Functions**   Suppose $f$ and $u$ are functions such that the image of $u$ is a subset of the domain of $f$. Then the function $g$ defined by the formula

$$g(x) = f(u(x))$$

has a domain equal to the domain of $u$ and is called the *composition of f with u*. We will usually denote such a function by simply writing $y = f(u(x))$, but the notation $f \circ u$ is also used. Hence

$$(f \circ u)(x) = f(u(x)).$$

The composition of $f$ and $u$ is referred to as a *composite function* or *function of a function*. The chain rule gives us a formula that expresses the derivative of the composite function in terms of the derivatives of the component functions.

The following examples and problems will familiarize the student with these ideas.

→   We will see in Example 3 that $h'(x) = 30(1 + x^2)^{29}2x$.

→   This section will enable the reader to differentiate what are known as "composite functions," such as $f(x) = \sin x^3$, $g(x) = \tan 2x$, or $h(x) = \sin (\tan x)$.

For example, if $f(x) = \sin x$ and $u(x) = x^3$, then

$$f(u(x)) = \sin u(x) = \sin (x^3).$$

→   It is important not to confuse $f(x)u(x)$ with $f(u(x))$. For the functions above we have

$$f(x)u(x) = (\sin x)x^3 \quad \text{and} \quad f(u(x)) = \sin (x^3).$$

These two functions are not the same.

→   The order in which the functions are composed is essential. For example, if $f(x) = \sin x$ and $u(x) = x^3$ we have

$$(f \circ u)(x) = \sin (x^3) \quad \text{but}$$
$$(u \circ f)(x) = u(f(x))$$
$$= f(x)^3 = (\sin x)^3,$$

and these functions are not equal.

→   The composition of more than two functions can also be defined. Given three functions $f$, $g$, and $h$ we define $(f \circ g \circ h)(x) = f(g(h(x)))$. Composition, like multiplication, is associative, and we have

$$(f \circ g) \circ h = f \circ (g \circ h).$$

---

*Example 1*   Suppose

$$f(x) = \cos x$$
$$u(x) = x^3 + 4$$
$$v(x) = 3x.$$

Then

a) $f(u(x)) = \cos (u(x)) = \cos (x^3 + 4)$
b) $(f \circ v)(x) = \cos (v(x)) = \cos (3x) = \cos 3x$
c) $u(v(x)) = (v(x))^3 + 4 = (3x)^3 + 4$
$$= 27x^3 + 4$$
d) $v(v(t)) = 3(v(t)) = 3 \cdot 3t = 9t$
e) $f(v(u(x))) = \cos (v(u(x))) = \cos (3u(x))$
$$= \cos (3(x^3 + 4))$$
$$= \cos (3x^3 + 12)$$
f) $(u \circ u)(1) = (u(1))^3 + 4 = (1^3 + 4)^3 + 4 = 129$
g) $v(f(\frac{\pi}{3})) = 3f(\frac{\pi}{3}) = 3 \cos (\frac{\pi}{3}) = \frac{3}{2}.$

---

*Problem 1*   Using the functions in Example 1, compute

a) $u(f(x))$       e) $u(u(x))$
b) $f(f(x))$       f) $f(u(0))$
c) $(v \circ u)(y)$     g) $(u \circ f)(\frac{\pi}{3})$
d) $f(u(v(x)))$    h) $v(v(21))$

i) $f\left(u\left(v\left(\frac{4^{1/3}}{-3}\right)\right)\right)$

j) $f\left(u\left(v\left(\left(\frac{\pi - 8}{54}\right)^{1/3}\right)\right)\right)$

*Example 2*   It is important to recognize when a function is the composite of two others. For example,

a) if $h(x) = (1 + x^2)^{30}$ then $h(x) = f(u(x))$ where $f(x) = x^{30}$ and $u(x) = 1 + x^2$.

b) if $h(x) = \sin \sqrt{1 + x^2}$ then $h(x) = g(v(u(x)))$ where $u(x) = 1 + x^2$, $v(x) = \sqrt{x}$, and $g(x) = \sin x$.

**B. The chain rule**   First we will state the theorem in a form in which it is convenient to prove and then restate it in a form suitable for applications.

**THEOREM 1**   *If*

$$g(x) = f(u(x)),$$

*then*

$$g'(x) = f'(u(x))u'(x).$$

*Proof*   The Newton Quotient is

$$\frac{g(x + h) - g(x)}{h}$$

$$= \frac{f(u(x + h)) - f(u(x))}{h}$$

$$= \frac{f(u(x + h)) - f(u(x))}{u(x + h) - u(x)} \cdot \frac{u(x + h) - u(x)}{h}.$$

We let $k = u(x + h) - u(x)$. Then, substituting

$$u(x + h) = k + u(x),$$

$$\frac{g(x + h) - g(x)}{h}$$

$$= \frac{f(u(x) + k) - f(u(x))}{k} \cdot \frac{u(x + h) - u(x)}{h}.$$

As $h \to 0$,

$$\frac{u(x + h) - u(x)}{h} \to u'(x)$$

and, by the Lemma in 2.5A,

$$k = u(x + h) - u(x) \to 0.$$

Hence,

$$\frac{f(u(x) + k) - f(u(x))}{k} \to f'(u(x))$$

*Problem 2*   Determine $f$ and $u$ so that $h = f \circ u$ where

a) $h(x) = (1 - x^2)^{1/3}$      b) $h(x) = \tan (3x^2 - 5)$.

Find $f$, $u$, and $v$ so that $h = f \circ u \circ v$ where

c) $h(x) = \sin^7 (x^2 + 3)$      d) $h(x) = \dfrac{1}{(1 + \sin x)^4}$

→   More precisely we should assume that $u$ is differentiable at $x$ and that $f$ is differentiable at $u(x)$. The conclusion is that $f \circ u$ is then differentiable at $x$ and

$$(f \circ u)'(x) = f'(u(x))u'(x).$$

→   There is a small flaw here because $k = u(x + h) - u(x)$ might be zero and it appears in the denominator. We will now give a proof which avoids this defect.

*Proof*   We define the quantity

$$A = \begin{cases} \dfrac{f(u(x) + k) - f(u(x))}{k} - f'(u(x)), & k \neq 0 \\ 0, & k = 0. \end{cases}$$

Since $f$ is differentiable at $u(x)$, we have for $k \neq 0$

$$\frac{f(u(x) + k) - f(u(x))}{k} - f'(u(x)) = A \to 0 \quad \text{as} \quad k \to 0.$$

Multiplying by $k$ and transposing gives

$$f(u(x) + k) - f(u(x)) = kf'(u(x)) + kA.$$

Notice that this equation is true even when $k = 0$ because both sides are then equal to zero. Now we also have

$$\frac{u(x + h) - u(x)}{h} - u'(x) = B \to 0 \quad \text{as} \quad h \to 0$$

since $u$ is differentiable at $x$. Hence

$$u(x + h) - u(x) = hu'(x) + hB.$$

Now letting $k = u(x + h) - u(x)$, we have      ↓

as $h \to 0$, so

$$g'(x) = \lim_{h \to 0} \frac{g(x + h) - g(x)}{h} = f'(u(x))u'(x) \quad \bullet$$

We now state Theorem 1 in the clearer Leibniz notation.

*THEOREM*   *Suppose* $y = f(u(x))$. *Then*

$$\frac{dy}{dx}\bigg|_x = \left(\frac{dy}{du}\bigg|_{u(x)}\right)\left(\frac{du}{dx}\bigg|_x\right)$$

*or, more simply,*

$$\frac{dy}{dx} = \frac{dy}{du}\frac{du}{dx}.$$

*Explanation*   This is essentially a mere change in notation. When we write $y = f(u(x))$ we are saying $y = g(x) = f(u(x))$, that is, $y$ is a function of $x$; hence

$$\frac{dy}{dx} = \frac{dg}{dx}(x) = g'(x).$$

On the other hand, when we write $y = f(u)$ we simply consider $y$ as a function of $u$ and ignore the fact that $u$ is a function of $x$. Thus

$$\frac{dy}{du} = f'(u) \quad \text{and} \quad \frac{dy}{du}\bigg|_{u(x)} = f'(u(x)).$$

Since $u'(x) = \dfrac{du}{dx} = \dfrac{du}{dx}\bigg|_x$, we have

$$\frac{dy}{dx} = g'(x) = f'(u(x))u'(x) = \frac{dy}{du}\bigg|_{u(x)}\frac{du}{dx}\bigg|_x;$$

that is,

$$\frac{dy}{dx} = \frac{dy}{du}\frac{du}{dx}.$$

## C. Examples

---

*Example 3*   The function $h(x) = (1 + x^2)^{30}$ can be written as

$$h(x) = f(u(x))$$

where

$$f(x) = x^{30} \quad \text{and} \quad u(x) = 1 + x^2.$$

$f(u(x + h)) - f(u(x))$
$\quad = f(u(x) + k) - f(u(x))$
$\quad = kf'(u(x)) + kA$
$\quad = (hu'(x) + hB)f'(u(x)) + (u(x + h) - u(x))A$
$\quad = hu'(x)f'(u(x)) + hBf'(u(x)) + (u(x + h) - u(x))A.$

Dividing by $h \neq 0$ then gives

$$\frac{f(u(x + h)) - f(u(x))}{h} = f'(u(x))u'(x)$$

$$+ Bf'(u(x)) + \left(\frac{u(x + h) - u(x)}{h}\right)A.$$

Now let $h \to 0$. Then $B \to 0$ and since $k \to 0$ as $h \to 0$ (2.5A Lemma) we also know that $A \to 0$. Since

$$\frac{u(x + h) - u(x)}{h} \to u'(x),$$

we then have

$$\lim_{h \to 0} \frac{f(u(x + h)) - f(u)x)}{h} = f'(u(x))u'(x) + 0 \cdot f'(u(x)) + u'(x) \cdot 0$$

$$= f'(u(x))u'(x) \quad \bullet$$

In Leibniz's notation the chain rule is easier to remember and clearer to write because of the suggestive

$$\frac{dy}{dx} = \frac{dy}{d\not{u}}\frac{d\not{u}}{dx}.$$

However, the student should realize that at this point we have only defined the quantities $\frac{dy}{dx}$, $\frac{dy}{du}$, and $\frac{du}{dx}$, not $dy$, $du$, and $dx$. In Section 3.5 $dy$, $du$, and $dx$ will be defined as separate quantities. They are called *differentials* and are quite useful.

More generally, if $h(x) = u(x)^n$ we have $h(x) = f(u(x))$ where $f(x) = x^n$. Since $f'(x) = nx^{n-1}$,

$$h'(x) = f'(u(x))u'(x) = nu(x)^{n-1}u'(x).$$

In Leibniz's notation we can write this as

$$\frac{d}{dx}u^n = nu^{n-1}\frac{du}{dx}.$$

Since $f'(x) = 30x^{29}$ and $u'(x) = 2x$, we have
$$h'(x) = f'(u(x))u'(x) = 30u(x)^{29}2x = 30(1 + x^2)^{29}2x.$$

---

**Example 4** The function $y = (3x^2 - 7x + 8)^{93}$ can be written as $y = u^{93}$ where $u = 3x^2 - 7x + 8$. Since $\frac{dy}{du} = 93u^{92}$ and $\frac{du}{dx} = 6x - 7$, we have

$$\frac{dy}{dx} = \frac{dy}{du}\frac{du}{dx} = 93u^{92}(6x - 7)$$

$$= 93(3x^2 - 7x + 8)^{92}(6x - 7).$$

---

**Example 5** Given $y = \sin(x^3)$, we have $y = \sin u$ where $u = x^3$. Since $\frac{dy}{du} = \cos u$, and $\frac{du}{dx} = 3x^2$, this gives

$$\frac{dy}{dx} = \frac{dy}{du}\frac{du}{dx} = (\cos u)3x^2$$

so

$$\frac{d}{dx}\sin(x^3) = 3x^2 \cos(x^3).$$

---

The derivatives of several composite functions are found in Example 6.

In a similar fashion, if $y = \sin u$ where $u$ is a function of $x$, then

$$\frac{d}{dx}\sin u = \frac{d}{du}\sin u \frac{du}{dx} = (\cos u)\frac{du}{dx}.$$

If $u$ itself is a composite function, say $u = (1 - x^2)^5$, we must also use the chain rule to compute $\frac{du}{dx}$. In this case we have

$$\frac{du}{dx} = 5(1 - x^2)^4(-2x) = -10x(1 - x^2)^4.$$

Hence,

$$\frac{d}{dx}\sin(1 - x^2)^5 = (\cos(1 - x^2)^5)(-10x(1 - x^2)^4).$$

After studying Example 6, go on to Problem 3. If you cannot do it in less than five minutes, then you are either not clear about the material in Chapter 2, or do not yet understand how to apply the chain rule.

(Jean)

---

**Example 6**

a) $\frac{d}{dx}(3 - x^2)^4 = 4(3 - x^2)^3(-2x)$

b) $\frac{d}{dx}(x^2 + a^2)^{-3} = -3(x^2 + a^2)^{-4}(2x)$

c) $\frac{d}{dx}\cos^7 x = 7\cos^6 x(-\sin x)$

d) If $y = \tan(2x^3 - 5x)$, then

$$y' = (\sec^2(2x^3 - 5x))(6x^2 - 5)$$

e) If $y = \sec 3x$, then $y' = (\sec 3x \tan 3x)3$

f) If $u = (2t^2 + 7t)^{10}$, $\frac{du}{dt} = 10(2t^2 + 7t)^9(4t + 7)$

g) $\frac{d}{dx}\sin(\cos x^5) = (\cos(\cos x^5))(-\sin x^5)(5x^4)$

---

**Problem 3**

a) $\frac{d}{dx}(2 - x + x^2)^5 =$

b) $\frac{d}{dx}(x^3 - 1)^{-4} =$

c) $\frac{d}{dx}\cot^{10} x =$

d) If $y = \csc(3x^2 - 2x + x^4)$, $y' =$

e) If $y = \sec\left(\frac{1}{x}\right)$, $y' =$

f) If $u = \left(\frac{1}{t^2} - \frac{3}{t} + t^4\right)^{10}$, $u' =$

g) $\frac{d}{dx}\cos^7(x^2 - 5x) =$

## D. Exercises

**B1.** Find $f \circ g$ and $g \circ f$ in each of the following cases.
  a) $f(x) = x^2$, $g(x) = 2x$
  b) $f(x) = \cos x$, $g(x) = x^3$
  c) $f(x) = 1 + \cos x$, $g(x) = \dfrac{x}{x - 1}$

**B2.** Given $f(x) = 1 + x + x^2$ and $g(x) = 1 - x$, find
  a) $(f \circ g)(0)$
  b) $(g \circ f)(0)$
  c) $(f \circ f)(1)(g \circ g)(1)$
  d) $(f \circ g \circ f)(2)$

**B3.** Find the derivatives of the following functions.
  a) $y = (1 + x^2)^{-8}$
  b) $y = \sin (\cos x^2)$
  c) $y = (x^{-2} - x^2)^{-2}$
  d) $y = \sec (a + bx)$

**B4.** Find the indicated derivatives:

  a) $\dfrac{dy}{dx}$ given $y = (\sin x)^5$

  b) $\dfrac{df}{dt}$ given $f(t) = \sin (t^5)$

  c) $\dfrac{dz}{dx}$ given $z = \dfrac{1}{1 + (1 + x^2)^2}$

  d) $\dfrac{dy}{dx}$ given $y = x(3x + 1)^{1/3}$

**B5.** Given $f(u) = 2u + 3u^2 - 4$ and $u(x) = 1 - 3x$,
  a) find $h(x) = f(u(x))$ and compute $\frac{dh}{dx}$ directly.
  b) use the chain rule to find $\frac{dh}{dx}$.

**B6.** Given $y = u^3$, $u = 3v + 4$, and $v = 5x - x^2$, find

  a) $\dfrac{dy}{du}\Big|_{u=1}$     d) $\dfrac{du}{dv}\Big|_{v=1}$

  b) $\dfrac{dy}{dv}\Big|_{v=1}$     e) $\dfrac{du}{dx}\Big|_{x=1}$

  c) $\dfrac{dy}{dx}\Big|_{x=1}$     f) $\dfrac{dv}{dx}\Big|_{x=1}$

**D7.** Given $f(x) = x^2$ and $g(x) = 1 - x$, find
  a) $f(g(0))$     e) $(f \circ g)(3)$
  b) $f(g(1))$     f) $(g \circ f)(\sqrt{3})$
  c) $g(f(2))$     g) $(f \circ f)(3)$
  d) $g(f(3))$     h) $(g \circ g)(78)$

  a) 1    b) 0    c) −3    d) −8    e) 4    f) −2
  g) 81    h) 78

**D8.** Find $\frac{dy}{dx}$ in each of the following.
  a) $y = (1 + x^2)^5$     e) $y = (1 + x^2)^{-3}$
  b) $y = (1 + x^4)^5$     f) $y = (1 + \sin x)^3$
  c) $y = 2(1 + x^3)^4$     g) $y = (1 + x^{-2})^{-3}$
  d) $y = 3(1 + 3x^5)^4$     h) $y = (1 + \cos x)^{-5}$

  a) $10x(1 + x^2)^4$     b) $20x^3(1 + x^4)^4$
  c) $24x^2(1 + x^3)^3$     d) $180x^4(1 + 3x^5)^3$
  e) $-6x(1 + x^2)^{-4}$     f) $3(\cos x)(1 + \sin x)^2$
  g) $6x^{-3}(1 + x^{-2})^{-4}$     h) $5(\sin x)(1 + \cos x)^{-6}$

**D9.** Find $\frac{dy}{dx}$ in each of the following.
  a) $y = \sin 3x$     e) $y = \sin^4 x$
  b) $y = \cos 6x$     f) $y = \sin^4 5x$
  c) $y = \tan (2x + 3)$     g) $y = \sin (1 + x^2)$
  d) $y = \sin x^4$     h) $y = \sin^3 (4x + 5x^2)$

  a) $3 \cos 3x$    b) $-6 \sin 6x$    c) $2 \sec^2 (2x + 3)$
  d) $4x^3 \cos x^4$    e) $4 \sin^3 x \cos x$
  f) $20 \sin^3 5x \cos 5x$    g) $2x \cos (1 + x^2)$
  h) $(12 + 30x) \sin^2 (4x + 5x^2) \cos (4x + 5x^2)$

**D10.** Determine the value of the constant that makes each of the following equations valid.

  a) $\dfrac{d}{dx}(1 + x^3)^4 = Ax^2(1 + x^3)^3$

  b) $\dfrac{d}{dx} \sin^2 x^4 = Bx^3 \sin x^4 \cos x^4$

  c) $\dfrac{d}{dx}(1 + \cos^2 x)^3 = C \sin x \cos x(1 + \cos^2 x)^2$

  d) $\dfrac{d}{dx} \sin^3 (4 + 5x^6)$
  $$= Dx^5 \sin^2 (4 + 5x^6) \cos (4 + 5x^6)$$

  a) 12    b) 8    c) −6    d) 90

**M11.** Each of the functions $h$ below has the form $h = f \circ u$. Determine $f$ and $u$. (There are many possible answers.)
  a) $h(x) = \sin^2 x$
  b) $h(x) = \sin x^2$
  c) $h(x) = \sqrt{3 - 2x + x^4}$

  d) $h(x) = \dfrac{1}{1 + x^2}$

**M12.** Each of the following functions has the form $f \circ f$. Find $f$. (There are many possible answers.)
  a) $x^4$     d) $4x - 1$
  b) $x^2$     e) $x$
  c) $\sin (\sin x)$

**M13.** Find the derivatives of the following functions.
a) $f(x) = (1 + x^2)^4(2 - x^3)^5$
b) $g(x) = (1 + x^2)^5 \sin^4 x$
c) $h(x) = \sin^3 (x \cos x)$

**M14.** Suppose $h(x) = f(u(x))$, $u(0) = 1$, $f'(1) = 3$, and $h'(0) = 4$. Find $u'(0)$.

**M15.** As a spherical balloon is inflated with gas, its radius and volume will change with time but will always satisfy the relation $V = \frac{4}{3}\pi r^3$.
a) Find $\frac{dV}{dt}$ as a function of $r$ and $\frac{dr}{dt}$.
b) If the balloon is inflated with gas at the rate of 50 ft³/min ($\frac{dV}{dt}$), how fast is the radius changing when the radius is 4 ft?

**M16.** The current $I$ in certain electrical circuits is given by a function of the form $I = A \sin (\omega t + \phi)$, where $A$, $\omega$, and $\phi$ are constants.
a) Find $\frac{dI}{dt}$.
b) Let $W = \frac{dI}{dt}$ and find $\frac{dW}{dt}$.
c) Show that $\frac{dW}{dt} + \omega^2 I = 0$ for all $t$.

**M17.** Prove that composition of functions is associative; that is, prove that $(f \circ g) \circ h = f \circ (g \circ h)$.

**M18.** Given three functions $f$, $g$, and $h$, prove that
$$(f + g) \circ h = f \circ h + g \circ h.$$
Give an example showing that $f \circ (g + h) \neq f \circ g + f \circ h$.

**M19.** Recall that a function $f$ is said to be *even* if $f(-x) = f(x)$ for all $x$, and *odd* if $f(-x) = -f(x)$ for all $x$. Prove that if $f$ is even and $g$ is odd then
a) $f \circ g$ is even
b) $g \circ g$ is odd
c) if $f$ is differentiable then $f'$ is odd
d) if $g$ is differentiable then $g'$ is even

**M20.** Given three functions $f$, $g$, and $h$, prove that
$$(f \circ g \circ h)'(x_0) = f'(g(h(x_0)))g'(h(x_0))h'(x_0)$$
assuming that $h$ is differentiable at $x_0$, $g$ is differentiable at $h(x_0)$, and $f$ is differentiable at $g(h(x_0))$.

**M21.** Prove that if $\lim_{x \to a} g(x) = L$ and if $f$ is continuous at $L$ then $\lim_{x \to a} f(g(x)) = f(\lim_{x \to a} g(x))$. Note that it follows that the composition of continuous functions is continuous.

## 3.2  *Differentiation of Implicitly Defined Functions*

The functions we have studied until now have always been explicitly given in the form

$$y = f(x).$$

However, functions can be given implicitly. In this → section we will learn about such functions and the techniques of differentiating them.

**A. Implicitly Defined Functions**  Before defining implicit functions we will give an illustration. Consider the equation

$$y^2(1 + y) = (x - y)^2 + 2xy.$$

Multiplying this out, we see that it is equivalent to → the equation

$$y = x^{2/3}.$$

For example, if
$$y = x^2 + 4$$
then $y$ is said to be an explicit function of $x$. However, the equation
$$x^2 - y + 4 = 0$$
is said to define $y$ implicitly as a function of $x$.

$$
\begin{array}{ll}
y^2(1 + y) = (x - y)^2 + 2xy & \text{or} \\
y^2 + y^3 = x^2 - 2xy + y^2 + 2xy & \text{or} \\
y^2 + y^3 = x^2 + y^2 & \text{or} \\
y^3 = x^2 & \text{or} \\
y = x^{2/3}
\end{array}
$$

In the original equation $y$ is given implicitly as a function of $x$, while in the second equation $y$ is explicitly defined as a function of $x$.

More generally, suppose we are given an equation involving $x$ and $y$. If it is possible to solve for $y$ in terms of $x$, the function so obtained is said to be a *function implicitly defined by the equation.*

A particular equation need not implicitly define any function (Example 1), or it could implicitly define several functions (Example 2).

---

*Example 1*　Consider the equation

$$x^2 + y^2 = -4.$$

Clearly there are no real numbers that satisfy this equation, since for any pair of real numbers $x$ and $y$ we have $x^2 + y^2 \geq 0$. This means that it is impossible to solve for $y$ in terms of $x$, and no function is implicitly defined.

---

*Example 2*　Consider the equation

$$x^2 + y^2 = 4$$

whose graph is shown in Figure 1a. Rewriting this equation we have

$$y^2 = 4 - x^2.$$

Now taking the positive square root gives the function (Fig. 1b)

$$y = (4 - x^2)^{1/2}.$$

The negative square root gives another function (Fig. 1c)

$$y = -(4 - x^2)^{1/2}.$$

Hence the original equation implicitly defines at least two functions.

---

Finding a function implicitly defined by a given equation often involves more computation than that shown in Example 2. Consider the following example.

The equation

$$y^2(1 + y) = (x - y)^2 + 2xy$$

is then said to implicitly define the function $y = x^{2/3}$.

A precise definition of "implicitly defined" requires the notion of a function of two variables.

Let $\phi$ be such a function and consider the equation

$$\phi(x, y) = 0.$$

If there is a function $y = u(x)$ that satisfies $\phi(x, u(x)) = 0$ for all $x$ in the domain of $u$, then the function $u$ is said to be implicitly defined by the equation.

It is not often clear whether an equation $\phi(x, y) = 0$ implicitly defines any function. However, the *Implicit Function Theorem* does give a criterion:

Suppose $(a, b)$ satisfies $\phi(a, b) = 0$. If, treating $x$ as a constant, $\frac{d\phi}{dy}(a, b) \neq 0$, then there exists a function $u$ defined in some open interval $I$ about $a$ satisfying $u(a) = b$ and $\phi(x, u(x)) = 0$ for $x \in I$.

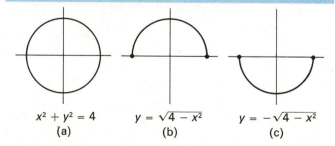

$$x^2 + y^2 = 4 \qquad y = \sqrt{4 - x^2} \qquad y = -\sqrt{4 - x^2}$$
$$\text{(a)} \qquad\qquad\quad \text{(b)} \qquad\qquad\quad \text{(c)}$$

**Figure 1**

Actually the equation $x^2 + y^2 = 4$ implicitly defines an infinite number of functions. Indeed, if $A$ is any subset of $[-2, 2]$, define

$$h_A(x) = \begin{cases} -(4 - x^2)^{1/2} & \text{if } x \text{ is in } A, \\ (4 - x^2)^{1/2} & \text{if } x \text{ is not in } A. \end{cases}$$

**Figure 2**

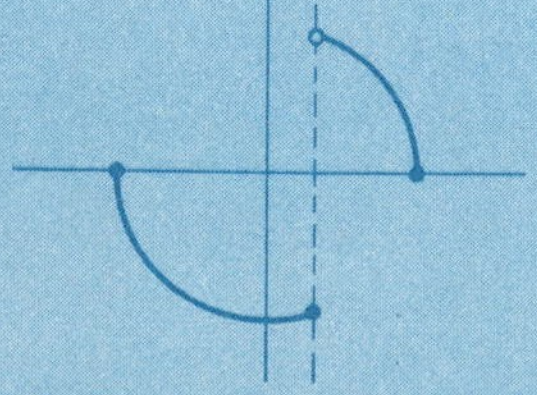

For example, the graph of $h_{[-2, 1/2]}$ is shown in Figure 2. Our interest will almost exclusively be with the functions $h_{[-2, -2]}$ and $h_{[-2, 2]}$ given in Example 2. They are the only *continuous* functions with domain $[-2, 2]$ implicitly defined by $x^2 + y^2 = 4$.

*Example 3*   Find two functions implicitly defined by the equation

$$y^2 - 2y + x - 1 = 0.$$

*Solution*   Using the Quadratic Formula (front cover), we solve for $y$ to obtain

$$y = \frac{-(-2) \pm \sqrt{(-2)^2 - 4(x-1)}}{2}$$

$$= \frac{2 \pm \sqrt{4 - 4x + 4}}{2}$$

$$= 1 \pm \sqrt{2 - x}.$$

Hence, each of the functions

$$\begin{aligned} y = f(x) &= 1 + \sqrt{2-x} \quad \text{and} \\ y = g(x) &= 1 - \sqrt{2-x} \end{aligned}$$

is implicitly defined by the original equation.

> As in Example 2, there are infinitely many functions implicitly defined by the given equation.

**B. Implicit Differentiation**   If we have the equation

$$y = x^2 + 2xu + u^3,$$

where $u$ is a function of $x$, its derivative is found to be

$$\frac{dy}{dx} = 2x + 2u + 2x\frac{du}{dx} + 3u^2\frac{du}{dx}.$$

Now take the similar looking equation

$$y = x^2 + 2xy + y^3.$$

If this equation implicitly defines $y$ as a function of $x$ we then have

$$\frac{dy}{dx} = 2x + 2y + 2x\frac{dy}{dx} + 3y^2\frac{dy}{dx}.$$

Hence

$$\frac{dy}{dx} - 2x\frac{dy}{dx} - 3y^2\frac{dy}{dx} = 2x + 2y$$

and solving for $\frac{dy}{dx}$ yields

$$\frac{dy}{dx} = \frac{2x + 2y}{1 - 2x - 3y^2}.$$

> Using the product and chain rules, we have
>
> $$\frac{dy}{dx} = \frac{d}{dx}(x^2 + 2xu + u^3) = \frac{d}{dx}x^2 + \frac{d}{dx}2xu + \frac{d}{dx}u^3$$
>
> $$= 2x + \left(\frac{d}{dx}2x\right)u + 2x\frac{du}{dx} + 3u^2\frac{du}{dx}$$
>
> $$= 2x + 2u + 2x\frac{du}{dx} + 3u^2\frac{du}{dx}.$$

> What we have done is to replace $u$ in the first equation by $y$, consider $y$ as a function of $x$, and then simply differentiate with respect to $x$.

This is an application of the technique of *implicit differentiation*. In general, if we are given an equation in $x$ and $y$ and the equation implicitly defines $y$ as a function of $x$, its derivative can be computed by implicit differentiation.

---

**Example 4** Differentiate the following equations implicitly, but do not solve for $\frac{dy}{dx}$. (It is typical of the technique of implicit differentiation that when one does solve for $\frac{dy}{dx}$ it is usually given in terms of both $x$ and $y$.)

a) Given $\quad y^2 = x, \quad 2y\dfrac{dy}{dx} = 1$

b) Given $\quad y^3 = x^2, \quad 3y^2\dfrac{dy}{dx} = 2x$

c) Given $\quad xy^4 = 8x^7 + 9, \quad x4y^3\dfrac{dy}{dx} + y^4 = 56x^6$

d) Given $\quad x^3y^4 = \sin y, \quad x^34y^3\dfrac{dy}{dx} + 3x^2y^4$

$$= (\cos y)\dfrac{dy}{dx}$$

e) Given $\quad x + \sin xy = y^2,$

$$1 + (\cos xy)\left(x\dfrac{dy}{dx} + y\right) = 2y\dfrac{dy}{dx}$$

f) Given $\quad \dfrac{1}{x} + \dfrac{1}{y} = 1, \quad -\dfrac{1}{x^2} - \dfrac{1}{y^2}\dfrac{dy}{dx} = 0$

g) Given $\quad (1 + x^4)^{33} = (1 + y^{33})^4,$

$$33(1 + x^4)^{32}4x^3 = 4(1 + y^{33})^333y^{32}\dfrac{dy}{dx}$$

h) Given $\quad y^2 + \tan y^2 = 1492,$

$$2y\dfrac{dy}{dx} + (\sec^2 y^2)\left(2y\dfrac{dy}{dx}\right) = 0$$

i) Given $\quad y = \sin(\cos(x^2 + y^2))$

$$\dfrac{dy}{dx} = (\cos(\cos(x^2 + y^2)))(-\sin(x^2 + y^2))$$

$$\cdot \left(2x + 2y\dfrac{dy}{dx}\right)$$

It is possible that an equation does not implicitly define any function. In such a case, we may still be able to differentiate implicitly, but the derivative is meaningless. For example, given $x^2 + y^2 + 4 = 0$, implicit differentiation yields $2x + 2y\frac{dy}{dx} = 0$ or $\frac{dy}{dx} = -\frac{x}{y}$. Since the initial equation does not implicitly define any function (Example 1), the formula $\frac{dy}{dx} = -\frac{x}{y}$ has no meaning.

Don't worry about whether an equation implicitly defines a function. The technique of implicit differentiation always applies and is correct whenever a function is defined.

(Dana)

You should try doing *Example 4* without pencil and paper. The equations get increasingly more involved, so be sure you understand a particular equation before going on to the next one. When you have finished the example try *Problem 1*.

(Beth)

---

**Problem 1** Match the answers to the problems.

| Given | | $\dfrac{dy}{dx}$ equals | |
|---|---|---|---|
| a) $y^4 = x^2 + 5$ | | A) $-\dfrac{x^2}{y^2}$ | |
| b) $x^2 = a^2 + y^2$ | | B) $\dfrac{-y}{2y + x}$ | |
| c) $x = -y - xy$ | | C) $-\dfrac{x}{y}$ | |
| d) $y^2 + xy = 3$ | | D) $\dfrac{x}{2y^3}$ | |
| e) $x^2 = a^2 - y^2$ | | E) $\dfrac{x}{y}$ | |
| f) $x^3 + y^3 - a^3 = 0$ | | F) $\dfrac{1}{(x + 1)^2}$ | |
| g) $x - y = xy$ | | G) $\dfrac{-y^2}{x^2}$ | |

*Answer:*    a) D      b) E      c) G      d) B      e) C
                 f) A      g) F

**Example 5** Find the slope of the tangent to the circle

$$x^2 + y^2 = 5$$

at the point $(1, 2)$.

*Solution* Differentiate implicitly to obtain

$$2x + 2yy' = 0$$

or

$$y' = \frac{-x}{y}.$$

Hence

$$y'\Big|_{(1,2)} = \frac{-1}{2}.$$

**Problem 2** Find the equation of the tangent line to

$$2x^2 - 3y^2 = 6$$

at the point $(3, 2)$.

*Answer:* $x - y = 1$

---

**Example 6** Show that the circles

$$x^2 + y^2 - 12x - 6y + 25 = 0$$

and

$$x^2 + y^2 + 2x + y - 10 = 0$$

are tangent at the point $(2, 1)$.

*Solution* Differentiating the first equation implicitly, we obtain

$$2x + 2yy' - 12 - 6y' = 0 \quad \text{or}$$
$$y' = \frac{12 - 2x}{2y - 6}.$$

The second equation gives

$$2x + 2yy' + 2 + y' = 0 \quad \text{or}$$
$$y' = \frac{-2 - 2x}{1 + 2y}.$$

Since $y'|_{(2,1)} = -2$ in both cases, the circles are tangent at $(2, 1)$.

---

We will now extend the formula $\frac{d}{dx}x^n = nx^{n-1}$ to apply to all rational exponents.

**THEOREM 1** If $y = x^r$, then $\dfrac{dy}{dx} = rx^{r-1}$ for all ⟶

rational numbers $r$.

**Problem 3** Show that the parabolas $y^2 = 1 - 2x$ and $y^2 = 2x + 1$ are perpendicular at each of their two points of intersection.

---

Using this result we may differentiate functions such as

$$y = (5 - x^2)^{1/2}.$$

Indeed, we have

$$\frac{dy}{dx} = \frac{1}{2}(5 - x^2)^{-1/2}(-2x) = \frac{-x}{(5 - x^2)^{1/2}}.$$

Notice that by replacing $(5 - x^2)^{1/2}$ by $y$ we have

$$\frac{dy}{dx} = -\frac{x}{y},$$

which is the same as the result obtained implicitly in Example 5.

*Proof* Since $r$ has the form $r = \frac{m}{n}$ for some integers $m$ and $n$, we have $y = x^{m/n}$ or $y^n = x^m$. Assuming that $y$ is differentiable, we differentiate implicitly, obtaining

$$ny^{n-1}\frac{dy}{dx} = mx^{m-1} \quad \text{or} \quad \frac{dy}{dx} = \frac{m}{n}\frac{x^{m-1}}{y^{n-1}}.$$

Now replacing $y$ by $x^{m/n}$, we have

$$\frac{dy}{dx} = \frac{m}{n}\frac{x^{m-1}}{x^{m(n-1)/n}} = r\frac{x^{m-1}}{x^{m-m/n}}$$
$$= rx^{m-1-m+m/n}$$
$$= rx^{r-1} \qquad \bullet$$

**C. Related Topics**   Even when a function is given explicitly as a function of $x$ it is sometimes more useful to differentiate implicitly.

---

*Example 7*   Find the derivative of
$$y = (3 + 6x + \sqrt{x^2 + 9})^{1/5} \quad \text{at } x = 4.$$

*Solution*   Raising each side of the equation to the fifth power we have
$$y^5 = 3 + 6x + \sqrt{x^2 + 9}.$$

Differentiating implicitly gives
$$5y^4 y' = 6 + \frac{x}{\sqrt{x^2 + 9}}.$$

When $x = 4$, $y = 2$, so we have
$$5 \cdot 16 y' = 6 + \tfrac{4}{5},$$
which gives $y'|_{x=4} = \frac{17}{200}$.

---

We may also find the slope of a curve represented parametrically. If $y = f(t)$ and $x = g(t)$ then, assuming we can solve for $y$ as a function of $x$, we have

$$\frac{dy}{dt} = \frac{dy}{dx} \cdot \frac{dx}{dt}, \quad \text{so} \quad \frac{dy}{dx} = \frac{dy}{dt} \Big/ \frac{dx}{dt} \quad \left(\frac{dx}{dt} \neq 0\right)$$

---

*Example 8*   Find the slope of the curve given by $x = t - \sin t, y = 1 - \cos t$ at the point corresponding to $t = \frac{\pi}{6}$.

*Solution*   The formula gives

Doing this directly we have
$$y' = \tfrac{1}{5}(3 + 6x + \sqrt{x^2+9})^{-4/5}\left(6 + \frac{x}{\sqrt{x^2+9}}\right),$$
so
$$y'|_{x=4} = \tfrac{1}{5}(32)^{-4/5}(6 + \tfrac{4}{5})$$
$$= \tfrac{1}{5} \cdot \tfrac{1}{16} \cdot \tfrac{34}{5} = \tfrac{17}{200}$$

The reader can decide for himself which method he prefers.

Even if the parametric equations do not represent $y$ as a function of $x$, for example when $y = t$, $x = \sin t$ (Fig. 3), the formula in the text can be applied to find the tangent.

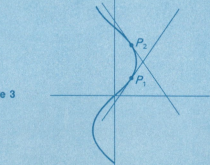

**Figure 3**

In this case
$$\frac{dy}{dx} = \frac{dy}{dt} \Big/ \frac{dx}{dt} = \frac{1}{\cos t}.$$

Hence, when $t = \frac{\pi}{4}$, the tangent at $P_1 = \left(\frac{\sqrt{2}}{2}, \frac{\pi}{4}\right)$ has slope
$\frac{1}{\cos(\pi/4)} = \sqrt{2}$. When $t = \frac{3\pi}{4}$ the tangent at $P_2 =$

$$\frac{dy}{dx} = \frac{dy}{dt} \bigg/ \frac{dx}{dt} = \frac{\sin t}{1 - \cos t},$$

so

$$\frac{dy}{dx}\bigg|_{t=\pi/6} = \frac{\sin \frac{\pi}{6}}{1 - \cos \frac{\pi}{6}} = \frac{\frac{1}{2}}{1 - \frac{\sqrt{3}}{2}}$$

$$= \frac{1}{2 - \sqrt{3}} = 2 + \sqrt{3}.$$

$\left(\frac{\sqrt{2}}{2}, \frac{3\pi}{4}\right) = \left(\frac{\sqrt{2}}{2}, \frac{3\pi}{4}\right)$ has slope $\frac{1}{\cos(3\pi/4)} = -\sqrt{2}.$

In all cases the expression $\frac{dy}{dt}\bigg|_{t=t_0} + \frac{dx}{dt}\bigg|_{t=t_0}$ unambiguously gives the slope of the tangent at the point corresponding to $t = t_0$ provided that $\frac{dx}{dt}\bigg|_{t=t_0} \neq 0$. However, as seen above, the expression $\frac{dy}{dx}\bigg|_{x=x_0}$ is ambiguous.

## D. Exercises

**B1.** Find two functions implicitly defined by the equation
$$x^2 - 4y + y^2 = 0.$$

**B2.** Show that the equation $11x^2 + 4y^2 - 12xy + 1 = 0$ does not implicitly define any function.

**B3.** Find $\frac{dy}{dx}$ by differentiating implicitly.
a) $3x^2 + 8xy - y^2 - 4 = 0$
b) $x \cos y + y \cos x = 78$

**B4.** Find the slope of the curve at the indicated point.
a) $y^3 + 3y^2 = 2x$, $(2, 1)$     b) $y = \tan \frac{\pi x y}{8}$, $(2, 1)$

**B5.** Find the equations of two tangents to the circle $x^2 + y^2 = 25$ corresponding to $x = 3$.

**B6.** Show that the curves whose equations are $x^2 + y^2 = 25$ and $3y^2 - 18y - 8x + 48 = 0$ are perpendicular at the point $(3, 4)$.

**B7.** Find $\frac{dy}{dx}\bigg|_{t=1}$ given the parametric equations
a) $y = t^3 + 2t^2 + 1$, $x = t^2 + t + 3$
b) $y = \sin^2 \frac{\pi t}{4}$, $x = \tan \frac{\pi t}{4}$

**B8.** Differentiate each of the following.
a) $y = x(x^2 + 1)^{1/2}$
b) $y = \left(\frac{x + 1}{x - 1}\right)^{2/3}$

**D9.** Show that no function $y = y(x)$ is defined by any of the equations below.
a) $2x^2 + 3y^2 + 4 = 0$     c) $\sin y^2 = 3 + \cos x$
b) $(x + y)^2 + 3 = 0$     d) $\sin xy = 1 + x^2$

**D10.** Find $\frac{dy}{dx}$ by differentiating implicitly. Your answer can be in terms of both $x$ and $y$.
a) $y^2 = x$                e) $x = \sin^2 y$
b) $2y^3 = 3x$              f) $x = \cos y^2$
c) $y^4 = \sin x$           g) $y^2 + y = x$
d) $x = \sin y$             h) $y^2 + y = \sin x$

a) $\frac{1}{2y}$   b) $\frac{1}{2y^2}$   c) $\frac{\cos x}{4y^3}$   d) $\sec y$   e) $\csc 2y$

f) $\frac{1}{-2y \sin y^2}$   g) $(1 + 2y)^{-1}$   h) $(1 + 2y)^{-1} \cos x$

**D11.** Find the slope of the following curves at the point $(1, 2)$.
a) $y^3 = 8x^2$                    c) $y^2 + 2y = x^2 + 7$
b) $2x^2 + 3y^4 = 50$              d) $2x = \sin^2 \frac{\pi y}{8}$

a) $\frac{4}{3}$   b) $\frac{-1}{24}$   c) $\frac{1}{3}$   d) $\frac{16}{\pi}$

**D12.** Find the derivatives of the following functions.
a) $f(x) = 2\sqrt{x}$              e) $f(x) = \sin \sqrt{x}$
b) $f(x) = 3x^{1/3}$              f) $f(x) = \sqrt{\cos x}$
c) $f(x) = 2x^{-1/2}$            g) $f(x) = \sqrt{1 + x^6}$
d) $f(x) = -3x^{-4/3}$           h) $f(x) = (1 + \sqrt{x})^2$

a) $\frac{1}{\sqrt{x}}$   b) $x^{-2/3}$   c) $-x^{-3/2}$   d) $4x^{-7/3}$

e) $\frac{1}{2\sqrt{x}} \cos \sqrt{x}$   f) $\frac{-\sin x}{2\sqrt{\cos x}}$   g) $\frac{3x^5}{\sqrt{1 + x^6}}$

h) $\frac{1 + \sqrt{x}}{\sqrt{x}}$

**D13.** Find $\frac{dy}{dx}$ at the indicated point for the following parametric equations.

a) $x = t^2$,   $y = 6t + 4$,   $t = 3$
b) $x = 3t + 4$,   $y = 6t^3 - 3$,   $t = 1$
c) $x = t^2 + t$,   $y = 2t^2 + \sin t$,   $t = 0$

d) $x = \tan t$,   $y = \sin t + \cos t$,   $t = \dfrac{\pi}{4}$

a) 1     b) 6     c) 1     d) 0

**M14.** Consider the equation $y^2 + axy + x^2 = 0$. Find:
 a) The values of $a$ for which no implicitly defined function exists.
 b) The values of $a$ for which more than one implicitly defined function exists.
 c) The values of $a$ for which exactly one implicitly defined function exists.

**M15.** Find the equation of the normal line at the indicated point.
 a) $x + x^2y^2 - y - 1 = 0$,   $(1, 1)$
 b) $4x^2 - 9y^2 = 28$,   $(4, -2)$

**M16.** Find $\dfrac{dy}{dx}\Big|_{t=1}$ given

 a) $y = t^3 + 2t^2 + 3t + 1$,   $x = t^2 + t + 3$
 b) $y = t^3 + 1$,   $x = t^3 + 3$

**M17.** Find the equation of the tangent to the curve given by $x = \sin 2t$, $y = \cos 3t$ at the point corresponding to $t = \frac{\pi}{3}$.

**M18.** Find $\frac{dy}{dx}$ given $\frac{x}{y} + \frac{y}{x} + \frac{x-y}{y-x} = 1$

**M19.** Suppose $x = f(2x + 3y)$ and $f'(5) = 3$. Find $\frac{dy}{dx}$ corresponding to the point $(1, 1)$.

**M20.** Find $y'$ given
 a) $3x^2 + 8xy - y^2 = 4$
 b) $\sin xy - x - y = 0$

**M21.** Show that the tangents at $(0, 0)$ to the curves given by the equations

$$5y - 2x + y^3 - x^2y = 0$$

and

$$5x + 2y + x^4 - x^3y^2 = 0$$

are perpendicular.

**M22.** Show that the equation of the tangent to the ellipse $b^2x^2 + a^2y^2 = a^2b^2$ at the point $(x_0, y_0)$ is given by $b^2x_0x + a^2y_0y = a^2b^2$.

**M23.** a) Prove that the curves $2x^2 + y^2 = 2$ and $y^2 = 3x$ are perpendicular at their points of intersection.
 b) Prove that the curves $2x^2 + y^2 = 2$ and $y^2 = cx$ are perpendicular at their points of intersection for any value $c > 0$.

**M24.** a) If $f(y) = x$, show that $\dfrac{dy}{dx} = \dfrac{1}{f'(y)}$.

 b) If $f(g(y)) = x$, show that $\dfrac{dy}{dx} = \dfrac{1}{f'(g(y))g'(y)}$.

**M25.** a) If $f(x) + g(y) = C$, $C$ equals constant, show that
$$\frac{dy}{dx} = \frac{-f'(x)}{g'(y)}.$$

 b) If $f(x)g(y) = C$, $C$ equals constant, show that
$$\frac{dy}{dx} = \frac{-g(y)f'(x)}{f(x)g'(y)}.$$

---

## 3.3   *Inverse Functions*

We will begin this section by discussing *monotone functions*. These will lead naturally to the conditions under which a function has a so-called "inverse." As we will see, the tangent and sine functions are important instances in which an inverse can be defined.

**A. Monotone Functions**   Suppose $f$ is a function defined on a set $D$. Then $f$ is said to be

→   This section was undoubtedly the most difficult one in the book to write. More than nine drafts were written before we finally were satisfied with the presentation. You should study this section carefully because there are a number of important concepts to get straight.

a) *strictly increasing on D* (Fig. 1a) if

$$x_1 < x_2 \quad (x_1, x_2 \in D) \quad \text{implies}$$
$$f(x_1) < f(x_2), \quad \text{and}$$

b) *strictly decreasing on D* (Fig. 1b) if

$$x_1 < x_2 \quad (x_1, x_2 \in D) \quad \text{implies} \quad f(x_1) > f(x_2).$$

In either (a) or (b) the function is said to be *strictly monotone on D*.

Also $f$ is said to be

c) *increasing on D* (Fig. 1c) if

$$x_1 < x_2 \quad (x_1, x_2 \in D) \quad \text{implies}$$
$$f(x_1) \le f(x_2), \quad \text{and}$$

d) *decreasing on D* (Fig. 1d) if

$$x_1 < x_2 \quad (x_1, x_2 \in D) \quad \text{implies} \quad f(x_1) \ge f(x_2).$$

In either (c) or (d) the function is said to be *monotone on D*.

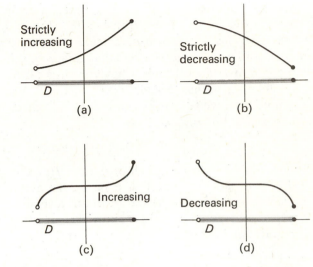

**Figure 1**

---

*Example 1* The function $f(x) = x^3$ (Fig. 2) is strictly increasing on $(-\infty, \infty)$ because whenever $x_1 < x_2$, $x_1^3 < x_2^3$.

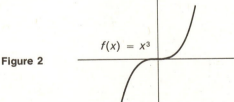

**Figure 2**

$f(x) = x^3$

**Figure 3**

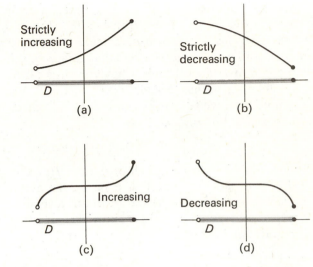

$f(x) = x^2$

---

Example 2 shows that a function which is not monotone can be made so by restricting its domain.

---

*Example 2* The function $f(x) = x^2$ (Fig. 3) is not monotone or strictly monotone on $(-\infty, \infty)$. However, the function $f(x) = x^2$ with domain $[0, \infty)$ (Fig. 4a) is strictly increasing, and the function $f(x) = x^2$ with domain $(-\infty, 0]$ is strictly decreasing.

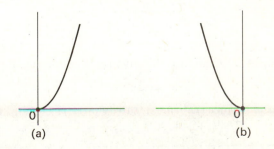

**Figure 4**

**B. Inverse Functions**    Let $f$ be a function with domain $D$ and image $T = \{f(a): a \in D\}$ (Fig. 5). Now suppose that for each $b$ in $T$ there is a *unique* $a$ in $D$ such that $f(a) = b$. The rule that assigns to the $b$ in $T$ the $a$ in $D$ then defines a function. It is called the *inverse of f* and has domain $T$ and image $D$. We will denote the inverse of $f$ by the symbol $f^{-1}$. Since $f(a) = b$ and $f^{-1}(b) = a$, we have

$$f(f^{-1}(b)) = b \quad \text{and} \quad f^{-1}(f(a)) = a.$$

Suppose $(a, b)$ is on the graph of $f$. Then $f(a) = b$, so $f^{-1}(b) = a$ and $(b, a)$ must then be on the graph of $f^{-1}$. Since the points $(a, b)$ and $(b, a)$ are related as in Figure 6, the graph of $f^{-1}$ is simply the reflection of the graph of $f$ about the line $y = x$ (Fig. 7).

→    Do not confuse $f^{-1}$ with the reciprocal $\frac{1}{f}$. The notation is admittedly ambiguous.

→    For example, if $f(x) = 2x + 1$ and $f^{-1}(x) = \frac{x-1}{2}$, then this equation is valid. (See Example 3.)

(Greg)

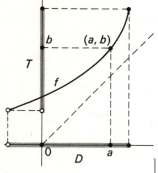

**Figure 5**

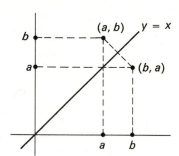

**Figure 6**

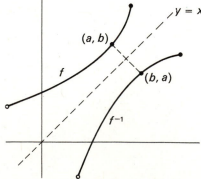

**Figure 7**

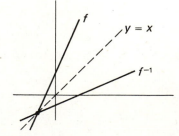

**Figure 8**

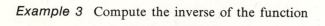

*Example 3*    Compute the inverse of the function

$$f(x) = 2x + 1.$$

*Solution*    Since $f^{-1}$ must satisfy $x = f(f^{-1}(x))$, we have $x = 2f^{-1}(x) + 1$, so $f^{-1}(x) = \frac{x-1}{2}$ (Fig. 8).

→    Notice that $f^{-1}(x)$ is *not* equal to $\frac{1}{2x+1} = \frac{1}{f(x)} = (f(x))^{-1}$.

Now suppose a function $f$ is strictly monotone. Then there could not be two distinct $a$'s, say $a_1 < a_2$, such that $f(a_1) = f(a_2) = b$. Namely, for each $b$ in its image there must be only one $a$ in its domain such that $f(a) = b$. This proves

**THEOREM 1**  *If a function $f$ is strictly monotone then $f$ has an inverse $f^{-1}$.*

→ The converse of this theorem is also true provided that $f$ is assumed to be nice enough. In particular, it can be shown that if $f$ is continuous on an interval $I$ and has an inverse $f^{-1}$ then $f$ is strictly monotone on $I$. The proof of this is left to the reader as an exercise (M20).

**Example 4**  Consider the function $y = f(x) = x^2$. If we restrict its domain to $(-\infty, 0]$ so that the function is strictly monotone (decreasing) (Fig. 9a) then $f$ will have an inverse $f^{-1}$. Its image is $[0, \infty)$, so the domain and image of $f^{-1}$ are $[0, \infty)$ and $(-\infty, 0]$ respectively. Since $f^{-1}$ must satisfy $x = f(f^{-1}(x)) = (f^{-1}(x))^2$, and the image of $f^{-1}$ is $(-\infty, 0]$, we must take the negative square root to obtain $f^{-1}(x) = -\sqrt{x}$ (Fig. 9b).

*All we are actually doing when computing inverses is solving for $x$ in terms of $y$, then switching letters. For example, if $y = x^2$ $(-\infty < x \le 0)$ then $x = -\sqrt{y}$, so the inverse function*
→ *is $y = -\sqrt{x}$ $(0 \le x < \infty)$.*

(Greg)

### C. The Inverse Tangent and Inverse Sine Functions

The function $y = \tan x$ (Fig. 10) does not have an inverse. However, the function

$$y = \tan x \quad (-\tfrac{\pi}{2} < x < \tfrac{\pi}{2}) \quad \text{(Fig. 11)}$$

is strictly increasing with image $(-\infty, \infty)$. Hence this function has an inverse denoted by

$$y = \mathrm{Tan}^{-1} x,$$

whose domain is $(-\infty, \infty)$ and image is $(-\tfrac{\pi}{2}, \tfrac{\pi}{2})$ (Fig. 12).

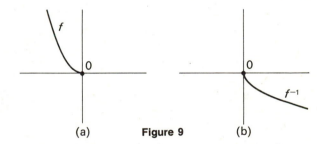

(a)      **Figure 9**      (b)

→ There are several notations for the inverse trigonometric functions. The most common are $\arctan x$, $\sin^{-1} x$, etc., but they are often considered to be multiple-valued. To avoid any confusion, we use the notation $\mathrm{Tan}^{-1} x$ and $\mathrm{Sin}^{-1} x$, introduced by E. E. Moise in his book *Calculus* (Reading, Mass.: Addison Wesley Publishing Company, 1967).

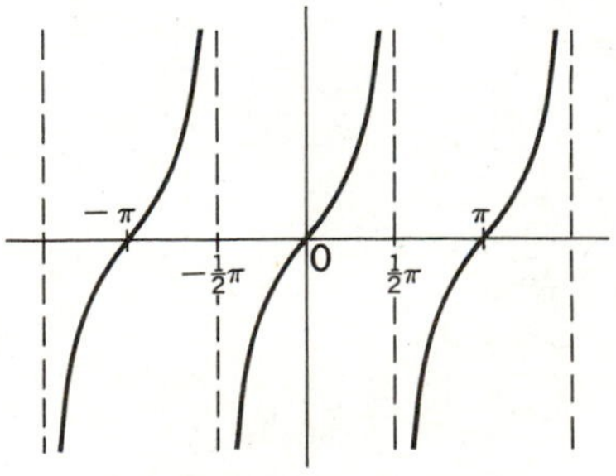

$y = \tan x \quad (-\infty < x < \infty)$

**Figure 10**

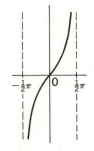

$y = \tan x \; \left(-\dfrac{\pi}{2} < x < \dfrac{\pi}{2}\right)$

**Figure 11**

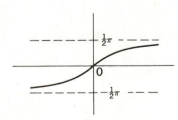

$y = \mathrm{Tan}^{-1} x \quad (-\infty < x < \infty)$

**Figure 12**

*THEOREM 1*

$$\frac{d}{dx}\,\text{Tan}^{-1}x = \frac{1}{1+x^2}$$

*Proof*  Letting $y = \text{Tan}^{-1}x$, we have

$$x = \tan y.$$

In this step we used the fact that $f^{-1}(x) = y$ if and only if
➡ $x = f(y)$.

Assuming that $y$ is differentiable, we differentiate implicitly, obtaining

➡ In Chapter 10 it is shown that if a function has a derivative different from zero on an interval then its inverse is differentiable.

$$\frac{d}{dx}x = \frac{d}{dx}\tan y,\quad \text{so}$$

$$1 = (\sec^2 y)\frac{dy}{dx},\quad \text{or}$$

$$\frac{dy}{dx} = \frac{1}{\sec^2 y} = \frac{1}{1 + \tan^2 y}.$$

Using the identity $\sin^2 y + \cos^2 y = 1$, we have

$$\sec^2 y = \frac{1}{\cos^2 y} = \frac{\cos^2 y + \sin^2 y}{\cos^2 y}$$

$$= \frac{\cos^2 y}{\cos^2 y} + \frac{\sin^2 y}{\cos^2 y}$$

Since $x = \tan y$, we have $\dfrac{dy}{dx} = \dfrac{1}{1+x^2}$ ●

$$= 1 + \tan^2 y.$$

*COROLLARY*  *If u is a function of x, then*

➡ This follows from the Chain Rule. We have

$$\frac{d}{dx}\,\text{Tan}^{-1}u = \frac{1}{1+u^2}\frac{du}{dx}.$$

$$\frac{d}{dx}\,\text{Tan}^{-1}u = \left(\frac{d}{du}\,\text{Tan}^{-1}u\right)\frac{du}{dx} = \frac{1}{1+u^2}\frac{du}{dx}.$$

---

*Example 5*

a) $\dfrac{d}{dx}\,\text{Tan}^{-1}3x = \dfrac{1}{1+(3x)^2}\cdot 3 = \dfrac{3}{1+9x^2}$

b) $\dfrac{d}{dx}\,\text{Tan}^{-1}(1+x^2) = \dfrac{1}{1+(1+x^2)^2}\cdot 2x$

$$= \dfrac{2x}{2 + 2x^2 + x^4}$$

*Problem 2*  Find

a) $\dfrac{d}{dx}\,\text{Tan}^{-1}\dfrac{1}{x}\bigg|_{x=1}$     b) $\dfrac{d}{dx}\,\text{Tan}^{-1}(\sin x)$

*Answer:*  a) $\dfrac{-1}{2}$     b) $\dfrac{\cos x}{1 + \sin^2 x}$

---

Now consider the function $y = \sin x$ (Fig. 13). On the interval $[-\frac{\pi}{2}, \frac{\pi}{2}]$ it is strictly increasing with image $[-1, 1]$, so the function

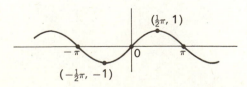

**Figure 13**          $y = \sin x\quad(-\infty < x < \infty)$

$$y = \sin x \quad \left(-\frac{\pi}{2} \le x \le \frac{\pi}{2}\right) \quad \text{(Fig. 14)}$$

has an inverse. We denote it by $y = \text{Sin}^{-1} x$, and it has domain $[-1, 1]$ and image $[-\frac{\pi}{2}, \frac{\pi}{2}]$ (Fig. 15).

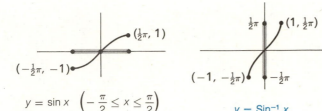

$y = \sin x \quad \left(-\frac{\pi}{2} \le x \le \frac{\pi}{2}\right)$

$y = \text{Sin}^{-1} x$

**Figure 14**          **Figure 15**

➡ We will have no use for the inverses of the remaining four trigonometric functions.

**THEOREM 2**

$$\frac{d}{dx} \text{Sin}^{-1} x = \frac{1}{\sqrt{1 - x^2}}$$

*Proof* Letting $y = \text{Sin}^{-1} x$, we have

$$x = \sin y.$$

Now differentiate implicitly:

$$\frac{d}{dx} x = \frac{d}{dx} \sin y,$$

so

$$1 = \cos y \frac{dy}{dx}$$

or

$$\frac{dy}{dx} = \frac{1}{\cos y} = \frac{1}{\sqrt{1 - \sin^2 y}}.$$

Since $x = \sin y$, we have $\dfrac{dy}{dx} = \dfrac{1}{\sqrt{1 - x^2}}$  ●

➡ Our replacing $\cos y$ by $\sqrt{1 - \sin^2 y}$ requires further comment. Since $\cos^2 y + \sin^2 y = 1$, we have $\cos^2 y = 1 - \sin^2 y$, so $\cos y = \pm\sqrt{1 - \sin^2 y}$. Because $y = \text{Sin}^{-1} x$ and the image of this function is $[-\frac{\pi}{2}, \frac{\pi}{2}]$, we must have $-\frac{\pi}{2} \le y \le \frac{\pi}{2}$. For these values of $y$, $\cos y \ge 0$, so we must take the positive square root to obtain $\cos y = \sqrt{1 - \sin^2 y}$.

**COROLLARY** *If $u$ is a function of $x$, then*

$$\frac{d}{dx} \text{Sin}^{-1} u = \frac{1}{\sqrt{1 - u^2}} \frac{du}{dx}.$$

➡ This follows from the chain rule as in the previous corollary.

*Example 6*

a) $\dfrac{d}{dx} \text{Sin}^{-1} x^2 = \dfrac{1}{\sqrt{1 - (x^2)^2}} 2x = \dfrac{2x}{\sqrt{1 - x^4}}$

b) $\dfrac{d}{dx}\operatorname{Sin}^{-1}\dfrac{x}{a} = \dfrac{1}{\sqrt{1-\left(\frac{x}{a}\right)^2}}\cdot\dfrac{1}{a} = \dfrac{1}{\sqrt{a^2-x^2}}$

when $a > 0$.

c) $\dfrac{d}{dx}\operatorname{Sin}^{-1}\dfrac{x}{-3} = \dfrac{1}{\sqrt{1-\left(\frac{x}{-3}\right)^2}}\cdot\dfrac{1}{-3}$

$\qquad = \dfrac{\sqrt{9}}{\sqrt{9-x^2}}\cdot\dfrac{1}{-3} = \dfrac{-1}{\sqrt{9-x^2}}$

d) $\dfrac{d}{dx}\operatorname{Sin}^{-1}\dfrac{3}{x^2} = \dfrac{1}{\sqrt{1-\left(\frac{3}{x^2}\right)^2}}\dfrac{d}{dx}\left(\dfrac{3}{x^2}\right)$

$\qquad = \dfrac{x^2}{\sqrt{x^4-9}}\cdot\dfrac{-6}{x^3} = \dfrac{-6}{x\sqrt{x^4-9}}.$

---

**Problem 3**  Evaluate

a) $\dfrac{d}{dx}\operatorname{Sin}^{-1}(3x-2)$  b) $\dfrac{d}{dx}\operatorname{Sin}^{-1}(\sqrt{1-x^2})$

c) $\dfrac{d}{dx}\operatorname{Sin}^{-1}(\cos x)$  $(0 \le x \le \pi)$

*Answer:*  a) $\dfrac{3}{\sqrt{12x-9x^2-3}}$  b) $\dfrac{-x}{\sqrt{x^2(1-x^2)}}$

c) $-1$

---

## D. Exercises

**B1.** What is the largest interval for which the function

$$y = x^2 - 4x + 5$$

a) is strictly increasing?
b) is strictly decreasing?

**B2.** Explain why each of these functions has an inverse and find the inverse.
a) $f(x) = \frac{x}{4} - 3$
b) $g(x) = x^{1/3}$

**B3.** Restrict the function $f(x) = \cos x$ to the interval $[0, \pi]$ and draw the graph of the inverse of this function.

**B4.** Show that
a) $\operatorname{Sin}^{-1}\left(\sin\dfrac{3\pi}{2}\right) = \dfrac{-\pi}{2}$

b) $\operatorname{Sin}^{-1}\left(\sin\dfrac{13\pi}{6}\right) = \dfrac{\pi}{6}$

c) $\operatorname{Tan}^{-1}\left(\tan\dfrac{5\pi}{4}\right) = \dfrac{\pi}{4}$

d) $\operatorname{Sin}^{-1}\left(\tan\dfrac{5\pi}{4}\right) = \dfrac{\pi}{2}$

**B5.** Differentiate the following.
a) $\operatorname{Tan}^{-1}2x$     c) $\operatorname{Sin}^{-1}(\sin x \cos x)$

b) $x^2\operatorname{Tan}^{-1}\dfrac{2}{x}$     d) $\operatorname{Tan}^{-1}(2\tan x)$

**B6.** Find $\dfrac{dy}{dx}$ given
a) $x^2 = \operatorname{Tan}^{-1}y$
b) $x = \operatorname{Sin}^{-1}(1-y)$

**D7.** Find the inverse function $f^{-1}(x)$ for each of the following functions $f(x)$.
a) $f(x) = 3x$          d) $f(x) = x^3$
b) $f(x) = x - 7$       e) $f(x) = x^3 - 2$
c) $f(x) = 3x - 7$      f) $f(x) = 4 + x^{1/3}$

a) $\dfrac{x}{3}$     b) $x + 7$     c) $\dfrac{x+7}{3}$     d) $x^{1/3}$

e) $(x+2)^{1/3}$     f) $(x-4)^3$

**D8.** Evaluate the following.
a) $\operatorname{Tan}^{-1}(-\sqrt{3})$     e) $\operatorname{Sin}^{-1}(-1)$

b) $\operatorname{Tan}^{-1}(-1)$     f) $\operatorname{Sin}^{-1}\left(\dfrac{-\sqrt{2}}{2}\right)$

c) $\operatorname{Tan}^{-1}\dfrac{\sqrt{3}}{3}$     g) $\operatorname{Sin}^{-1}\dfrac{\sqrt{2}}{2}$

d) $\operatorname{Tan}^{-1}\sqrt{3}$     h) $\operatorname{Sin}^{-1}1$

a) $\dfrac{-\pi}{3}$  b) $\dfrac{-\pi}{4}$  c) $\dfrac{\pi}{6}$  d) $\dfrac{\pi}{3}$  e) $\dfrac{-\pi}{2}$

f) $\dfrac{-\pi}{4}$  g) $\dfrac{\pi}{4}$  h) $\dfrac{\pi}{2}$

**D9.** Differentiate the following.

a) $y = \text{Tan}^{-1} 2x$     d) $y = \text{Tan}^{-1}(1 + x^2)$

b) $y = \text{Tan}^{-1} 2x^2$     e) $y = \text{Tan}^{-1}(\sin x)$

c) $y = \text{Tan}^{-1}\dfrac{x}{2}$     f) $y = \text{Tan}^{-1}\dfrac{1}{x}$

a) $\dfrac{2}{1 + 4x^2}$    b) $\dfrac{4x}{1 + 4x^4}$    c) $\dfrac{2}{4 + x^2}$

d) $\dfrac{2x}{x^4 + 2x^2 + 2}$    e) $\dfrac{\cos x}{1 + \sin^2 x}$    f) $\dfrac{-1}{1 + x^2}$

**D10.** Differentiate the following.

a) $y = \text{Sin}^{-1} 3x$     d) $y = \text{Sin}^{-1}(x - 1)$

b) $y = \text{Sin}^{-1} x^3$     e) $y = \text{Sin}^{-1}(2x - 3)$

c) $y = \text{Sin}^{-1}\dfrac{x}{3}$     f) $y = \text{Sin}^{-1}(\tan x)$

a) $\dfrac{3}{\sqrt{1 - 9x^2}}$    b) $\dfrac{3x^2}{\sqrt{1 - x^6}}$    c) $\dfrac{1}{\sqrt{9 - x^2}}$

d) $\dfrac{1}{\sqrt{2x - x^2}}$    e) $\dfrac{1}{\sqrt{3x - 2 - x^2}}$

f) $\dfrac{\sec^2 x}{\sqrt{1 - \tan^2 x}}$

**M11.** Given $f(x) = \frac{x-2}{x+k}$, find a value for $k$ so that $f = f^{-1}$.

**M12.** Determine conditions on $a$, $b$, $c$, and $d$ so that the function $f(x) = \frac{ax+b}{cx+d}$ is equal to its inverse function.

**M13.** Draw the graph of the function

$$f(x) = \begin{cases} x & \text{for } 0 \le x < 1 \\ -x + 3 & \text{for } 1 \le x \le 2 \end{cases}$$

and show that $f^{-1}(x) = f(x)$.

**M14.** Show that

a) $\dfrac{d}{dx}\text{Sin}^{-1}\dfrac{1}{x} = \dfrac{-1}{|x|\sqrt{x^2 - 1}}$

b) $\dfrac{dy}{dx} = \dfrac{y}{x}$ if $\dfrac{y}{x} = \text{Tan}^{-1}\dfrac{x}{y}$

**M15.** Given $y = \sin(\text{Tan}^{-1} x)$ show that $y = x\sqrt{1 - y^2}$. Then compute $\frac{dy}{dx}$ directly from $y = \sin(\text{Tan}^{-1} x)$ and implicitly from $y = x\sqrt{1 - y^2}$, and verify that they are the same.

**M16.** Rewrite the equation $y = \tan(\text{Sin}^{-1} x)$ without using any trigonometric or inverse trigonometric functions and compute $\frac{dy}{dx}$.

**M17.** Prove that

a) $\text{Tan}^{-1} x = \text{Sin}^{-1}\dfrac{x}{\sqrt{1 + x^2}}$

b) $\text{Sin}^{-1} y = \text{Tan}^{-1}\dfrac{y}{\sqrt{1 - y^2}}$

**M18.** The function $y = \cos x$ ($0 \le x \le \pi$) is strictly monotone and therefore has an inverse $y = \text{Cos}^{-1} x$ ($-1 \le x \le 1$). Prove that

$$\text{Cos}^{-1} x = \tfrac{\pi}{2} - \text{Sin}^{-1} x$$

and thus

$$\dfrac{d}{dx}\text{Cos}^{-1} x = -\dfrac{1}{\sqrt{1 - x^2}} \quad (-1 < x < 1).$$

**M19.** Although the function $y = \sec x$ is not strictly monotone on the set $S = \{x: 0 \le x \le \pi, x \ne \frac{\pi}{2}\}$ it has the property that for each $b$ in its image $T = \{y: y \ge 1$ or $y \le -1\}$ there is a unique $a$ in its domain $S$ such that $b = \sec a$. It therefore has an inverse $y = \text{Sec}^{-1} x$ with domain $T$ and image $S$. Prove that $\text{Sec}^{-1} x = \frac{\pi}{2} - \text{Sin}^{-1}(\frac{1}{x})$ and hence (see M14(a))

$$\dfrac{d}{dx}\text{Sec}^{-1} x = \dfrac{1}{|x|\sqrt{x^2 - 1}} \quad (x > 1 \quad \text{or} \quad x < -1).$$

**M20.** Prove that if $f$ is continuous on an interval $I$ and has an inverse then $f$ is strictly monotone on $I$. (Hint: Intermediate Value Theorem.)

## 3.4 *Exponentials and Logarithms*

Our discussion of exponential functions, though brief, will be sufficient for our purposes. Logarithm functions are then easily defined as inverses of exponential functions.

We will refer several times to the following laws of exponents.

For $a, b, x, y \in R$, $a, b > 0$

1. $a^{x+y} = a^x a^y$      4. $a^{-x} = \dfrac{1}{a^x}$

2. $(a^x)^y = a^{xy}$

3. $(ab)^x = a^x b^x$      5. $a^0 = 1$.

**A. The Definition of $a^x$**  Suppose $a$ is a positive real number. When $x = 0$ we define

$$a^x = a^0 = 1,$$

and for $x$ equal to a positive integer $n$,

$$a^n = a \cdot a \cdot \ldots \cdot a \quad (n \text{ factors}),$$

and

$$a^{-n} = \frac{1}{a^n}.$$

For $x = \frac{1}{q}$ where $q$ is a positive integer, let $a^{1/q} = b$, where $b$ is that positive real number satisfying $a = b^q$. →

Next, given any positive rational number $x = \frac{p}{q}$, we define

$$a^x = a^{p/q} = (a^{1/q})^p = (a^p)^{1/q}, \quad →$$

and when $x$ is a negative rational number we define

$$a^x = 1/a^{-x}.$$

Finally, when $x$ is irrational, choose an infinite set →
of rational numbers

$$r_1, r_2, r_3, \ldots$$

approaching $x$. Then the numbers

$$a^{r_1}, a^{r_2}, a^{r_3}, \ldots$$

are defined and it can be shown that they approach a unique number. That number is defined to be $a^x$. →

We have now defined $a^x$ for all real numbers $x$, and we assume without proof that the above laws of exponents are valid.

**B. The Exponential Functions**  Let $a$ be a positive real number not equal to one. The rule that assigns to the number $x$ the number $a^x$ is called an *exponential function*.

For example, if $a = 2$, we have $f(x) = 2^x$, and in order to find its graph (Fig. 1) we observe that the →
points

$$(0, 1), (1, 2), (2, 4), (3, 8), (-1, \tfrac{1}{2}), (-2, \tfrac{1}{4})$$

satisfy the equation. We assume that the function is increasing and complete the graph accordingly.

The graph of $a^x$ for $a > 1$ (Fig. 2) is similar to that →
of $2^x$.

---

*Problem 1*  Verify the above laws for $x$ and $y$ integers.

---

A proof of the existence of $b$ is difficult and will not be given until Chapter 10. The proof that $b$ is unique is given in Exercise M27.

More precisely, we should first prove $(a^{1/q})^p = (a^p)^{1/q}$ (Exercise M27) and then define $a^{p/q}$ as in the text.

The decimal expansion of $x$ provides such a set of rational numbers. For example, if $x = \pi = 3.14159\ldots$, let $r_1 = 3$, $r_2 = 3.1$, $r_3 = 3.14$, $r_4 = 3.141$, $r_5 = 3.1415$, $r_6 = 3.14159, \ldots$

We have not proven that the value of $a^x$ is independent of the particular set of rational numbers chosen. As usual, these subtleties will be discussed later.

Notice that $f(x) = a^x$ is greater than zero for all $x$, and the point $(0, 1)$ always lies on the graph of this function.

If $a = 1$, $f(x) = 1^x = 1$ for all $x$ and the function is a constant. The graph is then a horizontal line.

**Figure 1**

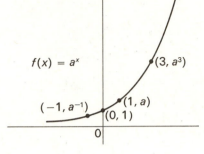

**Figure 2**

The graphs of $a^x$ and $(\frac{1}{a})^x = a^{-x}$ are the reflections of each other about the $y$-axis. This allows us to graph $g(x) = a^x$ if $0 < a < 1$. For example, the graphs of $g(x) = (\frac{1}{2})^x = 2^{-x}$ and $f(x) = 2^x$ are compared in Figure 3.

**Figure 3**

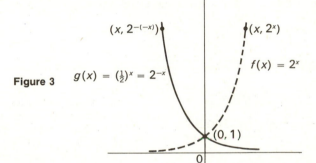

**C. The Logarithm Functions**  For $a > 0$ $(a \neq 1)$, the function $f(x) = a^x$ is strictly monotone with domain $R$ and image $(0, \infty)$. Hence it has inverse $f^{-1}$ with domain $(0, \infty)$ and image $R$. Since $x = f(f^{-1}(x))$, we have

$$x = a^{f^{-1}(x)}.$$

Denoting $f^{-1}(x)$ by $\log_a x$ (read "logarithm to the base $a$ of $x$") we have the defining equation

$$x = a^{\log_a x}.$$

The properties of logarithms (front cover) follow from those of exponentials and from this equation. In the following lemma we prove two identities to be used in the proof of Theorem 1.

Since $f(x) = \log_a x$ is the inverse of the function

$$g(x) = a^x,$$

its graph (Fig. 4) is the reflection of the graph of $a^x$ about the line $x = y$.

*The $\log_a x$ is defined to be the exponent to which $a$ is raised in order to obtain $x$. In high school I was often asked to solve equations of the type $\log_2 32 = y$, and usually simplified them to the form $2^y = 32$. Then it was clear that $y = 5$, so $5 = \log_2 32$. Problem 2, on the next page, can be solved in a similar manner.*

(Robin)

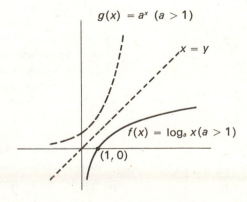

**Figure 4**

*LEMMA*

a) $\log_a \dfrac{x}{y} = \log_a x - \log_a y$    b) $\log_a x^r = r \log_a x$

*Proof*  a) Since $x = a^{\log_a x}$ and $y = a^{\log_a y}$, if we let $u = \log_a x$ and $v = \log_a y$, we have $x = a^u$ and $y = a^v$. Hence,

$$\frac{x}{y} = \frac{a^u}{a^v} = a^u a^{-v} = a^{u-v}.$$

Since $\frac{x}{y}$ also equals $a^{\log_a (x/y)}$, the exponents must be equal, so

$$\log_a \frac{x}{y} = u - v = \log_a x - \log_a y.$$

b) Let $u = \log_a x$, so $a^u = x$ and $(a^u)^r = x^r$, or $a^{ur} = x^r$. Hence,

$$a^{ur} = x^r = a^{\log_a x^r}.$$

Since the exponents must be equal, we have

$$ur = \log_a x^r \quad \text{or} \quad r \log_a x = \log_a x^r \quad \bullet$$

**Problem 2**   Find $x$.

a) $\log_2 16 = x$      e) $(\log_3 4)(\log_4 x) = 1$
b) $\log_3 x = 3$       f) $\log_a 1 = x$
c) $\log_x 625 = 4$     g) $\log_a a = x$
d) $\log_x \frac{1}{32} = 5$

*Answer:*    a) 4  b) 27  c) 5  d) $\frac{1}{2}$  e) 3  f) 0  g) 1

## *The Invention of Logarithms*

Seeing there is nothing (right well-beloved Students of Mathematics) that is so troublesome to mathematical practice, nor doth more molest and hinder calculators, than the multiplications, divisions, square and cubical extractions of great numbers, which besides the tedious expense of time are for the most part subject to many slippery errors, I began therefore to consider in my mind by what certain and ready art I might remove those hindrances.

John Napier (1550–1617)

Napier began these considerations around 1594, and, laying all other work aside, spent most of his next twenty years in a lonely Scottish castle preparing a manuscript on logarithms. His time was well spent. In 1914, at an international congress of mathematicians honoring Napier's work, Lord Moulton said, "The invention of logarithms came to the world as a bolt from the blue. No previous work had led up to it; nothing had foreshadowed it or heralded its arrival. It stands isolated, breaking in upon human thought abruptly, without borrowing from the work of other intellects or following proven lines of mathematical thought. It reminds one of those islands in the ocean which rise suddenly from great depths and which stand solitary, with deep water close around all their shores."

Prior to his invention of logarithms, Napier's work, though original, was not entirely successful. He devised a number of remarkably ingenious instruments of war, but none advanced beyond the drawing board. In 1593 he published a book in defense of Protestantism which was popular enough to warrant twenty-one editions. Here he proved, with a closely reasoned sequence of thirty-six propositions, that the world would end sometime between 1688 and 1700.

But Napier's work on logarithms was revolutionary and, with one notable exception, received instant acclaim. Johann Kepler (1571–1630) had been anxiously awaiting the publication of Napier's book, for he was in the midst of the monumental calculations that would eventually lead him to his third law of planetary motion. When Kepler received Napier's tables, an incomplete theoretical explanation accompanied them. Despite his need for them and his realization that they gave correct results, he refused to use them, saying:

. . . so far I have been unable to figure out how the tables are constructed, which makes me suspect that the inventor, on purpose, used some abstruse number as base to make it difficult, if not impossible, to check it. I regard it as unworthy of a mathematician to see with other people's eyes and to accept as true or as proven that for which he himself has no proof. For one may always doubt whether these computations, even though correct ten or a hundred times, may not one day provide wrong results.

Kepler was not long without logarithms, however, for he soon did figure out how the tables were constructed. He discovered Napier's fundamental formula, invented others, and proceeded to compute his own tables.

Napier's original tables were very cumbersome. They did not give the logarithm of $x$ but $10^7 \log_e (10^7 x^{-1})$. The Englishman Henry Briggs (1561–1631), upon seeing Napier's book, immediately recognized its importance and began working on logarithms himself. Within months he joined Napier in Scotland and convinced him that tables for $\log_{10} x$ would be more convenient. Together they began constructing them. After Napier's death in 1617, Briggs continued work on the tables, and they were published concurrently with Kepler's in 1624. By 1630 logarithms were in common use.

Today, when logarithms appear so easy, the student may have trouble understanding how they could ever have been difficult. But in Napier's time the notion of a function was unknown, and even the simple exponential notation $a^n = a \cdot a \cdot \cdots \cdot a$, introduced later by Descartes, did not exist. Napier invented logarithms to ease computation, and by so doing became known as the founder of applied mathematics. Although the computational efficiency of machines is now greater than that of logarithms, Napier's place in mathematics remains secure, for from his work came the logarithm functions and their inverses, the exponentials. These functions occupy a central role in the mathematical sciences.

**D. The Derivatives of $\log_a x$ and $a^x$**  In finding the formula for the derivative of $\log_a x$ the expression $(1 + k)^{1/k}$ will appear and its limit as $k \to 0$ must be evaluated. The limit does exist and is denoted by the letter $e = 2.71828\ldots$. In other words,

$$\lim_{k \to 0} (1 + k)^{1/k} = e.$$

*THEOREM 1*

$$\frac{d}{dx} \log_a x = \frac{1}{x} \log_a e$$

*Proof*  Let $f(x) = \log_a x$. Then

$$\frac{f(x + h) - f(x)}{h} = \frac{1}{h}(\log_a (x + h) - \log_a x).$$

Applying (a) of the lemma in 3.4C we rewrite this as

$$\frac{1}{h} \log_a \left(\frac{x + h}{x}\right) = \frac{1}{h} \log_a \left(1 + \frac{h}{x}\right)$$

$$= \frac{1}{x} \frac{x}{h} \log_a \left(1 + \frac{h}{x}\right).$$

We will not prove now that $\lim_{k \to 0} (1 + k)^{1/k}$ exists and equals the indicated number $e$, but a proof is given in Chapter 10 following an entirely different treatment of logarithms. If, like Kepler, the reader is reluctant to use an unverified result, the following table suggests that $(1 + k)^{1/k}$ approaches a value between 2 and 3.

The reader familiar with compound interest can compute that one dollar invested at 100% interest compounded $n$ times per year would be worth $(1 + \frac{1}{n})^n$ dollars at the end of the year. Hence $e = \lim_{n \to \infty} (1 + \frac{1}{n})^n$ dollars would be the value of one dollar invested at 100% interest compounded continuously throughout the year.

| $(1 + k)^{1/k}$ to five places corresponding to $\frac{1}{k}$ equals 1, 2, 3, 4, 10, 100, 1000, and 10,000. | |
| --- | --- |
| $(1 + 1)^1 = 2.00000$ | $(1 + \frac{1}{10})^{10} = 2.59374$ |
| $(1 + \frac{1}{2})^2 = 2.25000$ | $(1 + \frac{1}{100})^{100} = 2.70481$ |
| $(1 + \frac{1}{3})^3 = 2.37037$ | $(1 + \frac{1}{1000})^{1000} = 2.71691$ |
| $(1 + \frac{1}{4})^4 = 2.44141$ | $(1 + \frac{1}{10000})^{100000} = 2.71814$ |

By (b) of the same lemma this equals

$$\frac{1}{x}\log_a\left(1+\frac{h}{x}\right)^{x/h},$$

and letting $k=\frac{h}{x}$ we get

$$\frac{1}{x}\log_a(1+k)^{1/k}.$$

Now as $h \to 0$, $k=\frac{h}{x} \to 0$ and we have

$$f'(x) = \lim_{h\to 0}\frac{f(x+h)-f(x)}{h}$$

$$= \lim_{h\to 0}\frac{1}{x}\log_a\left(1+\frac{h}{x}\right)^{x/h}$$

$$= \lim_{k\to 0}\frac{1}{x}\log_a(1+k)^{1/k}$$

$$= \frac{1}{x}\log_a\left(\lim_{k\to 0}(1+k)^{1/k}\right)$$

$$= \frac{1}{x}\log_a e \quad\bullet$$

> We have used the fact that
> $$\lim_{k\to 0}\log_a(1+k)^{1/k} = \log_a\lim_{k\to 0}(1+k)^{1/k},$$
> which follows from the continuity of $\log_a x$.

**COROLLARY** *If $u$ is a function of $x$, then*

$$\frac{d}{dx}\log_a u = \left(\frac{1}{u}\log_a e\right)\frac{du}{dx}.$$

The proof is just an application of the chain rule.

> Note that we must have $u(x) > 0$ because the logarithm is only defined for positive numbers.
>
> $$\frac{d}{dx}\log_a u = \left(\frac{d}{du}\log_a u\right)\frac{du}{dx} = \left(\frac{1}{u}\log_a e\right)\frac{du}{dx} \quad\bullet$$

---

*Example 1*

a) If $f(x) = \log_a(3x^2+5x)$, then

$$f'(x) = \frac{1}{3x^2+5x}(\log_a e)(6x+5)$$

$$= \frac{6x+5}{3x^2+5x}\log_a e.$$

b) If $y = \log_{10}(1+t^2)$, then

$$\frac{dy}{dt} = \frac{2t}{1+t^2}\log_{10} e.$$

c) If $y = \log_2 \cos x$, then

$$\frac{dy}{dx} = \frac{1}{\cos x}(-\sin x)\log_2 e = (-\tan x)\log_2 e.$$

*Problem 3*   Find $\frac{dy}{dx}$ given

a) $y = \log_{10} x^5$

b) $y = \log_3\left(\frac{1}{1+x^2}\right)$

c) $y = \log_a \sqrt{1+x^2}$

*Answer:*    a) $\dfrac{5}{x}\log_{10} e$     b) $\dfrac{-2x}{1+x^2}\log_3 e$

c) $\dfrac{x}{1+x^2}\log_a e$

The constant $\log_a e$ appearing in Theorem 1 and its corollary is a constant source of annoyance, except when $a = e$, because then $\log_e e = 1$. Hence

$$\frac{d}{dx} \log_e x = \frac{1}{x} \log_e e = \frac{1}{x}.$$

→ When $a = 10$, $\log_{10} e = .434294$ to six places.

→ Indeed, $x = \log_e e$ means $e^x = e$, so $x = 1$.

We will introduce the notation

$$\ln x = \log_e x$$

to denote logarithms to the base $e$. In this notation the above corollary becomes

$$\frac{d}{dx} \ln u = \frac{1}{u} \frac{du}{dx} \quad \text{or} \quad (\ln u)' = \frac{u'}{u}.$$

→ The $\log_e x$ is called the *natural logarithm* of $x$, which suggests the notation ln $x$. With the exception of some exercises, our concern in what follows will almost always be with the natural logarithm.

The formula $\frac{d}{dx} x^r = rx^{r-1}$ can now be shown to be valid for all real numbers $r$.

**THEOREM 2**   *For any real number r*

$$\frac{d}{dx} x^r = rx^{r-1}.$$

→ In this theorem we assume $x > 0$ to insure that $x^r$ is defined and that ln $x$, appearing in the proof, makes sense.

*Proof*  Let $y = x^r$, so $\ln y = \ln x^r = r \ln x$. Assuming that $y$ is differentiable, we differentiate implicitly with respect to $x$, to obtain

$$\frac{1}{y} \frac{dy}{dx} = r \frac{1}{x}, \quad \text{so} \quad \frac{dy}{dx} = \frac{ry}{x}.$$

Replacing $y$ by $x^r$ then gives

$$\frac{d}{dx} x^r = \frac{rx^r}{x} = rx^{r-1} \quad \bullet$$

*Example 2*  In a similar fashion the derivative of $y = x^x$ can be found. Taking logarithms we have

$$\ln y = \ln x^x = x \ln x,$$

and differentiating implicitly gives

$$\frac{1}{y} y' = x \frac{1}{x} + \ln x = 1 + \ln x.$$

Hence

$$y' = y(1 + \ln x) = x^x(1 + \ln x).$$

**THEOREM 3**  $\dfrac{d}{dx} a^x = \dfrac{a^x}{\log_a e}$, *and when u is a function of x,*

$$\frac{d}{dx} a^u = \frac{a^u}{\log_a e} \frac{du}{dx}.$$

When $a = e$, $\log_e e = 1$, so these formulas become

$$\frac{d}{dx} e^x = e^x$$

$$\frac{d}{dx} e^u = e^u \frac{du}{dx}.$$

→ *Proof*  Let $y = a^x$. Taking logarithms to the base $a$,

$$\log_a y = \log_a a^x = x \log_a a = x.$$

Differentiating $\log_a y = x$ with respect to $x$ we have

$$\frac{1}{y} \frac{dy}{dx} \log_a e = 1, \quad \text{so} \quad \frac{dy}{dx} = \frac{y}{\log_a e} = \frac{a^x}{\log_a e}.$$

One of the most frequent mistakes one makes is to use the incorrect formula $\frac{d}{dx} e^x = x e^{x-1}$. This formula holds only when a variable $x$ is raised to a constant power and not the other way around.

(Alan)

**E. The Hyperbolic Functions** The following simple combinations of exponential functions are found to be quite useful. We define:

$$\sinh x = \frac{e^x - e^{-x}}{2} \quad \text{(hyperbolic sine)}$$

$$\cosh x = \frac{e^x + e^{-x}}{2} \quad \text{(hyperbolic cosine)}$$

$$\tanh x = \frac{\sinh x}{\cosh x} \quad \text{(hyperbolic tangent)}$$

$$\coth x = \frac{\cosh x}{\sinh x} \quad \text{(hyperbolic cotangent)}$$

$$\text{sech } x = \frac{1}{\cosh x} \quad \text{(hyperbolic secant)}$$

$$\text{csch } x = \frac{1}{\sinh x} \quad \text{(hyperbolic cosecant)}$$

The hyperbolic functions have properties that are quite similar to those of the trigonometric functions.

---

*Even though hyperbolic functions will not be dealt with extensively in this book, they are used frequently by mathematicians and engineers, and extensive tables of their values (similar to sine and cosine tables) exist.*

*(Vahan)*

*The circle $x^2 + y^2 = a^2$ is represented parametrically by $x = a \cos t$, $y = a \sin t$, while the hyperbola $x^2 - y^2 = a^2$ has the parametric representation*

$$x = a \cosh t, \quad y = a \sinh t,$$

*This is the reason for the word "hyperbolic."*

---

*Example 3* Show that

a) $\cosh^2 x - \sinh^2 x = 1$

b) $\dfrac{d}{dx} \sinh x = \cosh x$

*Solution*

$\cosh^2 x - \sinh^2 x$

$$= \left(\frac{e^x + e^{-x}}{2}\right)^2 - \left(\frac{e^x - e^{-x}}{2}\right)^2$$

$$= \left(\frac{e^{2x} + 2 + e^{-2x}}{4}\right) - \left(\frac{e^{2x} - 2 + e^{-2x}}{4}\right)$$

$$= \frac{4}{4}$$

$$= 1,$$

which proves (a). Since

$$\frac{d}{dx} \sinh x = \frac{d}{dx}\left(\frac{e^x - e^{-x}}{2}\right) = \frac{e^x - (-e^{-x})}{2}$$

$$= \cosh x,$$

equation (b) is valid.

---

*Problem 4* Verify the following.

a) $\dfrac{d}{dx} \cosh x = \sinh x$

b) $\dfrac{d}{dx} \tanh x = \text{sech}^2 x$

c) $\sinh (x + y) = \sinh x \cosh y + \sinh y \cosh x$

## F. Exercises

**B1.** Find $x$ in each of the following equations.
    a) $\log_3 81 = x$      c) $\log_x 243 = 5$
    b) $\log_{5/2}\left(\frac{25}{4}\right) = x$     d) $\log_2 x = 10$

**B2.** Show that
    a) $2\ln\sin x = \ln(1 - \cos x) + \ln(1 + \cos x)$
    b) $\ln(x + \sqrt{x^2 - 1}) = -\ln(x - \sqrt{x^2 - 1})$

**B3.** Evaluate the following.

    a) $\left.\dfrac{d}{dx}\dfrac{\ln x}{x^2}\right|_{x=e}$      b) $\left.\dfrac{d}{dx}\cos(\ln x)\right|_{x=e^{\pi/3}}$

**B4.** Differentiate the following functions and evaluate their derivatives at $x = 0$.
    a) $f(x) = e^{(e^x)}$
    b) $g(x) = e^{2x}\ln(1 + x^2)$
    c) $h(x) = \text{Sin}^{-1}(e^x - 1)$

**B5.** Differentiate each of the following functions.
    a) $y = \tanh e^x$
    b) $y = \cosh^2 5x$
    c) $y = e^{\sin 3x}$

**B6.** Find the coordinates of the point $P$ lying on the graph of $y = e^x$ which satisfy the condition that the line tangent to the curve at $P$ passes through the origin.

**D7.** Find $x$ in each of the following.
    a) $\log_2 x = 3$      e) $\log_x \frac{1}{9} = -2$
    b) $\log_3 27 = x$     f) $\log_{10} 1000 = x$
    c) $\log_x 128 = \frac{7}{3}$     g) $\log_x 125 = 3$
    d) $\log_{1/2}\frac{1}{16} = x$    h) $\log_2 x = 10$

    a) 8    b) 3    c) 8    d) 4    e) 3    f) 3    g) 5
    h) 1024

**D8.** Differentiate the following functions.
    a) $y = \ln 5x$      e) $y = \ln(1 + x^2)^2$
    b) $y = \ln 27x$     f) $y = \ln\sqrt{1 + x^2}$
    c) $y = \ln x^5$      g) $y = \ln\sec x$
    d) $y = \ln(1 + x^2)$   h) $y = \ln\sec^6 x$

    a) $\dfrac{1}{x}$    b) $\dfrac{1}{x}$    c) $\dfrac{5}{x}$    d) $\dfrac{2x}{1 + x^2}$    e) $\dfrac{4x}{1 + x^2}$

    f) $\dfrac{x}{1 + x^2}$    g) $\tan x$    h) $6\tan x$

**D9.** Differentiate the following functions.
    a) $y = e^{3x}$      d) $y = 2e^{2x}$
    b) $y = (e^x)^3$    e) $y = e^{\sin x}$
    c) $y = e^{x^2}$     f) $y = e^{\cos x}$

    g) $y = \sin e^x$    h) $y = \tan e^{2x}$

    a) $3e^{3x}$    b) $3e^{3x}$    c) $2xe^{x^2}$    d) $4e^{2x}$
    e) $(\cos x)e^{\sin x}$    f) $(-\sin x)e^{\cos x}$    g) $e^x\cos e^x$
    h) $2e^{2x}\sec^2 e^{2x}$

**M10.** The normal to the graph of $f(x) = \ln x$ at some point $P$ passes through the point $(0, 1 + e^2)$. Find the coordinates of the point $P$.

**M11.** Find the derivative of each of the following functions, first directly and then by taking logarithms.
    a) $y = (x(x + 1))^{1/2}$    $(x > 0)$

    b) $y = \left(\dfrac{x(x - 1)}{x^2 + 1}\right)^{2/3}$    $(x > 1)$

**M12.** Find the derivatives of the following functions.
    a) $y = 2^x$      c) $y = \log_{10} x^2\sin^2 x$
    b) $y = 3^{\sin x}$    d) $y = \log_2(\text{Tan}^{-1} x)$

**M13.** Find the derivatives of the following functions and simplify.
    a) $y = 10^{\log_{10} x^2}$                c) $y = \log_a b^x$
    b) $y = \log_{10}(\sin^2 x + \cos^2 x)$

**M14.** Find the derivatives of the following functions.
    a) $y = x^{(x^2)}$    b) $y = (x^x)^2$

**M15.** Prove the following identities.

    a) $\dfrac{d}{dx}\coth x = -\text{csch}^2 x$

    b) $2\ln(\cosh x) = \ln(1 + \sinh x) + \ln(1 - \sinh x)$

    c) $\dfrac{d}{dx}\text{sech}\, x = -\tanh x\,\text{sech}\, x$

**M16.** The number $N$ of bacteria in certain cultures as a function of time is given by a function of the form

$$N = Ae^{kt} \quad (k > 0)$$

where $A$ and $k$ are constants.
a) Show that $N$ satisfies the equation

$$\frac{dN}{dt} = kN \quad \text{for all } t.$$

This states that the rate of growth of the bacteria is proportional to the number of bacteria present.
b) Show that the time it takes for the bacteria to double in number is $\frac{\ln 2}{k}$.

**M17.** When a cable is suspended at two points it takes the shape of a curve of the form $y = a \cosh\left(\frac{x}{a}\right)$ called a *catenary* from the Latin word *catena,* meaning chain.
  a) Find $\frac{dy}{dx}$ and set $w = \frac{dy}{dx}$.
  b) Show that

$$\frac{dw}{dx} = \frac{1}{a}\sqrt{1 + w^2} \quad \text{for all } x.$$

**M18.** Using the fact that if $f(x) = a^x$ then $f'(x) = a^x \log_e a$, show that $\frac{a^h - 1}{h} \to \log_e a$ as $h \to 0$ by considering

$\frac{(f(x + h) - f(x))}{h}$. In particular, show that

$\frac{e^h - 1}{h} \to 1$ as $h \to 0$.

**M19.** Show that if $b > e^a$ then no tangent line to $f(x) = e^x$ passes through the point $(a, b)$.

**M20.** Find a tangent to $y = e^x$ that passes through the point $(1, 2)$.

**M21.** Prove that $2 \le (1 + \frac{1}{n})^n < 3$ for every positive integer $n$.

**M22.** Show that every point of the form $(a, b)$ with $b < 0$ lies on a unique tangent to $f(x) = e^x$.

**M23.** Properties of $\sinh^{-1} x$:

  a) For $0 < a < b$, prove that $\frac{a^2 - 1}{a} < \frac{b^2 - 1}{b}$.

  b) Using the result in (a) prove that $f(x) = \sinh x$ is strictly increasing on $(-\infty, \infty)$.
  c) Let $g(x) = \sinh^{-1} x$ be the inverse of $f(x) = \sinh x$ and show that $\frac{d}{dx}\sinh^{-1} x = \frac{1}{\sqrt{1 + x^2}}$.

**M24.** Solve the equation

$$x = \frac{e^y - e^{-y}}{2} = \sinh y$$

for $y$ in terms of $x$ to show that

$$y = \sinh^{-1} x = \ln(x + \sqrt{x^2 + 1}).$$

**M25.** Properties of $\tanh^{-1} x$:
  a) Show that $f(x) = \tanh x$ is strictly increasing on $(-\infty, \infty)$ and has image $(-1, 1)$.
  b) Let $g(x) = \tanh^{-1} x$ be the inverse of $f(x) = \tanh x$ and prove that $\frac{d}{dx}\tanh^{-1} x = \frac{1}{1 - x^2}$.

**M26.** Solve the equation

$$x = \frac{e^y - e^{-y}}{e^y + e^{-y}} = \tanh y$$

for $y$ in terms of $x$ to show that

$$y = \tanh^{-1} x = \frac{1}{2}\ln\left(\frac{1 + x}{1 - x}\right) \quad (-1 < x < 1).$$

**M27.** a) Suppose $a, b > 0$, $q$ is a positive integer, and $a^q = b^q$. Prove that $a = b$. (Hint: Divide $a^q - b^q$ by $a - b$.)

  b) Suppose $a > 0$ and $p$ and $q$ are positive integers. Using the definitions in 3.4A, show that $(a^{1/q})^p = (a^p)^{1/q}$ by first showing they have the same $pq$ power and then using (a).

**M28.** If an amount of money $A_0$ is deposited in a bank at an interest rate $r$, the amount of money $A$ accumulated after $t$ years is given by $A = A_0 e^{rt}$. Show that the rate $\frac{dA}{dt}$ at which the money is accumulating is proportional to $A$.

---

| 3.5 | ***Derivatives of Higher Order and Differentials*** |

We now introduce higher derivatives and their notations. These notions will be illustrated with a number of examples. The concept of differentials will also be introduced.

**A. Definition and Examples**  Given a function

$$y = f(x),$$

As an example, consider the function

$$y = x^3 + e^{2x} - 5 \ln x.$$

we will often refer to its derivative $\frac{dy}{dx}$ as the *first derivative*.

Its first derivative is

$$\frac{dy}{dx} = 3x^2 + 2e^{2x} - \frac{5}{x},$$

The *second derivative* is defined to be the derivative of the first derivative, that is, $\frac{d}{dx}\left(\frac{dy}{dx}\right)$.

its second derivative is

$$\frac{d}{dx}\left(\frac{dy}{dx}\right) = \frac{d}{dx}\left(3x^2 + 2e^{2x} - \frac{5}{x}\right) = 6x + 4e^{2x} + \frac{5}{x^2},$$

The *third derivative* is defined to be the derivative of the second derivative, that is, $\frac{d}{dx}\left(\frac{d}{dx}\left(\frac{dy}{dx}\right)\right)$.

and its third derivative is

$$\frac{d}{dx}\left(\frac{d}{dx}\left(\frac{dy}{dx}\right)\right) = \frac{d}{dx}\left(6x + 4e^{2x} + \frac{5}{x^2}\right) = 6 + 8e^{2x} - \frac{10}{x^3}.$$

In general, the *nth derivative* is defined to be the derivative of the $(n-1)$th derivative.

The notations for the second derivative are

The *n*th derivative ($n > 3$) is

$$2^n e^{2x} + \frac{(-1)^n 5(n-1)!}{x^n}.$$

$$\frac{d^2y}{dx^2}, \frac{d^2f}{dx^2}, y'', f'',$$

Notice that the expression $\frac{d}{dx}\left(\frac{dy}{dx}\right)$ suggests the notation

$$\frac{d^2y}{dx^2}.$$

and, in general, those for the *n*th derivative are

$$\frac{d^n y}{dx^n}, \frac{d^n f}{dx^n}, y^{(n)}, f^{(n)}.$$

The prime notation, $y''$, $y'''$, is convenient only for derivatives of low order. Observe that we use parentheses in the expressions $f^{(n)}$ and $y^{(n)}$ to distinguish them from the products $f^n = f \cdot f \cdot \ldots \cdot f$ and $y^n = y \cdot y \cdot \ldots \cdot y$. Also, we use the convention $f^{(0)} = f$.

Correspondingly, to denote the *n*th derivative at a point $x = a$, we write

$$\left.\frac{d^n y}{dx^n}\right|_a = \left.\frac{d^n f}{dx^n}\right|_{x=a} = y^{(n)}(a), \quad \text{and so on.}$$

## Example 1

a) $\left.\dfrac{d^2}{dx^2}\left(4x - \dfrac{1}{x}\right)\right|_{x=2} = \left.\dfrac{d}{dx}\left(4 + \dfrac{1}{x^2}\right)\right|_{x=2}$

$$= \left.\frac{-2}{x^3}\right|_{x=2} = \frac{-1}{4}$$

b) If $y = \tan x$, then $y' = \sec^2 x$, and $y'' = 2\sec x \sec x \tan x = 2\sec^2 x \tan x.$

c) If $f(x) = e^{2x}$, then $f^{(1)}(x) = 2e^{2x}$, $f^{(2)}(x) = 4e^{2x}$, $f^{(3)}(x) = 8e^{2x}$, and $f^{(4)}(x) = 16e^{2x}$.

d) Suppose $y = \dfrac{x-1}{x+2}$. Then

$$\frac{dy}{dx} = \frac{(x+2)\cdot 1 - (x-1)\cdot 1}{(x+2)^2} = \frac{3}{(x+2)^2},$$

$$\frac{d^2y}{dx^2} = \frac{-6}{(x+2)^3}, \frac{d^3y}{dx^3} = \frac{18}{(x+2)^4}, \quad \text{and}$$

$$\frac{d^4y}{dx^4} = \frac{-72}{(x+2)^5}.$$

*Be careful to compute the appropriate derivative first, and then evaluate it at the indicated point. Reversing the process will always give zero.*

*(Bob)*

**Problem 1** Find

a) $\left.\dfrac{d^3}{dx^3}(3x^4 - 6x + 8)\right|_{x=2}$

b) $y^{(6)}(0)$ given $y = e^{3x}$

c) $\left.\dfrac{d^2}{dx^2}(x^2 \ln x)\right|_{x=e}$

d) $f^{(2)}(\pi)$ given $f(x) = e^{\sin x}$

*Answer:* a) 144    b) 729    c) 5    d) 1

Higher derivatives of implicitly defined functions can also be found.

---

***Example 2*** Suppose $x^2 + y^2 = 9$. Find $y''$.

*Solution* Differentiating implicitly we have

$$2x + 2yy' = 0, \quad \text{so} \quad y' = \frac{-x}{y}.$$

Differentiating again we have

$$y'' = -\left(\frac{x}{y}\right)' = -\left(\frac{y - xy'}{y^2}\right).$$

Now replace $y'$ by $\frac{-x}{y}$ to obtain $y''$ in terms of $x$ and $y$. We have

$$y'' = -\frac{y - x\left(\frac{-x}{y}\right)}{y^2} = \frac{-y^2 - x^2}{y^3}.$$

Since $x^2 + y^2 = 9$ we have $y'' = \frac{-9}{y^3}$.

---

Notice that for functions $f$ and $g$ and constants $A$ and $B$, we generalize Theorem 1 of 2.3A to

$$(Af + Bg)^{(n)} = Af^{(n)} + Bg^{(n)}.$$

**B. Some Very High Derivatives** When $f(x) = e^x$, then

$$f'(x) = e^x, \quad f''(x) = e^x,$$

and in general for any positive integer $n$, $f^{(n)}(x) = e^x$. In the examples below, other cases are given in which the higher derivatives are very similar to the original function.

---

***Example 3*** Suppose $f(x) = e^{2x}$. Find $f^{(n)}(x)$.

*Solution* By successive differentiation we have $f'(x) = 2e^{2x}$, $f''(x) = 4e^{2x}$, $f^{(3)}(x) = 8e^{2x}$, etc. Each time we take another derivative an additional factor of 2 appears. Rewriting $f^{(3)}(x) = 8e^{2x}$ as

$$f^{(3)}(x) = 2^3 e^{2x},$$

it becomes clear that

$$f^{(n)}(x) = 2^n e^{2x}.$$

---

⟶ It is more convenient to use the prime notation $y'$ in place of Leibniz's notation $\frac{dy}{dx}$ in doing these types of problems.

---

***Problem 2*** Referring to Example 2, find (a) $y^{(3)}$ and (b) $y^{(4)}$ in terms of $x$ and $y$.

*Answer:* a) $\dfrac{-27x}{y^5}$   b) $-27\left(\dfrac{y^2 + 5x^2}{y^7}\right)$

---

⟶ The other rules for differentiation do not generalize so easily. For example, $(fg)'' = fg'' + 2f'g' + f''g$. (Do not make the mistake of writing $(fg)'' = fg'' + f''g$.)

(Kent)

A proof using mathematical induction (9.1C) of the formula

$$\frac{d^n}{dx^n} e^{2x} = 2^n e^{2x} \quad (n = 1, 2, \ldots)$$

is as follows:

*Step 1* For $n = 1$ the formula is correct since $\frac{d}{dx} e^{2x} = 2e^{2x}$.

*Step 2* Assume the formula is true for $n = k$. Hence $\frac{d^k e^{2x}}{dx^k} = 2^k e^{2x}$. Now take the derivative of this equation to obtain

$$\frac{d^{k+1}}{dx^{k+1}} e^{2x} = \frac{d}{dx}(2^k e^{2x}) = 2^k 2e^{2x} = 2^{k+1} e^{2x}.$$

This verifies the formula for $n = k + 1$.

We have proven the formula is valid for $n = 1$, and that it is valid for $n = k + 1$ whenever it is valid for $n = k$. Hence, by induction, it is true for all $n = 1, 2, \ldots$.

*Example 4*   Suppose $y = e^{-x}$. Find $y^{(n)}$.

*Solution*   Starting with the function, we have

$$y = e^{-x}, \quad y' = -e^{-x}, \quad y'' = e^{-x}, \quad y''' = -e^{-x}, \quad \text{etc.}$$

Each time we take another derivative the previous derivative is multiplied by $-1$. Hence

$$y' = y^{(3)} = y^{(5)} = \cdots = -e^{-x} \quad \text{and}$$
$$y'' = y^{(4)} = y^{(6)} = \cdots = e^{-x}.$$

These facts are neatly summarized by the single equation

$$y^{(n)} = (-1)^n e^{-x}$$

because when $n$ is even then $(-1)^n = 1$ and when $n$ is odd $(-1)^n = -1$.

---

*Example 5*   Suppose $y = \sin x$. Then

$$y' = \cos x$$
$$y'' = -\sin x$$
$$y^{(3)} = -\cos x$$
$$y^{(4)} = \sin x.$$

Since $y^{(4)} = \sin x = y$, we have $y^{(5)} = y'$, $y^{(6)} = y''$, $y^{(7)} = y^{(3)}$, etc. We may summarize these facts by writing

$$\left.\begin{array}{l} y^{(4n)} = \sin x \\ y^{(4n+1)} = \cos x \\ y^{(4n+2)} = -\sin x \\ y^{(4n+3)} = -\cos x. \end{array}\right\} n = 0, 1, 2, \ldots$$

---

**C. Differentials**   The symbols $dx$ and $dy$ that appear in the Leibniz notation for the derivative of $y = f(x)$ will now be defined as separate quantities. They are called *differentials* and are quite useful in the applications of calculus.

Suppose then that a function $y = f(x)$ is differentiable at $x$. We define $dx$, called the *differential of x,* to be a new variable and then define $dy$, called the *differential of y,* by the formula

$$dy = f'(x)\, dx.$$

Note that $dy$ is a function of $x$ and $dx$. At a fixed $x$

*Problem 3*   Given $f(x) = e^{-2x}$, find
a) $f^{(5)}(0)$
b) $f^{(78)}(39 \ln 2)$

*Answer:*   a) $-32$     b) $1$

---

*Problem 4*   Given $f(x) = \dfrac{1}{1 - x}$, prove that

$$f^{(n)}(x) = \frac{n!}{(1 - x)^{n+1}}.$$

---

It is a common error to forget to write the $dx$ at the end of $dy = f'(x)\, dx$. This often results in unnecessary confusion when applying differentials.

→   Note also that the differential notation is consistent in that if $y$ is the function $y = x$ then $dy = dx$. For, if $y = f(x) = x$ then $dy = f'(x)\, dx = 1 \cdot dx = dx$.

it is defined for all real numbers $dx$. Using the Leibniz notation for the derivative, it is clear that we can also write

$$dy = \frac{dy}{dx}\, dx.$$

---

**Example 6**  Given $y = x^3 + 2x^2 + 1$, evaluate $dy$ at $x = 2$ and $dx = 0.1$.

*Solution*

$$dy = \frac{dy}{dx}\, dx = \frac{d}{dx}(x^3 + 2x^2 + 1)\, dx$$

$$= (3x^2 + 4x)\, dx$$

and so

$$dy\Big|_{\substack{x=2 \\ dx=0.1}} = [3(2)^2 + 4(2)](0.1) = 2$$

---

If $dx \neq 0$, we can divide $dy = f'(x)\, dx$ by $dx$ obtaining

$$\frac{dy}{dx} = f'(x).$$

This should not be surprising, since we *defined dx* and *dy* so that this would be the case. What is useful however is that the equation $dy = f'(x)\, dx$ is valid when $x$ is a function of a third variable, say $t$, and $dx$ and $dy$ are the differentials of $x$ and $y$ in terms of $t$ and $dt$. Namely,

**THEOREM 1**  *Suppose $x = g(t)$ and $y = f(x) = f(g(t)) = h(t)$. Then the differential $dy = h'(t)\, dt$ is given by*

$$dy = f'(x)\, dx$$

*where $dx = g'(t)\, dt$.*

*Proof*  Using the Chain Rule to find the derivative of $h(t)$, we have

---

**Problem 5**  Given $y = x \ln x$, evaluate $dy$ at $x = 1$ and $dx = 2$.

*Answer:* 2

---

**Example 7**  Using differentials, find the slope of the curve $y = t + \sin t$, $x = t^3 + t$ at the point corresponding to $t = 0$.

*Solution*

$$dy = \frac{d}{dt}(t + \sin t)\, dt = (1 + \cos t)\, dt$$

$$dx = \frac{d}{dt}(t^3 + t)\, dt = (3t^2 + 1)\, dt$$

Assuming $dt \neq 0$ and hence $dx \neq 0$ we have

$$\frac{dy}{dx} = \frac{(1 + \cos t)\, dt}{(3t^2 + 1)\, dt} = \frac{1 + \cos t}{3t^2 + 1}.$$

The slope of the curve at the point corresponding to $t = 0$ is therefore

$$\frac{dy}{dx}\Big|_{t=0} = \frac{2}{1} = 2$$

$$dy = h'(t)\,dt = f'(g(t))g'(t)\,dt$$
$$= f'(x)g'(t)\,dt = f'(x)\,dx$$

where $dx = g'(t)\,dt$  ●

If in the above Theorem, $dx \neq 0$, we can divide $dy = f'(x)\,dx$ by $dx$, obtaining the *derivative* $dy/dx$ as a quotient of the differentials of $y$ and $x$ in terms of $t$ and $dt$, as shown in Example 7.

Higher order differentials can be defined but we have no need for them in this book.

**D. Approximations**  Suppose a function $y = f(x)$ is differentiable at $x$. Let $P$ be the point $(x, f(x))$, $T$ the tangent line at $P$, and $R$ the point on $T$ with $x$-coordinate equal to $x + dx$, where $dx$ is any real number. See Figure 1.

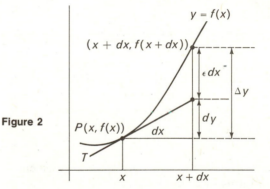

Figure 2

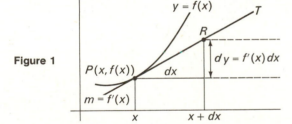

Figure 1

Since the slope of $T$ is $f'(x)$, it follows that the $y$-coordinate of $R$ minus the $y$-coordinate of $P$ is equal to $f'(x)$ times $dx$, the difference of the $x$-coordinates. But $f'(x)\,dx$ is the differential $dy$. Thus the differential

$$dy = f'(x)\,dx$$

represents the difference or "change" in the $y$-coordinates along the tangent to the graph of $f$ at $x$ corresponding to a change in the $x$-coordinates equal to $dx$.

Now suppose $x + dx$ is in the domain of $f$. See Fig. 2. Let $\Delta y$ denote the difference

$$\Delta y = f(x + dx) - f(x).$$

Of all lines passing through $P(x, f(x))$ the tangent $T$ is the one that best "fits" the graph of $f$ near $P$. Thus for small $dx$ we can expect that $\Delta y$, the change in $y$ along the graph of $f$, will be approximately equal to $dy$, the change along $T$. This is indeed the case, as shown in the following theorem.

**THEOREM 2**  *Let* $y = f(x)$ *be differentiable at* $x$. *Let* $\Delta y = f(x + dx) - f(x)$ *and* $dy = f'(x)\,dx$. *Then*

$$\Delta y = dy + \varepsilon \cdot dx,$$

*where* $\varepsilon \to 0$ *as* $dx \to 0$.

Thus we see that $dy$ differs from $\Delta y$ by $\varepsilon \cdot dx$, where $\varepsilon \to 0$ as $dx \to 0$. For small $dx$, this is a product of two small quantities, $\varepsilon$ and $dx$, and in general will be small as compared to $dy$ which is a product containing only one small quantity, $dx$.

We express the fact that $\Delta y$ is approximated by $dy$ by writing

$$\Delta y \approx dy,$$

read "$\Delta y$ is approximately $dy$." This approximation can also be written as

$$f(x + dx) - f(x) \approx f'(x)\,dx$$

or

$$f(x + dx) \approx f(x) + f'(x)\,dx.$$

Here the approximate value of $f$ itself is given at $x + dx$ in terms of $dx$ and $f$ and $f'$ at $x$.

*Proof of theorem:*  We define

$$\varepsilon = \begin{cases} \dfrac{\Delta y - dy}{dx}, & \text{for } dx \neq 0 \\[2mm] 0, & \text{for } dx = 0. \end{cases}$$

Note that $\varepsilon$ is defined for all $dx$ for which $x + dx$ is in the domain of $f$. Clearly, for $dx \neq 0$, it follows that

$$\Delta y = dy + \varepsilon \cdot dx.$$

But this is also valid for $dx = 0$, since $\Delta y = dy = 0$ when $dx = 0$. Finally

$$\lim_{dx \to 0} \varepsilon = \lim_{dx \to 0}\left[\frac{\Delta y - dy}{dx}\right]$$

$$= \lim_{dx \to 0}\left[\frac{f(x + dx) - f(x) - f'(x)\,dx}{dx}\right]$$

$$= \lim_{dx \to 0}\left[\frac{f(x + dx) - f(x)}{dx}\right] - \lim_{dx \to 0} f'(x)$$

$$= f'(x) - f'(x) = 0 \quad\bullet$$

---

**Example 8**  Using differentials, find the approximate increase in the volume of a spherical balloon as its radius increases from 20 ft to 20.1 ft.

*Solution*  The volume of a sphere as a function of its radius is

$$V = \tfrac{4}{3}\pi r^3.$$

The approximate change in $V$ as the radius changes from $r$ to $r + dr$ is

$$dV = \frac{d}{dr}\left(\frac{4}{3}\pi r^3\right) dr = 4\pi r^2\,dr.$$

Using $r = 20$ ft and $dr = 0.1$ ft gives

$$dV\bigg|_{\substack{r=20 \\ dr=0.1}} = 4\pi(20)^2(0.1)$$

$$= 160\pi \text{ ft}^3$$

---

**Example 9**  Using differentials, find $\sqrt[3]{7.98}$ approximately.

*Solution*  Here we use the formula

$$f(x + dx) \approx f(x) + f'(x)\,dx$$

for the function $f(x) = \sqrt[3]{x} = x^{1/3}$. Since $f'(x) = \frac{1}{3}x^{-2/3} = \dfrac{1}{3(\sqrt[3]{x})^2}$, we have

$$\sqrt[3]{x + dx} \approx \sqrt[3]{x} + \frac{1}{3(\sqrt[3]{x})^2}\,dx.$$

If we take $x = 8$ and $dx = -0.02$, then $x + dx = 7.98$, and so, substituting, gives

$$\sqrt[3]{7.98} \approx \sqrt[3]{8} + \frac{1}{3(\sqrt[3]{8})^2}(-0.02) = 2 - \frac{0.02}{12}$$

$$\approx 1.998.$$

## The Story of Little 0

Since

$$\frac{dy}{dx} = \lim_{\Delta x \to 0} \frac{\Delta y}{\Delta x} \quad \text{and} \quad \Delta y \to 0 \quad \text{as} \quad \Delta x \to 0,$$

it might seem reasonable to define the differentials $dx$ and $dy$ as $dx = \lim_{\Delta x \to 0} \Delta x$ and $dy = \lim_{\Delta x \to 0} \Delta y$. But it then follows that $dy = dx = 0$, and this leads to the meaningless expression $dy/dx = 0/0$ for the derivative. To avoid this difficulty Leibniz attempted to define $dx$ and $dy$ as very small quantities, greater than zero but less than any positive real number. Leibniz knew there were no such real numbers, but he considered them to exist in some ideal sense and used them with great success. He felt that those who wished could discard them at their own expense. The quantities were dubbed infinitesimals or "little zeros" and were the source of much confusion and controversy.

In the middle of the nineteenth century, Augustin Cauchy (1789–1857) defined them as we have, in a completely different manner than Leibniz had intended. With Cauchy the era of the little zero passed, being characterized as the "golden age of nothing." Then, since most mathematicians understood differentials quite precisely, they were often irritated by engineers and applied mathematicians who still thought of differentials as infinitely small nonzero quantities. In the twentieth century there were still many textbooks written containing a great deal of nonsense about infinitesimals. The applied scientists found them very useful, but the pure mathematicians often ridiculed them.

The situation changed completely in 1960 with Abraham Robinson's discovery of nonstandard analysis (see p. 413). Using certain ideas from mathematical logic, Robinson constructed a set that looks very much like the real numbers but has the additional feature that it contains quantities that are greater than zero but less than any positive real number. Moreover, Robinson has developed all the standard theorems of calculus in this setting. Hence, the infinitesimals, or little zeros, that Leibniz talked about almost 300 years ago indeed exist and behave quite exactly as he thought they would.

## E. Exercises

**B1.** Find the first and second derivatives of the following functions.

a) $f(x) = \dfrac{1 - x}{1 + x}$

b) $g(x) = \sin e^x$

c) $h(x) = x^2 \ln x^3$

**B2.** Given $f(x) = \dfrac{x}{x^2 + 1}$, find $f(2)$, $f'(2)$, and $f''(2)$.

**B3.** Find $dy$ in each of the following.

a) $y = 2x + 1$

b) $y = x^3 - 2x^2 + 3$

c) $y = \dfrac{x + 1}{(1 - x)^2}$

**B4.** Show that the given functions satisfy the indicated equations.

a) $y = e^{-2x}(\sin 2x + \cos 2x)$,

$y'' + 4y' + 8y = 0$

b) $y = A \sin \ln x + B \cos \ln x$,

$x^2 y'' + xy' + y = 0$

**B5.** Solve for $y''$ in terms of $y$, given

a) $b^2 x^2 + a^2 y^2 = a^2 b^2$

b) $y^3 = 1 - x^2$

**B6.** Evaluate

a) $\dfrac{d^{23}}{dx^{23}} e^{-x} \Big|_{x = -\ln 15}$

b) $\dfrac{d^7}{dx^7} \cos 2x \Big|_{x = \pi/12}$

c) $\dfrac{d^{10}}{dx^{10}} (x - x \ln x) \Big|_{x = 1}$

**B7.** Given $x^3 - y^3 = a^3$, differentiate implicitly to show that $y'' = \dfrac{-2a^3 x}{y^5}$.

**D8.** Verify that $f''(0) = 0$ in each of the cases below.
a) $f(x) = 8x - 11$
b) $f(x) = x^4 + \pi$
c) $f(x) = x^3 + x^{53} - 7x^{99}$
d) $f(x) = \sin 5x$
e) $f(x) = \tan x$
f) $f(x) = \text{Tan}^{-1} x$
g) $f(x) = e^{(x^3)}$
h) $f(x) = \ln e^x$

**D9.** For what value of $n$ is each of the following equations valid?
a) $(3x^3)^{(n)} = 18x$
b) $(x^4)^{(n)} = 24x$
c) $(e^{2x})^{(n)} = 8e^{2x}$
d) $(\sin 3x)^{(n)} = 81 \sin 3x$
e) $(\cos 4x)^{(n)} = -16 \cos 4x$
f) $(\ln x^3)^{(n)} = 6x^{-3}$
g) $(e^{-2x})^{(n)} = 64e^{-2x}$
h) $(xe^x)^{(n)} = (x + 3)e^x$

a) 2    b) 3    c) 3    d) 4    e) 2    f) 3    g) 6
h) 3

**M10.** If $x = 2t + 1$ and $y = x^2$, find $dy$ in terms of $t$ and $dt$.

**M11.** If the edges of a cube are increased in length from 4 to 4.05 inches, what is the approximate change in the volume?

**M12.** Find an approximate value for each of the following.
a) $\sqrt[3]{1001}$     b) $\sin (0.02)$     c) $\ln (0.96)$

**M13.** Find $y''$ in each of the following.
a) $y = e^{x \ln x}$
b) $y = xe^{-x^2}$
c) $y = x^x$

**M14.** Show that in each of the following cases $y^{(4)} = y$.
a) $y = a \sin x + b \cos x$
b) $y = a \sinh x + b \cosh x$
c) $y = ae^{-x} + be^x$

**M15.** a) Given $f(x) = \frac{1}{2}(x^3 - 4x^2 + 7x)$, verify that $f^{(n)}(n) = n$ for $n = 0, 1, 2, 3$.
b) Find a polynomial $p$ of degree 4 that satisfies $p^{(n)}(n) = n$ for $n = 0, 1, 2, 3, 4$.

**M16.** A polynomial $p$ of degree 2 satisfies
$$p(0) + p'(0) = 1,$$
$$2p(0) + p''(0) = 2,$$
and
$$p'(0) - p''(0) = 3.$$
Find $p$.

**M17.** Find those points where the second derivative of each of the following functions is equal to 0.
a) $y = x^2 - 3x + 6$
b) $y = x^3 - 3x^2 + 6$
c) $y = (x - 1)(x - 2)(x - 3)$

**M18.** Given $y^2 = ax^2 + bx + c$ show that
$$4y^3 y'' = 4ac - b^2.$$

**M19.** Find $f''(x)$ for each of the following.
a) $f(x + 3) = x^3 + 7x^2 + 3$
b) $f(x^2) = 1 + x^2 - 3x^6$
c) $f(ax) = \sin 3x$,   $a$ constant.

**M20.** Given $f(x) = a + bx + cx^2 + dx^3$, verify each of the following.
a) $f(x) = f(0) + f'(0)x + \frac{1}{2}f''(0)x^2 + \frac{1}{6}f'''(0)x^3$
b) $f(x) = f(1) + f'(1)(x - 1) + \frac{1}{2}f''(1)(x - 1)^2 + \frac{1}{6}f'''(1)(x - 1)^3$

**M21.** Given $y = ae^x + be^{-2x} + ce^{3x}$, where $a$, $b$, and $c$ are constants. Show that
$$y''' - 2y'' - 5y' + 6y = 0 \quad \text{for all } x.$$

**M22.** The displacement $s$ of a weight at the end of a spring as a function of time when it is set into motion is given by a function of the form
$$s = A \cos (\omega t + \phi)$$
where $A$, $\omega$, and $\phi$ are constants. Show that
$$\frac{d^2 s}{dt^2} + \omega^2 s = 0 \quad \text{for all } t.$$

**M23.** Prove that
a) $\dfrac{d^{n+1}}{dx^{n+1}} \ln (1 - x) = \dfrac{-n!}{(1 - x)^{n+1}}$

b) $\dfrac{d^{2n}}{dx^{2n}} \sin (3x + 10) = (-1)^n 3^{2n} \sin (3x + 10)$

**M24.** a) Prove that for any $n$ there is a polynomial $P_n$ of degree $n$ satisfying $\dfrac{d^{2n-1}}{dx^{2n-1}} \tan x = P_n(\sec^2 x)$.
b) Find $P_1$, $P_2$, and $P_3$.

**M25.** Prove that if a function $f$ is twice differentiable on $R$ then for any $x \in R$,
$$f''(x) = \lim_{h \to 0} \frac{f(x + h) + f(x - h) - 2f(x)}{h^2}.$$

**M26.** Show that if $f(x_0 + dx) = f(x_0) + a \cdot dx + \varepsilon \cdot dx$ where $\varepsilon \to 0$ as $dx \to 0$, then $f$ is differentiable at $x_0$ and $f'(x_0) = a$.

## 3.6 Sample Exams

### Sample Exam 1 (45–60 minutes)

1. Differentiate the following functions.
   a) $f(x) = (2 - 3x^2 + 2x^{-3})^8$

   b) $g(x) = \cos\left(\dfrac{x}{3} + 4\right)$

   c) $h(x) = \tan(\sin x)$

2. Given the equation $xy^2 + x^2y = 6$ find the slope of its graph at
   a) the point $(1, 2)$
   b) the point $(1, -3)$.

3. Find $\left.\dfrac{dy}{dx}\right|_{x=0}$ given
   a) $y = \text{Tan}^{-1} x + (\cot x)^{-1}$
   b) $y = \text{Sin}^{-1}(x^2) + (\text{Sin}^{-1} x)^2$

4. Find $dy'$ given
   a) $y = (e^{(x^2)})^2$
   b) $y = \ln(\ln(\ln(e^x)))$

5. a) Given $y = \dfrac{1}{x} + x^2$, find $\left.\dfrac{d^2y}{dx^2}\right|_{x=1}$

   b) Given $y = \ln(1 + x)$, find $\left.\dfrac{d^3y}{dx^3}\right|_{x=1}$

   c) Given $y = xe^x$, find $\left.\dfrac{d^4y}{dx^4}\right|_{x=0}$

### Sample Exam 2 (45–60 minutes)

1. Given $y = f(u)$ and $u = u(x)$, find $\dfrac{dy}{dx}$ in terms of $x$ (not $u$) in the case that
   a) $f(u) = (1 + u)^{-3}$, $u(x) = x^3$
   b) $f(u) = u^3$, $u(x) = (1 + x)^{-3}$
   c) $f(u) = e^u$, $u(x) = \ln x$

2. a) Find two functions $y = y(x)$ that are implicitly defined by the equation
   $$y^2 + 2xy - x^3 = 0.$$

   b) Show that there are no points $(x, y)$ whose coordinates satisfy the equation
   $$1 + x^2 + e^y = 0.$$

   c) Find all points $(x, y)$ whose coordinates satisfy the equation

$$1 + y^2 + \sin x = 0.$$

3. Consider the parametric equations
   $$x = \cos^3 t \quad \text{and} \quad y = \sin^3 t.$$

   Find:

   a) $\dfrac{dy}{dt}$  c) $\dfrac{dy}{dx}$  e) $\dfrac{d^2x}{dt^2}$

   b) $\dfrac{dx}{dt}$  d) $\dfrac{d^2y}{dt^2}$

4. Find the inverse function for each of the following functions; that is, find a function $f^{-1}$ satisfying
   $$f(f^{-1}(x)) = f^{-1}(f(x)) = x.$$

   a) $f(x) = 4x + 5$
   b) $f(x) = 2x^3 + 7$

5. Match the function to a property of its derivative.
   a) $y = \text{Tan}^{-1} 4x$          0) $y'(1) = 0$
   b) $y = \text{Sin}^{-1} \dfrac{x}{3}$          1) $y'(2\sqrt{2}) = 1$
   c) $y = e^{-2x} \sin 3x$          2) $y'(0) = 2$
   d) $y = \ln(x^2 - 2x + 2)$          3) $y'(0) = 3$
   e) $y = 2e^{\sin x}$          4) $y'(0) = 4$

### Sample Exam 3 (45–60 minutes)

1. Write down the derivatives of the following functions:
   a) $y = (1 + x + x^2)^{78}$
   b) $y = \sin e^x$
   c) $y = e^{\cos 3x}$
   d) $y = \text{Tan}^{-1} 4x^2$
   e) $y = \text{Sin}^{-1} \sqrt{x}$
   f) $y = \ln(e^x + e^{-x})$

2. Find a second degree polynomial $p(x)$ that satisfies $p(0) = 3$, $p'(2) = 10$, and $p''(10) = 4$.

3. Use implicit differentiation to find the slope of the tangent lines to the following curves at the indicated points.
   a) $x^2 + y^2 = 13$          at $(2, 3)$
   b) $x^3 - 4xy + y^3 = 0$   at $(2, 2)$

4. Consider the curve defined parametrically by the equations
   $$x = t^3 - 3t + 6, \quad y = t^3 - 12t + 9.$$

   Find the points $(x, y)$ on the curve where

a) the tangent is horizontal, and
b) the tangent is vertical.
(Remember to find the points $(x, y)$, not just the values of the parameter $t$ that correspond to these points.)

5. Find where the third derivative of $f(x)$ is zero if
   a) $f(x) = \frac{12}{5}x^5 - 6x^3 - \frac{27}{2}x^2 + \frac{8}{5}$
   b) $f(x) = \sin 2x$
   c) $f(x) = \cos 3x$

## Sample Exam 4 (45–60 minutes)

1. Find the derivatives of
   a) $f(x) = x^3 + 3^x$
   b) $g(x) = \text{Sin}^{-1} x + \sin(x^{-1}) + (\sin x)^{-1}$
   c) $h(x) = \log_{10} x + \log_x 10$

2. Suppose $f(x) = \sin^2 x$ and $u(x) = x^2$. Find
   a) $(f \circ u)'(\sqrt{\frac{\pi}{4}})$
   b) $(u \circ f)'(\frac{\pi}{4})$
   c) $(u \circ u)^{(3)}(\frac{1}{24})$

3. a) Find two functions of $x$ implicitly defined by the equation
   $$y^2 + 4x^2 - 18y - 16x - 3 = 0.$$
   b) Graph each of these two functions.

4. For each of the following functions find an open interval with the property that the function restricted to this interval has an inverse function.
   a) $y = 2 + \sqrt{1 - (x - 3)^2}$     $(2 < x < 4)$
   b) $y = \cos 2x$   $(\frac{\pi}{4} < x < \frac{3\pi}{4})$

5. Given $f(x) = 6 + 4x - 2x^2$ and $g(x) = \ln x$, show that
   a) $f'(x) = g'(x)$     for one value of $x$.
   b) $f''(x) = g''(x)$    for two values of $x$.
   c) $f'''(x) = g'''(x)$   for no values of $x$.

## Sample Exam 5 (45–60 minutes)

1. Given the function $f(x) = \sec x$:
   a) Find some (there are many of them) open interval of length $\frac{\pi}{2}$ on which this function has an inverse.
   b) Then compute the derivative of this inverse function.

2. Given $f(x) = x^3 + 3x^2 + 6x + 6$, find the points where
   a) $f(x) = f'(x)$
   b) $f(x) = f''(x)$
   c) $f'(x) = f''(x)$

3. Find $f'(x)$ in terms of $x$, $g$, and $g'$, given

a) $f(x) = g(x + g(x))$
b) $f(x) = g(xg(x))$
c) $f(x - 3) = g(x^3)$

4. Given the ellipse $2x^2 + 3y^2 = 14$ and the point $A = (2, 5)$, find the point $P = (x, y)$ on the ellipse in the first quadrant where the tangent at $P$ is perpendicular to the segment $AP$.

5. a) Given $f(x) = \dfrac{3x - 4}{x - 3}$, show that its inverse function
   $f^{-1}$ satisfies $f^{-1}(x) = f(x)$ for all $x \neq 3$.
   b) Given $g(x) = \dfrac{3x - 4}{x - 2}$ find two points $a$ where
   $$g^{-1}(a) = g(a).$$
   c) Show that the graph of $h(x) = \dfrac{x + 4}{3 - x}$ and that of its inverse function never intersect.

## Sample Exam 6 (Chapters 1, 2, and 3, approximately 2 hours)

1. Find the equation of the tangent line to the graph of $y = (3x^2 - 2x - 7)^4$ corresponding to the point $x = 2$.

2. Evaluate the indicated quantities.
   a) $\dfrac{d}{dx} e^{1+x^2} \Big|_{x=0}$
   b) $\dfrac{d^2}{dx^2} \text{Tan}^{-1} x \Big|_{x=1}$
   c) $\dfrac{d^3}{dx^3} \sin x \Big|_{x=\pi/3}$
   d) $\dfrac{d^4}{dx^4} e^{-2x} \Big|_{x=0}$

3. Given the circle $x^2 + y^2 = 25$, find the point of intersection of the tangent at $(4, 3)$ and the tangent at $(-3, 4)$.

4. Given $f(x) = 2x - 3$ and $g(x) = 3x - 2$, show that the graphs of the functions $f \circ g$ and $g \circ f$ never intersect.

5. Compute
   a) $\dfrac{d}{dx} e^{3x^2 - 2x^3}$
   b) $\dfrac{d}{dx} \csc^2 x^2$
   c) $\dfrac{d}{dx} \text{Tan}^{-1} \sqrt{1 + x^2}$

**6.** Find where $y'(x) = 0$, given
   a) $x = \ln xy$
   b) $x = (\ln x)(\ln y)$

**7.** Suppose $f(x) = x^2$, $g(x) = \cot x$, and $h(x) = e^x$. Find

$$(f \circ g \circ h)'(\ln \tfrac{\pi}{4}).$$

**8.** Find the values of the following.

   a) $\mathrm{Sin}^{-1}\dfrac{\sqrt{3}}{2} + \mathrm{Tan}^{-1}\dfrac{\sqrt{3}}{3}$

   b) $\lim\limits_{x \to 4} \dfrac{x - 4}{x^2 + x - 20}$

   c) $\ln \sqrt{e^{36}}$

**9.** Find $\dfrac{d^2y}{dx^2}$ in terms of $x$ and $y$, given

   a) $x^3 + y^3 = 23$
   b) $x = \tan y$

**10.** Given the function $f(x) = \dfrac{2x - 3}{3x - 2}$:

   a) What is its domain?
   b) Find its inverse function $f^{-1}(x)$.
   c) What is the domain of $f^{-1}(x)$?
   d) Compute $(f^{-1})'(1)$.

We are now in a position to apply the techniques of differential calculus. After establishing some basic principles, we will study the use of the derivative as an aid to graphing functions. The important class of maximum-minimum problems is then studied. This type of problem occurs in all branches of science, and its solution exhibits the power of differentiation. We next interpret the derivative as velocity, investigate moving bodies, and trace the ideas concerning falling objects from Aristotle to Galileo. More general rates of change along with other interpretations of the derivative are studied in the final section.

# Applications of Differentiation

4

## 4.1  Some General Theorems

We will now give some theoretical results that will be used in the remainder of the chapter.

**A. Some Important Points**  Given a function $f$, a point $c$ which satisfies $f'(c) = 0$ is called a *critical point* of $f$. Geometrically, this means that the tangent to $f$ at $(c, f(c))$ is horizontal (Fig. 1).

To find the critical points of a function $f$, compute $f'(x)$ and solve for $x$ in the equation

$$0 = f'(x).$$

→  *You should find the statements fairly clear but the proofs rather difficult. Also, if the terminology in this and the next section seems offensive, it would be worthwhile to spend some time in getting it straight.*

*(Beth)*

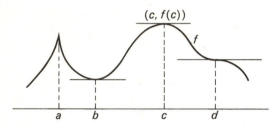

**Figure 1**  The points $b$, $c$, and $d$ are critical points but $a$ is not, since $f'(a)$ does not exist.

---

*Example 1*  Find the critical points of
a) $f(x) = x^3 - 3x^2 - 24x + 8$,

b) $y = 1 - \dfrac{1}{x^2}$.

*Solution*

a) The derivative of the function is

$$f'(x) = 3x^2 - 6x - 24,$$

so the critical points must satisfy the equation

$$0 = 3x^2 - 6x - 24 = 3(x^2 - 2x - 8).$$

Hence $0 = x^2 - 2x - 8 = (x - 4)(x + 2)$, so $x = 4$ and $x = -2$ are the critical points.

b) There are no critical points because in this case $\dfrac{dy}{dx} = \dfrac{2}{x^3}$ is never zero.

*Problem 1*  Compute the critical points of
a) $f(x) = x^3 - 3x + 7$
b) $y = 2x^3 + 6x^2 - 12x + 4$

*Answer:*  a) 1 and $-1$  b) $-1 + \sqrt{3}$ and $-1 - \sqrt{3}$

A number $f(p)$ is said to be a *local maximum* of a function $f$ at a point $p$ if

$$f(x) \leq f(p) \quad \text{for all} \quad x \quad \text{near} \quad p.$$

Similarly, $f(q)$ is a *local minimum* of $f$ at the point $q$ if

$$f(q) \leq f(x) \quad \text{for all} \quad x \quad \text{near} \quad q.$$

We illustrate these ideas below.

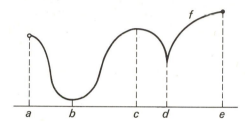

**Figure 2**

→ Many authors use the word "relative" in place of "local" in these definitions.

→ The word "near" is ambiguous, and precise definitions of local maxima and minima are as follows:

Suppose $f$ is a function with domain $D$. Then $f(p)$ is a *local maximum* at the point $p$ if there exists an open interval $I$ containing $p$ such that

$$f(x) \leq f(p)$$

for all $x$ in both $D$ and $I$. (A local minimum is defined similarly.) Note that this condition need not be satisfied for all intervals. For example, $f(c)$ (Fig. 2) is a local maximum because $f(x) \leq f(c)$ for all $x \in (b, d)$. The fact that $f(x) \leq f(c)$ is false for $x \in (b, e)$ does not contradict $f(c)$ being a local maximum.

The function $f$ (Fig. 2) has local maxima at the points $c$ and $e$ and has local minima at the points $b$ and $d$. At $a$ the function is not defined.

A number $f(r)$ is said to be the *absolute maximum* of $f$ at $r$ if $f(r) \geq f(x)$ for all $x$ in the domain of $f$. Similarly, $f(s)$ is the *absolute minimum* at $s$ if $f(s) \leq f(x)$ for all $x$ in the domain of $f$. In Figure 2, $f$ has an absolute maximum at $e$ and an absolute minimum at $b$.

Notice that a function may have an absolute maximum (or minimum) at more than one point. Also it may have no local maxima (or minima). We illustrate these cases in the following:

---

**Example 2** The function $y = x^2$ $(-\infty < x < \infty)$ has no local maxima (Fig. 3a). However, the function

$$y = x^2 \quad (-1 \leq x \leq 2)$$

has local maxima at the points $-1$ and $2$, and an absolute maximum at $2$ (Fig. 3b). Finally, the function

$$y = x^2 \quad (-1 \leq x \leq 1)$$

has an absolute maximum at both $-1$ and $1$ (Fig. 3c).

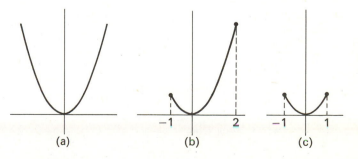

(a)    (b)    (c)

**Figure 3**

If $f$ has a local maximum or a local minimum at $p$, then $p$ need not be a critical point (Fig. 2, points $d$ and $e$). Conversely, if $p$ is a critical point, $f(p)$ need not be a local maximum or a local minimum (Fig. 1, point $d$). The following theorem simplifies the situation.

**THEOREM 1**  *Suppose $f$ is defined on an interval $I$ and*
  a) *$f$ is differentiable at a point $p$ in $I$*
  b) *$p$ is not an endpoint of $I$*
  c) *$f(p)$ is a local maximum or local minimum.*
*Then*

$$f'(p) = 0,$$

*in other words, $p$ is a critical point* (Fig. 4).

*Proof*  We assume $f$ has a local maximum at $p$. (The case of a local minimum is treated similarly.) Then for all $x$ near $p$ we have

$$f(x) \leq f(p),$$

so

$$f(p + h) - f(p) \leq 0$$

for all $h$ near zero. Since $p$ is not an endpoint of $I$, $p + h$ will be in $I$ for positive and negative values of $h$. When $h < 0$ (Fig. 5) we have

$$\frac{f(p + h) - f(p)}{h} \geq 0.$$

As $h \to 0, \dfrac{f(p + h) - f(p)}{h} \to f'(p)$, so it follows that

$f'(p) \geq 0$. When $h > 0$ (Fig. 6) we have

$$\frac{f(p + h) - f(p)}{h} \leq 0.$$

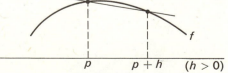

**Figure 6**

As $h \to 0, \dfrac{f(p + h) - f(p)}{h} \to f'(p)$, so it follows that
$f'(p) \leq 0$. Hence it must be true that $f'(p) = 0$ ●

We emphasize that every absolute maximum (or minimum) is a local maximum (or minimum). It is *not* true that every local maximum (or minimum) is an absolute maximum (or minimum).

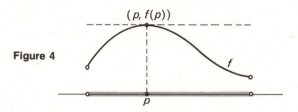

**Figure 4**

$(p, f(p))$

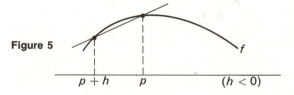

**Figure 5**

$p + h \qquad p \qquad (h < 0)$

It follows from this theorem that the local maximum or minimum points of a function are all found among (1) the critical points, (2) points where there is no derivative, such as corners, or (3) endpoints of intervals on which the function is defined.

## B. "All That Is Gold Does Not Glitter"

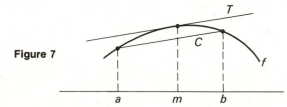

**Figure 7**

&#8594; From J. R. R. Tolkien's *Lord of the Rings*. The line is from a poem in the early part of the book describing the ranger Strider, whose shaggy appearance belied his nobility and gave no hint of the central role he was to play in the adventure to follow.

Figure 7 illustrates the *Mean Value Theorem*. A function $f$ and points $a$ and $b$, $a < b$, are given. The conclusion is that there exists a point $m$ between $a$ and $b$ with the property that the tangent $T$ at $(m, f(m))$ is parallel to the chord $C$ determined by the points $(a, f(a))$ and $(b, f(b))$. A precise statement is given next.

&#8594; A proof of this theorem is given below, but is based on the *Extreme Value Theorem*, whose proof will be given in 9.3A.

> **EXTREME VALUE THEOREM** *Suppose g is continuous on the closed interval* [a, b] ($-\infty < a < b < \infty$). *Then there are points c and d in* [a, b] *at which g has an absolute maximum and absolute minimum respectively.*

**THEOREM 2** *Suppose f is continuous on the closed interval* [a, b] ($-\infty < a < b < \infty$), *and differentiable in the open interval* (a, b). *Then there is a point* $m \in (a, b)$ *satisfying*

$$\frac{f(b) - f(a)}{b - a} = f'(m).$$

The hypothesis that $f$ be continuous on [a, b] is essential. For example, if the graph of $f$ is as in Figure 8, then $\dfrac{f(b) - f(a)}{b - a} = 0$, but there is no point $m \in (a, b)$ satisfying $f'(m) = 0$.

> **Proof of Theorem 2**   Let $g$ denote the function defined by
>
> $$g(x) = f(x) - f(a) - \frac{f(b) - f(a)}{b - a}(x - a).$$
>
> Then $g(b) = g(a) = 0$ and $g$ is continuous on [a, b] because $f$ is continuous on [a, b]. By the Extreme Value Theorem there are points, $m$, $M \in [a, b]$ such that
>
> $$g(m) \leq g(x) \leq g(M) \quad \text{for all } x \in [a, b].$$
>
> Now if $g(m) = g(M) = 0$ then $g(x) = 0$ for all $x \in [a, b]$, and this implies that
>
> $$f(x) = f(a) + \frac{f(b) - f(a)}{b - a}(x - a),$$
>
> or
>
> $$f'(x) = \frac{f(b) - f(a)}{b - a}.$$
>
> In this case the graph of $f$ is a straight line and the result is obvious. Therefore we may assume that either $g(m)$ or $g(M)$, say $g(m)$, is not zero. Since $g(a) = g(b) = 0$, $m$ cannot equal $a$ or $b$, so $m \in (a, b)$. Now $g$ has an absolute minimum at $m$. Since $g$ is differentiable at $m$, by Theorem 1, $m$ is a critical point for $g$. Hence $g'(m) = 0$, and from the defining equation we have
>
> $$g'(x) = f'(x) - \frac{f(b) - f(a)}{b - a} \cdot 1,$$
>
> so
>
> $$0 = g'(m) = f'(m) - \frac{f(b) - f(a)}{b - a}$$
>
> or
>
> $$f'(m) = \frac{f(b) - f(a)}{b - a} \quad \bullet$$

**Figure 8**    $(a, f(a))$        $f$        $(b, f(b))$

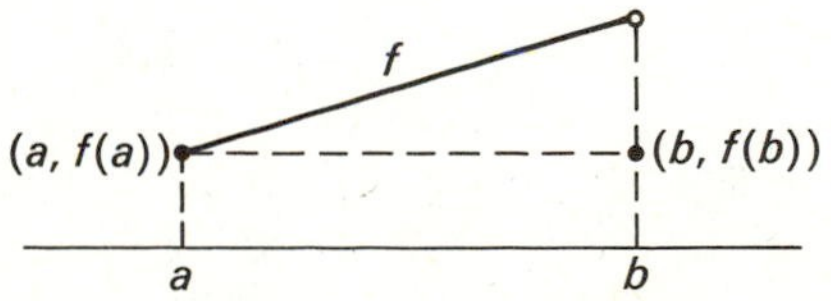

Also $f$ must be differentiable in $(a, b)$. The function whose graph is shown in Figure 9 is continuous on [a, b] and satisfies $\dfrac{f(b) - f(a)}{b - a} = 0$, but there is no point $m$ where $f'(m) = 0$.

**Figure 9**        $f$

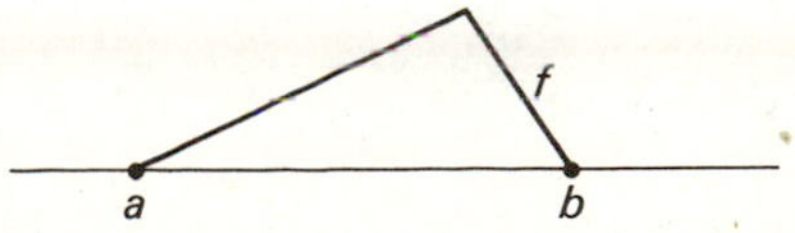

Finally, note that there may be more than one point $m$ satisfying the conclusion of the theorem (Fig. 10).

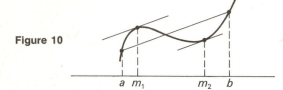

**Figure 10**

**C. Two Corollaries**  An earlier result stated that if $f$ is a constant function, then the derivative is zero. We will now prove the converse of this statement.

**COROLLARY 1**  *Suppose $f'(x) = 0$ for all $x$ on an interval $I$. Then $f$ is a constant function on $I$.*

*Proof*  Choose any point $a \in I$. We will show that $f(x) = f(a)$ for any other point $x \in I$. Assume that $x < a$. The proof when $x > a$ is similar. Since $f$ is continuous in $[x, a]$ and differentiable in $(x, a)$, we can apply the Mean Value Theorem to conclude that there exists a number $m \in (x, a)$ satisfying $f(x) - f(a) = (x - a)f'(m)$. Since $f'(m) = 0$, it follows that $f(x) - f(a) = 0$, or $f(x) = f(a)$. Hence $f$ is a constant function on $I$  ●

**COROLLARY 2**  *Suppose $f$ and $g$ are differentiable on any interval $I$ and $f'(x) = g'(x)$ for all $x \in I$. Then $f = g + K$ on $I$, where $K$ is a constant function.*

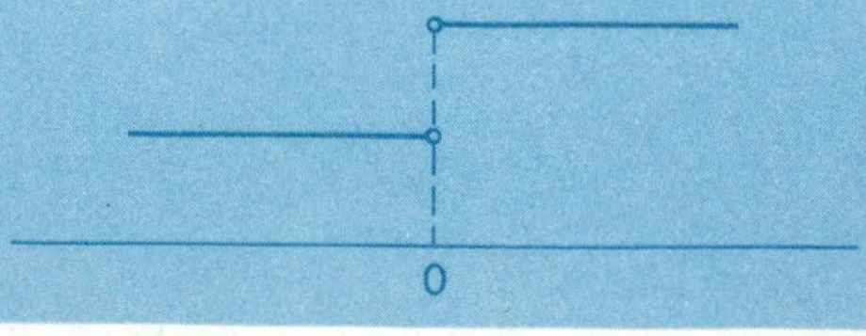

*Proof*  We have $(f - g)'(x) = f'(x) - g'(x) = 0$ for all $x$ in $I$. Hence, by Corollary 1, $f - g = K$ where $K$ is a constant function  ●

A direct proof of this is very easy. Suppose $f(x) = K$ is a constant function. Then

$$\frac{f(x + h) - f(x)}{h} = \frac{K - K}{h} = \frac{0}{h} = 0.$$

Hence

$$f'(x) = \lim_{h \to 0} \frac{f(x + h) - f(x)}{h} = 0.$$

The assumption that $I$ is an interval is essential. For example, suppose $f$ is defined on the two intervals $(-\infty, 0)$ and $(0, \infty)$ by

$$f(x) = \begin{cases} 1 & \text{for } x < 0 \\ 2 & \text{for } x > 0. \end{cases}$$

Then $f'(x) = 0$ for all $x \neq 0$, but $f$ is not a constant function.

Here also it is essential that $I$ is an interval. For example, let $f$ be the function shown in Figure 11, and define $g$ for all nonzero $x$ by the formula $g(x) = 0$. It follows that $f' = g'$, but $f - g$ is not a constant function.

**Figure 11**

**D. Exercises**

**B1.** Find the critical points of the following functions.
  a) $f(x) = x^2 - 4x$
  b) $g(x) = 2x^3 + 3x^2 - 36x + 7$
  c) $h(x) = \dfrac{6x^2 - 8x + 9}{x^2}$

**B2.** Find a point $m$ in $(a, b)$ such that $f'(m) = \dfrac{f(b) - f(a)}{b - a}$ given
  a) $f(x) = x^3, \quad a = 0, b = 1$
  b) $g(x) = \ln x, \quad a = 1, b = e$

**B3.** Find the critical points of the curve represented parametrically by $x = t - \sin t$, $y = 1 - \cos t$.

**B4.** Each of the following functions has either a local maximum or a local minimum at one point in the indicated interval. Find and describe these points.
  a) $f(x) = x^2 - x - 6, \quad (0, -6)$
  b) $g(x) = \dfrac{x + 1}{x^2 + 3}, \quad (-4, -2)$
  c) $h(x) = \dfrac{x + 1}{x^2 + 3}, \quad (0, 2)$

**B5.** Show that each of the following functions has exactly one critical point and the function has neither a local minimum nor a local maximum at that point.
  a) $f(x) = x^3$
  b) $h(x) = x^3 - 6x^2 + 12x - 7$

**D6.** Find the critical points for the following functions.
  a) $y = x^2 - 8x + 9$
  b) $y = 3x^2 - 12x + 8$
  c) $y = x^3 - 3x$
  d) $y = 2x^3 - 3x^2$
  e) $y = 2x^3 - 27x^2 + 120x$
  f) $y = 2x^3 + 27x^2 + 120x$
  g) $y = x^3 - 3x^2 - 189x + 283$
  h) $y = x^3 - 165x^2 + 3000x + 110$

  a) 4    b) 2    c) 1, −1    d) 0, 1    e) 4, 5
  f) −4, −5    g) −7, 9    h) 10, 100

**D7.** Show that the following functions have no critical points.
  a) $y = 7x$         e) $y = 2x + \cos x$
  b) $y = e^{3x}$        f) $y = \dfrac{x+3}{x+2}$
  c) $y = x^3 + x$     g) $y = \ln(x^3 + 2x)$
  d) $y = 3x - \sin x$    h) $y = x^3 + 6x^2 + 15x + 21$

**D8.** Find a point $m$ in the interval $(a, b)$ satisfying
$$(b - a)f'(m) = f(b) - f(a)$$
in each of the following cases.
  a) $f(x) = x^2 + 1$,   $a = 2$,     $b = 4$
  b) $f(x) = x^2 + 1$,   $a = -2$,   $b = 2$
  c) $f(x) = x^2 + 1$,   $a = -2$,   $b = 0$
  d) $f(x) = \sin x$,       $a = \pi$,      $b = 2\pi$

  a) $m = 3$    b) $m = 0$    c) $m = -1$    d) $m = \frac{3\pi}{2}$

**M9.** Prove that the function $f(x) = ax^3 + bx^2 + cx + d$ has exactly one critical point if and only if $b^2 = 3ac$.

**M10.** Suppose $f$ is twice differentiable on $(a, b)$ and $f(x_1) = f(x_2) = f(x_3) = 0$ for distinct points $x_1$, $x_2$, and $x_3$ in $(a, b)$. Prove that $f''(c) = 0$ for some $c \in (a, b)$ by using the Mean Value Theorem.

**M11.** In each of the following either verify that the Mean Value Theorem holds for the given function or give a reason why it does not.
  a) $f(x) = \dfrac{x+2}{x-1}$,   $a = 0$,     $b = 2$
  b) $f(x) = 1 - 3x$,   $a = 1$,     $b = 4$
  c) $f(x) = |x|$,        $a = -1$,   $b = 3$

**M12.** Prove, using the Mean Value Theorem, each of the following inequalities.
  a) $\dfrac{x-1}{x} < \ln x < x - 1$   for   $0 < x < 1$
  b) $\sqrt{x} < \dfrac{x+1}{2}$         for   $0 < x < 1$
  c) $x < \tan x$            for   $0 < x < \frac{\pi}{2}$

**M13.** Suppose $f$ is differentiable in $(a, b)$ and $|f'(x)| \le M$ for all $x \in (a, b)$. Prove that
$$|f(x_1) - f(x_2)| \le M|x_1 - x_2|$$
for all $x_1, x_2 \in (a, b)$.

**M14.** Using the result from Exercise M13 establish the following inequalities.
  a) $|\sin x_1 - \sin x_2| \le |x_1 - x_2|$
  b) $\dfrac{b-a}{b} < \ln \dfrac{b}{a} < \dfrac{b-a}{a}$   if   $0 < a < b$

**M15.** Establish the following identities.
  a) $\operatorname{Sin}^{-1} \dfrac{2x}{1+x^2} = 2\operatorname{Tan}^{-1} x$    $(|x| < 1)$
  b) $\sinh^{-1} x = \ln(x + \sqrt{1 + x^2})$
  c) $\tanh^{-1} x = \dfrac{1}{2}\ln\left(\dfrac{1+x}{1-x}\right)$

(Hint: Show first that the derivatives are the same.)

**M16.** Suppose $f$ is differentiable on $R$ and satisfies $f' = f$. Show that $f(x) = Ae^x$ for some constant $A$. Show first that the derivative of $f(x)e^{-x}$ is zero.

**M17.** Prove that a function of the form $p(x) = x^3 + ax + b$, $a, b > 0$, is strictly increasing by showing that $p'$ is never 0 and then using the Mean Value Theorem.

**M18.** Prove that if $f''(x) = 0$ for all $x \in R$ then $f$ has the form $f(x) = ax + b$ for constants $a$ and $b$.

**M19.** If $f^{(n)}(x) = 0$ for all $x \in R$, show that $f$ has the form
$$f(x) = a_0 + a_1 x + \cdots + a_{n-1}x^{n-1}.$$

**M20.** Prove that if the equation
$$x^n + a_{n-1}x^{n-1} + \cdots + a_1 x = 0$$
has a positive root $x = r$, then the equation
$$nx^{n-1} + (n-1)a_{n-1}x^{n-2} + \cdots + a_1 = 0$$
has a positive root less than $r$.

**M21.** Use Corollary 1 to show that
$$\sin^2 x + \cos^2 x = 1.$$

**M22.** The following generalization of the Mean Value Theorem is called the Extended Mean Value Theorem (see Section 10.1D): Suppose $f$ and $g$ are continuous in the closed interval $[a, b]$ ($-\infty < a < b < \infty$), and differentiable in the open interval $(a, b)$. Then there is a point $m$ in $(a, b)$ satisfying

$$(f(b) - f(a))g'(m) = (g(b) - g(a))f'(m).$$

Find all points $m$ in $(a, b)$ that satisfy this theorem given
a) $f(x) = x^3$, $g(x) = 2 - x$   $a = 0$, $b = 9$
b) $f(x) = 4x + 2$, $g(x) = 6x - 8$   $a = 1$, $b = 3$

**M23.** Find a function $g$ for which the Extended Mean Value Theorem reduces to Theorem 2.

---

## 4.2  *Graphing*

→   This section is long: "Proceed and faith shall be given to you." d'Alembert (1717–1783)

The derivative can be used to facilitate the graphing of functions. We will illustrate this with several examples.

**A. Using the First Derivative**  Consider the function $f$ whose graph is as shown in Figure 1. Notice that on the interval $[a, b]$ the function is increasing and that $f'(p) \geq 0$ for all $p \in [a, b]$. Similarly, on $[b, c]$ the function is decreasing and for each $q \in [b, c]$ we have $f'(q) \leq 0$. The precise relationship between the derivative and monotonicity is given in the following theorem.

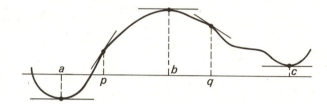

**Figure 1**

**THEOREM 1**  *Suppose $f$ is differentiable on an interval $I$. Then*

a) *$f$ is increasing on $I$ if and only if $f'(x) \geq 0$ for all $x \in I$, and*

b) *$f$ is decreasing on $I$ if and only if $f'(x) \leq 0$ for all $x \in I$.*

*Important*  In the above theorem the hypothesis that $f$ be defined on an interval is essential. For example, the function

$$f(x) = \frac{1}{x} \quad (x \neq 0) \qquad (\text{Fig. 2})$$

has derivative

$$f'(x) = \frac{-1}{x^2} \quad (x \neq 0).$$

Hence $f'(x) < 0$ for all $x \neq 0$ but the function is not

→   This result is intuitively plausible from Figure 1, and the proof is given below.

*Proof of* (a)  Assume $f$ is increasing, $x \in I$, and $x$ is not a right endpoint of $I$. Then for $h > 0$ and $x + h$ in $I$

$$f(x + h) \geq f(x),$$

so the Newton Quotient

$$\frac{f(x + h) - f(x)}{h} \geq 0.$$

Since this function of $h$ is greater than or equal to zero, its limit, $f'(x)$, must be greater than or equal to zero, or $f'(x) \geq 0$. If $x$ is a right endpoint of $I$, then for $h < 0$ and $x + h$ in $I$

$$f(x + h) \leq f(x)$$

so that again

$$\frac{f(x + h) - f(x)}{h} \geq 0$$

and thus $f'(x) \geq 0$.
   Conversely, assume that $f'(x) \geq 0$ for all $x \in I$ and pick $a$,

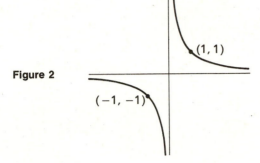

**Figure 2**

decreasing on its domain since $f(-1) < f(1)$. The theorem does apply, however, to each of the intervals $(-\infty, 0)$ and $(0, \infty)$.

---

**Example 1**   Show that the functions $f(x) = e^x$ $(x \in R)$ and $g(x) = \ln x$ $(x \in (0, \infty))$ are increasing on their domains.

*Solution*   Both functions are increasing on their domains because

$$f'(x) = e^x > 0 \quad \text{for all } x \in R,$$

and

$$g'(x) = \frac{1}{x} > 0 \quad \text{for all } x \in (0, \infty).$$

---

The theorems of this and the previous section help us to graph functions. Given a function $f$ (Fig. 3), we compute $f'$ and use it to find

   a) the points where $f'(x) = 0$ (critical points)
   b) the regions where $f'(x) \geq 0$ ($f$ increasing)
   c) the regions where $f'(x) \leq 0$ ($f$ decreasing).

We can now employ the *first derivative test* to determine whether a critical point yields a local maximum or local minimum.

Suppose $f'(a) = 0$ and $x$ is near $a$. Then:

Case (a)

If      $f'(x) \begin{cases} \leq 0 & \text{for} \quad x \leq a \\ \geq 0 & \text{for} \quad x \geq a \end{cases}$

$f$ has a local minimum at $a$ (Fig. 4a).

---

$b \in I$ with $a < b$. By the Mean Value Theorem (4.1B) there is a point $m \in (a, b)$ satisfying the equation

$$f(b) - f(a) = f'(m)(b - a).$$

Since $f'(m) \geq 0$ and $b - a > 0$, we have $f(b) - f(a) \geq 0$, or $f(a) \leq f(b)$. Hence $f$ is increasing on $I$.
   The proof of (b) is similar and will be omitted   ●

---

**Problem 1**   (a) Assume that $f'(x) > 0$ for all $x$ on an interval $I$ and show that $f$ is strictly increasing on $I$.
   b) Prove, by an example, that the converse to (a) is false.

→   Naturally we assume $f$ satisfies the hypotheses of Theorem 1.

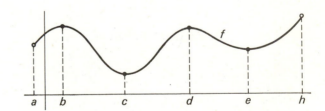

**Figure 3**   The function has critical points at $b$, $c$, $d$, and $e$. It is increasing on $(a, b]$, $[c, d]$, and $[e, h)$ and decreasing on $[b, c]$ and $[d, e]$.

Case (b)

If $f'(x) \begin{cases} \geq 0 & \text{for } x \leq a \\ \leq 0 & \text{for } x \geq a \end{cases}$

$f$ has a local maximum at $a$  (Fig. 4b).

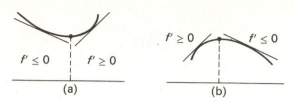

**Figure 4**

**Example 2**  Sketch the graph of $y = x^2 - 2x - 1$ using the first derivative test.

*Solution*  Since $\frac{dy}{dx} = 2x - 2$, we see that $x = 1$ is a critical point. Also

$$2x - 2 \begin{cases} \leq 0 & \text{if } x \leq 1 \\ \geq 0 & \text{if } x \geq 1 \end{cases}$$

so the function decreases when $x \leq 1$, and increases when $x \geq 1$. Hence the function must have an absolute minimum at $x = 1$ and the graph is as shown in Fig. 5.

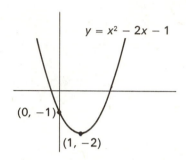

**Figure 5**  The points where the graph intersects the x-axis can be obtained by solving the equation $x^2 - 2x - 1 = 0$.

A more complicated example is the following:

**Example 3**  Sketch the graph of

$$f(x) = x^3 - 3x + 1.$$

*Solution*  Notice that

$$f'(x) = 3x^2 - 3 = 3(x^2 - 1)$$
$$= 3(x - 1)(x + 1).$$

Hence $x = -1$ and $x = 1$ are critical points. For $x \in (-1, 1)$ we have $x^2 < 1$, so

$$f'(x) = 3(x^2 - 1) < 0.$$

Hence $f$ must be decreasing. Similarly, $f'(x) = 3(x^2 - 1) > 0$ for

$$x \in (-\infty, -1) \quad \text{or} \quad x \in (1, \infty),$$

so $f$ increases on these intervals. Hence $f(1) = -1$ is a local minimum and $f(-1) = 3$ is a local maximum, so the graph appears as shown (Fig. 6).

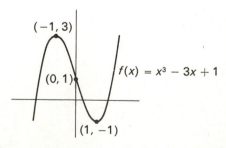

**Figure 6**  We will not find the points where the graph crosses the x-axis. This entails solving the cubic equation $x^3 - 3x + 1 = 0$.

**Example 4**  Show that the function

$$f(x) = x^3 + x + 1$$

has no critical points and is always increasing.

*Solution* The derivative is

$$f'(x) = 3x^2 + 1,$$

which is never zero. Because $f'(x) = 3x^2 + 1 > 0$ for all real numbers $x$, it follows from Theorem 1 that $f$ is increasing on $R$ (Fig. 7).

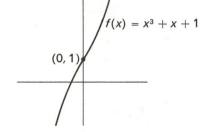

**Figure 7** (0, 1)

$f(x) = x^3 + x + 1$

**B. Vertical and Horizontal Asymptotes** We will investigate the behavior of the function

$$y = \frac{1}{x - 2}$$

in some detail to illustrate the notions of vertical and horizontal asymptotes.

For $x > 2$, $\frac{1}{x-2} > 0$, and

$$\frac{1}{x - 2} \to \infty \quad \text{as } x \to 2;$$

that is, $\frac{1}{x-2}$ becomes arbitrarily large as $x$ approaches 2. When $x < 2$, $\frac{1}{x-2} < 0$, and

$$\frac{1}{x - 2} \to -\infty \quad \text{as } x \to 2;$$

that is, $\frac{1}{x-2}$ remains negative and becomes arbitrarily large in absolute value. Hence the graph of $y = \frac{1}{x-2}$ for $x$ near 2 must be as shown (Fig. 8). The line $x = 2$ is called a *vertical asymptote* of the graph of the function.

Similarly, when $x > 2$, $\frac{1}{x-2} > 0$, and

$$\frac{1}{x - 2} \to 0 \quad \text{as } x \to \infty;$$

that is, as $x$ becomes larger, $\frac{1}{x-2}$ gets closer to zero.

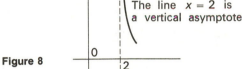

**Figure 8**

The line $x = 2$ is a vertical asymptote.

0
2

→ This is abbreviated by the notation

$$\lim_{\substack{x \to 2 \\ x > 2}} \frac{1}{x - 2} = \lim_{x \to 2^+} \frac{1}{x - 2} = +\infty.$$

→ This is also written

$$\lim_{\substack{x \to 2 \\ x < 2}} \frac{1}{x - 2} = \lim_{x \to 2^-} \frac{1}{x - 2} = -\infty.$$

→ In general, if $\lim_{x \to a^+} f(x)$ or $\lim_{x \to a^-} f(x)$ equals $\infty$ or $-\infty$, the line $x = a$ is called a *vertical asymptote* for $f$.

→ The notation for this is $\lim_{x \to \infty} \frac{1}{x - 2} = 0.$

When $x < 2$, $\frac{1}{x-2} < 0$,   and

$$\frac{1}{x-2} \to 0 \quad \text{as } x \to -\infty;$$

→ The notation for this is $\lim\limits_{x \to -\infty} \dfrac{1}{x-2} = 0$.

that is, as $x$ becomes large in absolute value but remains negative, $\frac{1}{x-2}$ gets closer to 0. In this case the line $y = 0$ is said to be a *horizontal asymptote,* and the graph of $y = \frac{1}{x-2}$ is completed as shown (Fig. 9).

→ In general if $\lim_{x \to \infty} f(x)$ or $\lim_{x \to -\infty} f(x)$ equals some number $b$, the line $y = b$ is called a *horizontal asymptote* for $f$.

There are types of asymptotes other than vertical and horizontal ones. For example, the lines $y = x$ and $y = -x$ are asymptotes for the hyperbola $x^2 - y^2 = 1$. However, we will not deal with this topic in the text.

**Figure 9**

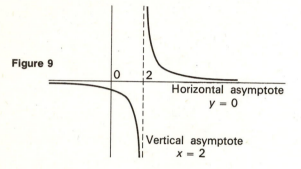

Horizontal asymptote
$y = 0$

Vertical asymptote
$x = 2$

When a function has the form $f(x) = \frac{u(x)}{v(x)}$ its vertical asymptotes, if any, are usually the lines through the points $a$ satisfying $v(a) = 0$. The horizontal asymptotes, if any, are found by studying the behavior of $f$ as $x \to \infty$ and as $x \to -\infty$.

For example, if

More exactly, the line $x = a$ is a vertical asymptote if $v(a) = 0$ and $u(a) \neq 0$. Without the assumption that $u(a) \neq 0$ there may not be a vertical asymptote. For example, if $f(x) = \dfrac{u(x)}{v(x)} = \dfrac{x^2}{x}$, then $v(0) = 0$ but the line $x = 0$ is not a vertical asymptote (Fig. 10).

$$f(x) = \frac{2x^2 - 3x + 4}{x^2 - x + 5}$$

we find $\lim\limits_{x \to \infty} f(x)$. To do so, we write

$f(x) = \dfrac{x^2}{x}$

**Figure 10**

$$\frac{2x^2 - 3x + 4}{x^2 - x + 5} = \frac{\dfrac{1}{x^2}(2x^2 - 3x + 4)}{\dfrac{1}{x^2}(x^2 - x + 5)}$$

$$= \frac{\left(2 - \dfrac{3}{x} + \dfrac{4}{x^2}\right)}{\left(1 - \dfrac{1}{x} + \dfrac{5}{x^2}\right)}.$$

Now as $x \to \infty$, $\dfrac{3}{x}$, $\dfrac{4}{x^2}$, $\dfrac{1}{x}$, and $\dfrac{5}{x^2}$ all approach 0. Hence,

$$\frac{\left(2 - \dfrac{3}{x} + \dfrac{4}{x^2}\right)}{\left(1 - \dfrac{1}{x} + \dfrac{5}{x^2}\right)} \to \frac{2}{1} = 2.$$

*To find the limit of an expression of the form $\frac{P(x)}{Q(x)}$ where $P(x)$ and $Q(x)$ are polynomials of degree p and q respectively, it is often convenient to divide $P(x)$ and $Q(x)$ by $x^r$ where r is the greater of the two numbers p and q.*

(Pat)

Also, as $x \to -\infty$, $f(x) \to 2$, so the line $y = 2$ is a horizontal asymptote for $f$.

**Example 5** Graph $f(x) = \dfrac{x}{x^2 - 4}$.

*Solution* Observe that the function has vertical asymptotes at $x = \pm 2$. Writing $f$ as

$$f(x) = \frac{1/x}{1 - (4/x^2)},$$

we see that $f(x) \to 0$ as $x \to \pm\infty$; hence, the $x$-axis is a horizontal asymptote. Since

$$f'(x) = \frac{x^2 - 4 - x \cdot 2x}{(x^2 - 4)^2} = \frac{-4 - x^2}{(x^2 - 4)^2} < 0$$

for all $x \neq \pm 2$, the function is strictly decreasing on the intervals $(-\infty, -2)$, $(-2, 2)$, and $(2, \infty)$. The graph passes through the origin and appears as shown in Figure 11.

**Problem 2** Find the indicated limits

a) $\displaystyle\lim_{x \to \infty} \sqrt{\dfrac{1 + x}{x}}$

c) $\displaystyle\lim_{x \to -\infty} \dfrac{3x^2 - 7x}{x^2}$

b) $\displaystyle\lim_{x \to \infty} \dfrac{x}{x^2 + 3x - 1}$

d) $\displaystyle\lim_{x \to \infty} \dfrac{2x}{\sqrt{9x^2 + 4}}$.

*Answer:* a) 1 b) 0 c) 3 d) $\frac{2}{3}$

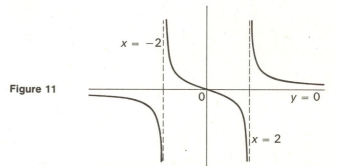

**Figure 11**

**C. Convexity** The functions shown in Figure 12a are said to be convex upward on the interval $[a, b]$, while those in Figure 12b are said to be convex downward on the interval $[a, b]$.

In general, a function is *convex upward* on an interval $I$ if the segment joining any two of its points lies "above" the graph, and *convex downward* on an interval $I$ if the segment joining any two of its points lies

A precise definition goes as follows:

Suppose $f$ is defined in an interval $I$. Then $f$ is said to be *convex upward* if, for all $x, y \in I$ $(x \neq y)$ and $0 < t < 1$,

$$f(tx + (1 - t)y) \leq tf(x) + (1 - t)f(y) \quad \text{(Fig. 13)}.$$

($f$ is said to be *strictly convex upward* if

$$f(tx + (1 - t)y) < tf(x) + (1 - t)f(y).)$$

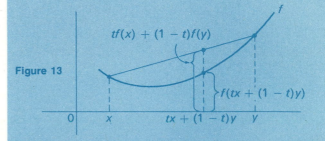

**Figure 13**

The term *convex downward* is defined analogously.
If $f(x) = ax + b$, then $f$ is convex upward and convex downward, but is not strictly convex in either direction.

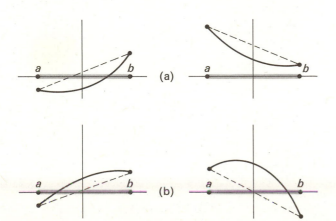

**Figure 12**

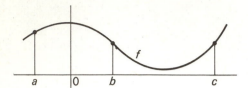

**Figure 14**

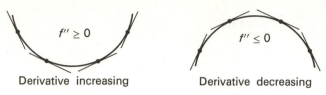

$f'' \geq 0$

Derivative increasing

**Figure 15**

$f'' \leq 0$

Derivative decreasing

**Figure 16**

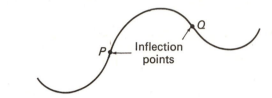

$P$ — Inflection points $Q$

**Figure 17**

"below" the graph. Notice that a function can be convex upward on one interval and convex downward on another. In Figure 14, $f$ is convex downward on $[a, b]$ and convex upward on $[b, c]$.

Using the second derivative, it is often easy to determine the types of convexity a function $f$ possesses. Indeed if $f$ is defined in an interval $I$, then

$f'' \geq 0$ in $I$ implies $f$ is convex upward on $I$.
$f'' \leq 0$ in $I$ implies $f$ is convex downward on $I$.

To see this, notice that $f'' \geq 0$ implies (using Theorem 1) that $f'$ is increasing. Hence the graph must be as shown in Figure 15. When $f'' \leq 0$ then $f'$ is decreasing, so the graph is as shown in Figure 16.

A point on the graph of a function is said to be an *inflection point* (Fig. 17) if the convexity changes as the graph passes the point.

If $(c, f(c))$ is an inflection point on the graph of $f$, then the second derivative must change signs at $c$. Assuming that $f''$ is continuous at $c$, we must have $f''(c) = 0$. However, not all points satisfying $f''(c) = 0$ are inflection points. For example, if $f(x) = x^4$ (Fig. 18), then

$$f''(x) = 12x^2 \geq 0 \quad \text{for all } x$$

so the curve is convex upward. The point $(0, 0)$ is not an inflection point even though $f''(0) = 0$ because $f''$ does not change sign at $0$.

These notions are illustrated in the following example.

---

*Example 6* Given the function $f(x) = e^{-x^2/2}$ discuss convexity and find its inflection points.

*Solution* The first derivative is equal to

$$f'(x) = -xe^{-x^2/2}$$

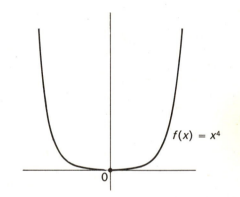

$f(x) = x^4$

**Figure 18**

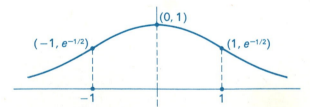

$(0, 1)$

$(-1, e^{-1/2})$ $(1, e^{-1/2})$

$-1$ $1$

**Figure 19**

To graph $f(x) = e^{-x^2/2}$ we use the following information along with our knowledge of convexity.

Since $f'(x) = -xe^{-x^2/2}$ the point $x = 0$ is a critical point. Also $f'(x) \geq 0$ for $x \leq 0$, so $f$ increases on $(-\infty, 0]$, and $f'(x) \leq 0$ for $x \geq 0$, so the function decreases on $[0, \infty)$. Hence $f$ has an absolute maximum at $x = 0$. Since $e^{-x^2/2} \to 0$ as $x \to \pm\infty$, the line $y = 0$ is a horizontal asymptote and the graph is as shown in Figure 19. Without the information about convexity, we would not have known about the inflection points.

and the second derivative is

$$f''(x) = (-x)(-x)e^{-x^2/2} + (-1)e^{-x^2/2}$$
$$= (x^2 - 1)e^{-x^2/2}.$$

Since $e^{-x^2/2} > 0$ for all $x$, $f''(x) = 0$ whenever $x^2 - 1 = 0$, that is, when $x = \pm 1$. Also, $f''(x) \leq 0$ whenever $x^2 - 1 \leq 0$. This occurs when $x^2 \leq 1$, or $-1 \leq x \leq 1$. Therefore $f$ is convex downward on $[-1, 1]$. Similarly, $f''(x) \geq 0$ whenever $x^2 \geq 1$, or equivalently, $f''(x) \geq 0$ when $x \in (-\infty, -1]$ or $x \in [1, \infty)$. Hence, in these intervals the graph is convex upward. Since $f''$ changes sign when $x = 1$ and $x = -1$, the points $(1, e^{-1/2})$ and $(-1, e^{-1/2})$ are inflection points.

---

**D. The Second Derivative Test**   If $f'(c) = 0$ the following criterion, known as the *second derivative test*, can often be used to determine whether $f$ has a local minimum or a local maximum at $c$. We have

$$f''(c) > 0 \text{ implies local minimum at } c,$$
and
$$f''(c) < 0 \text{ implies local maximum at } c.$$

An intuitive proof of this is very easy. If $f''(c) > 0$ then $f$ must be convex upward near $(c, f(c))$, and we have a local minimum (Fig. 20a). When $f''(c) < 0$, $f$ is convex downward near $(c, f(c))$, and we have a local maximum (Fig. 20b).

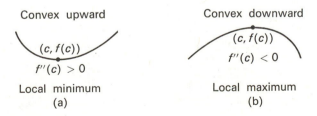

Convex upward
$(c, f(c))$
$f''(c) > 0$
Local minimum
(a)

Convex downward
$(c, f(c))$
$f''(c) < 0$
Local maximum
(b)

**Figure 20**

*Important*   When $f''(c) = 0$ the test *fails* and any possibility can occur. For example, let $a(x) = -x^4$, $b(x) = x^4$, and $c(x) = x^3$. Then 0 is a critical point for each of these functions, and the second derivative of each function is 0 at 0. But, as shown in Figure 21, all possibilities occur.

A precise statement and proof of this result, for local minima, is as follows:

*Theorem 2*   Suppose $f$ and $f'$ are defined on an interval containing $c$. If $f'(c) = 0$ and $f''(c) > 0$, then $f$ has a local minimum at $c$.

*Proof*   We will show that if $x$ is sufficiently near $c$ then $f(x) > f(c)$. Since $\dfrac{f'(x)}{x - c} = \dfrac{f'(x) - f'(c)}{x - c} \to f''(c) > 0$ as $x \to c$, we have $\dfrac{f'(x)}{x - c} > 0$ for $x$ near $c$. Hence $f'(x) > 0$ if $x - c > 0$ and $f'(x) < 0$ if $x - c < 0$. By the Mean Value Theorem (4.1B)

$$f(x) - f(c) = f'(m)(x - c)$$

where $m$ is between $c$ and $x$. If $x - c > 0$ then $f'(m) > 0$, so $f(x) - f(c) > 0$, or $f(x) > f(c)$. If $x - c < 0$ then $f'(m) < 0$, so $f(x) - f(c) > 0$, or $f(x) > f(c)$. Hence $f$ has a local minimum at $c$ ●

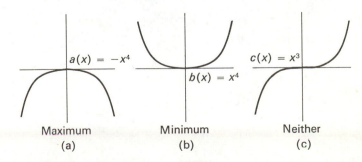

$a(x) = -x^4$
Maximum
(a)

$b(x) = x^4$
Minimum
(b)

$c(x) = x^3$
Neither
(c)

**Figure 21**

*Example 7*  Use the second derivative test to find the local maxima and minima for

$$y = 2x^3 - 3x^2 - 72x + 8.$$

*Solution*  Since

$$\frac{dy}{dx} = 6x^2 - 6x - 72 = 6(x^2 - x - 12)$$

$$= 6(x - 4)(x + 3),$$

the critical points are $x = 4$ and $x = -3$. Since

$$\frac{d^2y}{dx^2} = 12x - 6,$$

we have

$$\frac{d^2y}{dx^2}(4) = 12 \cdot 4 - 6 = 42 > 0,$$

so $x = 4$ yields a local minimum. Because

$$\frac{d^2y}{dx^2}(-3) = 12 \cdot (-3) - 6 = -42 < 0,$$

$x = -3$ yields a local maximum.

**E. The M₂I₂ ACIDS**  We will summarize these nine features of a graph in the following examples.

*Example 8*  Find, in detail, the graph (Fig. 22) of

$$y = \frac{x(x - 4)}{(x + 4)^2}.$$

*Solution*

*Intercepts*  When $x = 0$, $y = 0$ and when $y = 0$, $x(x - 4) = 0$ so $x = 0$ or $x = 4$. Hence the points $(0, 0)$ and $(4, 0)$ lie on the graph.

*Asymptotes*  Since

$$y = \frac{x(x - 4)}{(x + 4)^2} = \frac{\left(\dfrac{x(x - 4)}{x^2}\right)}{\left(\dfrac{(x + 4)^2}{x^2}\right)}$$

$$= \frac{\left(1 - \dfrac{4}{x}\right)}{\left(1 + \dfrac{4}{x}\right)^2} \to 1 \text{ as } x \to \pm\infty,$$

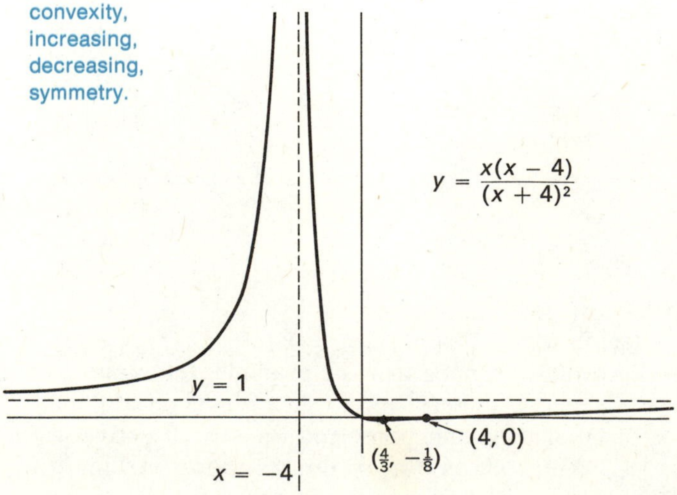

*Problem 3*  Find the local maxima and minima for the functions
a) $y = x + \frac{1}{x}$
b) $f(x) = 2x^3 - 3x^2 - 36x + 4$

*Answer:*
a) Local maximum at $x = -1$, and local minimum at $x = 1$.
b) Local maximum at $x = -2$, and local minimum at $x = 3$.

An acronym for the key features of a graph:

maximum,
minimum,
intercepts,
inflections,
asymptotes,
convexity,
increasing,
decreasing,
symmetry.

$$y = \frac{x(x - 4)}{(x + 4)^2}$$

$y = 1$

$x = -4$    $\left(\frac{4}{3}, -\frac{1}{8}\right)$    $(4, 0)$

**Figure 22**

the line $y = 1$ is a horizontal asymptote. Because the denominator is zero when $x = -4$ while the numerator is not, the line $x = -4$ is a vertical asymptote.

*Increasing, Decreasing*   We have $y' = \dfrac{4(3x - 4)}{(x + 4)^3} \geq 0$ when $3x - 4 \geq 0$ and $(x + 4)^3 > 0$ or when $3x - 4 \leq 0$ and $(x + 4)^3 < 0$. The first case gives $x \geq \frac{4}{3}$ and the second case gives $x < -4$. Therefore the function increases on $(-\infty, -4)$ and $[\frac{4}{3}, \infty)$. Also, we see that $y' \leq 0$ when $3x - 4 \leq 0$ and $(x + 4)^3 > 0$ or when $3x - 4 \geq 0$ and $(x + 4)^3 < 0$. In the first case, $y$ is decreasing on $(-4, \frac{4}{3}]$, while the second inequality is never satisfied.

*Maxima, Minima*   The first derivative is zero at $x = \frac{4}{3}$, and since $y'' = \dfrac{24(4 - x)}{(x + 4)^4}$ and $y''(\frac{4}{3}) > 0$, the function has a local minimum at $\frac{4}{3}$.

→   Here we are using the second derivative test.

*Convexity*   We have $y'' = \dfrac{24(4 - x)}{(x + 4)^4}$. The function is convex downward on $[4, \infty)$ because $y''(x) \leq 0$ for $4 \leq x < \infty$. Also, on each of the intervals $(-\infty, -4)$ and $(-4, 4]$ the function is convex upward because $y'' \geq 0$.

*Inflections*   Since $y'' = 0$ when $x = 4$, and the second derivative changes sign at that point, we have that $(4, 0)$ is an inflection point.

*Symmetry*   In this particular example there is no symmetry. (See the discussion to follow.)

---

The three simplest types of *symmetry* are:

a) with respect to the $y$-axis (Fig. 23a),
b) with respect to the $x$-axis (Fig. 23b), and
c) with respect to the origin (Fig. 23c).

In (a) a point $(x, y)$ lies on the graph if and only if the point $(-x, y)$ lies on the graph. A typical example of this is the graph of $y = x^2$.

In (b) a point $(x, y)$ lies on the graph if and only if the point $(x, -y)$ lies on the graph. A typical example of this is the graph of $y^2 = x$.

In (c) a point $(x, y)$ lies on the graph if and only if the point $(-x, -y)$ lies on the graph. A typical example of this is the graph of $y = x^3$.

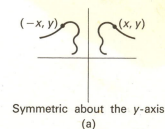

Symmetric about the $y$-axis
(a)

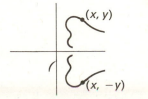

Symmetric about the $x$-axis
(b)

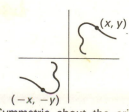

Symmetric about the origin
(c)

**Figure 23**

If one of these types of symmetry occurs it is only necessary to obtain the graph of the equation in at most two quadrants, because the graph can then be completed by using the symmetry that exists.

---

*Example 9*   The graphs of the following equations have the symmetry as indicated.

a) $y = x^4$   ($y$-axis)

b) $y^8 = x$   ($x$-axis)

c) $y^3 = x^5$   (origin)

d) $b^2x^2 \pm a^2y^2 = a^2b^2$   ($x$-axis, $y$-axis, and origin)

*Problem 4*   Prove that if any two of the three types of symmetry ($y$-axis, $x$-axis, origin) occur, then the third type also occurs.

---

## F. Exercises

**B1.** Find the regions where the following functions are increasing.

a) $f(x) = 2x^3 - 3x^2 - 36x + 6$

b) $g(x) = \sin 2x$

**B2.** Use the second derivative test to find the maxima and minima of the following functions.

a) $y = \dfrac{x}{1 + x^2}$     b) $y = \cos x + \sin x$

**B3.** Find the vertical and horizontal asymptotes for the following.

a) $f(x) = \dfrac{(x - 3)^2}{x^2 - 9}$

b) $g(x) = \dfrac{5(x^2 + 2x - 3)}{x^2 + 7x + 10}$

**B4.** Determine the regions of upward and downward convexity and the inflection points of these functions.

a) $g(x) = 2x^4 - 2x^3 - 6x^2 + 5x + 7$

b) $h(x) = x^4 - 8x^3 + 24x^2 + 8x + 5$

**B5.** Graph $f(x) = 2 \sin x + \cos x$ over the interval $[0, 2\pi]$, and discuss the $M_2I_2$ ACIDS features.

**B6.** Graph $g(x) = \dfrac{x^3}{(x - 2)^3}$ and discuss the $M_2I_2$ ACIDS features.

**D7.** Determine where the derivatives of the following functions are nonnegative.

a) $y = 4x + 8$     e) $y = 12x - x^3$

b) $y = -2x + 8$     f) $y = x^4 + 4$

c) $y = 3x^2 - 6x$     g) $y = x^4 + 4x$

d) $y = 12x - 3x^2$     h) $y = x^4 + 2x^2$

a) all $x$    b) no $x$    c) $x \geq 1$    d) $x \leq 2$
e) $-2 \leq x \leq 2$   f) $x \geq 0$   g) $x \geq -1$   h) $x \geq 0$

**D8.** Determine the vertical and horizontal asymptotes for each of the following functions.

a) $y = \dfrac{3}{x - 1}$     e) $y = \dfrac{2x}{x^2 + 1}$

b) $y = \dfrac{3 - 2x}{x - 1}$     f) $y = \dfrac{x - 2}{x^2 - 3x}$

c) $y = \dfrac{x - 3}{x^2 - 1}$     g) $y = \dfrac{x - 3}{x^2 - 3x}$

d) $y = \dfrac{x^3 - 3}{x^2 - 1}$     h) $y = \dfrac{x^2 + x - 6}{4 - x^2}$

a) $x = 1, y = 0$       b) $x = 1, y = -2$
c) $x = \pm 1, y = 0$    d) $x = \pm 1$
e) $y = 0$             f) $x = 0, x = 3, y = 0$
g) $x = 0, y = 0$      h) $x = -2, y = -1$

**D9.** Determine where the following functions are convex upward.

a) $y = 2x^2 + 8x + 3$     e) $y = 48 + 24x^2 - x^4$

b) $y = -x^2 + 8x + 4$     f) $y = x^5 + 27x + 13$

c) $y = x^3 - 3x^2 + 6x$     g) $y = x^5 + 10x^2 - 4x + 5$

d) $y = 63 + 6x^2 - x^3$     h) $y = 3x^5 + 10x^3 + 30x$

The answers are the same as those of Exercise D7.

**D10.** Use the second derivative test to find the $x$ yielding the relative minima of the following functions

a) $y = x^2 + 8$     e) $y = x^3 - 3x$

b) $y = x^2 - 2x + 4$     f) $y = 2x^3 + 3x^2 - 36x + 5$

c) $y = 2x - x^2$     g) $y = x^4 - 8x^2$

d) $y = x^3 - 3$     h) $y = x^4 + 8x^2$

a) $x = 0$   b) $x = 1$   c) none     d) none
e) $x = 1$   f) $x = 2$   g) $x = \pm 2$   h) $x = 0$

**D11.** Graph the following functions and discuss the $M_2I_2$ ACIDS features.
   a) $y = 3 + \sin x$      e) $y = \text{Tan}^{-1} 2x$
   b) $y = 4 \cos 2x$      f) $y = \text{Sin}^{-1} x^2$
   c) $y = 4 \cos^2 x$      g) $y = e^{-4x}$
   d) $y = \tan \frac{x}{2}$      h) $y = \ln (1 + x^2)$

**M12.** Show that every cubic polynomial of the form

$$ax^3 + bx^2 + cx + d \quad (a \neq 0)$$

must have exactly one inflection point.

**M13.** Show that $f(x) = ax^4 + bx^3 + cx^2 + dx + e \ (a \neq 0)$ has no inflection point whenever $3b^2 = 4ac$.

**M14.** Show that the graph of $f(x) = 2 \sin x + \cos 2x$ roughly has the same slope as the letter $M$ on the interval $(0, 2\pi)$. Find the maximum and minimum of the function on this interval.

**M15.** A ball thrown into the air follows the law $s = 60t - 16t^2$ where $s$ is the height in feet and $t$ is the time in seconds. How long does it take the ball to reach its maximum height? What is this maximum?

**M16.** The efficiency $E$ of a certain transformer is given by

$$E = \frac{3000I - (300 + 30I^2)}{3000I}$$

where $I$ is the current in amperes. At what current $I$ is the efficiency of the transformer a maximum? Use the second derivative test to verify that your value for $I$ yields a maximum. What is $E$ for this value of $I$?

**M17.** Show that the graph of $f(x) = \dfrac{x + 1}{x^2 + 1}$ has three inflection points and that these points lie on a line.

**M18.** Sketch the curve $y^3 = x^3 + 6x^2$ and show that the line $x + y = 2$ is an asymptote.

**M19.** Show that $f(x) = cx$ is strictly increasing for $c > 0$ and strictly decreasing for $c < 0$.

**M20.** Show that $f(x) = x^{2n+1}$ is strictly increasing on $(-\infty, \infty)$ for all integers $n \geq 0$.

**M21.** Show that $f(x) = x^{2n}$ has an absolute minimum at $x = 0$ for all integers $n \geq 1$.

**M22.** a) Show that for any real number $x$ there are rational numbers $r_n$ of the form $r_n = \dfrac{p_n}{2^{q_n}}$ ($p_n$ and $q_n \geq 0$ integers) satisfying $r_n \to x$ as $n \to \infty$.
   b) Prove that if $f$ is continuous and satisfies

$$f\left(\frac{x + y}{2}\right) \leq \frac{1}{2} (f(x) + f(y))$$

for all $x, y$, then $f$ is convex upwards.

---

## 4.3   *Maximum and Minimum Problems*

The class of problems dealt with in this section provides ideal examples in which calculus can be applied to practical situations. We will illustrate how these problems can be solved with a sequence of examples.

### A. The Method

a) *The problem*   This will often be in the form of a word problem arising from some physical situation. We will usually be asked to maximize or minimize a particular quantity $Q$.

*At first, we found it was quite difficult to solve the word problems in this section, for although we more or less understood the concepts, we had trouble translating the words into abstract symbols. In order to make the work easier for you, the method for solving these word problems is presented as explicitly as possible, with each step illustrated by a specific example.*

(Kent)

→ *Illustration*

a) A rectangular sheet of metal 8 inches wide and 100 inches long is folded along the center of its length to form a triangular trough. In order to maximize the carrying capacity of the trough, what should the width of the trough be at the top? What is the maximum carrying capacity of the trough?

b) *The figure*  It will be very helpful to draw a figure and label the pertinent features.

b)

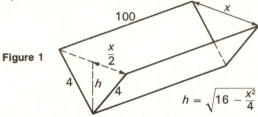

**Figure 1**

$$h = \sqrt{16 - \frac{x^2}{4}}$$

c) *The equation*  With the aid of the figure, find an equation giving the quantity $Q$ in terms of other variables. Use relations among the variables to obtain $Q$ in terms of a single variable $x$:

$$Q = Q(x).$$

c) We wish to maximize the volume $V$ of the trough. Using Figure 1, we can find $V$ in terms of $x$ and $h$:

$$V = \frac{xh}{2}\,100 = 50xh.$$

Writing $h$ in terms of $x$, $h = \sqrt{16 - \frac{1}{4}x^2}$, we have

$$V(x) = 50x\sqrt{16 - \tfrac{1}{4}x^2} = 25x\sqrt{64 - x^2}.$$

d) *Take the derivative*  Compute $Q'(x)$.

d) $V'(x) = \dfrac{-25x^2}{\sqrt{64 - x^2}} + 25\sqrt{64 - x^2}$

e) *Set the derivative equal to zero*  Let $0 = Q'(x)$ and solve to find critical points. The desired maximum or minimum usually occurs at a critical point.

e) $0 = \dfrac{-25x^2}{\sqrt{64 - x^2}} + 25\sqrt{64 - x^2}$ so

$$\frac{25x^2}{\sqrt{64 - x^2}} = 25\sqrt{64 - x^2} \quad \text{or} \quad x^2 = 64 - x^2.$$

This gives $x^2 = 32$ or $x = \pm 4\sqrt{2}$.

f) *Select the proper critical point*  There will often be more than one critical point, but it should be clear from the problem which critical point is the desired one. If it is not, then some test such as the second derivative test can be applied.

f) When $x = -4\sqrt{2}$ the problem has no meaning. Hence $x = 4\sqrt{2}$ is the desired width of the top. We see from the problem that $x = 4\sqrt{2}$ is not a minimum. The minimum occurs when $x = 0$ or $x = 8$ and the volume is then zero.

g) *The maximum or minimum*  In (f) we have determined the value of $x$ that either maximizes or minimizes $Q$. The desired value of $Q$ is obtained by simply finding $Q = Q(x)$ for this value of $x$.

g) Since the maximum occurs at $x = 4\sqrt{2}$ and since $V = 25x\sqrt{64 - x^2}$, the maximum volume is

$$V = 25(4\sqrt{2})\sqrt{64 - (4\sqrt{2})^2}$$
$$= 100\sqrt{2}\sqrt{32} = 800 \text{ in.}^3$$

## B. Examples

*Example 1*  Show that the square has the greatest area amongst all rectangles of a given perimeter.

*Solution*  Let

$$P = \text{the given perimeter (a constant)},$$

$x$ = the length, and
$y$ = the width

of the rectangle (Fig. 2). Then the area $A = xy$. Since

$$2x + 2y = P, \quad y = \frac{P - 2x}{2},$$

so

$$A(x) = x\left(\frac{P - 2x}{2}\right) = \frac{Px}{2} - x^2.$$

(Notice that $A(x)$ only makes physical sense for $0 \le x \le \frac{1}{2}P$.) We compute $A'(x) = \frac{1}{2}P - 2x$, and set it equal to zero:

$$0 = \frac{P}{2} - 2x.$$

Solving, we find $x = \frac{1}{4}P$ is the only critical point. When $x = \frac{1}{4}P$, $y = \frac{1}{4}P$ and the rectangle must be a square.

It is clear from the problem that the maximum area occurs at $x = \frac{1}{4}P$. To show this mathematically, notice that

$$A'(x) = \frac{P}{2} - 2x \begin{cases} \ge 0 & \text{if } x \le \dfrac{P}{4} \\[2mm] \le 0 & \text{if } x \ge \dfrac{P}{4}. \end{cases}$$

Hence $A(x)$ increases on $[0, \frac{1}{4}P]$ and decreases on $[\frac{1}{4}P, \frac{1}{2}P]$, which proves that the maximum occurs at $x = \frac{1}{4}P$.

---

**Example 2** Prove that, in order to construct a tin can of given volume with the least amount of metal, the diameter of the base should equal the height of the can (Fig. 4).

*Solution* Let

$V$ = the given volume (a constant),
$h$ = the height,
$d$ = diameter of base, and
$S$ = the surface area (amount of metal).

Then $\quad S = \text{top} + \text{bottom} + \text{side}$

$$= \frac{\pi d^2}{4} + \frac{\pi d^2}{4} + \pi dh \quad \text{(Fig. 5)},$$

or

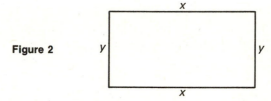

**Figure 2**

---

**Problem 1** A farmer has 100 ft. of fence and wishes to enclose a rectangular plot of land. The land borders a river and no fence is required on that side. What should the dimensions of the rectangle be in order that it include the largest possible area?

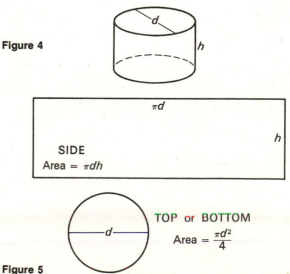

**Figure 3**

*Answer:* $x = 50, \quad y = 25$

---

**Figure 4**

SIDE
Area = $\pi dh$

TOP or BOTTOM
Area = $\frac{\pi d^2}{4}$

**Figure 5**

$$S = \frac{\pi d^2}{2} + \pi dh.$$

We will now eliminate $h$ from this equation. Since $V = \frac{\pi d^2}{4} h$ we have $h = \frac{4V}{\pi d^2}$, so

$$S = \frac{\pi d^2}{2} + \pi d \left( \frac{4V}{\pi d^2} \right) = \frac{\pi d^2}{2} + \frac{4V}{d}.$$

→ Remember that in this equation V is a constant and S and d are the variables.

(Vahan)

Hence $S'(d) = \pi d - \frac{4V}{d^2}$, and setting this equal to 0 gives

$$0 = \pi d - \frac{4V}{d^2}, \quad \text{so} \quad d = \left( \frac{4V}{\pi} \right)^{1/3}.$$

To find the corresponding $h$ we have

$$h = \frac{4V}{\pi d^2} = \frac{4V}{\pi \left( \left( \frac{4V}{\pi} \right)^{1/3} \right)^2} = \frac{4V}{\pi \left( \frac{4V}{\pi} \right)^{2/3}}$$

$$= \left( \frac{4V}{\pi} \right) \left( \frac{4V}{\pi} \right)^{-2/3} = \left( \frac{4V}{\pi} \right)^{1/3}.$$

Hence $h = d$, or the height equals the diameter of the base. Since it is clear from the problem that $S$ is a minimum, we will not prove it mathematically. It can be done as in Example 1.

Most cans in common use are not constructed to conserve metal. We wrote to several can companies to find out what factors went into their choice of dimensions. First of all, the price of the metal is almost negligible as compared to the price of its contents, and factors such as shelf advertising space and the aesthetic shape of the can are more important. Also, can manufacturers produce cans of various fixed diameters, and it is quite easy to vary the height of the can. To produce a can with a nonstandard diameter would entail significant costs on the manufacturer's part, and would thus → be reflected in the price of the product.

(Alan)

---

**Example 3** A strong swimmer is one mile off shore at point $P$ (Fig. 6). Assume that he swims at the rate of 3 mi. per hr. and walks at the rate of 5 mi. per hr. If he swims in a straight line from $P$ to $R$ and then walks from $R$ to $Q$, how far should $R$ be from $Q$ in order that he arrive at $Q$ in the shortest possible time?

We refer to Figure 6, labeling the features as indicated.

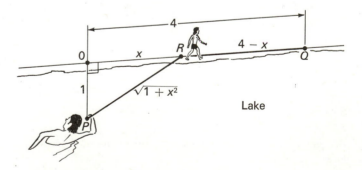

**Figure 6**

**Solution** Let

$$t_W = \text{his time in the water,}$$
$$t_L = \text{his time on the land, and}$$
$$T = t_W + t_L \text{ be the total time.}$$

Then

$$t_W = \frac{\sqrt{1 + x^2}}{3} \quad \text{and} \quad t_L = \frac{4 - x}{5},$$

so

→ We will use the basic formula

distance = (velocity)(time)

for the distance traveled at constant velocity during a certain interval of time. Here, though, we will solve for time in terms of our known quantities, distance and velocity.

(Lee)

$$T = \frac{\sqrt{1 + x^2}}{3} + \frac{4 - x}{5}.$$

Hence $\frac{dT}{dx} = \frac{x}{3\sqrt{1 + x^2}} - \frac{1}{5}$, and setting $0 = \frac{dT}{dx}$, we obtain $x = \pm\frac{3}{4}$. Since $x = -\frac{3}{4}$ is physically meaningless, we have $x = \frac{3}{4}$. Therefore the desired distance from $R$ to $Q$ is $3\frac{1}{4}$ miles.

## C. Further Examples

**Example 4** Assume that the ideal cigarette contains 9 cm.³ of tobacco. What should the dimensions of the cigarette be in order to use the least amount of paper?

*Solution*  Let $S$ denote the surface area of the paper. Then $S = xy$. Since the volume of the cigarette is 9 cm.³ and the radius is $\frac{x}{2\pi}$ (Fig. 7) we have

$$V = 9 = \pi r^2 y = \pi \left(\frac{x}{2\pi}\right)^2 y = \frac{x^2 y}{4\pi}.$$

Then  $\frac{36\pi}{x^2} = y$,   so  $S = x \cdot \frac{36\pi}{x^2} = \frac{36\pi}{x}.$

Hence $S'(x) = \frac{-36\pi}{x^2}$, and this is never zero. This means $S(x)$ has no critical points and therefore no minimum.

**Example 5** A rectangle is inscribed in the ellipse $\frac{x^2}{a^2} + \frac{y^2}{b^2} = 1$ as shown in Figure 9. What should the dimensions of the rectangle be so that its area is maximized?

Here is the calculation of this:

$$\frac{d}{dx}\left(\frac{\sqrt{1 + x^2}}{3} + \frac{4 - x}{5}\right) = \frac{1}{3} \cdot \frac{1}{2}(1 + x^2)^{-1/2}2x - \frac{1}{5}.$$

Then,   $0 = \frac{x}{3\sqrt{1 + x^2}} - \frac{1}{5}$,  so  $\frac{x^2}{9(1 + x^2)} = \frac{1}{25}$

and  $x^2 = \frac{9}{16}$,  or  $x = \pm\frac{3}{4}.$

The value $-\frac{3}{4}$ can be shown to be an extraneous root of the equation $\frac{x}{3\sqrt{1 + x^2}} - \frac{1}{5} = 0$. Hence the solution is $x = \frac{3}{4}$.

Note that $x = -\frac{3}{4}$ is also physically meaningless.
   We leave to the reader so disposed the exercise of proving mathematically that $T(x)$ indeed has a minimum at $x = \frac{3}{4}$.

Most of the time a minimum (or maximum) can be found by our methods. However, this example shows that a minimum does not always exist.

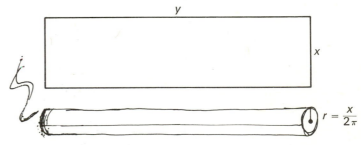

**Figure 7**

Observe that the graph (Fig. 8) of $S(x) = \frac{36\pi}{x}$ has no minimum. One might say that the minimum would be at $x$ infinitely large where $y$ is infinitely small. Unfortunately, this cigarette is a circle of infinite radius.

(Steve)

*Impossible to roll, light, or smoke.*

(Dana)

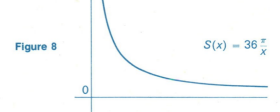

**Figure 8**                    $S(x) = 36\frac{\pi}{x}$

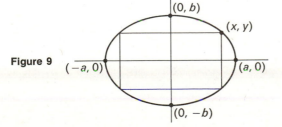

**Figure 9**   $(-a, 0)$           $(a, 0)$
   $(0, b)$   $(x, y)$
   $(0, -b)$

*Solution* Let $(x, y)$ be the point at the corner of the rectangle in the first quadrant. Then the area of the rectangle is

$$A = 2x \cdot 2y = 4xy.$$

The point $(x, y)$ is on the ellipse; hence $x$ and $y$ must satisfy

$$\frac{x^2}{a^2} + \frac{y^2}{b^2} = 1.$$

We will now use the technique of implicit differentiation to find $\frac{dA}{dx}$.

$$\frac{2x}{a^2} + \frac{2y}{b^2}\frac{dy}{dx} = 0$$

or

$$\frac{dy}{dx} = \frac{-xb^2}{a^2y}.$$

From $A = 4xy$ we have $\frac{dA}{dx} = 4y + 4x\frac{dy}{dx}$ or, substituting for $\frac{dy}{dx}$,

$$\frac{dA}{dx} = 4y + 4x\left(\frac{-xb^2}{a^2y}\right).$$

Setting this equal to zero gives

$$0 = \frac{dA}{dx} = 4y - \frac{4x^2b^2}{a^2y}$$

or

$$y^2 = \frac{b^2}{a^2}x^2,$$

so

$$y = \frac{\pm b}{a}x.$$

Since we assumed $(x, y)$ to be in the first quadrant, we must have $y = +\frac{b}{a}x$. Substituting this in the equation of the ellipse gives

$$\frac{x^2}{a^2} + \frac{1}{b^2}\left(\frac{bx}{a}\right)^2 = \frac{2x^2}{a^2} = 1,$$

so

$$x = \frac{a\sqrt{2}}{2}.$$

Hence the dimensions of the rectangle are

$$\text{length} = x = \frac{a\sqrt{2}}{2}$$

**Problem 2** Without using the result in Example 5, show directly that the rectangle of largest area that can be inscribed in a circle is a square.

We can also find $\frac{dA}{dx}$ directly by solving for $y$ in the equation of the ellipse. We then obtain

$$y = b\sqrt{1 - \frac{x^2}{a^2}}$$

and substituting this in the equation $A = 4xy$ gives

$$A = 4xb\sqrt{1 - \frac{x^2}{a^2}} = 4x\frac{b}{a}\sqrt{a^2 - x^2}.$$

Hence

$$\frac{dA}{dx} = \frac{4b}{a}\left(x\frac{1}{2}(a^2 - x^2)^{-1/2}(-2x) + \sqrt{a^2 - x^2}\right)$$

$$= \frac{4b}{a}\left(\frac{-x^2 + a^2 - x^2}{(a^2 - x^2)^{1/2}}\right).$$

Setting $\frac{dA}{dx} = 0$, we have

$$-2x^2 + a^2 = 0 \quad \text{or} \quad x = \frac{a\sqrt{2}}{2}.$$

and

$$\text{width} = y = \frac{b}{a}\left(\frac{a\sqrt{2}}{2}\right) = \frac{b\sqrt{2}}{2}.$$

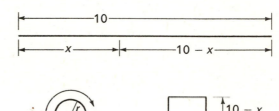

**Figure 10**

**Example 6**  A wire 10 inches long is to be cut into two pieces. A circle is formed from one piece and a square from the other. How should the wire be cut in order to maximize the total area enclosed by the two figures? How should it be cut in order to minimize the area?

*Solution*  If the wire is cut as shown in Figure 10, the radius of the circle is $\frac{x}{2\pi}$ and the side of the square is $\frac{10-x}{4}$. Hence the total area enclosed by the circle and square is

$$A = \pi\left(\frac{x}{2\pi}\right)^2 + \left(\frac{10-x}{4}\right)^2 = \frac{x^2}{4\pi} + \left(\frac{10-x}{4}\right)^2.$$

Therefore     $A' = \frac{x}{2\pi} - \left(\frac{10-x}{8}\right).$

Setting $A' = 0$ and solving we obtain

$$x = \frac{10\pi}{4+\pi}.$$

Now $A''(x) = \frac{1}{2\pi} + \frac{1}{8} > 0$ for all $x$; hence $x = \frac{10\pi}{4+\pi}$ yields a minimum. But we must also determine a maximum. Examining the graph of $A(x)$ (Fig. 11), we see that local maxima occur at 0 and at 10. Computing $A(0)$ and $A(10)$, we see that $A(10) > A(0)$, so $A(10)$ is the maximum. This means we should not cut the wire, but should use it all for the circle.

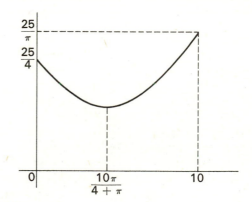

**Figure 11**

**Example 7**  A rectangle is inscribed in an isosceles triangle with one side along the base of the triangle. Show that the largest area such a rectangle can have is equal to one-half the area of the triangle. (See also Problems 3 and 4.)

*Solution*  Locate the triangle as shown in Figure 12 with its vertices as indicated. Let $x$ denote the distance from $(0,0)$ to the nearest corner $P$ of the rectangle. Then $P = (x, 0)$. The corner $Q$ that lies above $P$ lies on the line $y = \frac{b}{a}x$, so we have $Q = (x, \frac{b}{a}x)$. The other

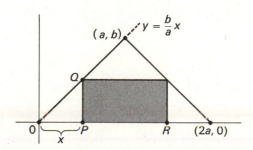

**Figure 12**

corner $R$ that lies on the $x$-axis has coordinates $R = (2a - x, 0)$. Therefore the area $A$ of the rectangle is

$$A = \text{(base)(altitude)} = (2a - x - x)\left(\frac{b}{a}x\right)$$

or

$$A = \frac{2b}{a}(ax - x^2).$$

The derivative is equal to $A'(x) = \frac{2b}{a}(a - 2x)$, and setting this equal to zero we obtain $0 = \frac{2b}{a}(a - 2x)$. Solving for $x$ we obtain $x = \frac{a}{2}$, so the area of the rectangle is

$$A = \frac{2b}{a}\left(a \cdot \frac{a}{2} - \left(\frac{a}{2}\right)^2\right) = \frac{2b}{a} \cdot \frac{a^2}{4} = \frac{ab}{2}.$$

Since this is one-half the area of the triangle, the argument is completed. (We leave to the reader the check that the above answer is indeed the maximum.)

**Problem 3 (Hard)** Prove that the rectangle of largest area that can be inscribed in any triangle has area equal to one-half the area of the triangle. (You must prove that one side of the rectangle lies along a side of the triangle.)

*Example 8* Suppose you bring electricity to your garage by plugging a long extension cord (with a 10 ampere rating and a resistance of 9 ohms) into a 120 volt outlet in your house. If you have ten 30 ohm electric heaters available and want to heat your garage, how many of the heaters should you plug in at the end of the extension cord, in order to bring as much heat as possible to the garage? What would be the maximum amount of heat?

The reader who has studied basic electrical circuits can show that the power radiated by the heaters equals

$$P = \frac{30n(120)^2}{(30 + 9n)^2}$$

where $n$ is the number of heaters plugged in.

*Solution* We consider $P$ to be a function defined for all real numbers $n$. We differentiate $P$ with respect to $n$, obtaining

$$\frac{dP}{dn} = \frac{(30 + 9n)^2 30(120)^2 - 30n(120)^2 2(30 + 9n)9}{(30 + 9n)^4}$$

Setting $\frac{dP}{dn} = 0$ and factoring, we obtain

→ *If you are interested in electrical engineering you should look at Example 8. It is a variation of the* impedance matching *problem which the power companies are constantly concerned with. Due to this concern, we obtain electricity in our homes at the proper impedance and, as a result, the power from the outlets seems to be limitless.*

(Vahan)

$$0 = 30(120)^2(30 + 9n)(30 - 9n).$$

It follows that $n = \pm\frac{10}{3}$. Taking the positive value and the nearest whole number, we obtain 3 for the number of heaters.

Substituting $n = 3$ gives the power for three heaters as

$$P = \frac{30(3)(120)^2}{(30 + 9(3))^2}, \text{ or about 400 watts.}$$

Notice that by plugging in more heaters you do not get more power radiated in the garage, you get less. In fact, if you plug in all 10 heaters, the heat radiated in the garage will be only 300 watts.

**Problem 4** (*Very easy*)   Prove that the triangle of largest area that can be inscribed in a rectangle has area equal to one-half the area of the rectangle.

## D. Exercises

**B1.** Consider the triangle in the first quadrant that is formed by the coordinate axes and a line $L$ through the point $(1, 2)$. Find the equation of the line $L$ which gives the least area for the triangle.

**B2.** A rectangle is situated inside a semicircle of radius $a$ with one side on the diameter. What is the largest area the rectangle can have?

**B3.** Show that of all isosceles triangles of a given perimeter the triangle of maximum area is the equilateral one.

**B4.** A sheet of metal is 10 feet long and 5 feet wide. It is bent lengthwise to form a rectangular trough 10 feet long. What should be the width of the trough to maximize the volume?

**B5.** What is the area of the isosceles triangle of maximum area whose vertex is at $(0, 0)$ and whose base, having endpoints on the ellipse $x^2 + 4y^2 = 4$, is parallel to the $x$-axis?

**D6.** Find the value of $x$ that maximizes the product $P = xy$, given that $x$ and $y$ are related as follows.
a) $x + y = 6$      d) $x^3 + y = 4$
b) $3x + y = 6$      e) $ye^x = 2$
c) $x^2 + y = 12$      f) $y + x^2y = 1$

a) $x = 3$     b) $x = 1$     c) $x = 2$     d) $x = 1$
e) $x = 1$     f) $x = 1$

**D7.** Among all rectangles with area 100 in.$^2$, determine the following.
a) The minimum perimeter
b) The maximum perimeter
c) The shortest diagonal
d) The shortest line from one vertex to the midpoint of a nonadjacent side

a) 40     b) no maximum     c) $10\sqrt{2}$     d) 10

**M8.** A rectangular swimming pool with square top is to be built with brick sides and bottom. If the volume is to be 500 ft.$^3$ what should the dimensions be in order to minimize the amount of brick used? (See Exercise M9.)

**M9.** A rectangular swimming pool with square top is to be built with brick sides and cement bottom. If the volume is to be 500 ft.$^3$ and the cost of brick is twice the cost of cement (per unit area), what should be the dimensions of the pool in order to minimize the cost? (See Exercise M8.)

**M10.** Find the point on the graph of $y = x^2 + 1$ that is nearest to the point $(3, 1)$.

**M11.** A rectangular poster is to contain 12 square inches of paint and is to have a margin of 1 inch on each side, 2 inches on the top and 2 inches on the bottom. Find the dimensions of the poster that will use the least amount of paper.

**M12.** A small manufacturing company makes a profit of $-x^3 + 9x^2 - 15$ dollars when it produces $x$ articles per day. If the company cannot manufacture more than 10 articles per day, determine the number that should be produced in order to maximize the profit.

**M13.** Suppose that the growth of a certain crop affects the amount of nitrogen in the soil according to the formula $y = -\dfrac{x^3}{3} + 15x^2 - 144x + 60$, where $y$ is the amount of nitrogen in the soil $x$ days after germination. How many days after germination should a farmer harvest that crop for maximum soil nitrogen content?

**M14.** Find the dimensions of the box of greatest volume which can be made from a rectangular piece of cardboard 10 in. wide by 16 in. long by cutting a square from each corner and bending up the sides.

**M15.** A window is in the shape of a rectangle capped by a semicircle. If the perimeter is $P$, what should the width be in order that the window admit the greatest possible amount of light?

**M16.** A building stands 64 feet from a wall 27 feet high. What is the length of the shortest ladder that will reach from the side of the building to the ground on the opposite side of the wall?

**M17.** Three sides of a trapezoid have length 10 inches each. What should the length of the fourth side be in order to maximize the area? (See Exercise M18.)

**M18.** A four-sided figure has three sides of length 10 inches each. What should the length of the fourth side be in order to maximize the area? (See Exercise M17.)

**M19.** Show that the square has the greatest area among all rectangles of a given diagonal.

**M20.** Show that the upper base of the trapezoid of largest area that can be inscribed in a semicircle of radius $r$ has length $r$, the lower base being a diameter.

**M21.** Show that the volume of the largest right circular cylinder that can be inscribed in a right circular cone of diameter $a$ and altitude $b$ is $\dfrac{\pi a^2 b}{27}$.

**M22.** Show that $\dfrac{4\pi r^3}{3\sqrt{3}}$ is the volume of the largest right circular cylinder that can be inscribed in a sphere of radius $r$.

**M23.** At time $t = 0$ a particle begins at $(0, 10)$ and moves to the right with a velocity of 3 units per second. Also, at $t = 0$ a particle starts at $(20, 0)$ and moves upward with a velocity of 2 units per second. When are the particles nearest to each other?

**M24.** It is an experimental fact that light travels in such a way as to traverse a distance in the shortest possible time and will travel in a particular medium in a straight line. Suppose the speed of light in air is $v_a$ and in water $v_w$. Given points $A$ (above water) and $B$ (under water), show that light will travel from $A$ to $B$ in such a way that

$$\frac{\sin \theta_a}{\sin \theta_w} = \frac{v_a}{v_w}$$

where $\theta_a$ is the angle formed with the normal $N$ to the surface by the path of the light above the water and $\theta_w$ is the angle formed with the normal by the path of the light under the water.

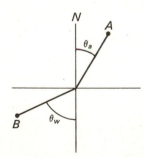

**M25.** The cost of producing a certain calculus book is $5.00 per volume. A publisher notices that the lower the price of the book the greater is the number of copies sold, the exact formula being given by $N = \dfrac{1080}{p-4} - 30$ where $p$ is the price of the book and $N$ is the number of copies sold. The publisher's total profit is $P = pN - 5N$, which equals the total income less the total cost. In order to maximize the publisher's profit, what should the selling price of the book be?

## 4.4   *Velocity and Acceleration*

The behavior of moving objects with respect to time has **always intrigued man**. However, it was calculus that provided a method for the systematic study of motion.

The Greeks understood constant motion, but beyond this had only primitive ideas about moving bodies. This was made evident by Zeno of Elea (495–435 B.C.), whose construction of four ingenious paradoxes showed that the Greek notions required drastic revision.

**A. Motion on a Line** If an object is moving on a line at constant velocity $v$, then the distance $s$ that is traveled in an interval of time $t$ is given by

$$s = vt.$$

Hence, in this case the velocity is

$$v = \frac{s}{t}.$$

When the motion is not constant this formula does *not* apply, and the velocity is defined as follows:

Suppose an object is moving on a line which we take to be the $x$-axis, and its position at time $t$ is given by the formula $x = x(t)$. We now *define* the *velocity* to be

$$v = \frac{dx}{dt}.$$

The number $\left.\dfrac{dx}{dt}\right|_{t=t_0}$ is called the *velocity of the object* *at time* $t_0$, or the *instantaneous* velocity at $t = t_0$. Notice that if $x = ct$ where $c$ is a constant, then

$$v = \frac{dx}{dt} = c \quad \text{which equals} \quad \frac{x}{t},$$

so that our definition of velocity reduces to the usual one for the case of constant motion.

The velocity can be negative as well as **positive**. When $\frac{dx}{dt} > 0$, then $x = x(t)$ is an increasing function of $t$ (4.2A), so the object is moving to the right. When $\frac{dx}{dt} < 0$, then $x = x(t)$ must be a decreasing function of $t$ (4.2A), so the object is moving to the left.

Having defined the velocity to be $v = \frac{dx}{dt}$, we now define the *acceleration a* to be the derivative of the

It is important to realize that the definition of velocity as $\frac{dx}{dt}$ is just that—a definition and not a theorem. It is possible, however, to appeal to physical intuition for justification. Suppose $x = x(t)$ and we have an object at position $x_0 = x(t_0)$ at time $t_0$. At a later moment in time, $t_0 + h$, the object is at position $x(t_0 + h)$. The quantity

$$v(t_0, h) = \frac{x(t_0 + h) - x(t_0)}{h}$$

is called the *average velocity* over the interval $[t_0, t_0 + h]$. That is, an object moving with constant velocity $v(t_0, h)$ would still cover the distance $x(t_0 + h) - x(t_0)$ in the time $h$. Hence $v(t_0, h)$ approximates $v(t_0)$ for $h$ small. If we then define

$$v(t_0) = \lim_{h \to 0} v(t_0, h) = \lim_{h \to 0} \frac{x(t_0 + h) - x(t_0)}{h} = \left.\frac{dx}{dt}\right|_{t=t_0}$$

we have our original definition. Notice that it is not necessary for $h$ to be positive for, although time does not move backwards, we must consider time increments on both sides of a fixed $t = t_0$. The derivative is not defined unless we do so. For example, if

$$x = x(t) = \begin{cases} t, & t \leq 1 \\ 2t - 1, & t \geq 1 \end{cases}$$

then $\frac{dx}{dt}$ is undefined at $t = 1$.

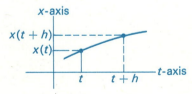

**Figure 1**

**Figure 2**

The *graph* of $x = x(t)$ when $\frac{dx}{dt} > 0$ is as shown in Figure 1; however, the *motion* of the object is as indicated in Figure 2.

velocity. Hence

$$a = \frac{dv}{dt} = \frac{d^2x}{dt^2}.$$

The acceleration of an object, as will be seen, plays an extremely important role in the study of moving objects.

---

**Example 1** Given $x = 3t^3 - 8t - 6$, find
a) the velocity when $t = 2$
b) the acceleration when $t = -3$
c) when the velocity equals 1.

*Solution*

a) $v = \dfrac{dx}{dt} = 9t^2 - 8$  so  $v|_{t=2} = 28$

b) $a = \dfrac{dv}{dt} = 18t$    so  $a|_{t=-3} = -54$

c) $v = 9t^2 - 8$ so, letting $v = 1$, we have $9t^2 = 9$, or $t^2 = 1$. Hence, at $t = 1$ and $-1$ the velocity is 1.

---

→  It is possible to coin names for even higher derivatives. For example, the term "jerk" (denoting $\dfrac{da}{dt} = \dfrac{d^2v}{dt^2} = \dfrac{d^3x}{dt^3}$ or the rate of change of acceleration with respect to time) has gained great popularity in recent years as a result of its importance in the design of rockets and space ships. During the vibration testing of space vehicle parts "jerkmeters" are employed to measure the amounts of jerk present. It is conceivable that even derivatives of jerk, yet unnamed, may become significant design parameters of the future.

(Vahan)

---

**Problem 1** Given $x = t^4 - 8t + 3$ find the following.
a) The velocity when $t = 3$
b) The acceleration when $t = -4$
c) The acceleration when $v = -4$

*Answer:*  a) 100  b) 192  c) 12

---

## B. Falling Objects

### A Revolution

Near the end of the sixteenth century, science was ripe for a revolution. For almost 2000 years, the views of Aristotle had dominated all branches of science. He had declared that the earth was the center of the solar system, and although Aristarchus, one of his contemporaries, disagreed with him, opposition was muffled and the views of Aristotle prevailed. Copernicus broke the spell of Aristotle's authority 18 centuries later with his theory of the solar system, but the revolution did not come from the heavens. It began with a single thud from 179 feet.

Aristotle believed that an object would fall with constant velocity dependent on its mass, that is $s = v_M t$ where $v_M$ is a constant determined by the mass of the substance being dropped. The formula is wrong, and the method by which he determined it is even more so. It was deduced as an exercise in pure thought and few questioned its validity or the method by which it was obtained. Leonardo da Vinci (1452–1519) took exception to Aristotle's result, and arrived at the formula $s = c(t^2 + t)$ with $c$ a constant. This formula is close to being correct and clearly recognizes the constant acceleration which is independent of the mass of the body. Da Vinci no doubt realized the importance of this result, but he did not pursue it. Perhaps he had more important matters to attend to.

Much later another Italian began his own investigations of falling bodies. In 1589 at the age of 25, Galileo Galilei (1564–1642) was appointed a lecturer in mathematics at Pisa. One day he went to the top of the leaning tower with two weights, one ten times that of the other, and dropped them simultaneously. Down below, a crowd of students, professors, and priests watched. According to Aristotle the heavier object would definitely strike the ground first, but the two objects struck the earth at the same time. Thus began the revolution and with it, its reception. Because of his insolence in disputing the accepted views of Aristotle, Galileo was forced to resign his position at Pisa.

He was fortunate in obtaining a position at the University in Padua, which had a tradition of encouraging new ideas. His work flourished, and enlightened men from all over Europe came to hear his lectures. Having accurately described the motion of falling bodies, he went on to lay the foundations for dynamics, the study of moving objects, thereby paving the way for Newton. His beautiful blending of ideas and theories with observations and experiments was to become known as the scientific method. Then in 1609 Galileo heard that Hans Lippershey, an apprentice to a spectacle-maker, had devised an instrument that would make objects appear larger. He immediately recognized the significance of the discovery and soon determined how the instrument must work. It was not long before Galileo was building the finest telescopes available. With their aid he made numerous astronomical discoveries and established the Copernican theory beyond all reasonable doubt.

In 1616 the Catholic Church outlawed the Copernican theory, and in 1630 Galileo published an outstanding book on dynamics. As soon as they understood it the authorities realized that the work supported the outlawed theory, and they reacted swiftly. On June 22, 1633 the 70 year old Galileo was brought before the Inquisition. Under threat of torture he was forced to renounce his "erring" ways and was sentenced to indefinite imprisonment. Because of his age he was allowed to remain a prisoner in his own home.

Reactions to Galileo's imprisonment were varied. Some scientists ignored the Church while others ignored Galileo. The very religious Descartes, who, prior to 1633, supported both the Copernican theory and the Catholic Church, was quite confused and withheld a publication of his own which clearly supported Galileo. However, his ingenuity eventually allowed him to somehow reconcile these opposing views. By the time of Newton, both the Copernican theory and Galileo were generally supported by the scientific community.

By making very careful observations, Galileo determined that any falling body has constant acceleration approximately equal to 32 ft./sec.$^2$ or 980 cm./sec.$^2$. This constant is called $g$.

The calculus will enable us to derive from $g$ the equation of motion of a falling body. Since our concern is with vertical motion, position will be measured along the $y$-axis. Because objects fall in the negative $y$ direction, their acceleration, according to Galileo, is given by

$$a = -g,$$

that is,

$$\frac{dv}{dt} = -g.$$

→ *In engineering applications we must be very careful about the units used, for they can unnecessarily complicate problems. At present, we will concentrate on the calculus involved in solving problems and choose simple and convenient units. You should note here that if the value of g is taken as 32 the answers will be in feet or feet per second, while if 980 is chosen the answers will be in centimeters and seconds.*

*Another practice in engineering is to initially assume ideal conditions; for example, the effects of air resistance or very small variations in gravity are neglected. First the idealized problem is solved; then the real one is considered with all its detailed complications.*

*(Vahan)*

But

$$\frac{d}{dt}(-gt) = -g;$$

hence

$$v = -gt + c$$

where $c$ is a constant. When $t = 0$, $v(0) = c$. If we denote $v(0)$ by $v_0$, we have $v_0 = c$; hence

$$v = -gt + v_0.$$

Now

$$\frac{dy}{dt} = v(t) = -gt + v_0,$$

and since

$$\frac{d}{dt}\left(\frac{-gt^2}{2} + v_0 t\right) = -gt + v_0,$$

we must have

$$y = \frac{-gt^2}{2} + v_0 t + C.$$

→   We are again using Corollary 2 (4.1C).

When $t = 0$, $y(0) = C$. Denoting $y(0)$ by $y_0$ yields →   We interpret $y_0$ as the starting position.

(*) $$y = \frac{-gt^2}{2} + v_0 t + y_0.$$

We now have a formula which describes the motion of any falling body.

→   Here we are using Corollary 2 of 4.1C, which states that if $f' = g'$ then $f = g + C$ where $C$ is a constant.

→   We interpret $v_0$ as the *initial velocity*. This quantity is often given in a problem.

---

**Example 3**   A ball is thrown upward from the top of a building 80 feet high with an initial velocity of 64 ft./sec.
a) When does the ball reach its maximum height?
b) How high above the ground does it rise?
c) When does it strike the ground?
d) At what time is the ball again at a height of 80 feet?

*Solution*

a) In this case $y_0 = 80$, $v_0 = 64$, $g = 32$. Hence our equation is

$$y(t) = -16t^2 + 64t + 80.$$

We seek to maximize $y$; hence we set

$$\frac{dy}{dt} = -32t + 64 = 0,$$

---

**Example 2**   A weight is dropped from a tower 179 feet high.
a) When does it strike the ground?
b) What is its velocity at that time?

*Solution*

a) Observe that $v(0) = v_0 = 0$ and $y(0) = y_0 = 179$. By Formula (*) above,

$$y(t) = \frac{-g}{2}t^2 + 0 \cdot t + 179 = -16t^2 + 179.$$

The weight will strike the ground when $y = 0$, that is when

$$-16t^2 + 179 = 0 \quad \text{or} \quad t = \tfrac{1}{4}\sqrt{179}.$$

b) Since $v(t) = \dfrac{dy}{dt} = -32t$ we have

$$v|_{t=\sqrt{179}/4} = -32 \cdot \tfrac{1}{4}\sqrt{179} = -8\sqrt{179} \text{ ft./sec.}$$

is the velocity when it hits the ground.

---

which occurs when $t = 2$ sec. At $t = 2$, $y$ is a maximum since

$$\frac{d^2y}{dt^2} = -32 < 0.$$

b) $y(2) = -16 \cdot 4 + 64 \cdot 2 + 80 = 144$ is the maximum height.

c) The ball strikes the ground when $y = 0$. Thus

$$y = -16t^2 + 64t + 80 = 0,$$

or

$$t^2 - 4t - 5 = 0;$$

hence $t = 5$ or $t = -1$. In the present case, only $t = 5$ makes sense.

d) The ball is at height 80 feet when

$$y(t) = -16t^2 + 64t + 80 = 80,$$

that is, when

$$-16t^2 + 64t = 0 \quad \text{or} \quad t = 0, 4.$$

Since $t = 0$ is the initial time, we therefore have that the ball is at a height of 80 feet for the second time when $t = 4$.

---

**C. Motion in the Plane** When a particle is moving in the plane, its position at time $t$ is given by the Cartesian coordinates

$$P(t) = (x(t), y(t)) \quad \text{(Fig. 3)}.$$

The derivatives $\frac{dx}{dt}$ and $\frac{dy}{dt}$ are called the **x and y components of the velocity,** respectively. The *speed of the particle* is defined by

$$s = \sqrt{\left(\frac{dx}{dt}\right)^2 + \left(\frac{dy}{dt}\right)^2}.$$

---

**Example 4** Suppose a particle is moving in the plane according to the formula

$$P = (t^2 - t, 1 - 4t).$$

Find

a) its position when $t = 4$
b) the $x$ component of the velocity when $t = 4$

---

**Problem 2** A ball is thrown vertically upward from the ground with initial velocity 80 ft./sec.
a) What is the velocity of the ball when it hits the ground?
b) How high will the ball go?
c) The ball is at a height of 96 ft. twice during its flight. What are the corresponding velocities?

*Answer:* a) $-80$ ft./sec.   b) 100 ft.
c) 16 ft./sec.   and   $-16$ ft./sec.

---

**Figure 3**     $(x(t_1), y(t_1))$          $(x(t_2), y(t_2))$

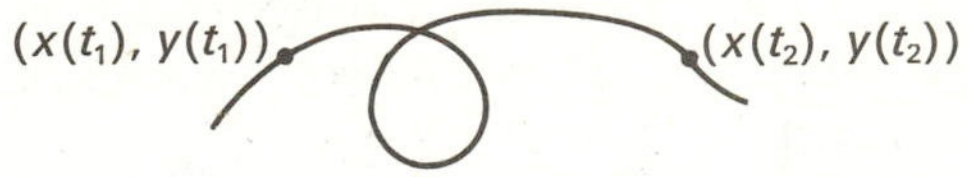

The velocity $v$ is defined by the equation

$$v = \frac{dP}{dt} = \left(\frac{dx}{dt}, \frac{dy}{dt}\right).$$

Observe that this is not a number, but a pair of numbers.

Notice that when the motion is confined to the $x$-axis, $y$ is always zero, so $\frac{dy}{dt} = 0$, and the speed equals

$$\sqrt{\left(\frac{dx}{dt}\right)^2} = \left|\frac{dx}{dt}\right|.$$

c) the $y$ component of the velocity when $t = -3$
d) the speed when $t = 2$
e) the time when the speed is a minimum

*Solution*

a) When $t = 4$, $P = (4^2 - 4, 1 - 4 \cdot 4) = (12, -15)$.

b) Since $\dfrac{dx}{dt} = 2t - 1$, $\left.\dfrac{dx}{dt}\right|_{t=4} = 2 \cdot 4 - 1 = 7$.

c) Since $\dfrac{dy}{dt} = -4$, $\left.\dfrac{dy}{dt}\right|_{t=-3} = -4$.

d) The speed is given by the formula

$$s = \sqrt{(2t - 1)^2 + (-4)^2} = \sqrt{4t^2 - 4t + 17}.$$

Hence, when $t = 2$,

$$s = \sqrt{4 \cdot 2^2 - 4 \cdot 2 + 17} = \sqrt{25} = 5.$$

e) To minimize $s$ we compute

$$\frac{ds}{dt} = \frac{4t - 2}{\sqrt{4t^2 - 4t + 17}}.$$

Setting $\frac{ds}{dt} = 0$ and solving for $t$ gives $t = \frac{1}{2}$.

---

*Example 5* A Colt 357 Magnum using 158 grain powder has a muzzle velocity (speed) of 1408 ft./sec. Assuming the pistol is fired at an angle of 30° from the horizontal (Fig. 4),

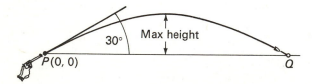

30° Max height

$P(0, 0)$     $Q$

**Figure 4**

a) how far horizontally does the bullet travel?
b) how high does the bullet rise?

*Solution* Assume that the pistol is fired from the point $P = (0, 0)$ at $t = 0$. The velocity can be broken down into horizontal and vertical components (Fig. 5a). Because these quantities lie on the sides of a 30°–60°–90° right triangle, we have $\frac{dx}{dt} = 704\sqrt{3}$ and $\frac{dy}{dt} = 704$ (Fig. 5b). By Formula (*) of 4.4B, since

---

*Problem 3* Suppose a particle is moving according to the formula

$$P = (3 \cos t, 3 \sin t).$$

a) At what instants is the particle not moving in a horizontal direction? (That is, when is $\frac{dx}{dt} = 0$?)
b) Show that the speed of the particle is constant.

*Answer:* a) $0, \pm\pi, \pm 2\pi, \ldots$

---

Since

$$\frac{ds}{dt} \text{ is } \begin{cases} <0 & \text{for } t < \frac{1}{2} \\ >0 & \text{for } t > \frac{1}{2} \end{cases}$$

it follows that there is indeed a minimum at $t = \frac{1}{2}$.

Our apologies to Colt Firearms of Hartford, Connecticut. The muzzle velocity of a Colt 357 Magnum using 158 grain powder is actually 1410 ft./sec., but the number 1408 yields a nicer answer.

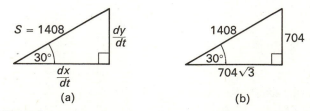

$S = 1408$   $\frac{dy}{dt}$

30°   $\frac{dx}{dt}$

(a)

1408   704

30°   $704\sqrt{3}$

(b)

**Figure 5**

Gravity affects only vertical motion, not horizontal motion. The bullet rises as if it were given an initial vertical velocity of 704, and moves horizontally with the constant velocity of $704\sqrt{3}$.

$$x(0) = y(0) = 0,$$

we have

$$x = 704\sqrt{3}t \quad \text{and} \quad y = -16t^2 + 704t.$$

a) Since $y = -16t^2 + 704t = -16t(t - 44)$, we have that $y$ is zero when $t = 0$ or $t = 44$. Hence the bullet falls to the ground when $t = 44$. At this time

$$x = 704\sqrt{3} \cdot 44 = 30{,}976\sqrt{3} \text{ ft.}$$

b) To maximize $y$ we compute $\frac{dy}{dt} = -32t + 704$ and, setting this equal to zero, obtain $t = 22$. At this time we have

$$y = -16(22)^2 + 704 \cdot 22 = 7744 \text{ ft.}$$

**Example 6** In order to maximize the horizontal distance the bullet travels, at what angle from the horizontal should the bullet be fired? What is the maximum horizontal distance the bullet travels?

*Solution* Let $\theta$ denote the angle at which the bullet is fired. Then the horizontal and vertical components of the initial velocity are

$$\frac{dx}{dt} = 1408 \cos\theta \quad \text{and} \quad \frac{dy}{dt} = 1408 \sin\theta,$$

and since $x(0) = y(0) = 0$, we have

$$x = t1408\cos\theta \quad \text{and} \quad y = -16t^2 + t1408\sin\theta.$$

The bullet returns to the ground when $y$ equals zero, so solving for nonzero $t$ in the equation $y = 0$ we obtain

$$t = 88 \sin\theta.$$

Putting this value for $t$ in the equation $x = t1408\cos\theta$ we obtain

$$x = (1408)(88)\sin\theta\cos\theta = (1408)(44)\sin 2\theta$$

as the horizontal distance traveled. To maximize this we compute $\frac{dx}{d\theta}$ and set it equal to zero. We have

$$0 = \frac{dx}{d\theta} = (1408)(88)\cos 2\theta,$$

so $\cos 2\theta = 0$ or $\theta = \tfrac{1}{4}\pi$ is the desired angle. The maximum distance traveled is then equal to

$$x = (1408)(44)\sin\left(2 \cdot \tfrac{1}{4}\pi\right)$$
$$= (1408)(44) = 61952 \text{ feet.}$$

## D. Exercises

**B1.** A particle moves along the $x$-axis according to the formula $x = t^2 - 6t$.
   a) When is the velocity equal to zero?
   b) Where is the particle when its velocity is zero?
   c) Where is the particle when its velocity is 4?

**B2.** A stone is thrown upward from a tower 160 feet high with an initial velocity of 32 feet per second.
   a) When does the stone reach its maximum height?
   b) What is the maximum height the stone reaches?
   c) When does the stone hit the ground?

**B3.** The position of a particle moving in the plane is given by $P = (t^2, t^2 - t)$.
   a) Find the position of the particle when $t = 4$.
   b) Find the speed when $t = 2$.
   c) At what time is the particle not making any vertical motion?

**B4.** Suppose a baseball is pitched horizontally with an initial velocity of 96 feet per second from a height of 7 feet. How high above the ground is the ball after it travels a horizontal distance of 60 feet?

**B5.** An arrow is shot from a bow at an angle of 30° from the horizontal with an initial velocity of 120 feet per second.
a) What is the maximum height of the arrow?
b) At what time does the arrow hit the ground?

**B6.** A particle moves along the $x$-axis according to the given formula. Determine when the particle is moving to the right.
a) $x = t^4 - 32t$
b) $x = 3t - t^3$

**D7.** A particle moves along the $x$-axis according to the formula

$$x = 2t^3 - 6t^2 + 18t.$$

Find
a) the velocity when $t = 3$
b) the acceleration when $t = 2$
c) the time when the acceleration is equal to zero
d) the position when the acceleration is zero
e) the velocity when the acceleration is zero

a) 36    b) 12    c) 1    d) 14    e) 12

**D8.** A particle moves along the $x$-axis according to the given formula. Determine when the particle is moving to the right.
a) $x = 4t^2$
b) $x = \ln(t^2 - 2t + 2)$
c) $x = t^3 + t^5 + t^7 + t^9$
d) $x = t^3 + 3t^2 - 9t + 10$

a) $t > 0$    b) $t > 1$    c) $t \neq 0$
d) $t < -3$ and $t > 1$.

**D9.** A ball is thrown upward with an initial velocity of $v_0$. Find the maximum height the ball attains, given the following values of $v_0$ (units in feet).
a) 32    b) 64    c) 160    d) 320

a) 16 ft.    b) 64 ft.    c) 400 ft.    d) 1600 ft.

**D10.** A particle moves in the plane according to the formula

$$P(t) = (x(t), y(t)).$$

Find the speed of the particle at the indicated time.
a) $x = 3t^2, y = 4t^2, t = 5$

b) $x = 4t^2, y = 2t^3, t = 1$
c) $x = 3 \cos t, y = 3 \sin t, t = 77$
d) $x = \cos t, y = 3 \sin t, t = \dfrac{\pi}{3}$
e) $x = e^{\sin t}, y = e^{\cos t}, t = 0$

a) 50    b) 10    c) 3    d) $\sqrt{3}$    e) 1

**M11.** Suppose that the position $x$ of a certain particle is given by $x = t^3 + 3t^2 - 9t + 10$; find
a) the position when $t = 3$
b) the velocity when $t = 1$
c) the acceleration when $t = 0$
d) the minimum velocity
e) the maximum acceleration

**M12.** A particle moves on the $x$-axis, its position $x$ at time $t$ being given by the formula

$$x = t^3 - 9t^2 + 24t + 4.$$

a) When is $x$ increasing?
b) When is the velocity increasing?
c) Starting at $t = 0$, how far does the particle move to the right before it reverses its motion?
d) What is the total distance the particle travels in the first 6 seconds?

**M13.** Two particles move on the $x$-axis according to the formulas

$$x_1 = t^2 - 2t + 3 \quad \text{and} \quad x_2 = -t^2 + t - 4.$$

a) How far apart are the particles after 3 seconds?
b) How close does the first particle come to the origin?
c) How close does the second particle come to the origin?
d) What is the least distance that occurs between the particles?

**M14.** A stone is dropped from a distance of $d$ feet above the ground. Show that it strikes the ground in $\sqrt{d}/4$ seconds.

**M15.** A stone is thrown downward from a height of $h$ feet above the ground with an initial velocity of $v_0$ feet per second. Show that it hits the ground in

$$\frac{\sqrt{v_0{}^2 + 64h} - |v_0|}{32} \text{ seconds.}$$

**M16.** Show that if a particle moves along the $x$-axis with velocity $v = \frac{dx}{dt} = f(x)$, then its acceleration is given by $a = f(x)f'(x)$.

**M17.** In his prime, Bobby Feller could throw a baseball with a speed of 100 miles per hour. Determine the maximum horizontal distance he could throw a baseball. (Ignore Feller's height and maximize over the angle at which the ball is thrown.)

**M18.** Assuming Jack Nicklaus can drive a golfball 960 feet

(ignore rolling), determine the minimum velocity it can have as it leaves the tee.

**M19.** Assuming a man can pass a football 224 feet, how fast must he be able to throw it? (Ignore the height of the passer.)

---

## 4.5  *Rates and Related Rates*

In the last section we saw that the derivative can be interpreted as velocity. In this section we will give several other interpretations and will then discuss related rate problems.

**A. Rates** Given any function $Q = Q(x)$, the derivative $\frac{dQ}{dx}$ is called the *rate of change of Q with respect to x*. For example, if $x = x(t)$ then the velocity is the rate of change of position with respect to time, and the acceleration is the rate of change of velocity with respect to time.

Most of the rates discussed below will be with respect to time, and in this case we will often refer to the rate of change of a certain quantity without explicitly referring to time.

---

*Example 1* A balloon is being filled with air. What is the relationship between the rate of change of the volume and the rate of change of the radius?

*Solution* Assuming the rates are with respect to time, let $V$ denote the volume of the balloon and $r$ denote its radius. Then we have

$$V = \tfrac{4}{3}\pi r^3.$$

Take the derivative with respect to time:

$$\frac{d}{dt}V = \frac{d}{dt}\left(\frac{4}{3}\pi r^3\right) = \frac{4\pi}{3}\frac{d}{dt}(r^3) = \frac{4\pi}{3}3r^2\frac{dr}{dt}.$$

Hence we have

$$\frac{dV}{dt} = 4\pi r^2\frac{dr}{dt}.$$

---

In economics the word "marginal" signifies a rate of change. For example, if $C(x)$ denotes the cost of producing $x$ items, then $\frac{dC}{dx}$ is called the *marginal cost*. However, in elementary economics courses the marginal cost is defined as $C(x + 1) - C(x)$, which is the cost of producing one additional item. When $x$ is large, then

$$\frac{C(x + 1) - C(x)}{1} \quad \text{is near} \quad \frac{dC}{dx}$$

and the elementary definition is practical. *Marginal price* and *marginal profit* are defined in similar ways.

If a charge $Q$ depends on time, then its rate of change with respect to time, $I = \frac{dQ}{dt}$, is called the *current*. The rate of change of work $W$ with respect to time $t$, or $\frac{dW}{dt}$, is defined to be *power*.

The density of a certain substance is its mass per unit volume, but the density can be defined even when the mass is distributed in a plane or a line. For example, if a mass $M$ is distributed along the $x$-axis, then $\frac{dM}{dx}$ is the *density*.

**B. Related Rate Problems** These problems are similar to maximum-minimum problems. We will give the method and illustrate it with an example.

→ *It might be amusing to make up your own "related rate" ratios. Our best one was:*

$$\frac{dDDT}{dt}\quad\text{(rate of contamination by insecticide with respect to time)}$$

*(Beth)*

### Method

→ **Illustration**

a) *The problem* It is usually a word problem. Two or more quantities are given and their rates of change are related. We are asked to find the rate of change of one of them when the other rates of change are specified.

a) A man is walking away from a wall 40 feet high at a rate of 3 feet per second. At what rate is his distance from the top of the wall changing when he is 30 feet from the wall? (We will assume that distance is measured from his feet to the top of the wall.)

b) *Figure* Use letters to describe each of the variables and draw a figure that exhibits them and the other data in the problem. Do not label a variable as though it were a constant. (For example, don't put 30 in place of $x$.)

b) As shown in Figure 1, let

$x =$ distance of man from the foot of the wall, and
$y =$ distance of man from the top of the wall.

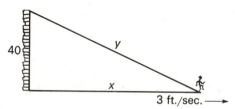

**Figure 1**

c) *Translation* Express the given and required information in mathematical terms.

c) We are given $\frac{dx}{dt} = 3$ and are asked to find $\frac{dy}{dt}$ when $x = 30$.

d) *Relations between the variables* This will usually be a single equation in the variables involved and the figure should help you find it.

d) By the Pythagorean Theorem we have
$$y^2 = x^2 + (40)^2.$$

e) *Differentiate and solve* The equation from (d) must now be differentiated with respect to the appropriate quantity (usually time). We then solve for the unknown and evaluate it at the specified values.

e) Differentiating with respect to time, we have
$$2y\frac{dy}{dt} = 2x\frac{dx}{dt}\quad\text{or}\quad\frac{dy}{dt} = \frac{x}{y}\frac{dx}{dt}.$$

Now $\frac{dx}{dt} = 3$, and when $x = 30$,
$$y^2 = (30)^2 + (40)^2 = 2500,$$
so $y = 50$. Hence, when $x = 30$,
$$\frac{dy}{dt} = \frac{30}{50}\cdot 3 = \frac{9}{5}\text{ ft./sec.}$$

*Example 2* An ice cream cone 5 inches high and 2 inches in diameter is leaking ice cream from a hole in the bottom at the rate of $\frac{1}{3}$ in.³/min. At what rate is the level of the ice cream falling when the height of the ice cream measures 3 inches? See Figure 2.

*Solution* Let $V$ denote the volume of the ice cream. We are given that $\frac{dV}{dt} = \frac{-1}{3}$. (This is negative because the volume is decreasing.) The problem is to find $\frac{dh}{dt}$ when $h = 3$. By similar triangles, we have

$$\frac{h}{5} = \frac{r}{1}.$$

Since $V = \frac{1}{3}\pi r^2 h$ (front cover), we have

$$V = \frac{1}{3}\pi\left(\frac{h}{5}\right)^2 h = \frac{1}{3}\pi\frac{h^3}{25}.$$

Hence

$$\frac{dV}{dt} = \frac{\pi}{25}h^2\frac{dh}{dt},$$

so

$$\frac{dh}{dt} = \frac{25}{h^2\pi}\frac{dV}{dt} = \frac{25}{h^2\pi}\left(\frac{-1}{3}\right).$$

Letting $h = 3$ gives

$$\frac{dh}{dt}\bigg|_{h=3} = \frac{25}{9\pi}\left(\frac{-1}{3}\right) = \frac{-25}{27\pi}.$$

The minus sign indicates that the level is decreasing, so we may say that the level is falling at the rate of $\frac{25}{27\pi}$ in./min.

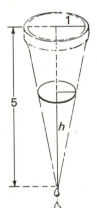

**Figure 2**

---

**Example 3** A man is walking away from a street light at the rate of 5 ft./sec. If the man is 6 feet tall and the light is 15 feet high, how fast is the man's shadow lengthening when he is 10 feet from the base of the street light?

*Solution* Referring to Figure 4, we let $x$ be the distance of the man from the foot of the street light and

---

**Problem 1** Two women start walking from the same point. One goes north at a rate of 4 miles/hr. and the other east at a rate of 5 miles/hr. How fast is the distance between them changing after 8 hours?

Hint: You are given $\frac{dx}{dt} = 5$ and $\frac{dy}{dt} = 4$ and are asked to find $\frac{dz}{dt}$ when $t = 8$ (Fig. 3).

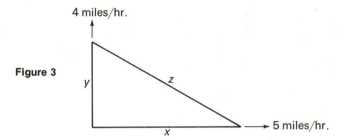

**Figure 3**

*Answer:* $\sqrt{41}$ mi./hr.

---

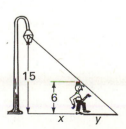

**Figure 4**

let $y$ be the length of his shadow. By similar triangles, we have

$$\frac{x + y}{15} = \frac{y}{6} \quad \text{so} \quad y = \frac{2}{3}x.$$

Differentiating the equation $y = \frac{2}{3}x$ with respect to time, we obtain

$$\frac{dy}{dt} = \frac{2}{3}\frac{dx}{dt}.$$

Since $\dfrac{dx}{dt} = 5$ we have $\dfrac{dy}{dt} = \dfrac{2}{3} \cdot 5 = \dfrac{10}{3}$.

Thus the man's shadow is always lengthening at a rate of $\frac{10}{3}$ ft./sec.

---

**Problem 2** A balloon is being filled with air at the rate of 2 in.³/min.
a) At what rate is the radius increasing after 18 minutes?
b) At what rate is the surface area increasing after 18 minutes?

*Answer:* a) $\dfrac{1}{18\pi^{1/3}}$ b) $\dfrac{4}{3}\pi^{1/3}$

---

## C. Exercises

**B1.** The radius of a circle begins to increase at a rate of 2 inches per minute.
   a) At what rate is the circumference increasing?
   b) At what rate is the area increasing when the radius is 6 inches?

**B2.** Water is pouring into a cylindrical bowl of height 10 feet and radius 3 feet at a rate of 5 cubic feet per minute.
   a) At what rate does the surface of the water rise?
   b) How long does it take to fill the bowl?

**B3.** A cone of height 10 inches and radius 3 inches is being filled with water at the rate of 2 cubic inches per second. At what rate is the surface of the water rising when the depth of the water is 6 inches?

**B4.** The slope of a line passing through the point $(6, 0)$ is increasing at the rate of $\frac{1}{3}$ inch per minute. At what rate is the $y$-intercept changing when the slope is 53?

**B5.** A point moves along the curve $6y = x^2$ in such a way that when $x = 8$ the $x$-coordinate is increasing at a rate of 3 feet per second. At what rate is the $y$-coordinate changing at that time?

**B6.** The side of an equilateral triangle increases at the rate of 3 inches per second. At what rate is the area increasing when the side is 10 inches long?

**B7.** In each case below, $x$ and $y$ are functions of time $t$. For each, find $\frac{dy}{dt}$ for the given conditions.
   a) $y = 3x^2$, when $\frac{dx}{dt} = 6$ and $x = 4$.
   b) $y^5 = x^2 + x + 2$, when $\frac{dx}{dt} = 5$ and $x = 5$.

**D8.** Consider the accompanying figure.

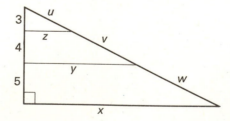

Express
a) $y$ in terms of $x$       e) $w$ in terms of $v$
b) $z$ in terms of $y$       f) $u$ in terms of $x$
c) $u$ in terms of $z$       g) $w$ in terms of $u$
d) $v$ in terms of $u$       h) $w$ in terms of $x$

a) $y = \frac{7x}{12}$   b) $z = \frac{3y}{7}$   c) $u = \sqrt{z^2 + 9}$
d) $v = \frac{4u}{3}$   e) $w = \frac{5v}{4}$   f) $u = \frac{1}{4}\sqrt{x^2 + 144}$
g) $w = \frac{5u}{3}$   h) $w = \frac{5}{12}\sqrt{x^2 + 144}$

**D9.** Referring to Exercise D7, suppose that $x$ increases at the rate of 3 inches per hour. When $x = 5$, find

a) $y$ and $\frac{dy}{dt}$        d) $v$ and $\frac{dv}{dt}$

b) $z$ and $\frac{dz}{dt}$        e) $w$ and $\frac{dw}{dt}$

c) $u$ and $\frac{du}{dt}$

a) $y = \frac{35}{12}, \frac{dy}{dt} = \frac{7}{4}$        b) $z = \frac{5}{4}, \frac{dz}{dt} = \frac{3}{4}$

c) $u = \frac{13}{4}, \frac{du}{dt} = \frac{15}{52}$        d) $v = \frac{13}{3}, \frac{dv}{dt} = \frac{5}{13}$

e) $w = \frac{65}{12}, \frac{dw}{dt} = \frac{25}{52}$

**D10.** In each case below, $x$ and $y$ are functions of time $t$. For each, find $\frac{dy}{dt}$ for the given conditions.

a) $y = 3x$,   when   $\frac{dx}{dt} = 6$.

b) $y = 3x^2$,   when   $\frac{dx}{dt} = 6$   and   $x = 4$.

c) $x^2 + y^2 = 4$,   when   $\frac{dx}{dt} = 3$   and   $x = y$.

d) $y^3 = 4x^2$,   when   $\frac{dx}{dt} = 6$, $x = 4$,   and   $y = 4$.

e) $y^5 = x^2 + x + 2$,   when   $\frac{dx}{dt} = 5$   and   $x = 5$.

f) $y^3 = 8t^2$,   when   $t = 1$.

a) 18    b) 144    c) $-3$    d) 4    e) $\frac{11}{16}$    f) $\frac{4}{3}$

**M11.** Show that the rate of change of the area of a circle with respect to the radius is equal to the circumference.

**M12.** Show that the rate of change of the volume of a sphere with respect to the radius is equal to the surface area.

**M13.** A boy is flying a kite at a height of 90 feet. If the kite is directly above the boy at $t = 0$ and moves horizontally away from him at a rate of 3 feet per second, how fast is the string being pulled out when the kite is 150 feet from the boy?

**M14.** A right triangle has a hypotenuse of constant length 5 inches. One leg of the triangle increases at the rate of 2 inches per second. When it is 3 inches long, what is the rate at which the other leg is decreasing?

**M15.** Two stones are dropped from the edge of a cliff, one 10 seconds after the other. What is the rate at which the distance between them changes before the first one hits the ground?

**M16.** A point moves along the line $y = 60$ with constant velocity. When the point is 100 feet from $(0, 0)$ it is measured that it is moving away from the origin at a rate of 8 feet per second. What is the velocity of the point?

**M17.** Two particles, initially coincident, begin to move apart along lines making an angle of $60°$. Suppose the first

particle moves according to the formula $d_1 = 4t$ and the second according to the formula $d_2 = 3t^2$. At what rate is the distance between the particles changing when $t = 3$?

**M18.** A ladder 12 feet long is resting flat against a building. The bottom of the ladder is then pulled along the ground at the constant rate of 2 feet per minute.
   a) How fast is the height of the top of the ladder decreasing when $t = 2$?
   b) How fast is the height of the midpoint of the ladder decreasing when $t = 2$?
   c) When is the height of the top of the ladder decreasing as fast as the midpoint was decreasing in part (b)?

**M19.** A girl sitting on a pier 16 feet above the water drops a stone. When the stone hits the water a circular wave is formed whose radius expands at the rate of 1 foot per second.
   a) What is the rate at which the area of the circle is increasing when the radius is 10 feet?
   b) At what rate is the area of the circle increasing 10 seconds after the girl drops the stone?

**M20.** When air expands adiabatically (without loss of heat), the pressure $P$ and volume $V$ satisfy $PV^{1.4} = C$, a constant. At a certain point in time the volume is 32 in.$^3$ and the pressure is 50 lb./in.$^2$ and is increasing at the rate of 8.75 lb./in.$^2$/sec. At what rate is the volume changing at this instant?

**M21.** The intensity of illumination at a point $P$ due to a candle is inversely proportional to the square of the distance $s$ from $P$ to the candle. If the candle moves away from $P$ at a constant rate, show that the rate of change of intensity at $P$ is eight times as great when $s = 10$ feet as when $s = 20$ feet.

**M22.** Show that when liquid is added to or removed from a container, the rate at which the volume $V$ of water is changing is the product of the area $A$ of the liquid surface and the rate at which the depth $d$ of the water is changing, given that $A$ is a continuous function of $d$.

**M23.** If the relation between the electric resistance $R$ of a wire at $T$ degrees centigrade and the resistance $R_0$ of the same wire at zero degrees centigrade is $R = R_0(1 + aT + b^3\sqrt{T})$ where $a$ and $b$ are constants, what is the rate of change of $R$ with respect to $T$?

# 4.6 Sample Exams

## Sample Exam 1 (45–60 minutes)

1. Find the critical points of
   a) $f(x) = 2x^3 + 9x^2 - 24x + 9$
   b) $g(x) = \frac{x^2}{2} - \ln(1 + x^2)$

2. Sketch the graph of $y = \frac{x}{x+3}$ and discuss the $M_2I_2$ ACIDS features.

3. Find the equation of the tangent to the graph of
   $$y = x^3 - 3x^2 + 5x + 2$$
   that has the smallest possible slope.

4. A ball is thrown upward with an initial velocity of 64 ft./sec.
   a) When does the ball reach its maximum height?
   b) How high does the ball rise?
   c) What is the velocity of the ball just before it returns to the ground?

5. One leg of a right triangle begins to increase at the rate of 2 inches per minute while the other leg remains at 8 inches. How fast is the hypotenuse increasing when the first leg is 6 inches?

## Sample Exam 2 (45–60 minutes)

1. Find the local maxima and local minima of the function
   $$f(x) = x^4 - 2x^2 + 3.$$

2. Sketch the graph of $y = \frac{x^2 + 4}{x^2 - 4}$ and discuss the $M_2I_2$ ACIDS features.

3. A window with perimeter 100 inches is in the shape of a rectangle (see figure) surmounted by an equilateral triangle. Find the dimensions of the rectangle for which the window admits the most light.

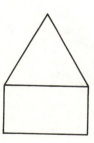

4. Find the regions where the graph of $y = \frac{x^2}{1 + x^2}$ is convex upward.

5. A conical paper cup is 4 inches high and its radius at the top is 2 inches. If water is poured into the cup at the rate of 1 in.$^3$/sec., how fast is the surface of the water rising after 6 seconds?

## Sample Exam 3 (45–60 minutes)

1. For $m_1, m_2, m_3 > 0$, show that the function
   $$f(x) = m_1(x - a_1)^2 + m_2(x - a_2)^2 + m_3(x - a_3)^2$$
   has a minimum when
   $$x = \frac{m_1a_1 + m_2a_2 + m_3a_3}{m_1 + m_2 + m_3}.$$

2. Find the maximum and minimum points and the points of inflection for each of the following functions.
   a) $y = x^3 - 6x^2$
   b) $y = 2x^3 + 3x^2 - 36x + 5$

3. Graph $y = x^3 + 6x^2$ and discuss the $M_2I_2$ ACIDS features.

4. Assume a particle moves on the $x$-axis according to the formula
   $$x = t^3 - 6t^2 + 9t + 5.$$
   Find
   a) the velocity when $t = 3$
   b) the acceleration when $t = 4$
   c) the times when the velocity is zero
   d) the place where the acceleration is zero
   e) the acceleration at each of the two moments when the particle is motionless.

5. From a tower 2000 feet high, a ball is thrown upward with a velocity of 64 ft./sec. at the same time a stone is dropped. At what rate is the distance between the ball and the stone changing when $t = 2$?

6. Find the base and altitude of the isosceles triangle of maximum area that can be inscribed in a circle of radius 12 inches. (See figure.)

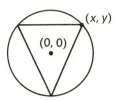

### Sample Exam 4 (45–60 minutes)

1. A rectangle with 16 inch perimeter is rotated about one of its sides to form a cylinder. What should the dimensions of the rectangle be in order to maximize the volume of the cylinder?

2. Show that a function of the form

$$f(x) = 2x^3 - 3x^2 + a^2x + b$$

where $a$ and $b$ are constants cannot vanish twice in the interval $[0, 1]$. (Hint: Use the Mean Value Theorem.)

3. A wire 17 inches long is cut into two pieces. One piece is bent to form a square and the other is bent to form a rectangle that is twice as wide as it is high. How should the wire be cut in order to minimize the total area of the square and rectangle?

4. Prove that if $a > 1$ then $1 + ax^2 < (1 + x^2)^a$ for all $x > 0$.

5. Consider the accompanying figure. If $x$ increases at the rate of 2 feet per hour, at what rate is $y$ increasing when $x = 6$ feet?

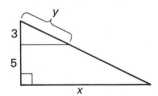

### Sample Exam 5 (Chapters 1, 2, 3, and 4, approximately 2 hours)

1. Let $f(x) = x - 1$. Find polynomials $p(x)$ and $q(x)$ of degree two such that
   a) $p(f(x)) = x^2 - 5x + 6$
   b) $f(q(x)) = x^2 - 5x + 6$

2. Find $\frac{dy}{dx}$ given
   a) $y = \tan^2 x + \tan x^2$
   b) $x = \ln(t^3 - 3t + 9), \quad y = t^3 - 3t + 9$
   c) $x = 3y + \sin y$

3. Consider the function $f(x) = \cos 2x + 2 \cos x$;
   a) find the critical points that lie in the interval $(0, 2\pi)$,
   b) sketch the graph of this function over the interval $(0, 2\pi)$, and
   c) label the various types of maxima and minima that occur.

4. The acceleration of a falling body on the moon is one-sixth that of a body on the earth. Suppose that from the top of a spaceship 576 feet high that is resting on the moon a rock is thrown upward with a velocity of 80 ft./sec.
   a) What is the equation of motion of the rock?
   b) What is the rock's maximum height?
   c) When does it hit the moon?

5. The base of a certain rectangle is increasing at the rate of 5 in./min. If the rectangle's height is 8 inches, how fast is the diagonal of the rectangle changing when the area is 48 in.²?

6. A triangle has vertices at $(1, 2)$, $(3, 4)$, and $(x, 0)$. Find the value of $x$ that minimizes the sum of the squares of the lengths of the sides.

7. Find two points on the graph of $x^2 - y^2 = 1$ that are nearest the point $(0, 6)$.

8. Find the equations of two lines through the origin that are tangent to the parabola $y = x^2 + 4$.

9. Find the 6th derivatives of
   a) $f(x) = xe^x$
   b) $g(x) = \ln(1 + 2x)$
   c) $h(x) = \sin 2x + \cos 3x$

### Sample Exam 6 (Chapters 1, 2, 3, and 4, approximately 3 hours)

1. Suppose $f(x) = \dfrac{1}{x^2 - 4}$ and $g(x) = \sqrt{\dfrac{1 + 4x}{x}}$.

   a) For what values of $x$ is $f(x)$ a real number?
   b) For what values of $x$ is $g(x)$ a real number?
   c) Show that $f(g(x)) = x$ for all $x > 0$.
   d) Show that $g(f(x)) = |x|$ for all $x \neq 2$.

2. Evaluate the following limits.

   a) $\lim\limits_{x \to 0} \dfrac{\sin x}{\sin 4x}$

   b) $\lim\limits_{x \to 2} \dfrac{x^2 + x - 6}{x^2 - 4}$

   c) $\lim\limits_{x \to \infty} \dfrac{3x^2 + 3x + 1}{3x^2 + 2}$

d) $\lim\limits_{x \to 0} \dfrac{(8 + x)^{1/3} - 8^{1/3}}{x}$

3. Given the equation    $y^3 = x^2 - x + 8$,
   a) find the equation of the tangent to its graph corresponding to the point $x = 0$, and
   b) find the equation of the normal to its graph corresponding to the point $x = 8$.

4. Find $a$, $b$, $c$, $d$, given that $f(x) = ax^3 + bx^2 + cx + d$ and $f(1) = 26$, $f'(0) = 6$, $f''(\frac{1}{2}) = 19$, and $f'''(18) = 6$.

5. Graph the equation $y = \dfrac{x + 1}{x}$ and discuss all the features summarized by the $M_2I_2$ ACIDS.

6. Show that the function $y = e^{3x} + \ln x$ satisfies the equation $x^2y'' - 3x^2y' + 3x + 1 = 0$.

7. Find $\frac{dy}{dx}$ given
   a) $e^x = e^y \sin x$
   b) $y = \sin(\sin x) + \sin x \sin x$
   c) $y = \text{Tan}^{-1} 3t^2$,   $x = \ln(1 + 9t^4)$

8. A rectangle is to have an area of 16 in.$^2$. Find its dimensions so that the distance from one corner to the midpoint of a nonadjacent side shall be a minimum.

9. The position of a particle moving on the $x$-axis at time $t$ is given by $x = 1 - \cos t$. The position of another particle moving on the $y$-axis is given by $y = \sin t$. Find the maximum distance between the two particles.

10. A swimming pool is 60 ft. long and 30 ft. wide. Its depth decreases uniformly from 20 ft. at one end to 4 ft. at the other end. If water is pumped into the pool at a rate of 10 ft.$^3$/min., how fast is the surface rising when the depth at the deepest end is 12 ft.?

11. The position of a particle moving on the $x$-axis at time $t$ is given by $x = t^2 - 4t$.
    a) When is the velocity of the particle equal to zero?
    b) Starting when $t = 0$, how far does the particle move before its velocity becomes zero?
    c) Where on the $x$-axis is the particle when its velocity is 8 ft./sec.?
    d) What is the acceleration when $t = 8$?

We will begin this chapter by looking at differentiation in reverse. This process, called antidifferentiation or integration, is intimately associated with the problem of finding area. The precise connection is expressed by the Fundamental Theorem of Calculus proven in the second section. We then study certain important techniques of integration (methods of finding antiderivatives). As will be seen, finding antiderivatives is not as easy as finding derivatives, and the techniques involved do not lend themselves to a systematic treatment. Nonetheless, the methods we will study cover the most important cases.

# Integration

## 5.1 Antidifferentiation

This section begins with the definition of an antiderivative and the introduction of some useful notations. A number of basic antidifferentiation formulas are then listed.

**A. The Indefinite Integral** Suppose $f$ is a given function. Then a function $F$ with domain equal to the domain of $f$ is called *an antiderivative* of $f$ if

$$F' = f$$

$$\left(\text{or in Leibniz notation } \frac{dF}{dx} = f\right).$$

For example, since

$$(\sin x)' = \cos x,$$

$F(x) = \sin x$ is an antiderivative of $f(x) = \cos x$.

The function $x^3$ is an antiderivative of $3x^2$ since

$$\frac{d}{dx}x^3 = 3x^2.$$

Now notice that if $C$ is any constant function, $x^3 + C$ is also an antiderivative of $3x^2$ since $\frac{d}{dx}(x^3 + C) = 3x^2$.

This fact is generalized in the following:

**THEOREM 1** *Suppose $f$ is a function defined on an interval $I$ and $F$ is an antiderivative of $f$. Then a function $G$ is also an antiderivative of $f$ if and only if it has the form*

$$G = F + C,$$

*where $C$ is a constant function.*

*Proof* If $G = F + C$, then $G' = (F + C)' = F' + C' = F' = f$, so $G$ is an antiderivative of $f$. Conversely, if $G$ is an antiderivative of $f$, then $G' = f = F'$. Recall that Corollary 2 (4.1C) states that two functions with the same derivative on an interval differ by a constant, so $G = F + C$ with $C$ a constant. ●

Suppose then $f$ is defined on an interval and $F$ is an antiderivative of $f$. By Theorem 1, we know that all antiderivatives of $f$ are of the form $F + C$, where $C$ is an arbitrary constant function. We therefore intro-

---

*Finding antiderivatives is harder, longer, and much more frustrating than finding derivatives, mainly because it obliges you to think "backwards" to differentiation, the latter being basically a forward process. Antidifferentiation often employs various nasty tricks which often can only be mastered by dedicated problem solving. Differentiation is a straightforward routine technique, but antidifferentiation is a nonlinear conglomeration of patience, foresight, and skill. Its mastery would be an artistic achievement.*

(Pat)

→ The words *primitive* and *integral* are synonyms for "antiderivative."

→ The equation $F' = f$ means $F'(x) = f(x)$ for all $x$. This means that $F$ must be differentiable but $f$ need not be.

→ Similarly, because $\frac{d}{dx}\text{Tan}^{-1}x = \frac{1}{1+x^2}$, $F(x) = \text{Tan}^{-1}x$ is an antiderivative of $f(x) = \frac{1}{1+x^2}$.

→ For example, $x^3 + \pi$, $x^3 - 4\sqrt{2}$, and $x^3 + 3019782$ are all antiderivatives of $3x^2$. In general, if a function has one antiderivative, it has infinitely many. This is in contrast to the fact that a function has at most one derivative.

→ The reason we require $f$ to be defined on an interval can be seen from the following: Suppose

$$f(x) = \frac{1}{x^2} \quad (x \neq 0).$$

(Recall that the set of nonzero reals ($x \neq 0$) is *not* an interval.) Define

$$F(x) = \frac{-1}{x} \quad (x \neq 0) \quad \text{and}$$

$$G(x) = \begin{cases} \frac{-1}{x} + 2 & (x > 0) \\ \frac{-1}{x} - 3 & (x < 0). \end{cases}$$

Then $G'(x) = F'(x) = f(x)$ for all $x \neq 0$, but $G - F$ is not a constant function.

duce the notation $\int f$ to denote $F + C$, the collection of all antiderivatives of $f$, and we write

$$\int f = F + C.$$

In the equation $\int f = F + C$ we call

$\int f$  the *indefinite integral of f*,

$C$  the *constant of integration*,
$f$  the *integrand*,  and

$\int$  the *integral sign*.

We will also use the notation

$$\int f(x)\, dx$$

in place of $\int f$. This will later prove very suggestive, and will help in remembering formulas.

**B. Basic Integration Formulas**  Each of the formulas given below can be immediately verified by showing that the derivative of the right-hand side of the equation equals the integrand on the left.

1. $\int x^r\, dx = \dfrac{x^{r+1}}{r+1} + C$   $(r \neq -1)$

2. $\int \cos x\, dx = \sin x + C$

3. $\int \sin x\, dx = -\cos x + C$

4. $\int e^x\, dx = e^x + C$

5. $\int \dfrac{dx}{1+x^2} = \text{Tan}^{-1}x + C$

6. $\int \dfrac{dx}{\sqrt{1-x^2}} = \text{Sin}^{-1}x + C$   $(-1 < x < 1)$

7. a) $\int \dfrac{dx}{x} = \ln x + C$   $(x > 0)$

   b) $\int \dfrac{dx}{x} = \ln(-x) + C$   $(x < 0)$

8. $\int (f+g) = \int f + \int g$

9. $\int cf = c\int f$   $(c$ a constant$)$

→ *When you see the expression $\int f$ you should ask yourself, "What must I differentiate to obtain f?"*

*(Beth)*

→ *When taking exams, be sure to remember to add the constant of integration to the indefinite integral formula, since it is common practice for the professor to take off points for failure to do this. For example*

$\int 3x^2 = x^3$  *is not correct, but*

$\int 3x^2 = x^3 + C$  *is correct.*

*(Lee)*

→ *The integral sign evolved from the summation sign $\Sigma$ (due to Euler) to $\mathcal{E}$, to $S$, and eventually to $\int$. In the next section we will see exactly why $\Sigma$ is an ancestor of $\int$, and in this respect the notation $\int f(x)\, dx$ will be especially helpful.*

→ *Be sure that the function $f(x) = x^r$ is defined on an interval. For example, if $f(x) = x^{-2}$, its domain must be an interval not containing zero.*

*The reason we have two formulas listed in (7) is that the logarithm is only defined for positive numbers. Equation (7b) is verified using the chain rule:*

$$\frac{d}{dx}(\ln(-x) + C) = \frac{1}{-x}\frac{d}{dx}(-x) = \frac{-1}{-x} = \frac{1}{x}.$$

*The formulas (7a) and (7b) are summarized by the single formula*

$$\int \frac{dx}{x} = \ln|x| + C$$

→ *where the function $f(x) = \frac{1}{x}$ is defined on either of the intervals $(-\infty, 0)$ or $(0, \infty)$, but not on both.*

→ *Since $\dfrac{d}{dx}\left(\int f + \int g\right) = \dfrac{d}{dx}\int f + \dfrac{d}{dx}\int g = f + g$*

→ *Since $\dfrac{d}{dx}c\int f = c\dfrac{d}{dx}\int f = cf$*

**False formula** $\int x f(x)\, dx = x \int f(x)\, dx$ (*Only constants* can be brought out from under the integral sign.)

**False formula** $\int f(x)g(x)\, dx = \int f(x)\, dx \int g(x)\, dx$ (The integral of a sum is the sum of the integrals (Formula 8), but the analogous formula for products is incorrect. See Problem 1.)

Using Formulas (8) amd (9), we may now integrate sums and constant multiples of the integrands given in Formulas (1) through (7). For example,

$$\int \left( x^2 - \frac{3}{x^3} + e^x - 2\sin x \right) dx$$

$$= \int x^2\, dx - 3 \int \frac{dx}{x^3} + \int e^x\, dx - 2 \int \sin x\, dx$$

$$= \frac{x^3}{3} + \frac{3}{2x^2} + e^x + 2\cos x + K.$$

The indefinite integral of any polynomial can now be found. Indeed, if

$$p(x) = a_0 + a_1 x + a_2 x^2 + \cdots + a_n x^n,$$

then

$$\int p(x)\, dx$$

$$= \int a_0\, dx + \int a_1 x\, dx + \int a_2 x^2\, dx + \cdots + \int a_n x^n\, dx$$

$$= a_0 x + \frac{a_1 x^2}{2} + \frac{a_2 x^3}{3} + \cdots + \frac{a_n x^{n+1}}{n+1} + C.$$

## C. Integrating the Chain Rule

**Example 1a (Easy)** Evaluate $\int (x^3 + 1)^6 3x^2\, dx$.

*Solution* Using the chain rule we have

$$\frac{d}{dx} \frac{(x^3 + 1)^7}{7} = (x^3 + 1)^6 3x^2.$$

It follows that

$$\int (x^3 + 1)^6 3x^2\, dx = \frac{(x^3 + 1)^7}{7} + C.$$

➡ For example, if $f(x) = x$, then

$$\int x f(x)\, dx = \int x^2\, dx = \frac{x^3}{3} + C,$$

but

$$x \int f(x)\, dx = x\left( \frac{x^2}{2} + C \right) \neq \frac{x^3}{3} + C.$$

---

**Problem 1** Show that it is true that the second "false formula" is indeed false by letting $f(x) = g(x) = x$.

---

➡ The four constants of integration, one from each of the four indefinite integrals, have been combined and replaced by the single constant $K$.

➡ Of course if the polynomial is not in standard form, then the integration might involve considerable computation. For example, to integrate $g(x) = (2 - 3x^2 + 7x^4)^{78}$, we must first multiply it out.

---

**Example 1b (Hard)** Find $\int (x^3 + 1)^6\, dx$.

*Solution* There is no way of evaluating this integral that avoids computation. One method is to write

$$(x^3 + 1)^6 = x^{18} + 6x^{15} + 15x^{12} + 20x^9$$
$$+ 15x^6 + 6x^3 + 1.$$

Now, integrating we obtain

$$\int (x^3 + 1)^6\, dx = \frac{x^{19}}{19} + \frac{3}{8} x^{16} + \frac{15}{13} x^{13} + 2x^{10}$$

$$+ \frac{15}{7} x^7 + \frac{3x^4}{2} + x + C.$$

*Example 2a (Still easy)*

$$\int (x^3 + 1)^{666} 3x^2 \, dx = \frac{(x^3 + 1)^{667}}{667} + C,$$

as can be seen by differentiating and using the chain rule.

→ *Example 2b (Absurd)*   Evaluate $\int (x^3 + 1)^{666} \, dx$.

We will not even attempt to solve this. One could expand $(x^3 + 1)^{666}$, using the binomial theorem, and then integrate. Fortunately, such problems do not occur very often in practice.

---

The difference between the examples on the left and right is the factor of $3x^2$. The importance of this factor is clarified by the more general discussion below.

Suppose $u = u(x)$ is a function of $x$. Using the chain rule, we have

$$\frac{d}{dx}\left(\frac{u^{n+1}}{n + 1}\right) = u^n \frac{du}{dx} \quad (n \neq -1).$$

It follows that

(\*) $$\int u^n \frac{du}{dx} \, dx = \frac{u^{n+1}}{n + 1} + C.$$

In Example 1, $u = x^3 + 1$ and the factor $3x^2$ equals $\frac{du}{dx}$. This factor is significant because it allows the integral $\int (x^3 + 1)^6 3x^2 \, dx$ to be cast in the form

$$\int u^6 \frac{du}{dx} \, dx.$$

This is easily evaluated by letting $n = 6$ in formula (\*). If in formula (\*) we interpret $dx$ as the differential of $x$ then the formula becomes

$$\int u^n \, du = \frac{u^{n+1}}{n + 1} + C$$

where $du = \frac{du}{dx} dx$ is the differential of $u$. In a similar way each basic integration formula in the paragraph above generalizes to a formula containing a function $u = u(x)$ and its differential $du = \frac{du}{dx} dx$ as shown below.

*Formulas*   (See also the back cover)

$$\int u^r \, du = \frac{u^{r+1}}{r + 1} + C \quad (r \neq -1)$$

$$\int \cos u \, du = \sin u + C$$

→   *This may not seem very difficult but, as the examples will soon show, it gets awfully hard sometimes to decide what u should be in order to agree with du. As the examples get more difficult, though, you will undergo alternating periods of frustration and elation as you struggle and finally solve the problem.*

                                  *(Ed)*

$$\int \sin u \, du = -\cos u + C$$

$$\int e^u \, du = e^u + C$$

$$\int \frac{du}{1 + u^2} = \text{Tan}^{-1} u + C$$

$$\int \frac{du}{\sqrt{1 - u^2}} = \text{Sin}^{-1} u + C \quad (-1 < u < 1)$$

$$\int \frac{du}{u} = \ln |u| + C$$

where $u \in (0, \infty)$ or $u \in (-\infty, 0)$ but not both.

Whenever an integral has one of the above forms, it is easily evaluated. One looks for a composite function $f(u(x))$ in the integrand and then tries the substitutions $u = u(x)$ and $du = (du/dx) \, dx$.

---

**Example 3**   Evaluate $\int (x^2 + a^2)^5 4x \, dx$.

*Solution*   Let $u = x^2 + a^2$. Then $du = (du/dx) \, dx = 2x \, dx$. The term $4x$ appears under the integral sign and we will write it as $2 \cdot 2x$ and bring the 2 outside the integral sign. Hence

$$\int (x^2 + a^2)^5 4x \, dx = 2 \int (x^2 + a^2)^5 2x \, dx$$

$$= 2 \int u^5 \, du = 2 \left( \frac{u^6}{6} + C \right)$$

$$= \frac{u^6}{3} + K = \frac{(x^2 + a^2)^6}{3} + K \qquad \longrightarrow \qquad \text{In this step we merely replaced the constant } 2C \text{ with the letter } K.$$

---

**Example 4**   Evaluate $\int \dfrac{dx}{\sqrt{1 - 3x^2}}$.

*Solution*   Let $u = x\sqrt{3}$, so $du = \dfrac{du}{dx} dx = \sqrt{3} \, dx$.

Then

$$\int \frac{dx}{\sqrt{1 - 3x^2}} = \int \frac{1}{\sqrt{1 - u^2}} \frac{du}{\sqrt{3}}$$

$$= \frac{1}{\sqrt{3}} \int \frac{1}{\sqrt{1 - u^2}} \, du$$

**Example 5**   Evaluate $\int \left( \dfrac{e^{\sqrt{x}} - 3}{\sqrt{x}} \right) dx$.

*Solution*   Let $u = \sqrt{x}$ so $du = \dfrac{du}{dx} dx = \dfrac{1}{2\sqrt{x}} dx$ and $dx = 2\sqrt{x} \, du$. Then

$$\int \frac{(e^{\sqrt{x}} - 3)}{\sqrt{x}} dx = \int \frac{(e^u - 3)}{u} 2\sqrt{x} \, du$$

$$= \int \frac{(e^u - 3)}{u} 2u \, du$$

$$= \frac{1}{\sqrt{3}} \operatorname{Sin}^{-1} u + K$$

$$= \frac{\operatorname{Sin}^{-1} x \sqrt{3}}{\sqrt{3}} + K.$$

$$= 2 \int (e^u - 3)\, du$$

$$= 2e^u - 6u + C$$

$$= 2e^{\sqrt{x}} - 6\sqrt{x} + C.$$

**D. Integrating the Trigonometric Functions** The integrals of $\sin x$ and $\cos x$ are already known to us, and those of the remaining four trigonometric functions all involve the logarithm.

*Example 6*   Show that

$$\int \tan x\, dx = -\ln |\cos x| + C \quad \text{or} \quad \ln |\sec x| + C.$$

*Solution*   $\int \tan x\, dx = \int \dfrac{\sin x}{\cos x}\, dx.$

Letting $u = \cos x$, we have $du = -\sin x\, dx$, so

$$\int \frac{\sin x\, dx}{\cos x} = -\int \frac{du}{u} = -\ln |u| + C$$
$$= -\ln |\cos x| + C.$$

Since $|\cos x| = |\sec x|^{-1}$, we have

$$-\ln |\cos x| = -\ln |\sec x|^{-1} = \ln |\sec x|.$$

*Example 7*   Prove that

$$\int \sec x\, dx = \ln |\sec x + \tan x| + C.$$

*Solution*   $\int \sec x\, dx = \int \sec x \left( \dfrac{\sec x + \tan x}{\sec x + \tan x} \right) dx$

$$= \int \frac{\sec x \tan x + \sec^2 x}{\sec x + \tan x}\, dx.$$

Now let $u = \sec x + \tan x$, so $du = (\sec x \tan x + \sec^2 x)\, dx$ and the integral becomes

$$\int \frac{du}{u} = \ln |u| + C = \ln |\sec x + \tan x| + C.$$

The formulas for the integrals of the trigonometric functions are summarized on the back cover.

*Problem 2*   Show that $\int \cot x\, dx = \ln |\sin x| + C.$

    →   *Here, Mr. Spivak, is one of the tricks* $\left(\text{letting } 1 = \dfrac{\sec x + \tan x}{\sec x + \tan x}\right)$ *that I was talking about in 2.5B.*
                             *(Robin)*

*Problem 3*   Prove that

$$\int \csc x\, dx = -\ln |\csc x + \cot x| + C.$$

## E. Exercises

Find the integrals given in Exercises B1–B7.

**B1. a)** $\int (x^2 + x^{-2} + 2x)\, dx$     **c)** $\int \frac{x^3 - 1}{x - 1}\, dx$

  **b)** $\int (2x + 3)^3\, dx$     **d)** $\int x(x + 3)\, dx$

**B2. a)** $\int (x^4 + 1)^8 x^3\, dx$     **b)** $\int \frac{(x^{-2} + 1)^{3/2}}{x^3}\, dx$

**B3. a)** $\int \frac{1}{x} \cos (\ln x)\, dx$     **b)** $\int \frac{1}{x} \sin (\ln x^2)\, dx$

**B4. a)** $\int e^{\tan 2x} \sec^2 2x\, dx$     **b)** $\int \frac{e^{\mathrm{Tan}^{-1} 2x}}{1 + 4x^2}\, dx$

**B5. a)** $\int \frac{x^2}{2x^3 - 4}\, dx$     **b)** $\int \frac{\ln x}{x}\, dx$

**B6. a)** $\int \frac{e^x}{1 + 4e^{2x}}\, dx$     **b)** $\int \frac{dx}{e^x + e^{-x}}$

**B7.** $\int x^2 \sin x^3 \cos x^3 (1 + \sin^2 x^3)^3\, dx$

**D8.** Find

  **a)** $\int 4x^3\, dx$     **d)** $\int 3e^x\, dx$

  **b)** $\int (\cos x - \sin x)\, dx$     **e)** $\int 2e^{\ln x}\, dx$

  **c)** $\int (\sec^2 x + \sec x \tan x)\, dx$     **f)** $\int \frac{7}{\sqrt{1 - x^2}}\, dx$

  a) $x^4 + C$     b) $\sin x + \cos x + C$
  c) $\tan x + \sec x + C$     d) $3e^x + C$
  e) $x^2 + C$     f) $7\,\mathrm{Sin}^{-1} x + C$

**D9.** Determine the constant $A$ that in each case makes the given equation valid, by differentiating the right side of the equation and comparing it with the integrand on the left.

  **a)** $A \int x^2 (1 + x^3)^7\, dx = (1 + x^3)^8 + C$

  **b)** $A \int \sin 2x \cos 2x\, dx = \sin^2 2x + C$

  **c)** $A \int \frac{dx}{x^2 + 9} = \mathrm{Tan}^{-1} \frac{x}{3} + C$

  **d)** $A \int \left(\frac{1}{x} + x^2\right) dx = \ln (x^3 e^{x^3}) + C$

  a) $A = 24$     b) $A = 4$     c) $A = 3$     d) $A = 3$

**D10.** Evaluate the following integrals.

  **a)** $\int (2 + 5x^3)^4 15x^2\, dx$

  **b)** $\int (1 + \sin x)^6 \cos x\, dx$

  **c)** $\int \tan^8 x \sec^2 x\, dx$

  **d)** $\int \frac{e^x}{1 + e^{2x}}\, dx$

  **e)** $\int 5x^4 \cos x^5\, dx$

  **f)** $\int -2e^{-2x} \sin e^{-2x}\, dx$

  **g)** $\int \frac{2x}{x^2 + 1}\, dx$

  **h)** $\int \frac{2x}{(x^2 + 1)^3}\, dx$

  a) $\frac{1}{5}(2 + 5x^3)^5 + C$     b) $\frac{1}{7}(1 + \sin x)^7 + C$
  c) $\frac{1}{9} \tan^9 x + C$     d) $\mathrm{Tan}^{-1} e^x + C$
  e) $\sin x^5 + C$     f) $-\cos (e^{-2x}) + C$
  g) $\ln (x^2 + 1) + C$     h) $\frac{-1}{2(x^2 + 1)^2} + C$

**M11.** Given the integral $\int \sin x \cos x\, dx$, we obtain

$$\int \sin x \cos x\, dx = \frac{\sin^2 x}{2} + C \text{ (letting } u = \cos x)$$

and

$$\int \sin x \cos x\, dx = -\frac{\cos^2 x}{2} + C \text{ (letting } u = \sin x).$$

Equating these answers gives

$$\frac{\sin^2 x}{2} + C = -\frac{\cos^2 x}{2} + C,$$

or

$$\sin^2 x + \cos^2 x = 0.$$

Since $\sin^2 x + \cos^2 x = 1$ we thus have that $0 = 1$. Where is the fallacy?

**M12.** Show that if $f$ is a function for which $f^{(n)}(x) = 0$ for some positive integer $n$ and all real numbers $x$, then $f$ is a polynomial.

**M13.** Find $y$ as a function of $x$ such that the following equations are satisfied.

a) $\dfrac{dy}{dx} = x^2 + 1$     c) $\dfrac{dy}{dx} = \left(2x + \dfrac{1}{x}\right)^2, \quad x > 0$

b) $\dfrac{dy}{dx} = (2x + 1)^3$

**M14.** Find $\displaystyle\int -\dfrac{1}{x}(\sin(\ln x))e^{\cos(\ln x)}\cos(1 + e^{\cos(\ln x)})\,dx.$

**M15.** Evaluate

a) $\displaystyle\int \sin 2x\, e^{\sin^2 x}\,dx$

b) $\displaystyle\int \dfrac{\cos x}{1 + \sin^2 x}\,dx$

c) $\displaystyle\int \dfrac{\sin 2x}{(1 + \sin^2 x)^2}\,dx$

## 5.2   *The Definite Integral*

In this section we will show how area can be defined as a limit of certain sums. We then prove the Fundamental Theorem of Calculus, which expresses the relation between area and the process of antidifferentiation.

**A. Area as the Limit of a Sum**   Let $f$ be a function continuous on a closed interval $[a, b]$, $a < b$. Suppose also that $f \geq 0$ on the interval. See Figure 1. Divide

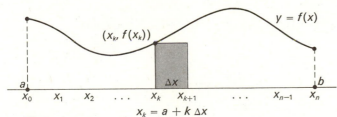

**Figure 1**

the interval into $n$ subintervals, each of length $\frac{b-a}{n} = \Delta x$ (read "delta $x$"). The endpoints of these subintervals are

$$a + 0\,\Delta x, \quad a + 1\,\Delta x, \quad a + 2\,\Delta x, \ldots, \quad a + n\,\Delta x$$

which we will label

$$x_0, \quad x_1, \quad x_2, \ldots, \quad x_n$$

⟶ Note that $x_0 = a$ and $x_n = b$.

respectively. The area of the rectangle shown in Figure 1 is clearly

$$f(x_k)\,\Delta x,$$

the base times the altitude. Now add the areas of all the $n$ rectangles (Fig. 2) to obtain

$$A_n = f(x_0)\,\Delta x + f(x_1)\,\Delta x + f(x_2)\,\Delta x \\ + \cdots + f(x_k)\,\Delta x + \cdots + f(x_{n-1})\,\Delta x,$$

or, in more convenient notation,

$$A_n = \sum_{k=0}^{n-1} f(x_k)\,\Delta x.$$

⟶   This is very common notation for finite sums such as $A_n$. Given $n$ numbers $a_1, a_2, \ldots, a_n$, we write

$$\sum_{k=1}^{n} a_k = a_1 + a_2 + \cdots + a_n.$$

The summation sign is read "sigma." In what follows, we will sometimes omit the limits of the sum and merely write $\Sigma a_k$ if no confusion will result. Below are some examples of the use of this notation.

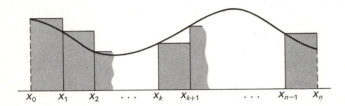

**Figure 2**

Notice now that if $n$ is large, then $\Delta x$ is small and there are many rectangles. In this case the sum $A_n$ is approximately equal to what one would consider intuitively to be the area $A$ between the graph of $f$, the $x$-axis, and the lines $x = a$ and $x = b$. In fact it appears that

$$A_n \to A \quad \text{as } n \to \infty \quad \text{(Fig. 3a, 3b, 3c)}$$

or, writing this out, that

$$\sum_{k=0}^{n-1} f(x_k)\,\Delta x \to A \quad \text{as } n \to \infty.$$

*Example 1*

a) $\displaystyle\sum_{i=1}^{4} b_i = b_1 + b_2 + b_3 + b_4$

b) $\displaystyle\sum_{j=0}^{3} a_j b_j{}^2 = a_0 b_0{}^2 + a_1 b_1{}^2 + a_2 b_2{}^2 + a_3 b_3{}^2$

c) $\displaystyle\sum_{k=1}^{n} d_k = \sum_{r=1}^{n} d_r = \sum_{i=1}^{n} d_i$

d) $\displaystyle\sum_{k=1}^{3} a_{2k} b_{2k+1} = a_2 b_3 + a_4 b_5 + a_6 b_7$

e) $\displaystyle\sum_{k=1}^{\infty} \frac{1}{k^2} = 1 + \frac{1}{2^2} + \frac{1}{3^2} + \cdots$

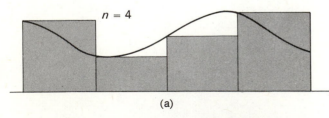

(a)

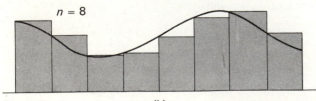

**Figure 3**                                    (b)

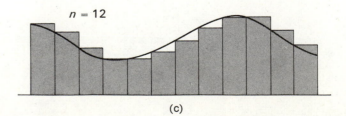

(c)

Since our notion of area under a general curve is only an intuitive one, we *define* the area $A$ under the curve $f$ between the points $a$ and $b$ to be the $\lim_{n \to \infty} A_n$, that is

$$A = \lim_{n \to \infty} \sum_{k=0}^{n-1} f(x_k)\,\Delta x.$$

→  In Chapter 9 a proof will be given that this limit exists. The proof will depend only on the continuity of $f$.

As an example of the above we will compute the area under the graph of the function $f(x) = 3x$ between $x = 0$ and $x = 2$. See Figure 4. We divide the interval $[0, 2]$ into $n$ equal subintervals. Then the length of each subinterval is

$$\Delta x = \frac{2-0}{n} = \frac{2}{n}$$

and the endpoints of the subintervals and altitudes of the rectangles are

$$x_0 = 0 + 0\,\Delta x = 0$$

$$x_1 = 0 + 1\,\Delta x = \frac{2}{n}$$

$$x_2 = 0 + 2\,\Delta x = 2\left(\frac{2}{n}\right)$$

$$x_3 = 0 + 3\,\Delta x = 3\left(\frac{2}{n}\right)$$

$$\vdots$$

$$x_{n-1} = 0 + (n-1)\,\Delta x = (n-1)\left(\frac{2}{n}\right)$$

and

$$f(x_0) = 3(0) = 0$$

$$f(x_1) = 3\left(\frac{2}{n}\right)$$

$$f(x_2) = 3(2)\left(\frac{2}{n}\right)$$

$$f(x_3) = 3(3)\left(\frac{2}{n}\right)$$

$$\vdots$$

$$f(x_{n-1}) = 3(n-1)\left(\frac{2}{n}\right)$$

respectively. The sum of the areas of the rectangles is

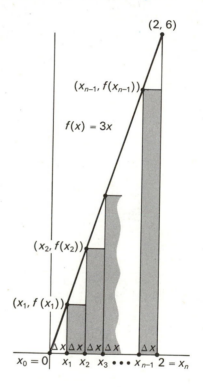

**Figure 4**

$$A_n := \sum_{k=0}^{n-1} f(x_k)\,\Delta x$$

$$= f(x_0)\,\Delta x + f(x_1)\,\Delta x + f(x_2)\,\Delta x$$
$$\qquad\qquad + f(x_3)\,\Delta x + \cdots + f(x_{n-1})\,\Delta x$$

$$= 0 + 3\left(\frac{2}{n}\right)\!\left(\frac{2}{n}\right) + 3(2)\left(\frac{2}{n}\right)\!\left(\frac{2}{n}\right)$$

$$\qquad + 3(3)\left(\frac{2}{n}\right)\!\left(\frac{2}{n}\right) + \cdots + 3(n-1)\left(\frac{2}{n}\right)\!\left(\frac{2}{n}\right)$$

$$= 3\left(\frac{2}{n}\right)^2 (1 + 2 + 3 + \cdots + (n-1)).$$

Now, the sum

$$1 + 2 + 3 + \cdots + (n-1) = \frac{(n-1)n}{2}.$$

Thus

$$A_n = 3\left(\frac{2}{n}\right)^2\left[\frac{(n-1)n}{2}\right] = \frac{6(n-1)}{n} = 6\left(1 - \frac{1}{n}\right).$$

And finally

$$A = \lim_{n\to\infty} A_n = \lim_{n\to\infty} 6\left(1 - \frac{1}{n}\right) = 6.$$

Observe that this agrees with the fact that the area of the triangle (Figure 4) with base equal to 2, the length of the interval $[0, 2]$, and altitude equal to $f(2) = 6$, is half the base times the altitude.

**B. The Definite Integral**  The sum $\sum_{k=0}^{n-1} f(x_k)\,\Delta x$ which was constructed for a nonnegative continuous function can be constructed for a continuous function in general and its limit as $n \to \infty$ has many applications in addition to calculating areas. Namely, let $f$ be any function continuous on a closed interval $[a, b]$, where $a < b$. See Figure 5. As before, we divide the interval $[a, b]$ into $n$ subintervals, each of length

$$\Delta x = \frac{b-a}{n}$$

and form the sum

$$S_n = f(x_0)\,\Delta x + f(x_1)\,\Delta x$$
$$\qquad\qquad + f(x_2)\,\Delta x + \cdots + f(x_{n-1})\,\Delta x$$

$$= \sum_{k=0}^{n-1} f(x_k)\,\Delta x$$

→ *Proof that* $1 + 2 + 3 + \cdots + (n-1) = \dfrac{(n-1)n}{2}$: We set

$$S_{n-1} = 1 + 2 + 3 + \cdots + (n-1).$$

Writing the sum backwards, gives

$$S_{n-1} = (n-1) + (n-2) + (n-3) + \cdots + 1.$$

Adding gives

$$2S_{n-1} = n + n + n + \cdots + n\ ((n-1)\ \text{times}) = (n-1)n.$$

Thus

$$S_{n-1} = 1 + 2 + 3 + \cdots + (n-1) = \frac{(n-1)n}{2}\quad\bullet$$

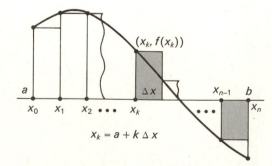

**Figure 5**

When $f(x_k) > 0$, the term $f(x_k)\,\Delta x$ is positive and equal to the area of the shaded rectangle in Figure 5. When $f(x_k) < 0$, the rectangle is below the x-axis and $f(x_k)\,\Delta x$ is the negative of the area of the rectangle. Thus $S_n$ equals the sum of the areas of the rectangles above the axis minus the sum of the areas of the rectangles below the axis.

where

$$x_0 = a + 0\,\Delta x, \quad x_1 = a + 1\,\Delta x,$$
$$x_2 = a + 2\,\Delta x, \ldots, \quad x_n = a + n\,\Delta x$$

are the endpoints of the subintervals.

Now, the limit of $S_n$ as $n$ approaches infinity is called the *definite integral of f from a to b* and is denoted by

$$\int_a^b f(x)\,dx = \lim_{n\to\infty} S_n = \lim_{n\to\infty} \sum_{k=0}^{n-1} f(x_k)\,\Delta x.$$

→ Observe the suggestive notation for the definite integral. The symbol $dx$ is used to represent $\Delta x$, the S-shaped symbol $\int$ is used in place of the summation sign $\Sigma$, and $a$ and $b$ replace the limits of the sum.

The function $f$ is called the *integrand* of the integral and the numbers $a$ and $b$ are called the *lower and upper limits* of the integral respectively. As in the case of the indefinite integral, we will often use $\int_a^b f$ in place of $\int_a^b f(x)\,dx$.

The following two theorems exhibit important and useful properties of the definite integral.

**THEOREM 1** *Let f be continuous on the closed interval [a, b] and let m and M be the absolute maximum and absolute minimum of f on [a, b] respectively. Then*

→ Recall that the Extreme Value Theorem states that every continuous function on a closed interval has an absolute maximum and absolute minimum.

$$m(b - a) \le \int_a^b f \le M(b - a).$$

→ Observe (Fig. 6) that if $f \ge 0$, the theorem simply states that the area under $f$ between $a$ and $b$, which is the integral $\int_a^b f$, is less than or equal to the area of the large rectangle with base equal to the length of $[a, b]$ and altitude equal to $M$, that is $M(b - a)$, and greater than or equal to the area of the smaller rectangle with the same base and with altitude equal to $m$.

*Proof:* Consider a subdivision of the interval $[a, b]$ into $n$ equal subintervals of length $\Delta x = \frac{b-a}{n}$ and having endpoints $x_k = a + k\,\Delta x$, $k = 0, 1, \ldots, n$. See Figure 6. For each $k$

$$m \le f(x_k) \le M.$$

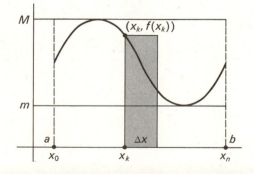

**Figure 6**

Since $\Delta x > 0$, also

$$m\,\Delta x \le f(x_k)\,\Delta x \le M\,\Delta x.$$

Adding over the $n$ subintervals gives

$$n(m\,\Delta x) \leq \sum_{k=0}^{n-1} f(x_k)\,\Delta x \leq n(M\,\Delta x).$$

Since $\Delta x = \frac{b-a}{n}$, we have

$$m(b-a) \leq \sum_{k=0}^{n-1} f(x_k)\,\Delta x \leq M(b-a).$$

Since this is true for each $n$, it follows that

$$m(b-a) \leq \lim_{n\to\infty}\sum_{k=0}^{n-1} f(x_k)\,\Delta x \leq M(b-a)$$

or

$$m(b-a) \leq \int_a^b f \leq M(b-a). \quad \bullet$$

The following important property of the definite integral, although geometrically obvious, is rather difficult to prove. The proof is given in Chapter 9.

**THEOREM 2**   *Let $f$ be a continuous function on $[a, c]$ and let $b \in (a, c)$. Then*

$$\int_a^b f + \int_b^c f = \int_a^c f.$$

See Figure 7.

Recall that the integral $\int_a^b f$ was defined for $a < b$. Actually the definition can be extended to include the case when $a = b$. In this case, for all $n$, $\Delta x = \frac{b-a}{n} = 0$. Hence $S_n = \sum_{k=0}^{n-1} f(x_k)\,\Delta x = 0$. And so

$$\int_a^a f = \lim_{n\to\infty} S_n = 0.$$

It is also convenient to have the integral $\int_a^b f$ defined for $a > b$. This we do by defining, for $a > b$,

$$\int_a^b f = -\int_b^a f.$$

With the above definition it can be shown that Theorem 2 is valid, regardless of the order of the numbers $a$, $b$ and $c$. For example, $\int_2^{-6} + \int_{-6}^{-4} = \int_2^{-4}$, since

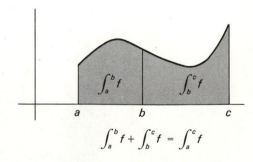

**Figure 7**

$$\int_{2}^{-6} + \int_{-6}^{-4} = -\int_{-6}^{2} + \int_{-6}^{-4}$$

$$= -\left[\int_{-6}^{-4} + \int_{-4}^{2}\right] + \int_{-6}^{-4} = -\int_{-4}^{2} = \int_{2}^{-4}.$$

Thus, regardless of the order of $a$, $b$, and $c$ and provided that $f$ is continuous on the largest interval involved, we have:

a) $\int_{a}^{a} f = 0$

b) $\int_{a}^{b} f = -\int_{b}^{a} f$

c) $\int_{a}^{b} f + \int_{b}^{c} f = \int_{a}^{c} f$

**C. The Fundamental Theorem of Calculus**  The direct evaluation of a definite integral as a limit of a sum, even for a simple function, is often a long and tedious process. It turns out, however, that the definite integral is intimately related to the process of anti-differentiation. When an antiderivative of a function can be found, the problem of evaluating a definite integral of the function reduces to a simple calculation. In fact, it will be shown that if $F$ is an antiderivative of $f$, that is if $\int f = F + C$, then

$$\int_{a}^{b} f = F\Big|_{a}^{b} = F(b) - F(a).$$

Here the expression $F|_{a}^{b}$ is a common notation for the value of $F$ at $b$ minus the value of $F$ at $a$.

This striking result is called the Fundamental Theorem of Calculus. Before proving this theorem we will consider several examples that illustrate this result.

→  Notice that the number $F(b) - F(a)$ does not depend on the particular antiderivative $F$ of $f$. Indeed, if $G$ is another antiderivative, then by Theorem 1, $G = F + C$ where $C$ is a constant function. Hence

$$G(b) - G(a) = (F + C)(b) - (F + C)(a)$$
$$= F(b) + C(b) - F(a) - C(a)$$
$$= F(b) + C - F(a) - C$$
$$= F(b) - F(a).$$

---

*Example 2*

a) Since $\int 3x^2 \, dx = x^3 + C$,

$$\int_{-2}^{5} 3x^2 \, dx = x^3\Big|_{-2}^{5} = 5^3 - (-2)^3 = 133.$$

b) Since $\int \cos x \, dx = \sin x + C$,

$$\int_{0}^{\pi/2} \cos x \, dx = \sin x\Big|_{0}^{\pi/2} = \sin\frac{\pi}{2} - \sin 0 = 1.$$

→  It is also true that $\int 3x^2 \, dx = x^3 + 78 + C$. Hence

$$\int_{-2}^{5} 3x^2 \, dx = x^3 + 78\Big|_{-2}^{5} = (5^3 + 78) - ((-2)^3 + 78)$$
$$= 5^3 + 78 - (-2)^3 - 78 = 5^3 - (-2)^3 = 133.$$

This answer is, of course, the same as the first one.

c) Since $\int \dfrac{dx}{x} = \ln |x| + C,$

$$\int_{-1}^{-2} \frac{dx}{x} = \ln |x| \Big|_{-1}^{-2} = \ln |-2| - \ln |-1| = \ln 2.$$

→ We may *not* write

$$\int_{-1}^{2} \frac{dx}{x} = \ln |x| \Big|_{-1}^{2} = \ln |2| - \ln |-1| = \ln 2$$

because the function $f(x) = \frac{1}{x}$ $(x \neq 0)$ is not defined on any interval containing both $-1$ and 2.

---

**THEOREM 3** Fundamental Theorem of Calculus
*Let $f$ be a continuous function defined on an interval $I$ and let $F$ be an antiderivative of $f$, i.e. $\int f = F + C$. Then for any $a$ and $b$ in $I$,*

$$\int_{a}^{b} f = F(b) - F(a).$$

*Proof:* We consider the function $H$ defined for all $x$ in $I$ by the integral

$$H(x) = \int_{a}^{x} f.$$

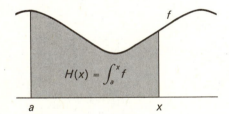

**Figure 8**

Observe in Figure 8 that if $f \geq 0$ and $a < x$ then $H(x)$ is the area under $f$ between $a$ and $x$. Now, we want to show that the function $H$ is differentiable on $I$, and in particular that its derivative is the function $f$, that is,

$$H'(x) = \frac{d}{dx} \int_{a}^{x} f = f(x).$$

Once this is established, the theorem itself will easily follow.

Keeping $x$ fixed, we consider the Newton quotient

$$\frac{H(x + h) - H(x)}{h} = \frac{1}{h} \left[ \int_{a}^{x+h} f - \int_{a}^{x} f \right].$$

Here $h$ is chosen small enough so that $x + h$ is in $I$. Since

We note that this result is very significant in itself, for it states that every continuous function $f$ defined on an interval in fact *has* an antiderivative and that this antiderivative can be obtained in terms of a definite integral of the function. For example, a function that frequently occurs in statistics is $f(x) = e^{-x^2}$. One can not find an antiderivative of this function in terms of our known functions. However, since this function is continuous, we know that such an antiderivative exists and can be given by $H(x) = \int_{0}^{x} e^{-t^2} \, dt$. Numerical integration methods can then be (and are being) used to provide us with tables of this function to any desired accuracy.

As another example of this result, it is not uncommon to *define* the natural logarithmic function as the integral

$$\ln x = \int_{1}^{x} \frac{1}{t} \, dt \quad (x > 0)$$

and then define the exponential function $e^x$ as the inverse of $\ln x$. Although that is perhaps not as natural a way of defining these functions as is the way we have defined them, one immediately obtains the result that

$$\frac{d}{dx} \ln x = \frac{d}{dx} \int_{1}^{x} \frac{1}{t} \, dt = \frac{1}{x}.$$

From this, and

$$\ln 1 = \int_{1}^{1} \frac{1}{t} \, dt = 0$$

one can obtain all the properties of the logarithmic and exponential functions. See problems M18 and M19, page 192.

$$\int_a^{x+h} - \int_a^x = \int_x^a + \int_a^{x+h} = \int_x^{x+h}$$

it follows that

$$\frac{H(x+h) - H(x)}{h} = \frac{1}{h}\int_x^{x+h} f.$$

Now, suppose that $h > 0$. See Figure 9. Since $f$ is continuous on $[x, x+h]$, it follows from Theorem 1 that

$$mh \le \int_x^{x+h} f \le Mh$$

where $m$ and $M$ are the absolute minimum and absolute maximum of $f$ on $[x, x+h]$ respectively. Dividing by $h$ gives

$$m \le \frac{1}{h}\int_x^{x+h} f \le M.$$

Thus the quantity $\frac{1}{h}\int_x^{x+h} f$, which is the Newton quotient of $H$ at $x$, is trapped between the absolute maximum and minimum of $f$ on $[x, x+h]$. The above inequalities are also valid when $h < 0$. In this case Theorem 1 gives

$$m(-h) \le \int_{x+h}^x f \le M(-h)$$

where $m$ and $M$ are the minimum and maximum respectively of $f$ on $[x+h, x]$. Dividing by $-h$, which we note is positive, gives

$$m \le \frac{1}{h}\int_x^{x+h} f \le M$$

as before.

Now, let $h \to 0$. Since $f$ is continuous at $x$, see Figure 9, the points on the graph of $f$ where the maximum and minimum are assumed approach the point $(x, f(x))$. Hence

$$m \to f(x) \quad \text{and} \quad M \to f(x) \quad \text{as } h \to 0.$$

Since the quantity $\frac{1}{h}\int_x^{x+h} f$ is squeezed between $m$ and $M$, it follows that

$$\frac{1}{h}\int_x^{x+h} f \to f(x) \quad \text{as } h \to 0$$

**Figure 9**

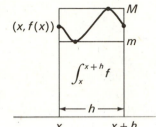

Proof that $M \to f(x)$ as $h \to 0$: Note that $M$ depends upon $h$. In particular, let $t_h$ denote a number in $[x, x+h]$ if $h > 0$ or $[x+h, x]$ if $h < 0$ at which $f$ takes on $M$, that is, for a given $h$, $f(t_h) = M$. Now, given an $\varepsilon > 0$ we want to find a $\delta > 0$ such that $|M - f(x)| < \varepsilon$ for $h \ne 0$ and satisfying $|h| < \delta$. Since $f$ is continuous at $x$, $\lim_{t \to x} f(t) = f(x)$. Thus for the given $\varepsilon > 0$ there is a $\delta^* > 0$ such that for $|t - x| < \delta^*$, we have $|f(t) - f(x)| < \varepsilon$. Now take $\delta = \delta^*$. Then for $|h| < \delta$, we have $|t_h - x| \le |h| < \delta = \delta^*$. Thus $|M - f(x)| = |f(t_h) - f(x)| < \varepsilon$. The proof that $m \to f(x)$ as $h \to 0$ is similar to the above and is left to the reader as an exercise.

and thus

$$H'(x) = \lim_{h \to 0} \frac{H(x + h) - H(x)}{h}$$

$$= \lim_{h \to 0} \frac{1}{h} \int_x^{x+h} f = f(x)$$

which is the required result.

The proof of the Fundamental Theorem now easily follows. For let $F$ be any antiderivative of $f$. Since $H$ is also an antiderivative of $f$, we have that for all $x$ in $I$,

$$H(x) = \int_a^x f = F(x) + C.$$

Now

$$H(a) = \int_a^a f = F(a) + C = 0.$$

Thus $C = -F(a)$. Substituting for $C$ above gives

$$H(x) = \int_a^x f = F(x) - F(a).$$

In particular, for $x = b$,

$$H(b) = \int_a^b f = F(b) - F(a) \quad \bullet$$

The following theorem gives us two important properties of the definite integral.

**THEOREM 4**   *Let f and g be continuous functions on a closed interval [a, b]. Then*

a) $\displaystyle\int_a^b (f + g) = \int_a^b f + \int_a^b g$

b) $\displaystyle\int_a^b kf = k \int_a^b f, \ k = constant.$

Now, suppose $f$ is continuous on $[a, b]$ and $f \le 0$ on $[a, b]$, as shown in Figure 10. Since $f \le 0$, the function $-f \ge 0$, as shown in Figure 11. Also it is intuitively clear that the area above $f$ between $a$ and $b$ is equal to the area below $-f$ between $a$ and $b$. Since $\int_a^b -f = -\int_a^b f$ (Theorem 4b), we are therefore led to define the area $A$ above $f$ between $a$ and $b$ to be equal to

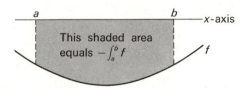

**Figure 10**

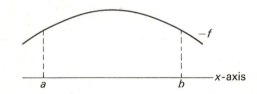

**Figure 11**

*Proof of (a):* Let $\int f = F + C_1$ and $\int g = G + C_2$. Then $\int (f + g) = F + G + C_3$ and so

$$\int_a^b (f + g) = (F + G) \Big|_a^b = (F(b) + G(b)) - (F(a) + G(a))$$

$$= (F(b) - F(a)) + (G(b) - G(a))$$

$$= F \Big|_a^b + G \Big|_a^b$$

$$= \int_a^b f + \int_a^b g.$$

$$A = -\int_a^b f.$$

Finally, suppose $f$ is a continuous function on $[a, b]$, as shown in Figure 12. For that part of the graph of $f$ above the axis we have

$$\int_a^c f = A_1 \quad \text{and} \quad \int_d^b f = A_3.$$

For that part of $f$ below the axis we have

$$-\int_c^d f = A_2 \quad \text{or} \quad \int_c^d f = -A_2.$$

It follows from Theorem 2 that

$$\int_a^b f = \int_a^c f + \int_c^d f + \int_d^b f = A_1 - A_2 + A_3.$$

This illustrates

**THEOREM 5**  *Suppose $f$ is a continuous function defined on a closed interval $[a, b]$. Then the definite integral*

$$\int_a^b f$$

*equals*
*the area of the region below the graph of $f$ and above the x-axis between the lines $x = a$ and $x = b$*
*minus*
*the area of the region above the graph of $f$ and below the x-axis between the lines $x = a$ and $x = b$.*

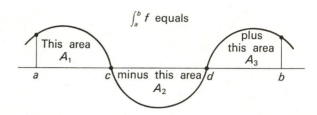

$\int_a^b f$ equals

This area $A_1$  minus this area $A_2$  plus this area $A_3$

**Figure 12**

We prove this result only in the above special case because the situation can become involved. For example, $f$ could have infinitely many regions above and below the x-axis, as does the function

$$f(x) = \begin{cases} x \sin \dfrac{1}{x} & (0 < x \le 1) \\ 0 & (x = 0) \quad \text{(Fig. 13)}. \end{cases}$$

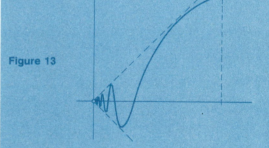

**Figure 13**

The theorem is true in this case, but the proof becomes complicated.

## The Invention of Calculus

The history of calculus is much more complex than that of analytic geometry, for the subject was widely anticipated and surrounded with controversy. Archimedes (287–212 B.C.) used a limiting procedure to find the area of certain regions, thereby anticipating integration by almost two thousand years. Later Kepler, using a crude form of the derivative, was able to solve a variety of maximum-minimum problems. Fermat, because he virtually wrote down the definition of the derivative, was credited by his countryman Lagrange with the creation of calculus. More significantly, Isaac Barrow (1630–1677), Newton's teacher, proved a theorem relating the problem of tangents to the problem of areas. Phrased in terms of derivatives and integrals, it later became known as the Fundamental Theorem of Calculus.

Despite the fact that so many men anticipated certain aspects of calculus, Isaac Newton and Gottfried Wilhelm von Leibniz are credited with its creation. Both men recognized that the current ideas concerning tangents, areas, limits and infinity were the genesis of something completely new. The Englishman and the German then fashioned them into the most powerful instrument science has ever known. Calculus not only changed the course of mathematics; the application of it revolutionized science.

### Isaac Newton (1642–1727)

Isaac Newton almost died at birth on Christmas Day in 1642, and is now considered by many historians to be the greatest genius of all time. He was supreme as a pure mathematician, applied mathematician, theoretical physicist, and experimental physicist. His father was a farmer and his childhood was spent on an isolated farm. Newton had few friends and spent much of his time inventing his own toys to play with. Although his mother wished him to become a farmer, he had no interest in this pursuit and was allowed to prepare for college.

Newton entered Trinity College at 18 and before he was 23 had proved the binomial theorem, worked out the fundamental ideas for calculus, revolutionized physical optics, and begun his considerations on gravity that were used to rewrite celestial mechanics. He was reluctant to publish his results, a lifelong characteristic, but his genius was recognized by his teacher Isaac Barrow.

When Newton was 26, Barrow resigned his professorship in favor of this remarkable student. The young scholar became the model for the absent-minded professor, often forgetting to eat or sleep and at times lecturing to empty classrooms.

In 1864 the astronomer Edmund Halley visited Newton, inquiring as to what curves the planets should describe in orbiting the sun. Newton informed Halley that he had long ago shown that the orbit had to be an ellipse and the excited Halley immediately wished to see the computations. Newton could not find his notes, which contained, among other things, the universal law of gravitation. With Halley's constant urging, Newton finally did write up notes in detail and they were published at Halley's expense. This work, known as the *Philosophiae Naturalis Principia Mathematica,* is considered to be the greatest contribution to science ever made by a single man.

At about the age of fifty, Newton's interests changed and he dropped out of the scientific community. His intellectual powers remained and his new interests were as varied as they were original. He reformed the British currency system, and intensely studied alchemy, occult philosophy, and theology. A great deal of his time was spent in studying the prophecies of the book of Daniel, and he considered his writings in theology to be far more important than his contributions to science and mathematics. Man became the focal point of his studies which included investigations concerning man's creation, his free will, and his ultimate destiny. To the orthodox scientific community Newton now appeared a bit cracked. He became somewhat paranoid and wrote bitter letters to friends, protesting wrongs that he imagined were being perpetrated against him. He was particularly angry with the philosopher John Locke who, Newton felt, ''attempted to embroil him with women.''

Newton's interests in mathematics never quite entirely died. In 1697 when Newton was 55, the mathematician Johann Bernoulli publicly challenged other mathematicians to solve an important problem whose solution he already knew, but had not published. Newton solved the problem in five hours and published the solution anonymously in the newspaper the following day. Bernoulli had no doubts about the author of this work and remarked simply, ''I can tell the lion by the imprint of his paw.'' The solution to the problem (called the *brachistochrone* problem) entailed finding a function that would minimize a certain expression. It initiated the subject of calculus of variations (calculus in infinite dimensional spaces), a branch of mathematics that is still actively studied today.

Near the end of his life, Newton offered his own estimation of himself:

I do not know what others may think of me; but to myself I seem to have been only like a boy playing on the seashore, and diverting myself in now and then finding a smoother pebble or a prettier shell than ordinary, whilst the great ocean of truth lay all undiscovered before me.

### Gottfried Wilhelm von Leibniz (1646–1716)

Leibniz had a highly intellectual childhood. His father was a philosopher, and Leibniz grew up surrounded by books. He showed his brilliance early and was reading at the age of five. By eight he knew Latin, and was composing verses in it at the age of twelve. When at age fifteen he entered the University of Leipzig, he was already bored with the classics and was passionately searching for some universal theory from which he could obtain all knowledge. His interests turned to philosophy, and by the age of seventeen he had studied and dismissed most of it.

Leibniz had an uncanny ability to rapidly absorb vast amounts of information, and his professors, jealous of his aptitude, denied him a law degree at the age of twenty on the pretense that he was too young. This setback did not damage his career, however, for he soon rose to eminence not only in law, but in history, literature, metaphysics, philosophy, and diplomacy. He excelled as well in mathematics, not only inventing the calculus, but contributing significantly to combinatorics, inventing a digital computing machine, and laying the foundations of mathematical logic.

This apparent diversity of interests sprang from Leibniz's desire to find

. . . a general method in which all truths of the reason would be reduced to a kind of calculation. At the same time this would be a sort of universal language or script, but infinitely different from all those projected hitherto; for the symbols and even the words in it would direct the reason; and errors, except for those of fact, would be mere mistakes in calculation.

At the time, Leibniz thought that, with a few able assistants, he would be able to carry out this program in five years. It was finally carried out by Bertrand Russell and A. N. Whitehead more than 200 years later.

Leibniz's serious study of mathematics began in 1672 when he met the physicist and mathematician Huygens. The latter immediately recognized the young man's genius, and with his encouragement, Leibniz invented calculus within six years. His first publication on the subject, a six-page paper, was printed in 1684, and this date marks the beginning in Europe of an era of great activity in mathematical analysis.

However, this work on calculus was not immediately understood, and it was the Bernoulli brothers, Jacob and Johann, who in 1687 came to assist Leibniz in developing his theory. By 1700 they had worked out all of elementary analysis and had built the foundations for several other branches of mathematics.

Leibniz wasted a great deal of time because of his desire to mingle with royalty. He spent the better part of forty years as the family librarian for the Duke of Brunswick, doubling as historian and diplomatic courier, and his dealings in this area led him to invent the time-honored concept of "balance of power."

Although Leibniz's metaphysical philosophy was not revolutionary, it is remembered from an historical point of view. His theory of monads was a crude atomic theory and reflected his eternal search for the ultimate truth. Despite Leibniz's multilateral contributions to human thought, he died alone and virtually forgotten. Since his death his fame has spread as his ideas have become better understood: his symbolic logic was not understood until this century, and, as will be seen in Chapter 10, much of what he wrote about calculus was not understood until this decade. Of himself he said,

I have so many ideas that may perhaps be of some use in time if others more penetrating than I go deeply into them someday and join the beauty of their minds to the labor of mine.

And he had this to say about his rival:

Taking mathematics from the beginning of the world to the time of Newton, what he has done is much the better half.

There is no doubt that Newton found calculus before Leibniz. But Newton only created what he felt necessary for his investigations in physics and astronomy. Leibniz pursued the subject far more deeply than did Newton, but both men independently discovered for themselves all the essential ideas of calculus.

Rarely is it so easy to give equal credit to two men for the same creation. This division of honors was not, however, accepted in the seventeenth century. A great controversy erupted between the British and German schools, and what ensued was an unfortunate chapter in the history of mathematics. The British school decided that Leibniz had plagiarized everything from Newton. This chauvinism caused British mathematics to suffer, for they refused to adopt the vastly superior version of calculus with its suggestive notation that Leibniz had created. British mathematics floundered for the next 100 years while continental mathematicians flourished.

Today, the controversy between the schools of Leibniz and Newton appears absurd. But the history of mathematics is punctuated with controversy. In fact, serious questions concerning the axioms and the validity of some methods of proof used in calculus have recently been raised. We will discuss some of these questions in Chapter 10.

## D. Exercises

**B1.** Evaluate the following definite integrals.

a) $\int_1^4 (x+4)^2 \, dx$

b) $\int_{\ln 3}^{\ln 8} 7e^x \, dx$

c) $\int_{-e^5}^{-e} \frac{dx}{x}$

d) $\int_{1/\sqrt 3}^1 \frac{6 \, dx}{1 + x^2}$

**B2.** Evaluate the following definite integrals.

a) $\int_0^1 e^{2 \ln (1+x)} \, dx$

b) $\int_{\pi/6}^{\pi/4} \sin x \cos x \csc x \sec x \, dx$

c) $\int_0^4 (9 + x^2)^{1/2} 2x \, dx$

d) $\int_0^{\sqrt \pi} 2x \sin x^2 \, dx$

**B3.** Find the area of the region in the plane bounded by the graph of $f(x) = 2x^{-3}$ and the lines $x = 1$, $x = 5$, and $y = 0$.

**B4.** Find the area of the region that lies above the $x$-axis and below the graph of $y = 4 - x^2$.

**B5.** Use the chain rule and the formula

$$\frac{d}{du} \int_0^u f(t)\, dt = f(u)$$

to prove that

$$\frac{d}{dx} \int_0^{g(x)} f(t)\, dt = f(g(x))g'(x).$$

**D6.** Evaluate the following definite integrals.

a) $\displaystyle\int_0^1 6x^5\, dx$

e) $\displaystyle\int_1^2 \frac{2\, dx}{x^2}$

b) $\displaystyle\int_0^{\pi/2} 4 \cos x\, dx$

f) $\displaystyle\int_{-1}^1 \ln\left(e^{3x^2 + 2x}\right) dx$

c) $\displaystyle\int_3^6 \frac{dx}{x}$

g) $\displaystyle\int_4^9 3\sqrt{x}\, dx$

d) $\displaystyle\int_{-6}^{-3} \frac{dx}{x}$

h) $\displaystyle\int_4^9 \frac{dx}{\sqrt{x}}$

a) 1    b) 4    c) $\ln 2$    d) $-\ln 2$    e) 1
f) 2    g) 38    h) 2

**D7.** Evaluate the expressions below.

a) $\displaystyle\int_0^x (2t + 3t^2)\, dt$

e) $\displaystyle\int_0^{x^2} (2t + 3t^2)\, dt$

b) $\displaystyle\int_3^x (2t + 3t^2)\, dt$

f) $\displaystyle\int_x^{x^2} (2t + 3t^2)\, dt$

c) $\displaystyle\int_0^{-x} (2t + 3t^2)\, dt$

g) $\displaystyle\int_{x^2}^{x^2} (2t + 3t^2)\, dt$

d) $\displaystyle\int_x^0 (2t + 3t^2)\, dt$

h) $\displaystyle\frac{d}{dx} \int_0^x \sin(2t + 3t^2)\, dt$

a) $x^2 + x^3$       b) $x^2 + x^3 - 36$
c) $x^2 - x^3$       d) $-x^2 - x^3$
e) $x^4 + x^6$       f) $-x^2 - x^3 + x^4 + x^6$
g) 0                 h) $\sin(2x + 3x^2)$

**M8.** Evaluate the following definite integrals.

a) $\displaystyle\int_{-1}^2 |4 - x^2|\, dx$

b) $\displaystyle\int_3^4 |4 - x^2|\, dx$

c) $\displaystyle\int_0^5 |4 - x^2|\, dx$

**M9.** Evaluate the following definite integrals.

a) $\displaystyle\int_{-\pi/4}^{\pi/3} |\sin x|\, dx$

b) $\displaystyle\int_{-\pi/4}^{\pi/3} \sin |x|\, dx$

**M10.** Differentiate the following functions.

a) $\displaystyle f(x) = \int_0^{x^2} \cos^4 t\, dt$

b) $\displaystyle g(x) = \int_{2x}^{3x} e^{-u^2}\, du$

c) $\displaystyle h(x) = \int_0^x \left(\int_0^u (t + \sin t)\, dt\right) du$

**M11.** Find the area of the region that lies under the curve $y = xe^{-x^2}$ between $x = 0$ and $x = A$ and above the $x$-axis. What happens to this area as $A \to \infty$?

**M12.** Find the area of the region bounded by the curves

$$y = 2x^2 - 5 \quad \text{and} \quad y = -x^2 + 7.$$

**M13.** Find $a$ and $b$ so that

$$\int_0^{3x} (at + b)\, dt = 9x^2 + 9x.$$

**M14.** Show that there is no function $f$ that satisfies the equation

$$\int_0^{2x} f(t)\, dt = x^4 + 6.$$

**M15.** Find $f$ given

$$\int_0^{x+5} f(t)\, dt = x^2 + 13x + 40.$$

**M16.** Solve the following equation for $a$.

$$\int_0^a e^{\cos x} \sin x\, dx = \int_0^{\pi/2} e^{\sin x} \cos x\, dx.$$

**M17.** Using the formula

$$\sum_{k=1}^n k^2 = \frac{n(n+1)(2n+1)}{6}$$

and the definition of the definite integral as a limit of a sum, show that

$$\int_0^1 x^2\, dx = \frac{1}{3}.$$

**M18.** Using only the definition $\ln x = \int_1^x \frac{1}{t}\,dt\ (x > 0)$ and hence that $\frac{d}{dx}\ln x = \frac{1}{x}$ and $\ln 1 = 0$, show that:
a) $\ln ab = \ln a + \ln b$ (Hint: Show that $\frac{d}{dx}\ln ax = \frac{1}{x}$.)
b) $\frac{d}{dx}e^x = e^x$, where $e^x$ is defined as the inverse of $\ln x$.
c) $e^a e^b = e^{a+b}$

**M19.** Using the definition $a^b = e^{b\ln a}(a > 0)$, show that:
a) $\hat{e}^x = e^x$, where $\hat{e} = e^1$
b) $\ln a^b = b\ln a$
c) $a^b a^c = a^{b+c}$
d) $(a^b)^c = a^{bc}$
e) $\frac{d}{dx}x^b = bx^{b-1}\quad (x > 0)$

---

## 5.3 Integration by Parts

The method of integration by parts is based on the Product Rule for differentiation. Not only is it an effective and versatile technique for integration, but it also has important theoretical applications.

→ *Most of us had more difficulty with integration by parts than with the chain rule technique. Don't feel bad if you have similar problems. Things will get better.*

*(Greg)*

**A. Integrating the Product Rule**  Suppose $u$ and $v$ are functions of $x$. Then

$$\int u\frac{dv}{dx}\,dx = uv - \int v\frac{du}{dx}\,dx$$

or more simply, using $du = \frac{du}{dx}\,dx$ and $dv = \frac{dv}{dx}\,dx$,

$$\int u\,dv = uv - \int v\,du.$$

For definite integration we have

$$\int_a^b u\frac{dv}{dx}\,dx = uv\Big|_a^b - \int_a^b v\frac{du}{dx}\,dx$$

or

$$\int_a^b u\,dv = uv\Big|_a^b - \int_a^b v\,du.$$

We will now use these formulas to evaluate a specific integral.

We will verify this formula using the product rule:

$$\frac{d}{dx}(uv) = \frac{du}{dx}v + u\frac{dv}{dx}.$$

→ Integrating both sides gives

$$uv = \int v\frac{du}{dx}\,dx + \int u\frac{dv}{dx}\,dx$$

or $\quad\int u\frac{dv}{dx}\,dx = uv - \int v\frac{du}{dx}\,dx.$

In our abbreviated notation, this is $\int u\,dv = uv - \int v\,du$.

→ This follows from the above formula by inserting the limits of integration.

### Method

1) Suppose we have an integral of the form

$$\int u\frac{dv}{dx}\,dx.$$

2) Compute $v$ and $\frac{du}{dx}\,dx$.

3) Provided we can evaluate $\int v\frac{du}{dx}\,dx$, the integration by parts formula gives

→ ### Illustration

1) Consider the integral $\int x\sin x\,dx$.

Letting $u = x$ and $\frac{dv}{dx}\,dx = \sin x\,dx$, observe that $\int x\sin x\,dx$ is in the form $\int u\frac{dv}{dx}\,dx$.

2) Since $u = x$, $\frac{du}{dx}\,dx = dx$, and since $\frac{dv}{dx}\,dx = \sin x\,dx$, we obtain by integration $v = -\cos x$.

3) $\int v\frac{du}{dx}\,dx = \int -\cos x\,dx = -\sin x + C$

4) $\int u \frac{dv}{dx} dx = uv - \int v \frac{du}{dx} dx$. Hence we have our integral expressed in terms of computable quantities.

5) A definite integral can be found directly from the above result.

## B. Examples

*Example 1*   Evaluate $\int x e^x \, dx$.

*Solution*   Let

$$u = x \quad \text{and} \quad dv = e^x \, dx.$$

Then

$$du = dx \quad \text{and} \quad v = e^x,$$

so

$$\int u \, dv = uv - \int v \, du = xe^x - \int e^x \, dx$$
$$= xe^x - e^x + C.$$

*Example 2*   Evaluate $\int \mathrm{Tan}^{-1} x \, dx$.

*Solution*   Let

$$u = \mathrm{Tan}^{-1} x \quad \text{and} \quad dv = 1 \cdot dx = dx$$

Then

$$du = \frac{1}{1 + x^2} dx \quad \text{and} \quad v = x.$$

Hence

$$\int u \, dv = uv - \int v \, du$$

$$= x \, \mathrm{Tan}^{-1} x - \int \frac{x \, dx}{x^2 + 1}$$

$$= x \, \mathrm{Tan}^{-1} x - \frac{1}{2} \int \frac{2x \, dx}{x^2 + 1}$$

$$= x \, \mathrm{Tan}^{-1} x - \frac{1}{2} \ln (x^2 + 1) + C.$$

## C. Some Complicated Examples
As will be seen below, it sometimes is necessary to integrate by parts more than once to solve a particular problem.

4) $\int x \sin x \, dx = -x \cos x + \sin x - C$

5) To evaluate, for example, $\int_0^{\pi/4} x \sin x \, dx$, we write

$$\int_0^{\pi/4} x \sin x \, dx = (-x \cos x + \sin x) \Big|_0^{\pi/4}$$

$$= \left( -\frac{\pi}{4} \cos \frac{\pi}{4} + \sin \frac{\pi}{4} \right) - (0)$$

$$= \frac{1}{\sqrt{2}} \left( 1 - \frac{\pi}{4} \right).$$

→   It is possible to transform the integral into a more complicated one by an indiscreet choice of $u$ and $dv$. For example, if in Example 1 we let $u = e^x$ and $dv = x \, dx$ we obtain $du = e^x \, dx$ and $v = x^2/2$. We then have

$$\int x e^x \, dx = \frac{x^2}{2} e^x - \int \frac{x^2 e^x}{2} dx.$$

This is a correct formula, but is of little use in evaluating the integral. In general, a first good choice for $u$ is the function which simplifies when differentiated, as for example, such functions as polynomials, $\ln x$, $\mathrm{Tan}^{-1} x$, and so on.

→   One might question our method of integrating $dv$. In Example 2 we had $dv = dx$ and we then wrote $v = x$, but we could have written $v = x + 2$ or $v = x + K$ for any constant $K$. This is true because if we replace $v$ by $v + C$ in the formula $\int u \, dv = uv - \int v \, du$, we obtain

$$\int u \, dv = u(v + C) - \int (v + C) \, du$$

$$= uv + uC - \int v \, du - \int C \, du$$

$$= uv - \int v \, du$$

and this is the original formula. Since $C$ can take any value, we generally let $C = 0$.

*Example 3* Find $\int x^2 e^x \, dx$.

*Solution* Let $u = x^2$ and $dv = e^x \, dx$. Then $du = 2x \, dx$ and $v = e^x$ and the Integration by Parts Formula gives

$$\int x^2 e^x \, dx = x^2 e^x - \int 2x e^x \, dx.$$

We must still evaluate $\int x e^x \, dx$. Letting $u = x$ and $dv = e^x \, dx$, we have $du = dx$ and $v = e^x$. Hence

$$\int x e^x \, dx = x e^x - e^x + C,$$

and substituting this into the previous equation gives

$$\int x^2 e^x \, dx = x^2 e^x - 2(x e^x - e^x + C)$$
$$= x^2 e^x - 2x e^x + 2e^x + K.$$

The method used in this example can be used to derive the formula:

$$\int x^n e^x \, dx = x^n e^x - n \int x^{n-1} e^x \, dx.$$

Letting $u = x^n$ and $dv = e^x \, dx$, we obtain $du = nx^{n-1} \, dx$ and $v = e^x$. The formula then follows using integration by parts.

The above equation is a typical example of what is called a *reduction formula*. Its repeated application will eventually lead to the solution of the problem.

Another method of applying the Integration by Parts Formula is illustrated in the following:

*Example 4* Show that

$$\int \cos^2 x \, dx = \frac{x}{2} + \frac{\sin x \cos x}{2} + C.$$

*Solution* Let

$$u = \cos x \quad \text{and} \quad dv = \cos x \, dx.$$

Then

$$du = -\sin x \, dx \quad \text{and} \quad v = \sin x.$$

By the Integration by Parts Formula we have

$$\int \cos^2 x \, dx = \cos x \sin x + \int \sin^2 x \, dx.$$

Now replace $\sin^2 x$ by $1 - \cos^2 x$. Then

$$\int \cos^2 x \, dx = \cos x \sin x + \int (1 - \cos^2 x) \, dx$$
$$= \cos x \sin x + x - \int \cos^2 x \, dx + C.$$

Notice that $\int \cos^2 x \, dx$ appears on both sides of the equals sign. Solving for $\int \cos^2 x \, dx$, we have

$$2 \int \cos^2 x \, dx = x + \cos x \sin x + C$$

This formula can also be derived by starting with the identity

$$\cos^2 x = \frac{1 + \cos 2x}{2}.$$

Hence

$$\int \cos^2 x \, dx = \frac{1}{2} \int (1 + \cos 2x) \, dx$$

$$= \frac{1}{2} \left( x + \frac{\sin 2x}{2} \right) + K = \frac{x}{2} + \frac{\sin 2x}{4} + K.$$

This answer can be put in the form of the answer given in Example 4 by using the identity

$$\sin 2x = 2 \sin x \cos x.$$

or

$$\int \cos^2 x \, dx = \frac{x}{2} + \frac{1}{2} \sin x \cos x + K.$$

---

The ideas of Examples 3 and 4 are combined in the following:

---

*Example 5*   Show that

$$\int e^{ax} \cos bx \, dx = \frac{e^{ax}(b \sin bx + a \cos bx)}{a^2 + b^2} + C.$$

*Solution*   Let

$$u = \cos bx \quad \text{and} \quad dv = e^{ax} \, dx.$$

Then

$$du = -b \sin bx \, dx \quad \text{and} \quad v = \frac{e^{ax}}{a}.$$

Hence

(\*)   $$\int e^{ax} \cos bx \, dx$$

$$= \frac{e^{ax}}{a} \cos bx + \frac{b}{a} \int (\sin bx) e^{ax} \, dx.$$

We now integrate the second integral by parts, letting

$$u = \sin bx \quad \text{and} \quad dv = e^{ax} \, dx.$$

Then

$$du = b \cos bx \, dx \quad \text{and} \quad v = \frac{e^{ax}}{a},$$

so we have

$$\int (\sin bx) e^{ax} \, dx = \frac{e^{ax}}{a} \sin bx - \frac{b}{a} \int (\cos bx) e^{ax} \, dx.$$

Substituting this in formula (\*) gives:

$$\int e^{ax} \cos bx \, dx = \frac{e^{ax} \cos bx}{a} + \frac{b}{a} \left( \frac{e^{ax}}{a} \sin bx \right)$$

$$+ \frac{b}{a} \left( \frac{-b}{a} \right) \int (\cos bx) e^{ax} \, dx.$$

Transposing the last term to the left we obtain

$$\left( 1 + \frac{b^2}{a^2} \right) \int e^{ax} \cos bx \, dx = \frac{e^{ax}}{a} \cos bx + \frac{b}{a^2} e^{ax} \sin bx:$$

thus

This example can be solved more easily using the complex exponential functions which will be introduced in 7.5C.

Example 6 shows that it is not always obvious what $u$ and $dv$ should be.

---

*Example 6*   Evaluate $\int \sec^3 t \, dt$.

*Solution*   Let

$$u = \sec t \quad \text{and} \quad dv = \sec^2 t \, dt.$$

Then

$$du = \sec t \tan t \, dt \quad \text{and} \quad v = \tan t,$$

so

$$\int \sec^3 t \, dt = \sec t \tan t - \int \sec t \tan^2 t \, dt$$

$$= \sec t \tan t - \int \sec t \, (\sec^2 t - 1) \, dt$$

$$= \sec t \tan t - \int \sec^3 t \, dt + \int \sec t \, dt.$$

Solving for $\int \sec^3 t \, dt$, we have

$$2 \int \sec^3 t \, dt = \sec t \tan t + \int \sec t \, dt$$

or

$$\int \sec^3 t \, dt = \tfrac{1}{2} \sec t \tan t + \tfrac{1}{2} \int \sec t \, dt.$$

Recall that

$$\int \sec t \, dt = \ln |\sec t + \tan t| + C \quad \text{(Example 7, 5.2C)};$$

hence

$$\int \sec^3 t \, dt = \tfrac{1}{2} \sec t \tan t + \tfrac{1}{2} \ln |\sec t + \tan t| + K.$$

*(It might be helpful to know that it took Bonic two hours to figure out the problem in Example 6 when he took freshman calculus.)*

(Lee)

$$\int e^{ax} \cos bx \, dx = \frac{e^{ax}}{a^2 + b^2} (a \cos bx + b \sin bx) + C.$$

(We added the constant of integration $C$ at the end. It was implicit in the above derivation.)

## D. Exercises

Evaluate the integrals given in Exercises B1–B8.

**B1.** $\int 3xe^{4x} \, dx$

**B2.** $\int x \cos 3x \, dx$

**B3.** $\int \ln 4x \, dx$

**B4.** $\int \text{Tan}^{-1} x \, dx$

**B5.** $\int x^2 e^x \, dx$

**B6.** $\int x^2 \sin x \, dx$

**B7.** $\int \cos (\ln x) \, dx$

**B8.** $\int \dfrac{x^3}{\sqrt{1 + x^2}} \, dx$

**D9.** Each of the following equations has the form

$$\int u \, dv = uv - \int v \, du.$$

Determine $u$ and choose the correct sign.

a) $\int x^2 \cos x \, dx = x^2 \sin x \pm \int 2x \sin x \, dx$

b) $\int \text{Sin}^{-1} x \, dx = x \, \text{Sin}^{-1} x \pm \int \dfrac{x}{\sqrt{1 - x^2}} \, dx$

c) $\int \cos^3 x \, dx = \cos^2 x \sin x \pm 2 \int \cos x \sin^2 x \, dx$

d) $\int \ln (1 + x^2) \, dx = x \ln (1 + x^2) \pm \int \dfrac{2x^2}{1 + x^2} \, dx$

e) $\int \sec^2 x \tan x \, dx = \sec^2 x \pm \int \sec^2 x \tan x \, dx$

f) $\int \sec^2 x \tan x \, dx = \tan^2 x \pm \int \sec^2 x \tan x \, dx$

g) $\int (2x^2 - x) \, dx = \dfrac{2x^3 - x^2}{2} \pm \int x^2 \, dx$

h) $\int (2x^2 - x) \, dx = x^3 - x^2 \pm \int (x^2 - x) \, dx$

a) $x^2,\ -$      b) $\text{Sin}^{-1} x,\ -$      c) $\cos^2 x,\ +$
d) $\ln (1 + x^2),\ -$      e) $\sec x,\ -$      f) $\tan x,\ -$
g) $2x - 1,\ -$      h) $x,\ -$

**D10.** Evaluate the following integrals.

a) $\int xe^{2x} \, dx$      d) $\int x \sec^2 x \, dx$

b) $\int xe^{-3x} \, dx$      e) $\int x \sec x \tan x \, dx$

c) $\int x \cos x \, dx$      f) $\int x \sin x \, dx$

a) $\dfrac{xe^{2x}}{2} - \dfrac{e^{2x}}{4} + C$      b) $\dfrac{-xe^{-3x}}{3} - \dfrac{e^{-3x}}{9} + C$
c) $x \sin x + \cos x + C$
d) $x \tan x + \ln |\cos x| + C$
e) $x \sec x - \ln |\sec x + \tan x| + C$
f) $-x \cos x + \sin x + C$

**M11.** Integrate by parts twice to evaluate the following integrals.

a) $\int \sin (\ln x) \, dx$      b) $\int e^{2x} \cos 3x \, dx$

**M12.** Find $\int \sin^3 x \, dx$.

**M13.** Evaluate $\int_0^1 x \sqrt{x + 1} \, dx$ using the integration by parts formula for definite integrals.

**M14.** Evaluate $\int_0^{\pi/2} \sin^2 x \cos^2 x \, dx$.

**M15.** a) Prove that $\int_0^{\pi/2} \sin^n x \, dx = \frac{n-1}{n} \int_0^{\pi/2} \sin^{n-2} x \, dx$.

b) Evaluate $\int_0^{\pi/2} \sin^6 x \, dx$.

**M16.** a) Express $\int (\ln x)^n \, dx$ in the form

$$f(x) + C \int (\ln x)^{n-1} \, dx.$$

b) Evaluate $\int_1^e (\ln x)^3 \, dx$.

**M17.** Find the following integrals.

a) $\int \frac{\ln (\ln x)}{x} \, dx$

b) $\int_1^e \sqrt{x} \ln x \, dx$

c) $\int x (\ln x)^2 \, dx$

**M18.** Given

$$\int \frac{1}{x} \, dx \quad \text{let} \quad u = \frac{1}{x} \quad \text{and} \quad dv = 1 \cdot dx.$$

Then $du = -\frac{dx}{x^2}$ and $v = x$, so that

$$\int \frac{1}{x} \, dx = \frac{1}{x} \cdot x - \int x \left(\frac{-dx}{x^2}\right) = 1 + \int \frac{1}{x} \, dx.$$

Subtracting $\int \frac{1}{x} \, dx$ from each side of the equation yields $0 = 1$. What is the fallacy?

**M19.** Evaluate the following definite integrals.

a) $\int_0^{1/\sqrt{2}} \mathrm{Sin}^{-1} x \, dx$     b) $\int_0^1 \mathrm{Tan}^{-1} x \, dx$

**M20.** Prove that

$$\int \sin^n x \, dx = -\frac{\sin^{n-1} x \cos x}{n} + \frac{n-1}{n} \int \sin^{n-2} x \, dx.$$

**M21.** Prove that

$$\int \cos^n x \, dx = \frac{\cos^{n-1} x \sin x}{n} + \frac{n-1}{n} \int \cos^{n-2} x \, dx.$$

**M22.** Prove that

$$\int x^n e^x \, dx = x^n e^x - n \int x^{n-1} e^x \, dx.$$

**M23.** Prove that

$$\int x^m (\ln x)^n \, dx$$

$$= \frac{x^{m+1} (\ln x)^n}{m+1} - \frac{n}{m+1} \int x^m (\ln x)^{n-1} \, dx.$$

**M24.** Prove that

$$\int \sec^n x \, dx = \frac{\sec^{n-2} x \tan x}{n-1} + \frac{n-2}{n-1} \int \sec^{n-2} x \, dx.$$

---

## 5.4  *Quotients of Polynomials*

In this section we study integrals of the form

$$\int \frac{p(x)}{q(x)} \, dx$$

where $p$ and $q$ are polynomials. We will confine ourselves to cases in which $p$ and $q$ are of low degree. →

The methods for polynomials of low degree that are given in this section generalize for high degree polynomials, but the mathematics is more complicated. We will indicate in the comments how the more general results are obtained.

**A. Linear Denominator**  Using the formula

$$\int \frac{1}{ax+b} \, dx = \frac{1}{a} \ln |ax + b| + C \quad (a \neq 0),$$

we can evaluate any integral of the form

→ We derive this as follows:

$$\int \frac{1}{ax+b} \, dx = \frac{1}{a} \int \frac{a}{ax+b} \, dx = \frac{1}{a} \ln |ax + b| + C.$$

$$\int \frac{p(x)}{ax + b} \, dx.$$

Since, by long division,

$$\frac{p(x)}{ax + b} = r(x) + \frac{c}{ax + b},$$

where $r$ is a polynomial and $c$ a constant, we have

$$\int \frac{p(x)}{ax + b} \, dx = \int r(x) \, dx + c \int \frac{1}{ax + b} \, dx.$$

Both of the integrals on the right side of the equation can be easily evaluated. As an illustration consider

⟶ This is a special case of the *division algorithm* for polynomials. In general, given polynomials $p$ and $q$, $q$ not a constant, we may write

$$\frac{p}{q} = r + \frac{s}{q},$$

where $r$ and $s$ are polynomials and the degree of $s$ is less than the degree of $q$.

---

*Example 1*  Let us evaluate

$$\int \frac{x^3 - 6x + 8}{x + 3} \, dx.$$

*Solution*  Using long division we obtain

$$\frac{x^3 - 6x + 8}{x + 3} = x^2 - 3x + 3 - \frac{1}{x + 3}.$$

Hence

$$\int \frac{x^3 - 6x + 8}{x + 3} \, dx$$

$$= \int (x^2 - 3x + 3) \, dx - \int \frac{dx}{x + 3}$$

$$= \frac{x^3}{3} - \frac{3x^2}{2} + 3x - \ln|x + 3| + C.$$

Here is the calculation of this.

$$
\begin{array}{r}
x^2 - 3x + 3 \\
x + 3 \overline{)\, x^3 \qquad - 6x + 8} \\
\underline{x^3 + 3x^2} \\
-3x^2 - 6x \\
\underline{-3x^2 - 9x} \\
3x + 8 \\
\underline{3x + 9} \\
-1
\end{array}
$$

Hence

$$\frac{x^3 - 6x + 8}{x + 3} = x^2 - 3x + 3 + \frac{-1}{x + 3}.$$

(Jean)

---

**B. Two Contrasting Formulas**  The two formulas given below are used frequently. We give the derivations alongside one another to contrast their differences.

When $a = 0$ each of these formulas is meaningless. In this

⟶ case we have $\int \frac{dx}{x^2 + 0^2} = \int x^{-2} \, dx = \frac{-1}{x} + C.$

(1)     $\int \frac{dx}{x^2 + a^2} = \frac{1}{a} \operatorname{Tan}^{-1} \frac{x}{a} + C$   $(a \neq 0)$

⟶ (2)   $\int \frac{dx}{x^2 - a^2} = \frac{1}{2a} \ln \left| \frac{x - a}{x + a} \right| + C$   $(a \neq 0)$

In order to show this, we write

$$\int \frac{dx}{x^2 + a^2} = \frac{1}{a^2} \int \frac{dx}{\frac{x^2}{a^2} + 1},$$

and we let $u = \frac{x}{a}$, so that $du = \frac{dx}{a}$. Substituting these into the above expression gives

Using the identity

$$\frac{1}{x^2 - a^2} = \frac{1}{(x - a)(x + a)}$$

$$= \frac{1}{2a} \left( \frac{1}{x - a} - \frac{1}{x + a} \right),$$

we have

$$\frac{1}{a^2} \int \frac{dx}{\frac{x^2}{a^2} + 1} = \frac{1}{a^2} \int \frac{a\, du}{u^2 + 1} = \frac{1}{a} \int \frac{du}{u^2 + 1}$$

$$= \frac{1}{a} \operatorname{Tan}^{-1} u + C$$

$$= \frac{1}{a} \operatorname{Tan}^{-1} \frac{x}{a} + C \quad \bullet$$

$$\int \frac{dx}{x^2 - a^2} = \frac{1}{2a} \int \left( \frac{1}{x - a} - \frac{1}{x + a} \right) dx$$

$$= \frac{1}{2a} \int \frac{dx}{x - a} - \frac{1}{2a} \int \frac{dx}{x + a}$$

$$= \frac{1}{2a} \ln |x - a| - \frac{1}{2a} \ln |x + a| + C$$

$$= \frac{1}{2a} \ln \left| \frac{x - a}{x + a} \right| + C \quad \bullet$$

Notice that both these formulas remain unchanged when $a$ is replaced by $-a$.

We will now show how an integral of the form

$$\int \frac{dx}{x^2 + ax + b}$$

can be reduced to form (1) or form (2) by completing the square and making a simple substitution. Two examples, each based on one of these formulas, will be given.

→ The fact that two integrals which appear so similar can result in completely different answers is an illustration of the contrast between integration and differentiation.

**Example 2** Show that

$$\int \frac{dx}{x^2 - 2x + 5} = \frac{1}{2} \operatorname{Tan}^{-1} \left( \frac{x - 1}{2} \right) + C.$$

Solution $\quad \displaystyle\int \frac{dx}{x^2 - 2x + 5} = \int \frac{dx}{x^2 - 2x + 1 + 4}$

$$= \int \frac{dx}{(x - 1)^2 + 4}$$

Now let $u = x - 1$, so $du = dx$. Then

$$\int \frac{dx}{(x - 1)^2 + 4} = \int \frac{du}{u^2 + 2^2} = \frac{1}{2} \operatorname{Tan}^{-1} \frac{u}{2} + C$$

$$= \frac{1}{2} \operatorname{Tan}^{-1} \left( \frac{x - 1}{2} \right) + C.$$

**Example 3** Show that

$$\int \frac{dx}{x^2 + 6x - 7} = \frac{1}{8} \ln \left| \frac{x - 1}{x + 7} \right| + C.$$

Solution

$$\int \frac{dx}{x^2 + 6x - 7} = \int \frac{dx}{x^2 + 6x + 9 - 16}$$

$$= \int \frac{dx}{(x + 3)^2 - 16}.$$

Now let $u = x + 3$, so $du = dx$. Then

$$\int \frac{dx}{(x + 3)^2 - 16} = \int \frac{du}{u^2 - 4^2} = \frac{1}{8} \ln \left| \frac{u - 4}{u + 4} \right| + C$$

$$= \frac{1}{8} \ln \left| \frac{x + 3 - 4}{x + 3 + 4} \right| + C$$

$$= \frac{1}{8} \ln \left| \frac{x - 1}{x + 7} \right| + C.$$

When the integral has the form

$$\int \frac{dx}{px^2 + qx + r}$$

where $p \neq 0$, it can be reduced to form (1) or (2) by first dividing out $p$. This simply means writing

$$\int \frac{dx}{px^2 + qx + r} = \frac{1}{p} \int \frac{dx}{x^2 + \frac{q}{p} x + \frac{r}{p}}.$$

**Problem 1** Show that

$$\int \frac{5\, dx}{2x^2 + 8x + 26} = \frac{5}{6} \operatorname{Tan}^{-1} \left( \frac{x + 2}{3} \right) + C.$$

## C. General Quadratic Denominator

When the integral has the form

$$\int \frac{p(x)\,dx}{x^2 + ax + b},$$

it can be reduced to one of the previous cases by using long division. We have

$$\frac{p(x)}{x^2 + ax + b} = r(x) + \frac{cx + d}{x^2 + ax + b}$$

where $r(x)$ is a polynomial. Now we do some arithmetic to put $\frac{cx + d}{x^2 + ax + b}$ in a more convenient form.

Let $u = x^2 + ax + b$. We would like to write $\frac{cx + d}{x^2 + ax + b}$ in the form $\frac{A}{u}\frac{du}{dx} + \frac{B}{u}$ where $A$ and $B$ are constants. To do so, let

$$cx + d = A\frac{du}{dx} + B = A(2x + a) + B$$

and solve for $A$ and $B$. Thus

$$cx + d = 2Ax + (aA + B),$$

which gives

$$c = 2A \quad \text{and} \quad d = aA + B,$$

so

$$A = \frac{c}{2} \quad \text{and} \quad B = \frac{2d - ac}{2}.$$

Therefore

$$\frac{cx + d}{x^2 + ax + b} = \frac{c}{2}\frac{1}{u}\frac{du}{dx} + \left(\frac{2d - ac}{2}\right)\frac{1}{u},$$

so

$$\int \frac{cx + d}{x^2 + ax + b}\,dx$$

$$= \frac{c}{2}\int \frac{du}{u} + \frac{2d - ac}{2}\int \frac{1}{x^2 + ax + b}\,dx$$

$$= \frac{c}{2}\ln|u| + \frac{2d - ac}{2}\cdot\int \frac{1}{x^2 + ax + b}\,dx.$$

The latter integral can be evaluated by the methods of 5.4B.

→ Of course $p(x)$ is a polynomial. Also, when the coefficient of $x^2$ in the denominator is not 1, divide it out to obtain the indicated form.

→ For example, suppose $p(x) = 2x^4 - 15x^3 + 45x^2 - 44x + 12$ and $x^2 + ax + b = x^2 - 6x + 13$. Then

$$
\begin{array}{r}
2x^2 - 3x + 1 \\
x^2 - 6x + 13\overline{)\,2x^4 - 15x^3 + 45x^2 - 44x + 12} \\
2x^4 - 12x^3 + 26x^2 \\
\hline
-3x^3 + 19x^2 - 44x \\
-3x^3 + 18x^2 - 39x \\
\hline
x^2 - 5x + 12 \\
x^2 - 6x + 13 \\
\hline
x - 1
\end{array}
$$

Hence

$$\frac{2x^4 - 15x^3 + 45x^2 - 44x + 12}{x^2 - 6x + 13}$$

$$= 2x^2 - 3x + 1 + \frac{x - 1}{x^2 - 6x + 13}.$$

(Jean)

**Example 4**  We will evaluate

$$\int \frac{(2x^4 - 15x^3 + 45x^2 - 44x + 12)}{x^2 - 6x + 13}\, dx.$$

*Solution*  Using the division computation given above, we may write the integral as

$$\int (2x^2 - 3x + 1)\, dx + \int \left(\frac{x - 1}{x^2 - 6x + 13}\right) dx.$$

The first integral is

$$\int (2x^2 - 3x + 1)\, dx = \tfrac{2}{3}x^3 - \tfrac{3}{2}x^2 + x + C.$$

The second integral is

$$\int \frac{x - 1}{x^2 - 6x + 13}\, dx$$

and can be written as

$$\frac{1}{2}\int \frac{2x - 2}{x^2 - 6x + 13}\, dx$$

$$= \frac{1}{2}\int \frac{(2x - 6)}{x^2 - 6x + 13}\, dx + \frac{1}{2}\int \frac{4}{x^2 - 6x + 13}\, dx \longrightarrow$$

$$= \frac{1}{2}\ln|x^2 - 6x + 13| + \text{Tan}^{-1}\left(\frac{x - 3}{2}\right) + K.$$

Adding this to the first integral we see that the original integral is equal to

$$\tfrac{2}{3}x^3 - \tfrac{3}{2}x^2 + x + \tfrac{1}{2}\ln|x^2 - 6x + 13|$$

$$+ \text{Tan}^{-1}\left(\frac{x - 3}{2}\right) + L.$$

**D. Partial Fractions**  An integral of the form

$$\int \frac{(ax + b)\, dx}{(x - c)(x - d)}$$

could be integrated as in 5.4C, but in this case the method of *partial fractions* is more efficient. The two cases that occur are illustrated in Examples 5 and 6.  $\longrightarrow$

**Example 5**  Integrate  $\displaystyle\int \frac{(7x - 13)\, dx}{(x - 3)(x + 1)}.$

---

**Problem 2**  Evaluate $\displaystyle\int \frac{x^2}{x^2 - 4}\, dx.$

*Answer:*  $x + \ln\left|\dfrac{x - 2}{x + 2}\right| + C$

---

The integral $\dfrac{1}{2}\displaystyle\int \frac{4}{x^2 - 6x + 13}\, dx$ is evaluated by completing the square in the denominator and thus reducing the integral to either 5.4B(1) or 5.4B(2). Specifically we have

$$\frac{1}{2}\int \frac{4}{x^2 - 6x + 13}\, dx = 2 \int \frac{1}{(x - 3)^2 + 2^2}\, dx$$

$$= 2 \left(\frac{1}{2}\,\text{Tan}^{-1}\left(\frac{x - 3}{2}\right)\right) + C$$

$$= \text{Tan}^{-1}\left(\frac{x - 3}{2}\right) + C.$$

Any integrand of the form $\dfrac{ax + b}{(x - c)(x - d)}$ does have the form $\dfrac{ax + b}{x^2 + px + q}$, but the converse is false. The latter form can only be factored when the roots of $x^2 + px + q$ are real.

These cases are when $c \neq d$ and $c = d$. In the first ($c \neq d$), set

$$\frac{ax + b}{(x - c)(x - d)} = \frac{A}{x - c} + \frac{B}{x - d}$$

$$= \frac{A(x - d) + B(x - c)}{(x - c)(x - d)}.$$

$\downarrow$

*Solution*  We set

$$\frac{7x - 13}{(x - 3)(x + 1)} \text{ equal to } \frac{A}{x - 3} + \frac{B}{x + 1}.$$

We then have

$$\frac{7x - 13}{(x - 3)(x + 1)} = \frac{A(x + 1) + B(x - 3)}{(x - 3)(x + 1)}$$

$$= \frac{(A + B)x + (A - 3B)}{(x - 3)(x + 1)}.$$

Equating corresponding parts, we therefore have

$$A + B = 7$$

and

$$A - 3B = -13.$$

Solving these equations gives $A = 2$ and $B = 5$. Hence

$$\int \frac{(7x - 13)\,dx}{(x - 3)(x + 1)} = \int \frac{2\,dx}{x - 3} + \int \frac{5\,dx}{x + 1}$$

$$= 2 \ln |x - 3| + 5 \ln |x + 1| + C.$$

Below is an example of the case when $c = d$.

**Example 6**  Integrate $\int \frac{(2x - 3)}{(x - 4)^2}\,dx.$

*Solution*  Write

$$\frac{2x - 3}{(x - 4)^2} = \frac{A}{(x - 4)} + \frac{B}{(x - 4)^2}$$

$$= \frac{A(x - 4) + B}{(x - 4)^2}.$$

Then we have

$$Ax + (B - 4A) = 2x - 3.$$

Hence $A = 2$ and $B - 4A = -3$, which gives $B = 5$. Therefore

$$\int \frac{2x - 3}{(x - 4)^2}\,dx = \int \frac{2\,dx}{x - 4} + \int \frac{5\,dx}{(x - 4)^2}$$

$$= 2 \ln |x - 4| - \frac{5}{x - 4} + C.$$

Since the numerators must be equal, we have

$$ax + b = (A + B)x - (dA + cB).$$

It follows that $A + B = a$ and $dA + cB = -b$, and solving these equations for $A$ and $B$ we have

$$A = \frac{b + ac}{c - d} \text{ and } B = \frac{b + ad}{d - c}.$$

In the second case ($c = d$), set

$$\frac{ax + b}{(x - c)^2} = \frac{A}{x - c} + \frac{B}{(x - c)^2}.$$

Multiplying this out, as above, we obtain

$$ax + b = A(x - c) + B.$$

Hence $A = a$ and $-Ac + B = b$, which yields

$$A = a \text{ and } B = b + ac.$$

**Problem 3**  Find $\int_4^6 \frac{-2\,dx}{(x - 2)(x - 3)}.$

*Answer:*  $\ln \frac{4}{9}$

This technique can be extended to denominators of higher degree.

---

*Example 7* Find $\int \dfrac{6x^2 - 11x - 14}{(x - 3)(x - 2)(x + 4)}\,dx$.

*Solution* Set

$$\dfrac{6x^2 - 11x - 14}{(x - 3)(x - 2)(x + 4)}$$

$$= \dfrac{A}{(x - 3)} + \dfrac{B}{(x - 2)} + \dfrac{C}{(x + 4)}$$

$$= \dfrac{A(x - 2)(x + 4) + B(x - 3)(x + 4) + C(x - 2)(x - 3)}{(x - 2)(x - 3)(x + 4)}$$

$$= \dfrac{(A + B + C)x^2 + (2A + B - 5C)x + (-8A - 12B + 6C)}{(x - 2)(x - 3)(x + 4)}$$

Equating corresponding parts, we have

$$A + B + C = 6, \quad 2A + B - 5C = -11,$$

and

$$-8A - 12B + 6C = -14.$$

Solving these equations gives

$$A = 1, \quad B = 2, \quad \text{and} \quad C = 3.$$

Therefore

$$\int \dfrac{6x^2 - 11x - 14}{(x - 3)(x - 2)(x + 4)}\,dx$$

$$= \int \dfrac{1}{x - 3}\,dx + 2\int \dfrac{1}{x - 2}\,dx + 3\int \dfrac{1}{x + 4}\,dx$$

$$= \ln|x - 3| + 2\ln|x - 2| + 3\ln|x + 4| + C.$$

---

→ In general given polynomials $p$ and $q$, we first divide to obtain

$$\dfrac{p}{q} = r + \dfrac{s}{q}$$

where $r$ is a polynomial and the degree of $s$ is less than the degree of $q$. Hence, to find $\int \dfrac{p}{q}$ it suffices to find $\int \dfrac{s}{q}$.

If $q$ can be put in the form

$$q(x) = (x - c_1)(x - c_2) \cdots (x - c_r)$$

and the roots are real and distinct, we then set

$$\dfrac{s}{q} = \dfrac{A_1}{x - c_1} + \dfrac{A_2}{x - c_2} + \cdots + \dfrac{A_r}{x - c_r}$$

and solve for $A_1, \ldots, A_r$.

If there are repeated roots, say for example $c_1 = c_2 = c_3$, write

$$\dfrac{A_1}{x_1 - c_1} + \dfrac{A_2}{(x - c_1)^2} + \dfrac{A_3}{(x - c_1)^3}$$

in place of

$$\dfrac{A_1}{x_1 - c_1} + \dfrac{A_2}{x - c_1} + \dfrac{A_3}{x - c_1}.$$

The case where the roots are complex numbers can also be treated with some slight variations.

---

## E. Exercises

Find the integrals given in Exercises B1–B8.

**B1.** a) $\int \dfrac{dx}{x - 3}$ b) $\int \dfrac{dx}{2x - 3}$ c) $\int \dfrac{dx}{(2x - 3)^2}$

**B2.** a) $\int \dfrac{dx}{x + 2}$ b) $\int \dfrac{x\,dx}{x + 2}$ c) $\int \dfrac{x^2\,dx}{x + 2}$

**B3.** a) $\int \dfrac{3\,dx}{x^2 + 16}$ b) $\int \dfrac{3x\,dx}{x^2 + 16}$ c) $\int \dfrac{3x^2\,dx}{x^2 + 16}$

**B4.** a) $\int \dfrac{2\,dx}{x^2 - 9}$ b) $\int \dfrac{2x\,dx}{x^2 - 9}$ c) $\int \dfrac{2x^2\,dx}{x^2 - 9}$

**B5.** a) $\int \dfrac{dx}{x^2 - 4x + 20}$ b) $\int \dfrac{dx}{x^2 - 4x - 12}$

**B6.** a) $\displaystyle\int \frac{dx}{(x-3)(x+4)}$     b) $\displaystyle\int \frac{2\,dx}{x^2+x-12}$

**B7.** $\displaystyle\int \frac{(3x^2-7)\,dx}{(x-1)(x-2)(x+3)}$

**B8.** $\displaystyle\int \frac{(x^3+3x^2-5x-6)\,dx}{x^2-2x+5}$

**D9.** Given polynomials $p$ and $q$, find polynomials $r$ and $s$ so that $\dfrac{p}{q} = r + \dfrac{s}{q}$ where the degree of $s$ is less than the degree of $q$.

a) $p = x^2 + 3x$,  $q = x + 1$
b) $p = 4x^2 - 8x$,  $q = 2x + 3$
c) $p = x^4$,  $q = x$
d) $p = x^2 - 3x - 4$,  $q = x + 1$
e) $p = x^3 - 2$,  $q = x - 1$
f) $p = x^3 - 1$,  $q = x + 1$
g) $p = x^3 - 1$,  $q = x^2 - 1$
h) $p = x^2 - 1$,  $q = x^3 + 1$

a) $r = x + 2$,  $s = -2$     b) $r = 2x - 7$,  $s = 21$
c) $r = x^3$,  $s = 0$         d) $r = x - 4$,  $s = 0$
e) $r = x^2 + x + 1$,  $s = -1$
f) $r = x^2 - x + 1$,  $s = -2$
g) $r = x$,  $s = x - 1$     h) $r = 0$,  $s = x^2 - 1$

**D10.** Each of the following expressions can be put in the form $\dfrac{A}{x-a} + \dfrac{B}{x-b}$. Find $A$, $B$, $a$, and $b$.

a) $\dfrac{3x-10}{x^2-7x+12}$     d) $\dfrac{-6x-21}{x^2+7x+12}$

b) $\dfrac{5-x}{x^2-1}$     e) $\dfrac{10}{x^2-x-6}$

c) $\dfrac{7x-3}{x^2-x}$     f) $\dfrac{3x-15}{x^2-9x+20}$

a) $A = 1$,  $B = 2$,  $a = 3$,  $b = 4$
b) $A = 2$,  $B = -3$, $a = 1$,  $b = -1$
c) $A = 3$,  $B = 4$,  $a = 0$,  $b = 1$
d) $A = -3$, $B = -3$, $a = -3$, $b = -4$
e) $A = 2$,  $B = -2$, $a = 3$,  $b = -2$
f) $A = 3$,  $B = 0$,  $a = 4$,  $b = 5$

**M11.** Evaluate $\displaystyle\int \frac{(x^3-4x+3)}{(x-1)(x+3)}\,dx$.

**M12.** Evaluate $\displaystyle\int \frac{x^3\,dx}{(x-1)^2}$.

**M13.** Evaluate $\displaystyle\int_3^4 \frac{dx}{(x-1)(x+2)}$.

**M14.** Evaluate $\displaystyle\int_0^1 \frac{(x^2-x+10)\,dx}{x^3-2x^2-4x+8}$.

**M15.** Evaluate

a) $\displaystyle\int_0^1 \frac{x+4}{(x+2)(x+3)}\,dx$

b) $\displaystyle\int_0^1 \frac{(x+2)(x+3)}{x+4}\,dx$

**M16.** Evaluate $\displaystyle\int_0^1 \frac{2x^3+4x^2-8x+16}{x^4-16}\,dx$.

**M17.** Find

$$\int \frac{(2\sin x + 7)\cos x\,dx}{(3+\sin x)(4+\sin x)}$$

by observing that it can be put in a form to which the partial fractions technique can be applied by letting $u = \sin x$.

**M18.** Evaluate $\displaystyle\int \frac{\sec x\,\tan x\,dx}{9+4\sec^2 x}$.

**M19.** Evaluate $\displaystyle\int \frac{x+1}{x^2-4x+8}\,dx$.

**M20.** Evaluate $\displaystyle\int \frac{2x+3}{9x^2-12x+8}\,dx$.

**M21.** Evaluate $\displaystyle\int \frac{2-x}{4x^2+4x-3}\,dx$.

**M22.** Prove that

$$\int \frac{dx}{a^2-x^2} = \frac{1}{2a}\ln\left|\frac{a+x}{a-x}\right| + C.$$

**M23.** Use Problem M22 to evaluate $\displaystyle\int \frac{dx}{4x-x^2}$.

**M24.** A theorem in algebra states that every polynomial with real coefficients can be factored as a product of linear and quadratic factors with real coefficients, and such that the quadratic factors cannot be further factored into linear factors with real coefficients (a linear factor is one of the form $ax + b$; a quadratic factor is one of the form $cx^2 + dx + e$). Hence, if we are given a quotient of polynomials $A(x)/B(x)$ with the degree of $A(x)$ less than the degree of $B(x)$, we may write it in the form

$$\frac{A(x)}{L_1(x)\cdots L_n(x)Q_1(x)\cdots Q_m(x)}$$

where the $L_i$ are linear factors and the $Q_i$ are quadratic factors. A modification of the method of partial fractions may then be applied. We write

$$\frac{A(x)}{L_1(x) \cdots L_n(x) Q_1(x) \cdots Q_m(x)}$$

$$= \frac{A_1}{L_1(x)} + \frac{A_2}{L_2(x)} + \cdots + \frac{A_n}{L_n(x)} + \frac{C_1 x + D_1}{Q_1(x)}$$

$$+ \frac{C_2 x + D_2}{Q_2(x)} + \cdots + \frac{C_m x + D_m}{Q_m(x)}$$

where the $A_i$, $C_i$, and $D_i$ are constants for which we wish to solve. We have implicitly assumed that the $L_i$ are distinct and the $Q_i$ are distinct. We already know what to do if two of the $L_i$ are equal; if two of the $Q_i$ are equal, say $Q_1 = Q_2$, then we write

$$\frac{C_1 x + D_1}{Q_1(x)} + \frac{C_2 x + D_2}{(Q_1(x))^2} + \cdots + \frac{C_m x + D_m}{Q_m(x)}.$$

Thus we have a method for simplifying any quotient of polynomials in order to facilitate integration. Solve the following problems.

a) $\int \dfrac{1}{(x-3)(x^2+1)}\, dx$    b) $\int \dfrac{x^4 + 10x + 25}{(x^2+5)^2(x-1)}\, dx$

c) $\int \dfrac{(4x^2 + 11x + 7)\, dx}{(2x^2 + 5x + 3)(x+1)}$

(Steve)

**M25.** Evaluate $\int \dfrac{x^3 + x^2 + x + 2}{x^4 + 3x^2 + 2}\, dx$.

**M26.** Evaluate $\int \dfrac{2x^3 + x^2 + 4}{(x^2+4)^2}\, dx$.

---

## 5.5 Substitution Techniques

A number of integrals can be simplified by making a suitable change of variable. This method, generally referred to as the substitution technique, will be applied in a variety of ways in the examples below. When dealing with definite integrals it will be necessary to also change the limits of integration.

→ *A comment telling you how difficult substitution is would be both redundant and useless. Let it be known, then, that we all found substitution aggravating, so may this comment serve as a relief from the throes of calculus and no more.*

*(Robin)*

**A. Examples** We start with the following very simple one.

*Example 1* Find $\int \dfrac{dx}{\sqrt{a^2 - x^2}}$ $(a > 0)$.

*Solution* Let $u = \frac{x}{a}$, so $x = au$ and $dx = a\, du$. We then have

$$\int \frac{dx}{\sqrt{a^2 - x^2}} = \int \frac{a\, du}{\sqrt{a^2 - a^2 u^2}} = \int \frac{du}{\sqrt{1 - u^2}}$$

$$= \mathrm{Sin}^{-1} u + C = \mathrm{Sin}^{-1} \frac{x}{a} + C.$$

→ This solution is very similar to that used in 5.4B where we derived the formula

$$\int \frac{dx}{x^2 + a^2} = \frac{1}{a} \mathrm{Tan}^{-1} \frac{x}{a} + C.$$

*Example 2* Show that

$$\int \frac{dx}{x\sqrt{x^2 - a^2}} = \frac{-1}{a} \operatorname{Sin}^{-1} \frac{a}{x} + C \quad (0 < a < x).$$

*Solution* We could of course verify this by differentiation, but instead we will reduce it to Example 1 by a suitable substitution.

Let $u = \dfrac{a}{x}$. Then $x = \dfrac{a}{u}$, $dx = \dfrac{-a\,du}{u^2}$,

and we have

$$\int \frac{dx}{x\sqrt{x^2 - a^2}} = \int \frac{1}{\dfrac{a}{u}\sqrt{\dfrac{a^2}{u^2} - a^2}} \left(\frac{-a\,du}{u^2}\right)$$

$$= \frac{-1}{a} \int \frac{du}{\sqrt{1 - u^2}}$$

$$= \frac{-1}{a} \operatorname{Sin}^{-1} u + C$$

$$= \frac{-1}{a} \operatorname{Sin}^{-1} \frac{a}{x} + C.$$

*Example 3* Find $\int x^3 \sqrt{x^2 + 4}\, dx$.

*Solution* Let $u = \sqrt{x^2 + 4}$. Then

$$du = \frac{x\,dx}{\sqrt{x^2 + 4}} = \frac{x\,dx}{u} \quad \text{and} \quad u^2 = x^2 + 4.$$

Hence $dx = \dfrac{u\,du}{x}$, $x^2 = u^2 - 4$, and we have

$$\int x^3 \sqrt{x^2 + 4}\, dx$$

$$= \int x^2 \cdot x \sqrt{x^2 + 4}\, dx$$

$$= \int (u^2 - 4)xu \left(\frac{u\,du}{x}\right) = \int (u^2 - 4)u^2\, du$$

$$= \frac{u^5}{5} - \frac{4u^3}{3} + C$$

$$= \frac{(x^2 + 4)^{5/2}}{5} - \frac{4}{3}(x^2 + 4)^{3/2} + C.$$

→ Many authors introduce the inverse secant function, $\operatorname{Sec}^{-1} x$, and show that

$$\frac{d}{dx} \operatorname{Sec}^{-1} x = \frac{1}{x\sqrt{x^2 - 1}} \quad (x > 0);$$

hence $\int \dfrac{1}{x\sqrt{x^2 - 1}}\, dx = \operatorname{Sec}^{-1} x + C.$

The main use of the inverse trigonometric functions is in integration but, as we see from Example 2, the inverse sine function can be used in place of the inverse secant function in this respect. Therefore we will speak no more of the inverse secant function.

→ Since $\operatorname{Cos}^{-1} y + \operatorname{Sin}^{-1} y = \frac{\pi}{2}$, this formula may also be written as

$$\int \frac{dx}{x\sqrt{x^2 - a^2}} = \frac{1}{a} \operatorname{Cos}^{-1} \frac{a}{x} + K.$$

However, we prefer to express everything in terms of the inverse sine and inverse tangent functions.

→ It is instructive to note how much more involved the integral $\int x^2 \sqrt{x^2 + 4}\, dx$ is. In fact

$$\int x^2 \sqrt{x^2 + 4}\, dx = \frac{x}{4}(x^2 + 4)^{3/2} - \frac{x}{2}(x^2 + 4)^{1/2}$$

$$- 2 \ln |x + \sqrt{x^2 + 4}| + C.$$

This is obtained by making the substitution $x = 2 \tan u$ (see 5.5C) which transforms the integral to $16\int(\sec^5 u - \sec^3 u)\, du$. We then evaluate this integral using integration by parts as shown in Example 6(5.3C). A great deal of computation is involved.

*Example 4*  Evaluate

a) $\displaystyle\int \frac{\sin x \, dx}{1 + \cos^2 x}$

b) $\displaystyle\int \frac{\sin 2x \, dx}{1 + \cos^2 x}$.

*Solution*   a) Letting  $u = \cos x$  we have  $du = -\sin x \, dx$  so

$$\int \frac{\sin x \, dx}{1 + \cos^2 x} = \int \frac{-du}{1 + u^2} = -\text{Tan}^{-1} u + C$$
$$= -\text{Tan}^{-1} (\cos x) + C.$$

b) In this case we let $u = 1 + \cos^2 x$. Then

$$du = -2 \cos x \sin x \, dx = -\sin 2x \, dx,$$

so we have

$$\int \frac{\sin 2x \, dx}{1 + \cos^2 x} = \int \frac{-du}{u} = -\ln |u| + C$$
$$= -\ln (1 + \cos^2 x) + C.$$

*Problem 1*  Evaluate $\displaystyle\int \frac{x^2 \, dx}{x + 4}$ by making the substitution $u = x + 4$.

*Answer:*  $\displaystyle\frac{(x + 4)^2}{2} - 8(x + 4) + 16 \ln |x + 4| + C$

*Problem 2*  Evaluate $\int x^2 \sqrt{x + 1} \, dx$ by making the substitution $u = \sqrt{x + 1}$.

*Answer:*
$\frac{2}{7}(x + 1)^{7/2} - \frac{4}{5}(x + 1)^{5/2} + \frac{2}{3}(x + 1)^{3/2} + C$

## B. Changing the Limits of Integration   Consider the definite integral

$$\int_0^3 x \sqrt{1 + x} \, dx.$$

We will evaluate this in two ways.

*First method (Limits unchanged)*  We first compute the indefinite integral.
   Let $u = \sqrt{1 + x}$, so $u^2 = 1 + x$, $u^2 - 1 = x$, and $2u \, du = dx$. Then

$$\int x \sqrt{1 + x} \, dx = \int (u^2 - 1) u (2u \, du)$$
$$= \int (2u^4 - 2u^2) \, du$$
$$= \frac{2u^5}{5} - \frac{2u^3}{3} + C$$
$$= \frac{2}{5}(1 + x)^{5/2} - \frac{2}{3}(1 + x)^{3/2} + C.$$

We then have

$$\int_0^3 x \sqrt{1 + x} \, dx = \frac{2}{5}(1 + x)^{5/2} - \frac{2}{3}(1 + x)^{3/2} \Big|_0^3$$

→ *Second method (Limits changed)*  For this method we again let $u = \sqrt{1 + x}$. The original limits are $x = 0$ and $x = 3$. Notice that when $x = 0$, $u = \sqrt{1 + 0} = 1$, and when $x = 3$, $u = \sqrt{1 + 3} = 2$. Hence we may write

$$\int_0^3 x \sqrt{1 + x} \, dx = \int_1^2 (2u^4 - 2u^2) \, du = \frac{2}{5}u^5 - \frac{2}{3}u^3 \Big|_1^2$$
$$= (\tfrac{2}{5} \cdot 2^5 - \tfrac{2}{3} \cdot 2^3) - (\tfrac{2}{5} - \tfrac{2}{3}) = \tfrac{116}{15}.$$

(This method is justified by the following theorem.)

$$= \left(\tfrac{2}{5} \cdot 2^5 - \tfrac{2}{3} \cdot 2^3\right) - \left(\tfrac{2}{5} - \tfrac{2}{3}\right)$$
$$= \tfrac{62}{5} - \tfrac{14}{3} = \tfrac{116}{15}.$$

The second method given, involving a change in the limits of integration corresponding to a change of variable, can be used to advantage in a variety of problems.

Typically we start with a definite integral

$$\int_c^d f(x)\, dx,$$

and then make a change of variable $x = \phi(u)$. This converts the integral to one of the form

$$\int_a^b g(u)\, du,$$

and we have

$$\int_c^d f(x)\, dx = \int_a^b g(u)\, du$$

where $c = \phi(a)$ and $d = \phi(b)$. We emphasize that the limits of integration change with a change of variable.

---

**Example 5**  Evaluate $\int_0^{\sqrt{\pi/4}} x \sin x^2\, dx$.

*Solution*  Letting $u = x^2$ we have $du = 2x\, dx$. Moreover, when $x = 0$ then $u = 0$ and when $x = \sqrt{\tfrac{\pi}{4}}$, $u = \tfrac{\pi}{4}$. Hence

$$\int_0^{\sqrt{\pi/4}} x \sin x^2\, dx = \int_0^{\pi/4} \sin u\, \frac{du}{2} = \left. \frac{-\cos u}{2} \right|_0^{\pi/4}$$

$$= \frac{-1}{2}\cos\frac{\pi}{4} + \frac{1}{2}\cos 0 = \frac{2 - \sqrt{2}}{4}.$$

---

Now we will apply this technique to derive the formula for the area of a circle.

---

**Example 6**  Prove that the area of a circle of radius $a$ is $\pi a^2$.

*Solution*  Let $A$ denote the area of the circle of radius $a$ (Fig. 1).

**Theorem 1**  (Change of Variable Theorem)

Suppose $x = \phi(u)$ is differentiable on an interval containing $[a, b]$ and $\phi'$ is continuous on $[a, b]$. If $f$ is continuous on an interval containing the image of $\phi$ then

$$\int_{\phi(a)}^{\phi(b)} f(x)\, dx = \int_a^b f(\phi(u))\phi'(u)\, du.$$

*Proof*  Take a function $F$ with $F' = f$ and let $G(u) = F(\phi(u))$. Using the chain rule we have

$$G'(u) = F'(\phi(u))\phi'(u) = f(\phi(u))\phi'(u).$$

By the Fundamental Theorem of Calculus we have

$$\int_{\phi(a)}^{\phi(b)} f(x)\, dx = F(\phi(b)) - F(\phi(a)) = G(b) - G(a)$$

$$= \int_a^b G'(u)\, du = \int_a^b f(\phi(u))\phi'(u)\, du \quad \bullet$$

**Figure 1**

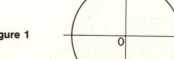

$x^2 + y^2 = a^2$

$(a, 0)$

Referring to Figure 2 we clearly have $A = 4\int_0^a \sqrt{a^2 - x^2}$ $dx$. To evaluate this integral we will make use of a trigonometric substitution.

Let $x = a \sin u$, so $dx = a \cos u\, du$. Since $x = 0$ when $u = 0$ and $x = a$ when $u = \frac{\pi}{2}$, we have

$$A = 4 \int_0^a \sqrt{a^2 - x^2}\, dx$$

$$= 4 \int_0^{\pi/2} \sqrt{a^2 - a^2 \sin^2 u}\; a \cos u\, du$$

$$= 4a^2 \int_0^{\pi/2} \sqrt{1 - \sin^2 u}\; \cos u\, du$$

$$= 4a^2 \int_0^{\pi/2} \cos^2 u\, du.$$

Using the identity $\cos^2 u = \dfrac{1 + \cos 2u}{2}$ and integrating, we see that this equals

$$4a^2 \left( \frac{u}{2} + \frac{\sin 2u}{4} \right) \Big|_0^{\pi/2} = 4a^2 \left( \frac{\pi}{4} - 0 \right) = \pi a^2.$$

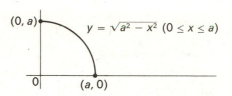

**Figure 2**

With almost no extra work we may compute the area enclosed by the ellipse $\dfrac{x^2}{a^2} + \dfrac{y^2}{b^2} = 1$. Referring to Figure 3,

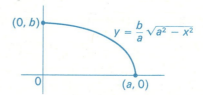

**Figure 3**

we have

$$A = 4 \int_0^a \frac{b}{a} \sqrt{a^2 - x^2}\, dx = \frac{4b}{a} \int_0^a \sqrt{a^2 - x^2}\, dx$$

$$= \frac{4b}{a} \cdot \frac{\pi a^2}{4} = \pi ab.$$

---

**C. More on Trigonometric Substitutions** When the integrand contains a term of the form

$$\sqrt{a^2 - x^2}, \qquad \sqrt{a^2 + x^2}, \qquad \sqrt{x^2 - a^2},$$

the substitutions

$$x = a \sin u, \qquad x = a \tan u, \qquad x = a \sec u,$$

respectively, will often transform the integrand to one that is more easily evaluated.

Making these substitutions and assuming $\cos u$, $\sec u$, $\tan u \geq 0$,

$$\sqrt{a^2 - x^2} = \sqrt{a^2 - a^2 \sin^2 u} = \sqrt{a^2 \cos^2 u} = a \cos u$$
$$\sqrt{a^2 + x^2} = \sqrt{a^2 + a^2 \tan^2 u} = \sqrt{a^2 \sec^2 u} = a \sec u$$
$$\sqrt{x^2 - a^2} = \sqrt{a^2 \sec^2 u - a^2} = \sqrt{a^2 \tan^2 u} = a \tan u$$

In the examples which follow, and in similar problems, the reader is advised to check his answers by differentiation. This is necessary because of the difficulty one has with signs when extracting the square root.

---

**Example 7** Find $\int \sqrt{a^2 - x^2}\, dx$   $(a > 0)$.

*Solution* Let

$$x = a \sin u \qquad \left( -\frac{\pi}{2} \leq u \leq \frac{\pi}{2} \right).$$

Restrict $u$ to satisfy $\frac{-\pi}{2} \leq u \leq \frac{\pi}{2}$. The reason for this will be clear in a moment.

Then $dx = a \cos u\, du$, and we have

$$\int \sqrt{a^2 - x^2}\, dx$$

$$= \int \sqrt{a^2 - a^2 \sin^2 u}\, a \cos u\, du$$

$$= \int a^2 \sqrt{\cos^2 u}\, \cos u\, du = a^2 \int \cos^2 u\, du$$

$$= a^2 \int \left(\frac{1 + \cos 2u}{2}\right) du$$

(since $\cos^2 u = \tfrac{1}{2}(1 + \cos 2u)$)

$$= \frac{a^2}{2} \int (1 + \cos 2u)\, du = \frac{a^2}{2}\left(u + \frac{\sin 2u}{2}\right) + C$$

$$= \frac{a^2}{2}(u + \sin u \cos u) + C.$$

Notice that in writing $\sqrt{\cos^2 u} = \cos u$ we assumed that $\cos u \geq 0$. This is valid since we have restricted $u$ to the interval $[\frac{-\pi}{2}, \frac{\pi}{2}]$.

Now we substitute $x$ back into the equation. Since

$$x = a \sin u \qquad (-\tfrac{1}{2}\pi \leq u \leq \tfrac{1}{2}\pi),$$

we have $u = \mathrm{Sin}^{-1}\frac{x}{a}$. Also, $\sin u = \frac{x}{a}$, so

$$\cos u = \frac{\sqrt{a^2 - x^2}}{a}.$$

To find $\cos u$, notice that, since $-\tfrac{1}{2}\pi \leq u \leq \tfrac{1}{2}\pi$, $\cos u$ is always positive. Hence

$$\cos u = +\sqrt{1 - \sin^2 u} = \sqrt{1 - \frac{x^2}{a^2}} = \frac{\sqrt{a^2 - x^2}}{a}.$$

The value of $\cos u$ may also be found by constructing a right triangle. Since $\sin u = \frac{x}{a}$, the sides are labeled (using the Pythagorean Theorem) as shown in Figure 4. Hence

$$\cos u = \frac{\sqrt{a^2 - x^2}}{a}.$$

We conclude that

$$\int \sqrt{a^2 - x^2}\, dx = \frac{a^2}{2}(u + \sin u \cos u) + C$$

$$= \frac{a^2}{2}\left(\mathrm{Sin}^{-1}\frac{x}{a} + \frac{x}{a}\frac{\sqrt{a^2 - x^2}}{a}\right) + C$$

$$= \frac{a^2}{2}\mathrm{Sin}^{-1}\frac{x}{a} + \frac{x}{2}\sqrt{a^2 - x^2} + C.$$

**Figure 4**

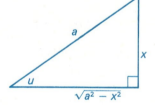

We leave to the reader the exercise of checking the answer by differentiation.

**Example 8** Find $\int \sqrt{x^2 + a^2}\, dx$.

**Solution** Letting $x = a \tan u$, we have $dx = a \sec^2 u\, du$. Hence

Assume that $u$ satisfies $-\tfrac{1}{2}\pi < u < \tfrac{1}{2}\pi$. The reason for this is similar to that given in Example 7.

$$\int \sqrt{x^2 + a^2}\, dx$$

$$= \int \sqrt{a^2 + a^2 \tan^2 u}\, a \sec^2 u\, du$$

$$= \int a^2 \sqrt{\sec^2 u}\, \sec^2 u\, du = a^2 \int \sec^3 u\, du.$$

Since $-\tfrac{1}{2}\pi < u < \tfrac{1}{2}\pi$, $\cos u > 0$, so $\sec u = \frac{1}{\cos u} > 0$. Hence $\sqrt{\sec^2 u} = \sec u$.

Using the formula for $\int \sec^3 u\, du$ (Example 6, 5.3C), we have

$$\int \sqrt{x^2 + a^2}\, dx$$

$$= \frac{a^2}{2}(\sec u \tan u + \ln|\sec u + \tan u|) + C.$$

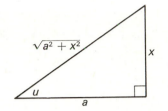

**Figure 5**

Now $\tan u = \frac{x}{a}$, and consulting Figure 5, we see that

$$\sec u = \frac{\sqrt{x^2 + a^2}}{a}.$$

Hence

$$\int \sqrt{x^2 + a^2}\, dx$$

$$= \frac{a^2}{2}\left(\frac{\sqrt{x^2 + a^2}}{a}\frac{x}{a} + \ln\left|\frac{\sqrt{x^2 + a^2}}{a} + \frac{x}{a}\right|\right) + C$$

$$= \frac{x\sqrt{x^2 + a^2}}{2} + \frac{a^2}{2}\ln\left|\frac{\sqrt{x^2 + a^2} + x}{a}\right| + C.$$

---

**Example 9**  Integrate $\int \sqrt{x^2 + 2x + 5}\, dx$.

*Solution*  We will reduce the integral to the form

$$\int \sqrt{u^2 + a^2}\, du$$

by completing the square. Indeed

$$x^2 + 2x + 5 = x^2 + 2x + 1 + 4$$
$$= (x + 1)^2 + 4.$$

Letting $u = x + 1$ we have $du = dx$, so

$$\int \sqrt{x^2 + 2x + 5}\, dx = \int \sqrt{u^2 + 4}\, du.$$

Using Example 8 and replacing $x$ with $u$ and $a$ with 2 gives

$$\int \sqrt{u^2 + 4}\, du$$

$$= \frac{u\sqrt{u^2 + 4}}{2} + 2\ln\left|\frac{\sqrt{u^2 + 4} + u}{2}\right| + C$$

$$= \frac{(x + 1)\sqrt{x^2 + 2x + 5}}{2}$$

$$+ 2\ln\left|\frac{\sqrt{x^2 + 2x + 5} + x + 1}{2}\right| + C.$$

**Problem 3**  Find $\int \sqrt{5 + 4x - x^2}\, dx$.

*Answer:*

$$\frac{9}{2}\operatorname{Sin}^{-1}\left(\frac{x - 2}{3}\right) + \frac{x - 2}{2}\sqrt{5 + 4x - x^2} + C$$

## D. Exercises

Find the integrals given in Exercises B1–B5.

**B1.** a) $\int \sqrt{4 - x^2}\, dx$     b) $\int x\sqrt{4 - x^2}\, dx$

**B2.** a) $\int \dfrac{dx}{\sqrt{4 - x^2}}$     b) $\int \dfrac{x\, dx}{\sqrt{4 - x^2}}$

**B3.** a) $\int \dfrac{(x + 3)\, dx}{\sqrt{x}}$     b) $\int \dfrac{dx}{(x + 3)\sqrt{x}}$

**B4.** a) $\int \sqrt{x}\,(x - 4)\, dx$     b) $\int x\sqrt{x - 4}\, dx$

**B5.** a) $\int 2x \sin x\, dx$     b) $\int \sin \sqrt{x}\, dx$

**B6.** Find $\displaystyle\int_0^8 15x\sqrt{1 + x}\, dx$   using the substitution

$$u = \sqrt{1 + x}.$$

**B7.** Find $\displaystyle\int_4^9 \dfrac{e^{\sqrt{x}}}{\sqrt{x}}\, dx$   using the substitution $u = \sqrt{x}$.

**B8.** Find $\displaystyle\int_0^{\ln \sqrt{3}} \left(\dfrac{1}{e^x + e^{-x}}\right) dx$   using   the   substitution $u = e^x$.

**D9.** Each of the following integrals is given in the form $\displaystyle\int_a^b f(x)\, dx$ and can be written in the form $\displaystyle\int_c^d g(u)\, du$ by making the indicated substitution. Find $g(u)$.

a) $\displaystyle\int_{-1}^4 x\sqrt{x + 5}\, dx, \quad u = \sqrt{x + 5}$

b) $\displaystyle\int_{-1}^4 x\sqrt{x + 5}\, dx, \quad u = x + 5$

c) $\displaystyle\int_0^4 \dfrac{\sqrt{x}\, dx}{x + 1}, \quad u = \sqrt{x}$

d) $\displaystyle\int_4^9 \dfrac{dx}{x + \sqrt{x}}, \quad u = \sqrt{x}$

e) $\displaystyle\int_{-2}^0 x^5\sqrt{1 - x^3}\, dx, \quad u = \sqrt{1 - x^3}$

f) $\displaystyle\int_3^4 (8x^2 + 16x^3 + 64x^5)\, dx, \quad u = 2x$

a) $g(u) = 2u^4 - 10u^2$     b) $g(u) = u^{3/2} - 5u^{1/2}$

c) $g(u) = \dfrac{2u^2}{u^2 + 1}$     d) $g(u) = \dfrac{2}{1 + u}$

e) $g(u) = \dfrac{-2u^2 + 2u^4}{3}$     f) $g(u) = u^2 + u^3 + u^5$

**D10.** Referring to Exercise D9 find the limits $c$ and $d$ for the transformed integral $\displaystyle\int_c^d g(u)\, du$.

a) $c = 2, d = 3$     b) $c = 4, d = 9$
c) $c = 0, d = 2$     d) $c = 2, d = 3$
e) $c = 3, d = 1$     f) $c = 6, d = 8$

**M11.** If $x = 2\,\mathrm{Tan}^{-1} u$ show that

$$\sin x = \frac{2u}{1 + u^2}, \quad \cos x = \frac{1 - u^2}{1 + u^2},$$

and

$$dx = \frac{2\, du}{1 + u^2}.$$

**M12.** Use the substitution given in Exercise M11 to evaluate the integrals.

a) $\displaystyle\int \dfrac{dx}{1 + \sin x - \cos x}$     b) $\displaystyle\int \dfrac{dx}{5 + 4\cos x}$

Find the integrals in Exercises M13–M18.

**M13.** $\displaystyle\int \dfrac{x\, dx}{\sqrt{x - 1}}$

**M14.** $\displaystyle\int 5x^3\sqrt{36 - x^2}\, dx$

**M15.** $\displaystyle\int \dfrac{dx}{2 + \tan x}$

**M16.** $\displaystyle\int \dfrac{dx}{\sqrt{1 + e^x}}$

**M17.** $\displaystyle\int (1 + x^2)^{-3/2}\,\mathrm{Tan}^{-1} x\, dx$

**M18.** $\displaystyle\int x^x(1 + \ln x)\, dx$

Evaluate the definite integrals given in Exercises M19–M22.

**M19.** $\displaystyle\int_4^9 \dfrac{dx}{x + \sqrt{x}}$

**M20.** $\displaystyle\int_1^{\sqrt{3}} \dfrac{dx}{x\sqrt{1 + x^2}}$

**M21.** $\displaystyle\int_1^e \dfrac{\ln^2 x}{x + x\ln x}\, dx$

**M22.** $\displaystyle\int_0^3 x^3\sqrt{9 - x^2}\, dx$

**M23.** Evaluate $\displaystyle\int \frac{dx}{\sqrt{20 + 8x - x^2}}$.

**M24.** Evaluate $\displaystyle\int \frac{x + 3}{\sqrt{5 - 4x - x^2}}\, dx$.

**M25.** a) Prove that

$$\int \frac{dx}{\sqrt{x^2 + a^2}} = \ln\left(x + \sqrt{x^2 + a^2}\right) + C.$$

b) Use the above result to evaluate

$$\int \frac{x + 2}{\sqrt{x^2 + 9}}\, dx.$$

**M26.** a) Prove that

$$\int \frac{dx}{\sqrt{x^2 - a^2}} = \ln\left|x + \sqrt{x^2 - a^2}\right| + C.$$

b) Use the above result to evaluate

$$\int \frac{x + 2}{\sqrt{x^2 + 2x - 3}}\, dx.$$

## 5.6   *Improper Integrals*

The definite integrals considered earlier all had the form

$$\int_a^b f$$

and we assumed that

$$-\infty < a < b < \infty$$

and that

$f$ was continuous on $[a, b]$.

We will relax these assumptions and give five examples of *improper integrals*.

### A. The Function "Blows Up" at Some Point

---

*Example 1*   Evaluate $\displaystyle\int_0^1 \frac{dx}{\sqrt{x}}$.

*Solution*   At first sight, this appears to be an ordinary definite integral, but notice that the function

$$y = \frac{1}{\sqrt{x}} \quad \text{(Fig. 1)}$$

is not defined at $x = 0$, and $y \to \infty$ as $x \to 0^+$.

We do not give the general definition of an improper integral in the text because the examples serve to illustrate the concept well enough. Moreover, a systematic discussion of these and more general types of integrals requires considerable preparation. The subject is usually treated in an upper division mathematics course on integration theory.

Consider first the case where $f$ is continuous on an interval of the form $[a, b)$, but not on $[a, b]$. Then for any $a < c < b$ the integral $\int_a^c f(x)\, dx$ is defined. We then define

$$\int_a^b f(x)\, dx = \lim_{c \to b^-} \int_a^c f(x)\, dx \quad (a < c < b).$$

If the limit exists and is a real number, the integral is called *convergent*. Otherwise, it is said to be *divergent*. The case when $f$ is continuous on $(a, b]$ is similarly treated.

When $f$ is continuous on $[a, b]$ *except* at some point $m \in (a, b)$ the situation is slightly more involved. One first computes $\int_a^m f(x)\, dx$ and $\int_m^b f(x)\, dx$ separately by the above limit procedure. If *both* of these limits exist then $\int_a^b f(x)\, dx$ is defined to be

$$\int_a^m f(x)\, dx + \int_m^b f(x)\, dx.$$

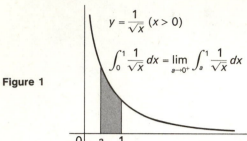

**Figure 1**

$$y = \frac{1}{\sqrt{x}} \ (x > 0)$$

$$\int_0^1 \frac{1}{\sqrt{x}} \, dx = \lim_{a \to 0^+} \int_a^1 \frac{1}{\sqrt{x}} \, dx$$

To compute the integral we first compute the ordinary integral $\int_a^1 \frac{dx}{\sqrt{x}}$ for $0 < a < 1$ and then let $a \to 0^+$. In other words, we define

$$\int_0^1 \frac{dx}{\sqrt{x}} = \lim_{a \to 0^+} \int_a^1 \frac{dx}{\sqrt{x}} \quad (0 < a < 1).$$

Since

$$\int_a^1 \frac{dx}{\sqrt{x}} = 2\sqrt{x}\,\Big|_a^1 = 2 - 2\sqrt{a},$$

and $2 - 2\sqrt{a} \to 2$ as $a \to 0^+$, we have

$$\int_0^1 \frac{dx}{\sqrt{x}} = 2.$$

This is an example of a *convergent* improper integral.

---

**Example 2** Evaluate $\int_{-1}^0 \frac{dx}{x^2}$.

*Solution* In this case $y = 1/x^2$ is continuous on $[-1, 0)$, but not on the closed interval $[-1, 0]$. To evaluate the integral we compute for $-1 < b < 0$

$$\int_{-1}^b \frac{dx}{x^2} = \frac{-1}{x}\,\Big|_{-1}^b = \frac{-1}{b} - 1.$$

As $b \to 0^-$ the expression $\frac{-1}{b} - 1$ tends to $\infty$. Hence no limit exists and the integral is said to be *divergent*.

---

**B. Infinite Domains** In this case the function is continuous, but the interval over which it is to be integrated is infinite, such as $[a, \infty)$, $(-\infty, b]$, or $(-\infty, \infty)$.

It is interesting to note that $\int_a^b f(x)\,dx$ is not defined as

$$\lim_{c \to m} \left[ \int_a^c f(x)\,dx + \int_c^b f(x)\,dx \right].$$

Example 3 (5.1D) dealing with $\int_{-3}^2 \frac{dx}{x^2}$ illustrates this. Using the above formula would give

$$\int_{-3}^2 \frac{dx}{x^2} = \lim_{c \to 0} \left( \int_{-3}^c \frac{dx}{x^2} + \int_c^2 \frac{dx}{x^2} \right) = \lim_{c \to 0} \left( \frac{-1}{x}\Big|_{-3}^c + \frac{-1}{x}\Big|_c^2 \right)$$

$$= \lim_{c \to 0} \left( \left(\frac{-1}{c}\right) - \left(\frac{-1}{-3}\right) + \left(\frac{-1}{2}\right) - \left(\frac{-1}{c}\right) \right)$$

$$= \lim_{c \to 0} \left( -\frac{1}{3} - \frac{1}{2} \right) = -\frac{5}{6}.$$

This result is clearly absurd since $f(x) = \frac{1}{x^2} > 0$ for $x \neq 0$. The correct procedure is to evaluate the limits separately. Thus

$$\int_{-3}^2 \frac{dx}{x^2} = \lim_{c \to 0^-} \int_{-3}^c \frac{dx}{x^2} + \lim_{c \to 0^+} \int_c^2 \frac{dx}{x^2}$$

$$= \lim_{c \to 0^-} \frac{-1}{x}\Big|_{-3}^c + \lim_{c \to 0^+} \frac{-1}{x}\Big|_c^2$$

$$= \lim_{c \to 0^-} \left( \frac{-1}{c} - \frac{-1}{-3} \right) + \lim_{c \to 0^+} \left( \frac{-1}{2} + \frac{1}{c} \right).$$

Since neither limit exists (Example 2), $\int_{-3}^2 \frac{dx}{x^2}$ is divergent.

*Example 3*   Evaluate $\int_1^\infty \dfrac{dx}{x^2}$.

*Solution*   The procedure in this case is similar to that of Example 1. See Figure 2. We evaluate $\int_1^b \dfrac{dx}{x^2}$ for $1 < b < \infty$ and then let $b \to \infty$. Since

$$\int_1^b \frac{dx}{x^2} = \frac{-1}{x}\bigg|_1^b = \frac{-1}{b} - \frac{-1}{1} = \frac{-1}{b} + 1$$

and $\frac{-1}{b} + 1$ clearly approaches 1 as $b \to \infty$, we see that $\int_1^\infty \dfrac{dx}{x^2} = 1$. The integral is called convergent.

*Example 4*   Evaluate $\int_1^\infty \dfrac{dx}{x}$.

*Solution*   We have $\int_1^b \dfrac{dx}{x} = \ln x \bigg|_1^b = \ln b - \ln 1 = \ln b$. Since $\ln b \to \infty$ as $b \to \infty$, the expression $\int_1^b \dfrac{dx}{x}$ does not have a limit as $b \to \infty$. We often write

$$\int_1^\infty \frac{dx}{x} = \infty.$$ This integral is divergent.

*Example 5*   Evaluate $\int_{-\infty}^\infty \dfrac{1}{1 + x^2}\, dx$.

*Solution*   Since both the upper and lower limits of integration are infinite, we split the integral into two parts. We have

$$\int_{-\infty}^\infty \frac{1}{1 + x^2}\, dx = \int_{-\infty}^0 \frac{1}{1 + x^2}\, dx + \int_0^\infty \frac{1}{1 + x^2}\, dx$$

and

$$\int_{-\infty}^0 \frac{1}{1 + x^2}\, dx = \lim_{b \to -\infty} \int_b^0 \frac{1}{1 + x^2}\, dx$$

$$= \lim_{b \to -\infty} (\mathrm{Tan}^{-1} x)\bigg|_b^0 = \lim_{b \to -\infty} (-\mathrm{Tan}^{-1} b)$$

$$= -\left(-\frac{\pi}{2}\right) = \frac{\pi}{2}.$$

Also

$$\int_0^\infty \frac{1}{1 + x^2}\, dx = \lim_{b \to \infty} \int_0^b \frac{1}{1 + x^2}\, dx$$

$$= \lim_{b \to \infty} (\mathrm{Tan}^{-1} x)\bigg|_0^b = \lim_{b \to \infty} (\mathrm{Tan}^{-1} b) = \frac{\pi}{2}.$$

Hence

$$\int_{-\infty}^\infty \frac{1}{1 + x^2}\, dx = \frac{\pi}{2} + \frac{\pi}{2} = \pi.$$

**Figure 2**

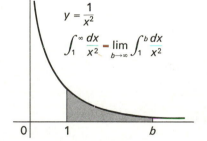

$$y = \frac{1}{x^2}$$

$$\int_1^\infty \frac{dx}{x^2} = \lim_{b \to \infty} \int_1^b \frac{dx}{x^2}$$

## C. Exercises

a) 4    b) 4    c) 6    d) 4    e) −27  f) 6

**B1.** Evaluate the following improper integrals.

a) $\int_4^8 \dfrac{1}{\sqrt{x-4}}\,dx$

b) $\int_{-3}^1 \dfrac{1}{\sqrt{1-x}}\,dx$

**B2.** Show that $\int_0^1 \dfrac{dx}{\sqrt{1-x^2}} = \dfrac{\pi}{2}$.

**B3.** Evaluate the following improper integrals.

a) $\int_1^\infty \dfrac{dx}{1+x^2}$

b) $\int_0^\infty xe^{-x}\,dx$   (Hint: $xe^{-x} \to 0$  as  $x \to \infty$)

**B4.** Show that the integrals below are divergent improper integrals.

a) $\int_0^{16} \dfrac{dx}{x^{5/4}}$

b) $\int_2^\infty \dfrac{x\,dx}{x^2+2}$

c) $\int_0^\infty \dfrac{dx}{\sqrt{x+1}}$

**B5.** The value of $\int_0^2 \dfrac{5x^2\,dx}{(x^3-1)^2}$ does *not* equal $\dfrac{-40}{21}$. Explain why and what error may cause this faulty answer to arise.

**D6.** Evaluate the following improper integrals.

a) $\int_2^\infty \dfrac{dx}{x^2}$    d) $\int_1^\infty \dfrac{dx}{x^4}$

b) $\int_4^\infty \dfrac{dx}{x^2}$    e) $\int_{-1}^\infty \dfrac{dx}{(x+2)^2}$

c) $\int_1^\infty \dfrac{dx}{x^3}$    f) $\int_{-\infty}^{-1} \dfrac{dx}{x^3}$

a) $\frac{1}{2}$    b) $\frac{1}{4}$    c) $\frac{1}{2}$    d) $\frac{1}{3}$    e) 1    f) $-\frac{1}{2}$

**D7.** Evaluate the following improper integrals.

a) $\int_0^4 \dfrac{dx}{\sqrt{x}}$       d) $\int_1^5 \dfrac{dx}{\sqrt{x-1}}$

b) $\int_0^4 \dfrac{dx}{\sqrt{4-x}}$     e) $\int_{-27}^0 \dfrac{2\,dx}{x^{1/3}}$

c) $\int_0^8 \dfrac{dx}{x^{1/3}}$        f) $\int_2^{10} \dfrac{dx}{(x-2)^{2/3}}$

**M8.** Evaluate $\int_{-3}^0 \dfrac{dx}{\sqrt{9-x^2}}$.

**M9.** Evaluate $\int_0^{3/4} (9-16x^2)^{-1/2}\,dx$.

**M10.** Evaluate $\int_{-\infty}^\infty \dfrac{1}{e^x+e^{-x}}\,dx$.

**M11.** Evaluate $\int_0^\infty e^{-x}\sin x\,dx$.

$\left(\text{Hint: } e^{-x}\dbinom{\cos x}{\sin x} \to 0 \quad \text{as } x \to \infty.\right)$

**M12.** For what values of $a$ are the following improper integrals convergent?

a) $\int_0^1 x^a\,dx$    b) $\int_1^\infty x^a\,dx$

**M13.** Show that

a) $\int_0^4 \dfrac{dx}{(x+4)\sqrt{x}} = \dfrac{\pi}{4}$

b) $\int_0^\infty \dfrac{dx}{(x+4)\sqrt{x}} = \dfrac{\pi}{2}$

**M14.** Evaluate $\int_0^{\pi/2} \dfrac{\cos x}{\sqrt{1-\sin x}}\,dx$.

**M15.** Evaluate $\int_{-2}^2 \dfrac{dx}{\sqrt{4-x^2}}$.

**M16.** Evaluate $\int_2^\infty \dfrac{dx}{x\ln^2 x}$.

**M17.** Evaluate $\int_{-\infty}^\infty \dfrac{dx}{1+4x^2}$.

**M18.** Evaluate $\int_1^\infty \dfrac{e^{-\sqrt{x}}}{\sqrt{x}}\,dx$.

**M19.** Evaluate $\int_{-\infty}^0 x^3 e^x\,dx$.

(Hint: $x^n e^x \to 0$  as $x \to -\infty$.)

**M20.** Determine whether or not each of the following integrals is convergent.

a) $\int_0^1 \dfrac{\cos t}{t^2}\,dt$

$\left( \text{Hint: Use the fact that } \dfrac{\cos t}{t} \to \infty \quad \text{as} \quad t \to 0. \right)$

b) $\displaystyle\int_{1}^{2} \dfrac{x}{\ln x}\, dx$   (Hint: Use a substitution.)

**M21.** Evaluate $\displaystyle\int_{1}^{\infty} \dfrac{\ln x}{x^2}\, dx$.

**M22.** Evaluate

   a) $\displaystyle\int_{0}^{1} \ln x \, dx$    b) $\displaystyle\int_{0}^{1} x \ln x \, dx$    c) $\displaystyle\int_{0}^{1} x^2 \ln x \, dx$

**M23.** Evaluate $\displaystyle\int_{0}^{\infty} x^n e^{-x}\, dx$    for    $n = 1, 2, 3, \ldots$.

---

## 5.7   Sample Exams

### Sample Exam 1 (45–60 minutes)

**1.** Find the following indefinite integrals.

   a) $\displaystyle\int (4 + 2x - x^3)^4 (2 - 3x^2)\, dx$

   b) $\displaystyle\int e^{\sin x} \cos x \, dx$    c) $\displaystyle\int \dfrac{8x}{4x^2 + 5}\, dx$

**2.** Find the area of the region bounded by the graph of $y = x\sqrt{x-1}$ and the lines $y = 0$, $x = 1$, and $x = 2$.

**3.** Evaluate the following definite integrals.

   a) $\displaystyle\int_{0}^{4} \dfrac{x}{x^2 + 9}\, dx$    b) $\displaystyle\int_{-8}^{0} \dfrac{2}{x + 9}\, dx$

**4.** Find $\displaystyle\int \dfrac{5x + 1}{(x-1)(x+2)}\, dx$.

### Sample Exam 2 (45–60 minutes)

**1.** Find the following.

   a) $\displaystyle\int (3x^2 - 5)^8 3x \, dx$

   b) $\displaystyle\int \dfrac{1}{\sqrt{9 - x^2}}\, dx$    c) $\displaystyle\int 2(x^2 + 2)^2 \, dx$

**2.** Use the substitution $u = \sqrt{1 - x^2}$ to find

$$\int x^3 \sqrt{1 - x^2}\, dx.$$

**3.** Integrate by parts to find $\int (\ln x)^2 \, dx$.

**4.** Evaluate $\displaystyle\int_{0}^{\ln\sqrt{3}} \dfrac{e^x}{1 + e^{2x}}\, dx$.

**5.** Evaluate $\displaystyle\int_{0}^{\pi/2} \cos^2 2x \, dx$.

### Sample Exam 3 (45–60 minutes)

**1.** Evaluate the following definite integrals.

   a) $\displaystyle\int_{0}^{1/2} \dfrac{\operatorname{Sin}^{-1} x}{\sqrt{1 - x^2}}\, dx$    b) $\displaystyle\int_{0}^{1} \dfrac{\operatorname{Tan}^{-1} x}{1 + x^2}\, dx$

**2.** Use the substitution $u = a \tan x$ to find

$$\int \dfrac{1}{a^2 \sin^2 x + \cos^2 x}\, dx.$$

**3.** Find $\displaystyle\int \dfrac{2}{3x^2 + 6x + 15}\, dx$.

**4.** Find $\displaystyle\int e^{\cos x} \sin x \cos x \, dx$.

   (Hint: Begin by letting $u = \cos x$.)

**5.** Given $g(x) = \displaystyle\int_{x}^{3x} f(t)\, dt$ and $f'(0) = 2$, find $g''(0)$.

### Sample Exam 4 (45–60 minutes)

**1.** Find

   a) $\displaystyle\int \tan x \, dx$    b) $\displaystyle\int \tan^2 x \, dx$    c) $\displaystyle\int \tan^3 x \, dx$

**2.** Use the substitution $u = \tan \frac{x}{2}$ to evaluate

$$\int \dfrac{dx}{2 - 3\cos x}.$$

**3.** Evaluate $\displaystyle\int_{0}^{\pi} e^x \sin x \, dx$.

**4.** Prove that there is no function $f$ that satisfies the equation

$$\int_{1}^{x} f(u)\, du = x^2 + x.$$

(Hint: Start by differentiating each side of the equation.)

**5.** Find $f$ given

$$\int_0^{2x-3} f(u)\, du = 2x^2 - 4x + \tfrac{3}{2}.$$

(Hint: Start by differentiating each side of the equation.)

## Sample Exam 5 (Chapters 1–5, approximately 2 hours)

**1.** Find $\frac{dy}{dx}$ given
   a) $y = (9 + x^{-4})^4$
   b) $y = \mathrm{Tan}^{-1} e^{2x}$
   c) $y = \sin(\ln x^3)$

**2.** Find $y'$ in terms of $x$ and $y$ given
$$x^2 + y^2 = \ln(1 + y^2).$$

**3.** Find $y''$ in terms of $x$ and $y$ given
$$x^2 - 4y^2 = 10.$$

**4.** Determine the region where the function $f(x) = \frac{1}{x} \ln x$ is increasing.

**5.** The side of an equilateral triangle increases at a rate of 2 in./hr. At what rate is the area changing when the length of the side is 4 inches?

**6.** Find numbers $x$ and $y$ which satisfy $2x + 3y = 8$ and whose product is a maximum.

**7.** Find

   a) $\int \dfrac{\mathrm{Tan}^{-1} 2x}{1 + 4x^2}\, dx$        b) $\int (\sin x)(\ln \cos x)\, dx$

**8.** Find $\int x \ln \sqrt{x}\, dx$.

**9.** Find $\int \dfrac{x - 6}{(x - 1)(x + 4)}\, dx$.

**10.** Compute

   a) $\dfrac{d}{dx}\left(\displaystyle\int_0^{x^2} e^{-t^2}\, dt\right)\Big|_{x=1}$        b) $\displaystyle\int_0^1 \dfrac{d}{dt}(\mathrm{Sin}^{-1} t\, \mathrm{Tan}^{-1} t)\, dt$

## Sample Exam 6 (Chapters 1–5, approximately 3 hours)

**1.** Given $f(x) = \sin x$, $g(x) = \cos x$, and $h(x) = x^2$, find
   a) $h(f(x)) + h(f'(x))$        c) $\int(h \circ g)(x)\, dx$
   b) $(f \circ h)'(x)$        d) $\int(g \circ h \circ g)(x) f(2x)\, dx$

**2.** Given $u(x) = e^x$, $v(x) = \ln x$, and $h(x) = x^2$, find
   a) $(u \circ v)(x) + (v \circ u)(x)$
   b) $(h \circ u \circ v)'(x)$
   c) $(u \circ h \circ v)'(x)$
   d) $\int(v \circ u)(x)(h \circ u)(x)\, dx$

**3.** Find $f'(x)$ given
   a) $f(x) = \displaystyle\int_0^{x^2} (1 + x^3)^{10}\, dx$
   b) $f(x) = \displaystyle\int_0^{x^2}\left(\int_0^u y^2\, dy\right) du$
   c) $f''(x) = (1 + x^3)^{10} x^2$

**4.** Given the function
$$f(x) = 3x^4 + 4x^3 - 12x^2 + 20.$$
   a) Find its critical points.
   b) Which critical points are local maxima and which critical points are local minima?

**5.** A box, open at the top, is made from a rectangular piece of cardboard that is 8 inches by 5 inches by cutting away square corners and then turning up the sides. What is the maximum possible volume for such a box?

**6.** The Otto and Sorani Canals in Venice are 8 feet and 27 feet wide respectively and cross at right angles. What is the length of the largest gondola that can pass from the Otto Canal into the Sorani Canal?

**7.** An isosceles triangle is inscribed in the circle $(x - 3)^2 + y^2 = 9$ with the vertex opposite the unequal side located at $(0, 0)$. Find the largest area such a triangle can have.

**8.** A conical paper cup of radius 2 inches and height 6 inches is filled with water. The water begins to leak out of a small hole in the bottom at a rate of 2 in.$^3$ per minute. How fast is the level of the water falling when the volume of water in the cup is $\pi/8$ in.$^3$?

**9.** Find the following integrals.
   a) $\displaystyle\int \dfrac{dx}{(1 + x^2)\, \mathrm{Tan}^{-1} x}$
   b) $\displaystyle\int (\mathrm{Sin}^{-1} x)^{-1}(1 - x^2)^{-1/2}\, dx$

**10.** Find
   a) $\displaystyle\int \dfrac{1}{x^2 + x - 2}\, dx$
   b) $\displaystyle\int \dfrac{x^4 + x^3 - 2x^2 + 1}{x^2 + x - 2}\, dx$

The formulation of the definite integral as a limit of sums is used to derive formulas for volumes, arc lengths, and surface areas. We also define several important physical quantities in terms of integrals and then study some applications. In the final section we give some techniques that can be used to approximate the value of a definite integral when an antiderivative of the integrand is difficult to find.

# Applications of Integration

## 6.1　Another Look at Area

**A. The Area Between Two Curves**　Suppose $f$ and $g$ are continuous functions on $[a, b]$ and suppose $f \geq g$ as shown in Figure 1. We want to derive a formula for the area $A$ between $f$ and $g$ from $x = a$ to $x = b$. We divide the interval $[a, b]$ into $n$ equal subintervals of length $\Delta x = \frac{b-a}{n}$. The area of the rectangle shown in Figure 1 is

$$(f(x) - g(x)) \Delta x,$$

the base times the altitude. The area we are looking for is given approximately by

$$\sum (f(x) - g(x)) \Delta x,$$

the sum of the areas of all such rectangles. See Figure 2. Now let $\Delta x \to 0$, or equivalently let $n \to \infty$. Then the sum approaches the required area. But by definition of the integral

$$\lim_{n \to \infty} \sum (f(x) - g(x)) \Delta x = \int_a^b (f(x) - g(x)) \, dx.$$

Thus the formula is

$$A = \int_a^b (f(x) - g(x)) \, dx.$$

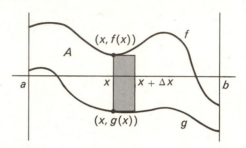

**Figure 1**

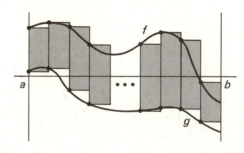

**Figure 2**

*Example 1*　Find the area of the finite region bounded by the graphs of $y = x^2$ and $y = x + 2$.

*Solution*　Referring to the graphs of the two functions (Fig. 3), we observe that the region is between the vertical lines passing through the points of intersection of the two curves. These points of intersection are $(-1, 1)$ and $(2, 4)$, and the desired area is given by

$$A = \int_{-1}^2 (x + 2 - x^2) \, dx = \left( \frac{x^2}{2} + 2x - \frac{x^3}{3} \right) \Big|_{-1}^2$$

$$= (2 + 4 - \tfrac{8}{3}) - (\tfrac{1}{2} - 2 + \tfrac{1}{3}) = \tfrac{9}{2}$$

→　*In problems of this type, the better your figure is, the less chance there is of making some ridiculous mistake. At least sketch a rough figure.*

*(Fred)*

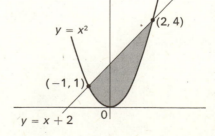

**Figure 3**

Given two functions $f$ and $g$ with $f(x) \geq g(x)$ on $[a, b]$, the area between them bounded by the lines $x = a$ and $x = b$ is *always* given by $\int_a^b (f(x) - g(x)) \, dx$

whether or not the functions themselves are nonnegative. When there is more than one region between the curves (Example 3) one must be careful to find the areas separately.

**Example 2**  Find the area of the finite region bounded by the graphs of $f(x) = -x^2 - 1$ and $g(x) = -2x - 1$.

*Solution*  Referring to the graphs of the two functions, we observe that the region is between the vertical lines $x = 0$ and $x = 2$ (Fig. 4).

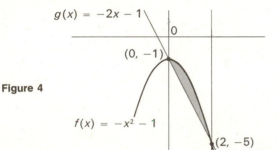

Figure 4

Notice that $f \geq g$ on $[0, 2]$. Hence the area is equal to

$$\int_0^2 (f - g) = \int_0^2 (2x - x^2)\, dx$$

$$= x^2 - \frac{x^3}{3}\Big|_0^2 = 4 - \frac{8}{3} = \frac{4}{3}.$$

**Example 4**  Find the area of the region bounded by the graphs of $f(x) = \sin x$, $g(x) = \cos x$ and the lines $x = \frac{\pi}{2}$ and $x = \frac{5\pi}{4}$.

*Solution*  Graph the functions as shown in Figure 6 with $x$ restricted to the interval $[\frac{\pi}{2}, \frac{5\pi}{4}]$. The desired area is then the area of the region that lies below the graph of $\sin x$ and above the graph of $\cos x$. Since in this interval, $\sin x \geq \cos x$, the area between the curves equals

$$\int_{\pi/2}^{5\pi/4} (\sin x - \cos x)\, dx = -\cos x - \sin x \Big|_{\pi/2}^{5\pi/4}$$

**Example 3**  Find the total area of the regions enclosed by the graphs of the functions $f(x) = 2x$ and $g(x) = 6x - x^3$.

*Solution*  Graphing the functions, we see that there are two distinct regions involved. In one region $f \geq g$ and in the other region $g \geq f$. The regions lie between the vertical lines that pass through the points of intersection, which are $(-2, -4)$, $(2, 4)$, and $(0, 0)$ (Fig. 5).

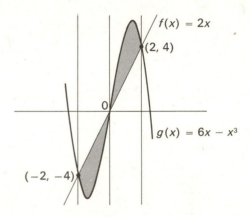

Figure 5

The areas must be computed separately; thus

$$A = \int_{-2}^0 (f(x) - g(x))\, dx + \int_0^2 (g(x) - f(x))\, dx$$

$$= \int_{-2}^0 (2x - 6x + x^3)\, dx + \int_0^2 (6x - x^3 - 2x)\, dx$$

$$= -2x^2 + \frac{x^4}{4}\Big|_{-2}^0 + 2x^2 - \frac{x^4}{4}\Big|_0^2$$

$$= 4 + 4 = 8.$$

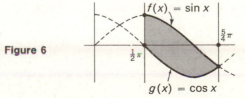

Figure 6

$$= \left(-\cos\frac{5\pi}{4} - \sin\frac{5\pi}{4}\right) - \left(-\cos\frac{\pi}{2} - \sin\frac{\pi}{2}\right)$$

$$= \left(-\left(\frac{-1}{\sqrt{2}}\right) - \left(\frac{-1}{\sqrt{2}}\right)\right) - (0 - 1)$$

$$= \frac{2}{\sqrt{2}} + 1 = 1 + \sqrt{2}.$$

**Problem 1** Find the total area of the regions enclosed by the graphs of the functions $f(x) = x^3 - 4x$ and $g(x) = -x^2 + 2x$.

*Answer:* $\frac{253}{12}$

---

**Example 5** Find the area of the region enclosed by the graphs of $y^2 = 4x$ and $y = 2x - 4$.

*Solution* In this problem it is more convenient to obtain the area by integrating with respect to $y$ rather than with respect to $x$. Solving the equations for $x$ in terms of $y$ we obtain $x = f(y) = \frac{y+4}{2}$ and $x = g(y) = y^2/4$. Graph the equations as shown in Figure 7. Observe that the region is between the horizontal lines $y = -2$ and $y = 4$ and bounded by the graphs of the functions $f(y)$ and $g(y)$ on the sides, where $f \geq g$ for $-2 \leq y \leq 4$. The area is now given by

**Figure 7**

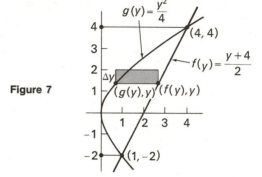

$$\int_{-2}^{4} (f(y) - g(y))\, dy = \int_{-2}^{4} \left(\frac{y+4}{2} - \frac{y^2}{4}\right) dy$$

$$= \int_{-2}^{4} \left(-\frac{y^2}{4} + \frac{y}{2} + 2\right) dy$$

$$= \left(-\frac{y^3}{12} + \frac{y^2}{4} + 2y\right)\Big|_{-2}^{4} = 9$$

**B. Area in Terms of Polar Coordinates** We will now derive a formula for the area enclosed by a given curve when the curve is defined in terms of polar coordinates.

Suppose $r = f(\theta)$. We will find the area of the region from the line $\theta = \theta_1$ to the line $\theta = \theta_2$ that is bounded by the graph of $r = f(\theta)$ (Fig. 8).

Consider the small region cut off by the angle $\Delta\theta$ (Fig. 9). The circular sector in Figure 9 has area

$$\frac{f(\theta)^2 \Delta\theta}{2}$$

as we know from 1.5B. Hence the area we are looking for is given approximately by

**Figure 8**

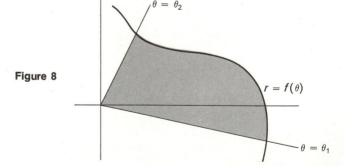

**Figure 9**

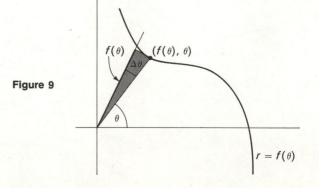

$$\sum \frac{f(\theta)^2}{2} \Delta\theta$$

(Fig. 10). We assume here that the angle determined by $\theta = \theta_1$ and $\theta = \theta_2$ is divided into equal subdivisions $\Delta\theta$ and the sum is extended over them.

If we let $\Delta\theta \to 0$, the sum approaches the area. But

$$\sum \frac{f(\theta)^2}{2} \Delta\theta \to \int_{\theta_1}^{\theta_2} \frac{f(\theta)^2}{2} \, d\theta.$$

The area is therefore given by the formula

$$A = \int_{\theta_1}^{\theta_2} \frac{f(\theta)^2}{2} \, d\theta.$$

As a first example, we will again derive the formula for the area of a circle.

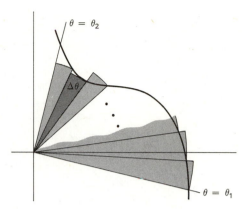

**Figure 10**

---

*Example 6*  Prove that the area of a circle with radius $a$ is $\pi a^2$.

*Solution*  The equation of the circle of radius $a$ centered at the origin is $r = a$. According to the above formula the area is

$$A = \int_0^{2\pi} \frac{f(\theta)^2}{2} \, d\theta = \int_0^{2\pi} \frac{a^2}{2} \, d\theta = \frac{a^2}{2} \theta \Big|_0^{2\pi} = \pi a^2. \quad \longrightarrow$$

---

We have not gained anything from this example. Indeed, the formula

$$\int_{\theta_1}^{\theta_2} \frac{f(\theta)^2}{2} \, d\theta$$

was derived from the formula $a^2\theta/2$ for the area of a sector of radius $a$ and angle $\theta$. This formula is in turn based on the formula $\pi a^2$ for the area of a circle of radius $a$. Hence the example does not provide us with an independent derivation for the area of a circle.

*In other words, our reasoning has been circular.*

(Estelle)

---

*Example 7*  Find the area enclosed by one loop of the curve $r = a \sin 2\theta$.

*Solution*  The graph of $r = a \sin 2\theta$ for $0 \le \theta \le \frac{\pi}{2}$ is shown in Figure 11. The area of the loop is therefore given by

$$A = \int_0^{\pi/2} \frac{f(\theta)^2}{2} \, d\theta = \int_0^{\pi/2} \frac{a^2}{2} \sin^2 2\theta \, d\theta$$

$$= \frac{a^2}{2} \int_0^{\pi/2} (\sin^2 2\theta) \, d\theta$$

$$= \frac{a^2}{4} \left( \theta - \frac{\sin 4\theta}{4} \right) \Big|_0^{\pi/2} = \frac{a^2\pi}{8}.$$

**Figure 11**

$(a, \frac{1}{4}\pi)$

$r = a \sin 2\theta$

## Continuity, Area, and Rigor

The notion of continuity slowly evolved from the intuitive idea of an unbroken curve to a rigorous definition, given by Karl Weierstrass (1815–1897), who phrased the notion in terms of $\epsilon$ and $\delta$. (See 9.3A for this definition.) The modern definition of continuity in terms of open sets did not appear until the twentieth century.

Paralleling the evolution of the notion of continuity was that of the area of a region. This concept also began as an intuitively evident notion. During the nineteenth century, it too was investigated carefully and was found to be imprecise.

By current standards the foundations of mathematics were in a pathetic state through the eighteenth century. Mathematicians used the words "function," "limit," "infinitesimal," "continuity," and "area," but could not give a precise definition of any of them. Newton and Leibniz had been aware of these difficulties but only halfheartedly attempted to correct them. In the eighteenth century, mathematicians began to consider these problems, but did little to rectify them. The inadequacies of the foundations of mathematics were most clearly pointed out by an amateur mathematician, George Berkeley (1685–1753), later to become the Bishop of Cloyne, Ireland. That Berkeley attempted to make fools of the mathematicians is evidenced by the title of his work: *The analyst: or, a discourse addressed to an infidel mathematician. Wherein it is examined whether the object, principles, and inferences of the modern analysis are more distinctly conceived, or more evidently deduced, than religious mysteries or points of faith.* But Berkeley, to his great annoyance, was virtually ignored by mathematicians, who were not yet ready to deal with subtleties. In the nineteenth century they finally attacked the foundations, and their debt to the past was great. As E. T. Bell (*The Development of Mathematics,* McGraw-Hill, 1945) has said, "The mistakes and unresolved difficulties of the past in mathematics have always been the opportunities of its future; and should analysis ever appear without flaw or blemish its perfection might only be that of death."

The nineteenth century mathematicians did resolve most of the difficulties. In addressing the second international congress of mathematicians in 1900, Henri Poincaré (1854–1912), the greatest mathematician of the time, stated, "Now, in analysis today, if we take the pains to be rigorous there are only syllogisms or appeals to the intuition of pure number that could possibly deceive us. We may say today that absolute rigor has been attained."

In Chapters 9 and 10 everything done in the earlier chapters will be made rigorous. As we will see, our appeals "to the intuition of pure number" are not "without flaw or blemish."

### D. Exercises

**B1.** Find the area of the region bounded by the graph of $y = x^2$ and the lines
  a) $x = 2$, $x = 4$, and $y = 0$.
  b) $x = 2$, $x = 4$, and $y = 4$.

**B2.** Find the area of the region that lies above the $x$-axis and below the graph of $y = -x^2 + 4x - 3$.

**B3.** Determine the area of the region that lies between the parabolas $y = 6x - x^2$ and $y = x^2 - 2x$.

**B4.** Find the area of the region bounded by the graphs of $y^2 = 2x$ and $y = x - 4$. (First make a sketch of the desired region.)

**B5.** Find the area of the region bounded by the graphs of $y = \sin x$ and $y = \cos x$ between $x = 0$ and $x = \frac{\pi}{4}$.

**B6.** Find the area of the region enclosed by the graph of $r = 2 - \cos \theta$ using polar coordinates.

**D7.** Find the area of the regions bounded by the graphs of the given curves.
  a) $y = x^3$, $y = 0$, $x = 0$, $x = 3$

b) $y = x^3,\quad y = 0,\quad x = 2,\quad x = 4$
c) $y = 1 + x^2,\quad y = 1,\quad x = 0,\quad x = 2$
d) $y = 4 - x^2,\quad y = 0,\quad x = -2,\quad x = 2$
e) $y = 4 - x^2,\quad y = 0,\quad x = -1,\quad x = 1$

f) $y = \cos x,\quad y = 0,\quad x = \frac{\pi}{6},\quad x = \frac{\pi}{2}$

g) $y = \sec^2 x,\quad y = 0,\quad x = 0,\quad x = \frac{\pi}{4}$

h) $y = e^x,\quad y = 0,\quad x = 0,\quad x = \ln 4$

a) $\frac{81}{4}$     b) $60$     c) $\frac{8}{3}$     d) $\frac{32}{3}$     e) $\frac{22}{3}$

f) $\frac{1}{2}$     g) $1$     h) $3$

**D8.** Find the area of the regions bounded by the graphs of the given curves using polar coordinates.
a) $r = 4,\quad \theta = 0,\quad \theta = 2\pi$
b) $r = a,\quad \theta = 0,\quad \theta = 2\pi$
c) $r = 2,\quad r = 3,\quad \theta = 0,\quad \theta = \pi$
d) $r = \sec \theta,\quad \theta = 0,\quad \theta = \frac{1}{4}\pi$
e) $r = 2 \sec \theta,\quad \theta = 0,\quad \theta = \frac{1}{4}\pi$
f) $r = \sin \theta,\quad \theta = 0,\quad \theta = \frac{1}{2}\pi$
g) $r = \cos \theta,\quad \theta = 0,\quad \theta = \frac{1}{2}\pi$
h) $r = \sqrt{\cos \theta},\quad \theta = 0,\quad \theta = \frac{1}{2}\pi$

a) $16\pi$     b) $\pi a^2$     c) $\frac{5}{2}\pi$     d) $\frac{1}{2}$     e) $2$
f) $\frac{1}{8}\pi$     g) $\frac{1}{8}\pi$     h) $\frac{1}{2}$

**M9.** Find the area of the regions bounded by the graphs of
a) $y = \sqrt{x},\quad y = x$     c) $y = \sqrt{x},\quad y = x^2$
b) $y = x^2,\quad y = x$

**M10.** Find the area bounded by the graph of $y^2 = 2y - x$ and the $y$-axis.

**M11.** Find the area bounded by the graph of $y^2 = x^2 - x^4$.

**M12.** Find the area of the region that is common to the circles $x^2 + y^2 = 9$ and $(x - 3)^2 + y^2 = 9$.

**M13.** Find the area of the region that lies inside $r = 1 + \cos \theta$ and outside $r = 1$.

**M14.** Find the area of the region inside the graph of $r^2 = 2 \cos 2\theta$.

**M15.** a) Show that $\lim\limits_{x \to 0} x \ln x = 0$.

b) Find the area of the region that lies above the graph of $y = \ln x\ (0 < x \le 1)$ and below the $x$-axis. (This gives rise to an improper integral.)

**M16.** a) Find the area of the region below the graph of $y = e^{-x}$, above the $x$-axis, and to the right of the line $x = 0$.

b) Explain why the area in (a) is the same as that in Exercise M15b.

**M17.** Find the area of the region bounded by the graphs of $y = x^3$ and $y = x^{1/3}$.

**M18.** Find the area of the region bounded by the graphs of $y = e^{x/2}$ and $y = 1/x^2$, and the lines $x = 2$ and $x = 3$.

**M19.** Find the area of the region bounded by the graph of $y = \dfrac{x}{\sqrt{2x^2 + 1}}$, the $x$-axis, and the lines $x = 0$ and $x = 2$.

**M20.** Find the value of $k\ (k > 0)$ for which the area bounded by the graphs of $y = x^2 - k^2$ and $y = k^2 - x^2$ is 9.

---

## 6.2  *Volume*

The regions considered in this section are very special. To find volumes of more general regions and their higher dimensional analogs, the notion of a multiple integral is required.

We will now find formulas, expressed in integrals, for the volumes of certain solids in space.

**A. Regions Obtained by Revolution**  Consider the function $y = f(x)$ on the interval $[a, b]$, and let $A$ denote the shaded region (Fig. 1). Rotating this region about the $x$-axis generates the solid region shown in

**Figure 1**

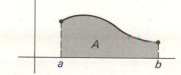

Figure 2. We will now derive a formula for its volume $V$.

Take the points $x$ and $x + \Delta x$ as shown in Figure 3a and denote the volume of the slice by $\Delta V$. The disc in Figure 3b has volume equal to $\pi(f(x))^2 \Delta x$ and, for $\Delta x$ small, approximates $\Delta V$.

Now divide the interval $[a, b]$ into $n$ equal intervals of length $\Delta x$ (Fig. 4a).

Then $\sum \pi(f(x))^2 \Delta x$ is the sum of the volumes of the discs (Fig. 4b). As $\Delta x$ gets smaller, this sum approximates the volume $V$. Hence we have that

$$\sum \pi(f(x))^2 \Delta x \to V \quad \text{as} \quad \Delta x \to 0.$$

On the other hand, we have

$$\sum \pi(f(x))^2 \Delta x \to \int \pi(f(x))^2 \, dx \quad \text{as} \quad \Delta x \to 0,$$

and this gives the formula

(*) 
$$V = \int_a^b \pi(f(x))^2 \, dx.$$

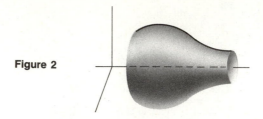

**Figure 2**

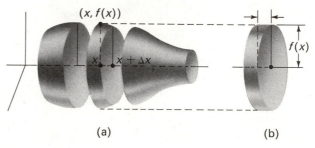

(a)           (b)

**Figure 3**

---

**Example 1** We will show that the volume of a sphere of radius $a$ is equal to $\frac{4}{3}\pi a^3$.

*Solution* Rotate the semicircle $y = \sqrt{a^2 - x^2}$ $(-a \le x \le a)$ about the $x$-axis (Fig. 5) to obtain the sphere of radius $a$. According to formula (*),

$$V = \int_{-a}^a \pi y^2 \, dx = \int_{-a}^a \pi(\sqrt{a^2 - x^2})^2 \, dx$$

$$= \pi \int_{-a}^a (a^2 - x^2) \, dx = \pi \left( a^2 x - \frac{x^3}{3} \right) \Big|_{-a}^a$$

$$= \pi \left( a^3 - \frac{a^3}{3} \right) - \pi \left( -a^3 + \frac{a^3}{3} \right)$$

$$= \frac{4\pi a^3}{3}.$$

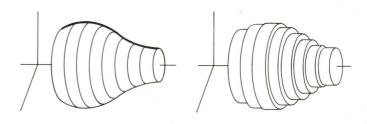

(a)           (b)

**Figure 4**

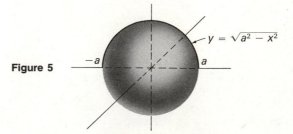

**Figure 5**

---

**Example 2** We will now find the volume determined by rotating the region bounded by the curves

$$y = x^2 - 4x + 5 \quad \text{and} \quad y = x + 1$$

about the $x$-axis.

*Solution* The region bounded by these curves is between the lines $x = 1$ and $x = 4$ because the points of intersection are $(1, 2)$ and $(4, 5)$ (Fig. 6).

Rotating this region about the $x$-axis then gives the region shown in Figure 7. The integral $\int_1^4 \pi(x + 1)^2 \, dx$ is the volume in Figure 8 while $\int_1^4 \pi(x^2 - 4x + 5)^2 \, dx$ represents the volume in Figure 9. Hence the desired volume is

$$V = \int_1^4 (\pi(x + 1)^2 - \pi(x^2 - 4x + 5)^2) \, dx.$$

Now, after an annoying amount of arithmetic, we arrive at the answer of $V = \frac{117}{5}\pi$.

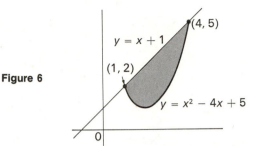

**Figure 6**

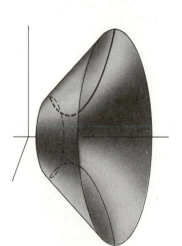

**Figure 7**

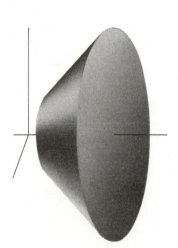

**Figure 8**

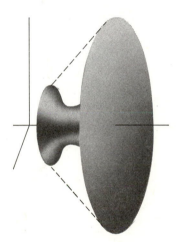

**Figure 9**

The following example provides an interesting case where the reader's intuition may be slightly troubled.

*Example 3* Find the volume of the region obtained by rotating about the $x$-axis the graph of $y = \frac{1}{x}$ between $x = 1$ and $x = \infty$ (Fig. 10).

*Solution* This is an improper integral. The volume $V(b)$ between $x = 1$ and $x = b$ is given by the formula

$$V(b) = \int_1^b \pi y^2 \, dx = \pi \int_1^b \frac{1}{x^2} \, dx$$

$$= \pi \left. \frac{-1}{x} \right|_1^b = \pi \left( 1 - \frac{1}{b} \right).$$

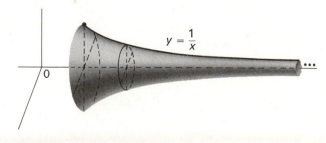

**Figure 10**

Now letting $b \to \infty$, we see that

$$V(b) = \pi\left(1 - \frac{1}{b}\right) \to \pi.$$

Hence the desired volume is given by

$$V = \pi.$$

Starting with a function defined as $x = g(y)$ and rotating about the $y$-axis between $y = c$ and $y = d$ gives the corresponding formula (Fig. 12)

$$(\#) \qquad V = \int_c^d \pi(g(y))^2 \, dy.$$

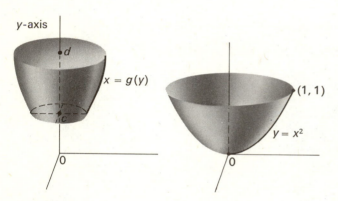

y-axis

x = g(y)

(1, 1)

y = x²

**Figure 12**     **Figure 13**

---

*Example 4*   Rotate the curve given by $y = x^2$ $(0 \le x \le 1)$ about the $y$-axis. Determine the volume enclosed by the resulting surface (Fig. 13).

*Solution*   We may rewrite the equation of the curve as $x = y^{1/2}$ $(0 \le y \le 1)$ and, using formula $(\#)$, we find the volume is

$$V = \int_0^1 \pi(y^{1/2})^2 \, dy = \int_0^1 \pi y \, dy = \frac{\pi y^2}{2}\Big|_0^1 = \frac{\pi}{2}.$$

---

**B. When the Cross Section Is Known**   When a solid figure can be cut up into parallel cross sections of known areas, then its volume can be computed by integration.

Now consider the area under the curve $y = \frac{1}{x}$ from 1 to $\infty$ (Fig. 11). It is given by $\int_1^\infty \frac{dx}{x}$ and is infinite.

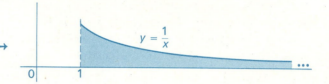

$y = \frac{1}{x}$

0     1

**Figure 11**

Hence a paint can having the shape described in Example 3 would hold a finite amount of paint. On the other hand, since the surface area of the can is infinite (the surface area from 1 to $b$ is greater than the area under the curve from 1 to $b$), we could not paint the outside of the can. But the inside is already painted, since the can is assumed filled with paint. Moreover, the walls of the can are infinitely thin.

---

*Problem 1*   Explain away the above apparent paradox by describing, in general, how an infinite area could be painted with one quart of paint. (Hint: Ignore details about the atomic theory of matter and start by using $\frac{1}{2}$ quart for the first square foot.)

---

**Figure 14**

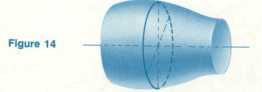

(Circular cross section)

A typical example of this is the case of rotations (6.2A), in which the cross sections are circles (Fig. 14).

Suppose we are given a solid object and for each $x$ the cross section through $x$ and perpendicular to the $x$-axis is known to be $A(x)$ (Fig. 15). If the solid lies in the region between the planes $x = a$ and $x = b$, divide the interval $[a, b]$ into subintervals of length $\Delta x$. Consider the subinterval $[x, x + \Delta x]$. The portion of the solid bounded by the cross sections through these points has a volume $\Delta V$ which can be approximated by $A(x) \Delta x$ (Fig. 16). Adding these volumes, we obtain

$V = \Sigma \, \Delta V$ which is approximately equal to

$$\sum A(x) \, \Delta x.$$

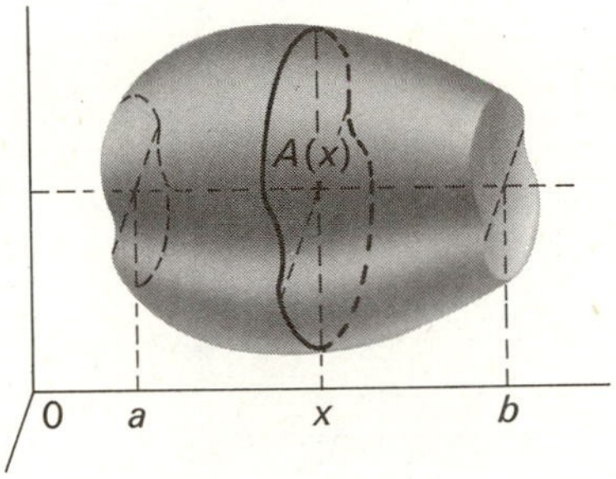

When $\Delta x$ is small, the approximation becomes more accurate and, letting $\Delta x \to 0$, we find that the exact volume is given by

$$V = \int_a^b A(x) \, dx.$$

**Figure 15**

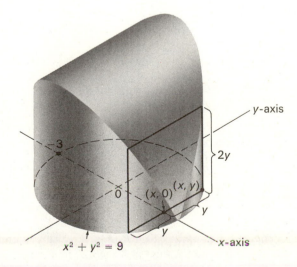

For example, when the solid is as in Figure 14, then

$$A(x) = \pi y^2 = \pi(f(x))^2$$

and we have our old formula

$$V = \int_a^b \pi(f(x))^2 \, dx.$$

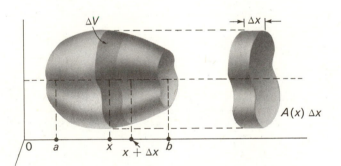

**Figure 16**

*Example 5*  Assume that the base of a solid is the circle

$$x^2 + y^2 = 9$$

and on each chord of the circle parallel to the $y$-axis there is erected a square (Fig. 17). Find the volume of the resulting solid.

*Solution*  The square passing through $(x, 0)$ has area $(2y)^2$, or $4y^2$. Since $x^2 + y^2 = 9$, we have

$$4y^2 = 4(9 - x^2) = 36 - 4x^2.$$

Hence

**Figure 17**

$$A(x) = 36 - 4x^2$$

and the volume is given by the formula

$$V = \int_{-3}^{3} A(x)\,dx = \int_{-3}^{3} (36 - 4x^2)\,dx$$

$$= 36x - \frac{4x^3}{3}\Big|_{-3}^{3} = 2\left(36 \cdot 3 - \frac{4(3)^3}{3}\right) = 144.$$

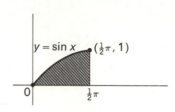

**Figure 18**

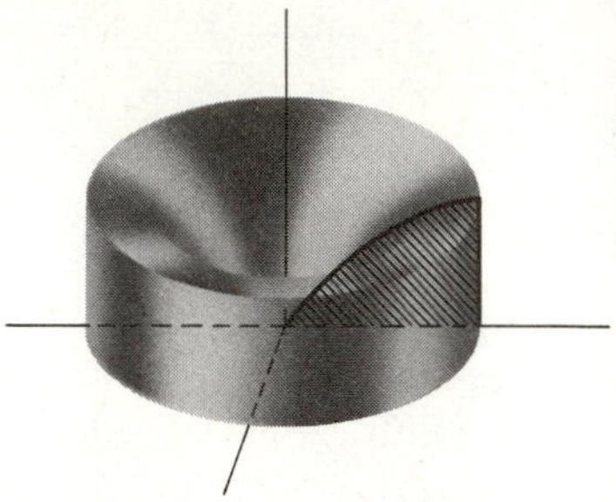

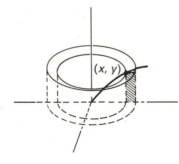

**Figure 20**

**Figure 19**

**Figure 21**

**C. The Method of Cylindrical Shells** The ideas here are similar to those of the previous section, but the technique is slightly different. They can be applied to problems where the method given in 6.2A is difficult to use. We will illustrate with an example.

*Example 6* Find the volume obtained by rotating the area bounded by $y = \sin x$, the $x$-axis, the line $x = 0$, and the line $x = \frac{\pi}{2}$ (Fig. 18) about the $y$-axis (Fig. 19).

*Solution* Consider the small rectangle with base $\Delta x$ and height $y$ (Fig. 20). When this rectangle is rotated about the $y$-axis it generates a cylindrical shell (Fig. 21). Observe that the surface area of the inside cylinder with radius $x$ is given by

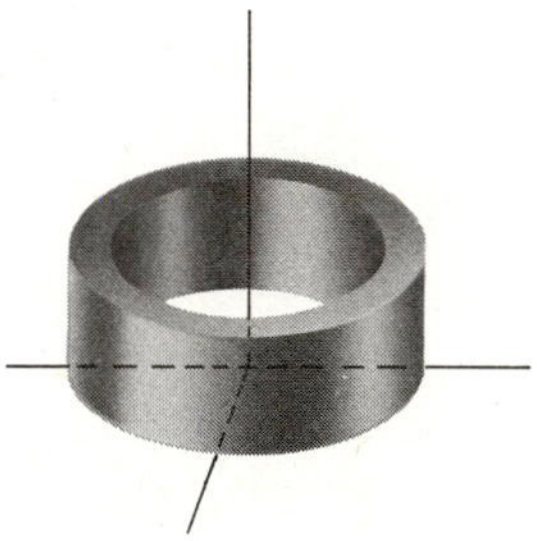

$$2\pi xy,$$

so the volume of the cylindrical shell is approximately

$$2\pi xy\,\Delta x.$$

The volume of the solid is then approximated by

$$\sum 2\pi xy\,\Delta x,$$

and as $\Delta x \to 0$ we find that the volume is

$$\int_{0}^{\pi/2} 2\pi xy\,dx.$$

Since $y = \sin x$, this is just

$$\int_{0}^{\pi/2} 2\pi x \sin x\,dx = 2\pi.$$

In general, suppose a region $R$ (Fig. 23) is given and for each $x$ the distance $h(x)$ is known. Then, rotating

*The reader who is not convinced that this approximates the volume of the cylindrical shell may roll it out to obtain a rectangular slab (Fig. 22).*

→

*The height is y, the thickness $\Delta x$, and, although the ends are not square, the length is approximately $2\pi x$. The volume of the slab is then approximately $2\pi xy\,\Delta x$.*

(Alan)

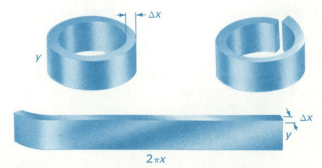

**Figure 22**

this region about the $y$-axis, we obtain a solid figure whose volume is given by the formula

$$V = \int_a^b 2\pi x h(x)\, dx.$$

In a similar way we obtain a formula for the volume generated by revolving the region $R$ (Fig. 24) about the $x$-axis. Indeed we have

$$V = \int_c^d 2\pi y k(y)\, dy.$$

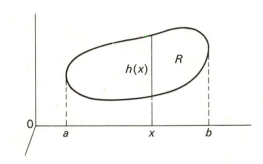

**Figure 23**

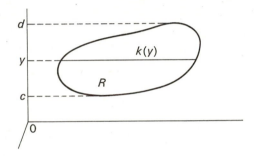

**Figure 24**

---

*Example 7*   Suppose the region bounded by $y = x^2$, the $x$-axis, and the line $x = 1$ (Fig. 25) is rotated about the $x$-axis. Find the volume.

**Figure 25**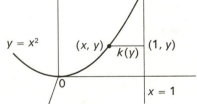

*Solution*   Since $k(y) = 1 - x = 1 - \sqrt{y}$, we have

$$V = \int_0^1 2\pi y(1 - \sqrt{y})\, dy = 2\pi \int_0^1 (y - y^{3/2})\, dy$$

$$= 2\pi \left( \frac{y^2}{2} - \frac{2y^{5/2}}{5} \right)\Big|_0^1 = \frac{\pi}{5}.$$

---

*Problem 2*   Rotate the region described in Example 7 about the $y$-axis and find the volume.

*Answer:*   $\dfrac{\pi}{2}$

## D. Exercises

**B1.** Find the volume of the solid obtained by rotating the region bounded by $y = 1 - x^2$ and $y = 0$ about the $x$-axis.

**B2.** Find the volume of the solid obtained by rotating the region bounded by $y = \sqrt{x}$ and $y = x^2$ about the $x$-axis.

**B3.** Rotate the region described in Exercise B2 about the $y$-axis and find the volume of the resulting solid.

**B4.** Use the method of cylindrical shells to determine the volume of the solid obtained by rotating the region bounded by $y = \frac{1}{x} \sin x$, $y = 0$, and $x = \frac{\pi}{2}$ about the $y$-axis.

**B5.** Assume the base of a solid is the circle $x^2 + y^2 = 9$ and on each chord parallel to the $y$-axis there is an equilateral triangle. What is the volume of the solid?

**B6.** The region bounded by $y = 6 - x^2$ and $y = 2$ is rotated about the $x$-axis. What is the volume of the solid obtained?

**D7.** In each of the cases below sketch the region bounded by the graphs of the given curves and then find the volume of the solid obtained by rotating this region about the $x$-axis.
a) $y = 3x$, $y = 0$, $x = 0$, $x = 3$
b) $y = 5x^2$, $y = 0$, $x = 0$, $x = 2$
c) $y = 5x^2$, $y = 0$, $x = 1$, $x = 2$
d) $y = x$, $y = x^2$, $x = 0$, $x = 1$
e) $y = \sqrt{16 + x^2}$, $y = 0$, $x = 0$, $x = 3$
f) $y = e^x$, $y = 0$, $x = 0$, $x = \ln 3$

a) $81\pi$    b) $160\pi$    c) $155\pi$    d) $\dfrac{2\pi}{15}$    e) $57\pi$

f) $4\pi$

**D8.** Use the method of cylindrical shells to find the volume of solids obtained by rotating the regions described in Exercise D7 about the $y$-axis.

a) $54\pi$    b) $40\pi$    c) $\dfrac{75\pi}{2}$    d) $\dfrac{\pi}{6}$    e) $\dfrac{122\pi}{3}$

f) $6\pi \ln 3 - 4\pi$

**M9.** Rotate the region bounded by $y = x^2 + 2x$ and $y = 0$ about the $x$-axis and find the volume of the solid obtained.

**M10.** Find the volume of the solid obtained by rotating the region bounded by $x = 3y - y^2$ and $x = 0$ about the $x$-axis.

**M11.** Find the volume of the solid obtained by rotating the ellipse $\dfrac{x^2}{a^2} + \dfrac{y^2}{b^2} = 1$ about the $x$-axis.

**M12.** The base of a solid is an ellipse given by $4x^2 + y^2 = 1$. Every cross section is a semicircle perpendicular to the $x$-axis. Find the volume of the solid.

**M13.** Assume the base of a solid is the circle $x^2 + y^2 = r^2$ and each plane perpendicular to the $x$-axis that meets this solid cuts the solid in a square. What is the volume of the solid?

**M14.** Find the volume of the solid obtained by rotating the area bounded by

$$y = \cos x \quad \left(\frac{3\pi}{2} \le x \le 2\pi\right) \quad \text{and} \quad y = 0$$

around the line $x = 2\pi$.

In Exercises M15–M20 consider the region bounded by the graphs of $y = x^2 + 2$ and $y = 3x$. Find the volume obtained by rotating this region as follows:

**M15.** About the $x$-axis

**M16.** About the $y$-axis

**M17.** About the line $x = 1$

**M18.** About the line $x = 2$

**M19.** About the line $y = 3$

**M20.** About the line $y = 6$

**M21.** A wedge is cut from a cylinder of radius 10 inches by two half-planes, one perpendicular to the axis of the cylinder. The second plane meets the first plane at an angle of 45° along a diameter of the circular cross section made by the first plane. What is the volume of the wedge?

## 6.3  *Arc Length and Surface Area*

We will now derive the formulas for the length $L$ of a curve and the area $S$ of a surface obtained by rotating a curve about an axis.

$$L = \int_a^b \sqrt{1 + \left(\frac{dy}{dx}\right)^2}\, dx$$

$$S = \int_a^b 2\pi y \sqrt{1 + \left(\frac{dy}{dx}\right)^2}\, dx$$

**A. Arc Length**  Given a function $f$ defined on $[a, b]$, divide the interval into $n$ equal intervals of length $\Delta x = \frac{b-a}{n}$ (Fig. 1). Now consider the two successive points (Fig. 2)

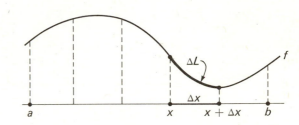

Here we assume that $f'$ is continuous on $[a, b]$.

$$(x, f(x)) \quad \text{and} \quad (x + \Delta x, f(x + \Delta x)).$$

Using the distance formula we find that the length of the dotted segment between these two points is

$$\sqrt{((x + \Delta x) - x)^2 + (f(x + \Delta x) - f(x))^2}.$$

Letting $\Delta y = f(x + \Delta x) - f(x)$, this equals

$$\sqrt{(\Delta x)^2 + (\Delta y)^2} = \sqrt{1 + \left(\frac{\Delta y}{\Delta x}\right)^2}\, \Delta x.$$

Now, for $\Delta x$ small

$$\Delta L \quad \text{is near} \quad \sqrt{1 + \left(\frac{\Delta y}{\Delta x}\right)^2}\, \Delta x.$$

Since the length $L$ of the curve is the sum of the lengths of the curve over the subintervals we see that

$$L \quad \text{is near} \quad \sum \sqrt{1 + \left(\frac{\Delta y}{\Delta x}\right)^2}\, \Delta x.$$

**Figure 1**

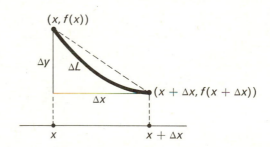

**Figure 2**

Now letting $\Delta x \to 0$ we have $\frac{\Delta y}{\Delta x} \to \frac{dy}{dx}$, so

$$\sum \sqrt{1 + \left(\frac{\Delta y}{\Delta x}\right)^2}\, \Delta x \to \int_a^b \sqrt{1 + \left(\frac{dy}{dx}\right)^2}\, dx.$$

The reason $f'$ was assumed continuous was to insure the continuity of the function $\sqrt{1 + (dy/dx)^2}$.

Since the sum also approaches $L$ as $\Delta x \to 0$, we have the formula

$$L = \int_a^b \sqrt{1 + \left(\frac{dy}{dx}\right)^2}\, dx.$$

When the curve is given parametrically,

$$x = x(t), \quad y = y(t),$$

the formula for its length between the points

$(x(t_0), y(t_0))$ and $(x(t_1), y(t_1))$ $(t_1 \geq t_0)$

is

$$L = \int_{t_0}^{t_1} \sqrt{\left(\frac{dx}{dt}\right)^2 + \left(\frac{dy}{dt}\right)^2}\, dt.$$

The derivation of this is analogous to the one given above. $\rightarrow$

---

**Example 1**  Find the length of the curve defined by

$$y = (a^{2/3} - x^{2/3})^{3/2} \quad (0 \leq x \leq a).$$

*Solution*  Since $y = (a^{2/3} - x^{2/3})^{3/2}$ we have

$$y' = \tfrac{3}{2}(a^{2/3} - x^{2/3})^{1/2}(-\tfrac{2}{3}x^{-1/3}),$$

so

$$(y')^2 = (a^{2/3} - x^{2/3})(x^{-2/3}) = a^{2/3}x^{-2/3} - 1.$$

Therefore

$$(1 + (y')^2)^{1/2} = (a^{2/3}x^{-2/3})^{1/2} = a^{1/3}x^{-1/3}.$$

Now the arc length $L$ is given by

$$L = \int_0^a (1 + (y')^2)^{1/2}\, dx$$

$$= \int_0^a a^{1/3}x^{-1/3}\, dx = \tfrac{3}{2}a^{1/3}x^{2/3}\Big|_0^a = \tfrac{3}{2}a.$$

---

In our next example the curve is parametrically defined. $\rightarrow$

---

**Example 2**  Find the length of the arc defined by

$$x = e^{-t}\cos t, \quad y = e^{-t}\sin t$$

from $t = 0$ to $t = \pi$.

*Solution*  Using the notation $\dot{x} = \frac{dx}{dt}$ and $\dot{y} = \frac{dy}{dt}$, we have

$$(\dot{x}^2 + \dot{y}^2)^{1/2} = ((-e^{-t}\sin t - e^{-t}\cos t)^2$$
$$+ (e^{-t}\cos t - e^{-t}\sin t)^2)^{1/2}$$
$$= e^{-t}((\sin t + \cos t)^2 + (\cos t - \sin t)^2)^{1/2}$$
$$= e^{-t}\sqrt{2}.$$

Hence we have

$$L = \int_0^\pi \sqrt{\dot{x}^2 + \dot{y}^2}\, dt = \sqrt{2}\int_0^\pi e^{-t}\, dt$$

---

We have, as before, that $\Delta L$ is near $\sqrt{(\Delta x)^2 + (\Delta y)^2}$. Since

$$\sqrt{(\Delta x)^2 + (\Delta y)^2} = \sqrt{\left(\frac{\Delta x}{\Delta t}\right)^2 + \left(\frac{\Delta y}{\Delta t}\right)^2}\,\Delta t,$$ 

we see that $L$ is near

$$\sum \sqrt{\left(\frac{\Delta x}{\Delta t}\right)^2 + \left(\frac{\Delta y}{\Delta t}\right)^2}\,\Delta t.$$

This sum approaches

$$\int_{t_0}^{t_1} \sqrt{\left(\frac{dx}{dt}\right)^2 + \left(\frac{dy}{dt}\right)^2}\, dt \quad \text{as} \quad \Delta t \to 0,$$

and we have the desired result.

This formula is superior to the original one since it generalizes naturally to higher dimensions. A curve is described there by the formula

$$P(t) = (x_1(t), x_2(t), \dots, x_n(t)),$$

and the formula for its length between the points $P(t_0)$ and $P(t_1)$ can be shown to be

$$\int_{t_0}^{t_1} (\dot{x}_1{}^2 + \dot{x}_2{}^2 + \cdots + \dot{x}_n{}^2)^{1/2}\, dt \quad \left(\dot{x}_k = \frac{dx_k}{dt}\right).$$

---

The examples and problems dealing with arc length must be chosen with great care. If not, the integrand $\sqrt{1 + (y')^2}$ is difficult to integrate by elementary methods. For example, to find the length of the curve

$$y = (a^{3/2} - x^{3/2})^{2/3}$$

between $x = 0$ and $x = a$, we obtain the integral

$$L = \int_0^a (1 + x(a^{3/2} - x^{3/2})^{-2/3})^{1/2}\, dx.$$

This is most easily handled with one of the approximation techniques in 6.6.

$$= \sqrt{2}(-e^{-t})|_0^\pi = \sqrt{2}(1 - e^{-\pi}).$$

---

**Example 3**  Find the arc length of a circle of radius $r$.

*Solution*  The parametric equations of the circle are

$$x = r \cos t, \qquad y = r \sin t \quad (0 \le t \le 2\pi).$$

Hence

$$\left(\left(\frac{dx}{dt}\right)^2 + \left(\frac{dy}{dt}\right)^2\right)^{1/2} = (r^2 \sin^2 t + r^2 \cos^2 t)^{1/2}$$

$$= (r^2)^{1/2} = r.$$

The arc length of the circle is $\displaystyle\int_0^{2\pi} r \, dt = 2\pi r.$

**Problem 1**  Using the formula

$$L = \int_a^b \sqrt{1 + \left(\frac{dy}{dx}\right)^2} \, dx,$$

prove that the circumference of a circle of radius $r$ is $2\pi r$.

---

**B. Surface Area**  In a manner similar to the way the arc length formula was obtained, we will now derive a formula for the area of the surface obtained by rotating a curve about an axis.

Suppose $f$ is defined on $[a, b]$, and $S$ denotes the area of the surface obtained by rotating the graph of $f$ about the $x$-axis (Fig. 3). Consider the strip of surface which forms a ring concentric about the subinterval of length $\Delta x$ (Fig. 3 and Fig. 4).

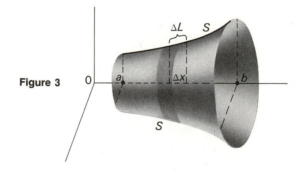

**Figure 3**

**Figure 4**

Since the length of the arc $\Delta L$ is approximately

$$\sqrt{(\Delta x)^2 + (\Delta y)^2}$$

and the circumference of the strip is $2\pi y$, the area of the strip is approximately

$$2\pi y \sqrt{(\Delta x)^2 + (\Delta y)^2} = 2\pi y \sqrt{1 + \left(\frac{\Delta y}{\Delta x}\right)^2} \, \Delta x.$$

Summing the areas of the strips, we find that the area of the surface is approximately

$$\sum 2\pi y \sqrt{1 + \left(\frac{\Delta y}{\Delta x}\right)^2}\, \Delta x.$$

Letting $\Delta x \to 0$ we obtain the formula

$$S = \int_a^b 2\pi y \sqrt{1 + \left(\frac{dy}{dx}\right)^2}\, dx.$$

When the curve is given parametrically, $x = x(t)$ and $y = y(t)$, where $t_0 \le t \le t_1$, the formula becomes

$$S = \int_{t_0}^{t_1} 2\pi y \sqrt{\left(\frac{dx}{dt}\right)^2 + \left(\frac{dy}{dt}\right)^2}\, dt.$$

**Example 4** Show that the surface area $S$ of a sphere of radius $r$ is $4\pi r^2$.

*Solution* Consider the arc

$$y = \sqrt{r^2 - x^2} \quad (-r \le x \le r)$$

and rotate it about the $x$-axis (Fig. 7).

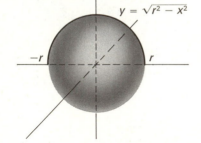

**Figure 7**

We have

$$S = \int_{-r}^{r} 2\pi y \sqrt{1 + (y')^2}\, dx.$$

Now

$$y' = \frac{-x}{\sqrt{r^2 - x^2}}, \quad \text{so} \quad (y')^2 = \frac{x^2}{r^2 - x^2}.$$

Therefore

$$y\sqrt{1 + (y')^2} = \sqrt{r^2 - x^2}\sqrt{1 + \frac{x^2}{r^2 - x^2}}$$

$$= \sqrt{r^2 - x^2}\,\frac{r}{\sqrt{r^2 - x^2}}$$

The estimate for the area of the strip given in the text is derived using the formula

$$(\#) \qquad A = \pi u(y_1 + y_2)$$

for the area of a section of a right circular cone (shaded region of Figure 5).

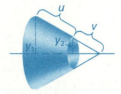

**Figure 5**

*Derivation of formula $(\#)$* Cut the cone and lay it flat as shown in Figure 6.

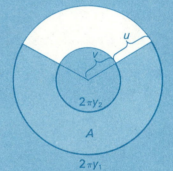

**Figure 6**

The area of the large disc is $\pi(u + v)^2$ and that of the small disc is $\pi v^2$ so the lateral area of the ring is

$$\pi(u + v)^2 - \pi v^2 = \pi(u^2 + 2uv).$$

Now using proportions we have that

$$\frac{\text{area of strip}}{\text{area of ring}} = \frac{\text{outer perimeter of strip}}{\text{outer perimeter of ring}},$$

or

$$\frac{A}{\pi(u^2 + 2uv)} = \frac{2\pi y_1}{2\pi(u + v)}$$

which gives

$$(*) \qquad A = \pi y_1 \left(\frac{u^2 + 2uv}{u + v}\right).$$

Referring to Figure 6, we also have $\frac{v}{2\pi y_2} = \frac{u + v}{2\pi y_1}$, and solving this for $v$ gives $v = \frac{y_2 u}{y_1 - y_2}$. Substituting this in formula $(*)$ for $v$ then gives $A = \pi u(y_1 + y_2)$ ●

Returning now to Figure 4, we see that the area of the strip is approximately

$$= r, \quad \text{and}$$

$$S = \int_{-r}^{r} 2\pi r \, dx = 2\pi r x \Big|_{-r}^{r} = 4\pi r^2.$$

**Example 5** Rotate the curve defined parametrically by

$$x = 2t^2, \quad y = \tfrac{1}{2}t^6 \quad (0 \le t \le 1)$$

about the $x$-axis and find the area of the generated surface.

*Solution* We have

$$S = \int_0^1 2\pi y \sqrt{\left(\frac{dx}{dt}\right)^2 + \left(\frac{dy}{dt}\right)^2} \, dt$$

$$= \int_0^1 \pi t^6 \sqrt{(4t)^2 + (3t^5)^2} \, dt$$

$$= \int_0^1 \pi t^7 \sqrt{16 + 9t^8} \, dt$$

$$= \frac{\pi}{108}(16 + 9t^8)^{3/2} \Big|_0^1 = \frac{\pi}{108}(125 - 64)$$

$$= \frac{61}{108}\pi.$$

$\pi(y + y + \Delta y)\sqrt{(\Delta x)^2 + (\Delta y)^2}$
$$= 2\pi y \sqrt{(\Delta x)^2 + (\Delta y)^2} + \pi \Delta y \sqrt{(\Delta x)^2 + (\Delta y)^2}.$$

Now adding over all the subintervals of length $\Delta x$ we find that the area of the surface is approximately

$$\sum 2\pi y \sqrt{1 + \left(\frac{\Delta y}{\Delta x}\right)^2} \, \Delta x + \sum \pi \Delta y \sqrt{1 + \left(\frac{\Delta y}{\Delta x}\right)^2} \, \Delta x.$$

As $\Delta x \to 0$ the first sum approaches $\int_a^b 2\pi y \sqrt{1 + \left(\frac{dy}{dx}\right)^2} \, dx$ and the second sum approaches zero since $\Delta y \to 0$.

**Problem 2** Rotate the graph of $y = x^3/16$ $(0 \le x \le 2)$ about the $x$-axis and find the area of the generated surface.

*Answer:* Same as for Example 5. (Explain why.)

## C. Exercises

**B1.** Find the length of the curve $y = x^{3/2}$ from $x = 0$ to $x = 5$.

**B2.** Determine the entire length of the curve
$$x^{2/3} + y^{2/3} = 1.$$

**B3.** Find the length of the curve $y = \ln \cos x$ from $x = 0$ to $x = \frac{\pi}{3}$.

**B4.** Find the length of the curve defined parametrically by $x = 2t^3$, $y = 3t^2$ $(0 \le t \le \sqrt{3})$.

**B5.** Rotate the curve $y = \sqrt{x}$ $(6 \le x \le 12)$ about the $x$-axis and find the area of the surface that is generated.

**B6.** Consider the curve defined parametrically by

$x = \cos^3 t, y = \sin^3 t$ $(0 \le t \le \pi)$. Determine the area of the surface that is generated when this curve is rotated about the $x$-axis.

**D7.** Find the lengths of the curves defined over the indicated intervals.
a) $y = \frac{2}{3}x^{3/2}$ $(0 \le x \le 3)$
b) $y = \frac{2}{3}x^{3/2}$ $(8 \le x \le 15)$
c) $y = (1 - x^{2/3})^{3/2}$ $(0 \le x \le 1)$

d) $y = \dfrac{e^x + e^{-x}}{2}$ $(0 \le x \le \ln 2)$

e) $y = \dfrac{e^x + e^{-x}}{2}$ $(\ln 2 \le x \le \ln 4)$

f) $y = e^{x/2} + e^{-x/2}$ $(\ln 4 \le x \le \ln 9)$

a) $\frac{14}{3}$    b) $\frac{74}{3}$    c) $\frac{3}{2}$    d) $\frac{3}{4}$    e) $\frac{9}{8}$    f) $\frac{7}{6}$

**D8.** Rotate each of the following curves about the $x$-axis and find the area of the surface generated.
a) $y = x$ $\quad(0 \le x \le 4)$
b) $y = 1 + 2x$ $\quad(1 \le x \le 3)$
c) $y = 2\sqrt{x}$ $\quad(3 \le x \le 8)$
d) $y = \sqrt{x}$ $\quad(2 \le x \le 6)$
e) $3y = x^3$ $\quad(0 \le x \le 1)$
f) $y = 3$ $\quad(3 \le x \le 7)$

a) $16\pi\sqrt{2}$　b) $20\pi\sqrt{5}$　c) $\dfrac{152\pi}{3}$　d) $\dfrac{49\pi}{3}$

e) $\dfrac{\pi}{9}(2^{3/2} - 1)$　f) $24\pi$

In Exercises M9–M14 find the lengths of the indicated curves.

**M9.** $y = \ln \cos x$ $\quad(\frac{\pi}{6} \le x \le \frac{\pi}{3})$

**M10.** $y = \ln \sin x$ $\quad(\frac{\pi}{6} \le x \le \frac{\pi}{2})$

**M11.** $y = x^2/2$ $\quad(0 \le x \le 1)$

**M12.** $x = e^t \cos t$, $y = e^t \sin t$ $(0 \le t \le \ln 2)$

**M13.** $x = \frac{1}{4}t^4$, $y = \frac{1}{6}t^6$ $(0 \le t \le 1)$

**M14.** $x = \ln \sqrt{1 + t^2}$, $y = \mathrm{Tan}^{-1} t$ $\quad(0 \le t \le 1)$

**M15.** Find the length of the curve $x = \theta - \sin \theta$, $y = 1 - \cos \theta$ $(0 \le \theta \le \frac{\pi}{2})$.

**M16.** Rotate the graph of $y = \sqrt{3 + x}$ $(3 \le x \le 9)$ about the $x$-axis, and find the area of the surface generated.

**M17.** Rotate the graph of $y = \dfrac{e^x + e^{-x}}{2}$ $(-1 \le x \le 1)$ about the $x$-axis, and find the area of the surface that is generated.

**M18.** Find the area of the surface generated by rotating the graph of

$$3y = 3 + x^3 \quad (0 \le x \le 8^{1/4})$$

about the line $y = 1$.

**M19.** Consider the curve defined parametrically by

$$x = 3t, \quad y = 2t^2 \quad (0 \le t \le 1).$$

Find the area of the surface generated by rotating this curve about
a) the $x$-axis,
b) the $y$-axis.

**M20.** Rotate the curve defined parametrically by

$$x = t - \sin t, \quad y = 1 - \cos t \quad (0 \le t \le 2\pi)$$

about the $x$-axis to find the area of the surface generated.

**M21.** Find the area of the surface of a zone cut from a sphere of radius $r$ by two parallel planes, each at a distance $\frac{a}{2}$ from the center.

**M22.** Find the area of the surface cut from a sphere of radius $r$ by a circular cone of half angle $\alpha$ with its vertex at the center of the sphere.

---

## 6.4 *Physical Applications*

→ The material in this section is optional and may be omitted by the non-science major.

The idea of expressing the integral as a limit of finite sums can also be used to define certain physical quantities and develop some of their properties.

**A. Work** Suppose that a force $F = F(x)$ is given. Then the *work* $W$ done by the force $F$ when applied from position $x = a$ to position $x = b$ is defined to be

$$W = \int_a^b F(x)\, dx.$$

→ Mathematically we simply are given a function $F$. The physical meaning of force will not be discussed.

*Motivation* The elementary definition of work,

work = (force) · (distance),

→ applies to the case of a constant force. When a variable

*Example 1*   Find the work done in moving a particle under the influence of the force $F(x) = x^2$ from $x = 1$ to $x = 5$.

*Solution*   We have

$$W = \int_1^5 F(x)\, dx = \int_1^5 x^2\, dx = \left.\frac{x^3}{3}\right|_1^5 = \frac{124}{3}.$$

*Example 2*   Two particles of masses $m_1$ and $m_2$ attract each other with a force equal to

$$F = \frac{km_1m_2}{x^2} \quad (k \text{ a constant})$$

where $x$ is the distance between them. Assuming the particles are 1 unit apart, how much work is required to separate them by a distance of 1000 units?

*Solution*   Place the particles as shown in Figure 2. The work involved is then equal to

$$W = \int_1^{1000} \frac{km_1m_2}{x^2}\, dx = km_1m_2 \int_1^{1000} \frac{dx}{x^2}$$

$$= km_1m_2 \left.\left(\frac{-1}{x}\right)\right|_1^{1000} = km_1m_2 \left(1 - \frac{1}{1000}\right)$$

$$= .999km_1m_2.$$

*Example 3*   A parabolic bowl with cross section $y = x^2$ is filled with a fluid of constant density $\rho$   → (Fig. 3). If the bowl has height 4, how much work must be done to remove the fluid from the bowl?

force $F$ is defined on $[a, b]$ we divide the interval into equal subintervals of length $\Delta x$ as shown in Figure 1.

**Figure 1**

If $\Delta x$ is small, then $F$ is almost constant on $[x, x + \Delta x]$, so the work accomplished in going from $x$ to $x + \Delta x$ is approximately equal to

$$F(x)\, \Delta x.$$

Hence the total work done in going from $a$ to $b$ is approximately

$$\sum F(x)\, \Delta x.$$

Since $\Sigma F(x)\, \Delta x \to \int_a^b F(x)\, dx$, it is natural to define the work by this integral.

**Figure 2**

*We can also compute the work required to move one of the particles to infinity. This gives the improper integral*

$$W = \int_1^\infty \frac{km_1m_2}{x^2}\, dx.$$

→ *Evaluating it we obtain $W = km_1m_2$.*
   *Comparing this with the answer in Example 2 shows that for some practical purposes infinity is not too far away.*
                                    (Vahan)

→   The word *density* will always refer to mass per unit volume. When near the earth's surface we multiply the mass of an object by $g$ (the constant acceleration due to gravity) to obtain the force due to gravity (its weight).
   We will use the fact that mass = (density)·(volume) and hence weight = $g$·(density)·(volume).

**Figure 3**

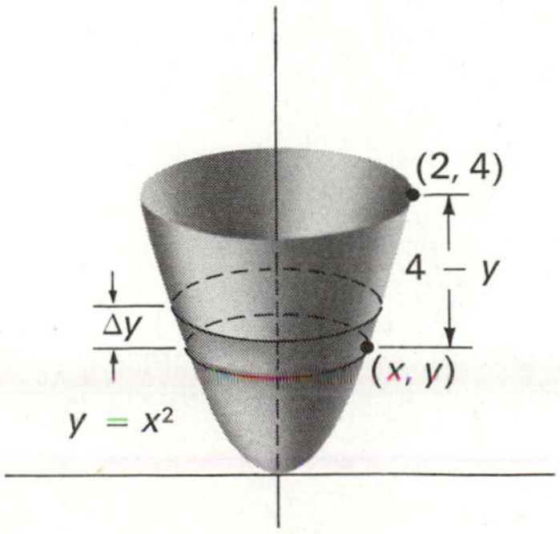

*Solution*  Consider a thin slice of the fluid of thickness $\Delta y$ as shown in Figure 3. Its volume is approximately

$$\pi x^2 \, \Delta y = \pi y \, \Delta y,$$

so its mass is approximately $\rho \pi y \, \Delta y$. Hence the force due to gravity, $\Delta F$, is approximately $g\rho\pi y \, \Delta y$, so the work done in removing this slice from the bowl is approximately

$$\Delta W = (4 - y)g\rho\pi y \, \Delta y.$$

The total work is then approximately

$$\sum (4 - y)g\rho\pi y \, \Delta y,$$

and now letting $\Delta y \to 0$ we obtain

$$W = \int_0^4 (4 - y)g\rho\pi y \, dy$$

$$= g\rho\pi \int_0^4 (4y - y^2) \, dy$$

$$= g\rho\pi \left( 2y^2 - \frac{y^3}{3} \right) \Big|_0^4$$

$$= \frac{32}{3} g\rho\pi.$$

*Problem 1*  Referring to Example 3 assume now that the container is a cylindrical can with radius 2 and height 4, and compute the work done in removing the water from it.

*Answer:*  $32 \, g\rho\pi$

According to Hooke's Law, an ideal spring exerts a pulling force proportional to the distance it is stretched (Fig. 4).

We write $F = kx$ ($k$ a constant) to represent the force which must be exerted to keep the spring stretched a distance of $x$ units. The work performed in stretching the spring from position zero to position $a$ is computed by noticing that the force exerted by the spring at each $x \in [0, a]$ is $F = kx$; hence

$$W = \int_0^a kx \, dx = \frac{k}{2} x^2 \Big|_0^a$$

$$= \frac{k}{2} a^2.$$

*Example 4*  A spring follows Hooke's Law with Hooke's constant 2. What work is required to stretch

⟶  No real spring is an ideal one, so Hooke's Law is only approximately valid. It works well as long as the spring is not stretched too far (beyond its elastic limit). The proportionality constant is known as the modulus of elasticity, and is determined by empirical observation.

(Alan)

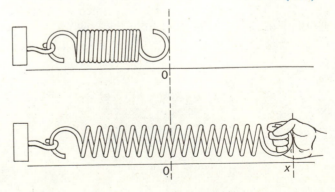

**Figure 4**

it 3 inches from its rest position? How much *more* work is required to stretch it 2 more inches?

*Solution*  We use the formula to obtain

$$W = \int_0^3 2x \, dx = x^2 \Big|_0^3 = 9 \text{ in. lbs.}$$

The work done in stretching the spring 2 more inches is given by

$$W = \int_3^5 2x \, dx = x^2 \Big|_3^5 = 16 \text{ in. lbs.}$$

---

**B. Pressure**  Given a fluid of constant density $\rho$, the pressure $p$ at a point $h$ feet below the surface is defined by the formula

$$p = g\rho h \quad \text{(Fig. 5a)}.$$

If an object has negligible thickness and is submerged so that it is parallel to the earth's surface (Fig. 5b), the pressure at all points in the object is the same and the total force on one side of the object is $\Delta F = g\rho h \, \Delta A$.

Moreover, if $\Delta A$ is the area of a piece of an object and $\Delta h$ is small, the force $\Delta F$ on one side of that piece is approximately given by

$$\Delta F = g\rho h \, \Delta A \quad \text{(Fig. 5c)}.$$

To find the total force on an object submerged in the fluid, it is necessary to integrate.

---

*Example 6*  Suppose the face of a dam is a square with side equal to 200 ft. (Fig. 6). What is the total force against the dam, given that the density of the water is $\rho$?

*Solution*  Consider the force on the strip shown in Figure 6. Since its area is $200 \, \Delta y$ and the pressure at that depth is

$$g(200 - y)\rho$$

(approximately), the force on the strip is

$$g(200 - y)\rho 200 \, \Delta y.$$

Adding over all strips we obtain

*Example 5*  It requires 8 inch-pounds of work to stretch a certain spring 2 inches from its rest position. Assuming that the spring follows Hooke's Law, what is its Hooke's constant $k$?

*Solution*  The formula yields

$$\int_0^2 kx \, dx = 8 \quad \text{so} \quad \frac{kx^2}{2}\Big|_0^2 = 8 \quad \text{or} \quad 2k = 8 \text{ in. lbs.}$$

Hence $k = 4$.

---

→  Observe that density $\rho$ is mass per unit volume, so $g\rho$ is weight per unit volume and $g\rho h$ multiplies this according to the depth of the fluid.

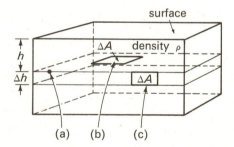

**Figure 5**  (a) Pressure at this point is $g\rho h$.
(b) Force on this area is $g\rho h \, \Delta A$.
(c) Force on this area is approximately $g\rho h \, \Delta A$.

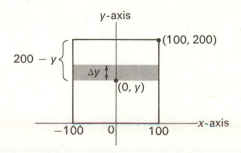

**Figure 6**

$$\sum g(200 - y)\rho 200\,\Delta y.$$

Now letting $\Delta y \to 0$, we obtain the total force

$$F = \int_0^{200} g(200 - y)\rho 200\, dy$$

$$= g200\rho \left(200y - \frac{y^2}{2}\right)\Big|_0^{200}$$

$$= 4 \cdot 10^6 \rho g.$$

---

**Example 7** A can of radius 1 and height 3 is submerged in a fluid of density $\rho$ as shown in Figure 8. Find the total force exerted on the can.

*Solution* Since the area of the top is $\pi \cdot 1^2$ and the pressure at that depth is $2\rho g$, the

$$\text{force on top} = 2\rho g\pi.$$

The depth at the bottom is 5, so the

$$\text{force on bottom} = 5\rho g\pi.$$

Now consider the side in terms of strips of width $\Delta y$. A strip at $y$ has area $2\pi \cdot 1\,\Delta y$ and the pressure at that depth is $g(5 - y)\rho$. Hence the force on the strip is

$$g(5 - y)\rho 2\pi\,\Delta y.$$

Summing, we find the total force on the side to be approximately

$$\sum g(5 - y)\rho 2\pi\,\Delta y.$$

Letting $\Delta y \to 0$, we have

$$\text{force on side} = \int_0^3 2\pi g\rho(5 - y)\,dy = 21\pi g\rho.$$

Hence the total force is

$$2\pi g\rho + 5\pi g\rho + 21\pi g\rho = 28\pi g\rho.$$

---

**C. Gravitation** Given two particles, assumed to be points, of masses $m_1$ and $m_2$ (Fig. 9), the force between them is given by *Newton's law of gravitation,*

---

**Problem 2** Assume that the face of a dam is a semicircle of radius 10 (Fig. 7). If the density of the water is $\rho$, find the total force on the dam.

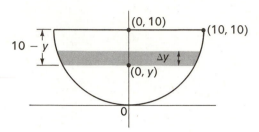

**Figure 7**

*Answer:*   $\frac{2000}{3} g\rho$

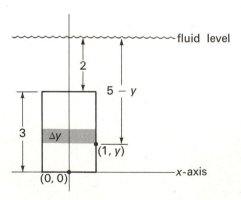

**Figure 8**

Newton first found his law for the particular case of mass points, and realized that before he could apply it to situations other than mass points (for example, an apple falling to the earth or the moon circling it), the law would have to be generalized. He eventually did generalize it, and in particular, proved the important theorem and corollary given below.

$$F = \frac{km_1 m_2}{d^2},$$

where $d$ is the distance between the particles and $k$ the gravitational constant. Using integration we can generalize this law to bodies that are not necessarily points.

**Figure 9**

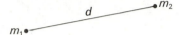

---

*Example 8* We will show that the force between a uniform rod of length $l$ and mass $M$, and a particle of mass $m$ located on the line of the rod $d$ units from the nearer end (Fig. 10) is given by

$$F = \frac{kmM}{d(d + l)}.$$

*Solution* Divide the rod into small segments of length $\Delta x$ and take a typical one as shown in Figure 10. Denoting its mass by $\Delta M$, we have

$$\frac{\Delta M}{M} = \frac{\Delta x}{l}, \quad \text{so} \quad \Delta M = \frac{M \Delta x}{l}.$$

Assuming $\Delta x$ is small and applying Newton's law for point masses, we see that the force between the particle and this small section of the rod is approximately

$$\Delta F = \frac{km \, \Delta M}{x^2} = \frac{kmM \, \Delta x}{lx^2}.$$

Now adding up these forces over all the pieces, we find that the approximate force is

$$\sum \frac{kmM \, \Delta x}{lx^2}.$$

Then letting $\Delta x \to 0$, we obtain the formula

$$F = \int_d^{d+l} \frac{kmM}{lx^2} \, dx = \frac{-kmM}{lx} \Big|_d^{d+l}$$

$$= \frac{-kmM}{l(d + l)} + \frac{kmM}{ld} = \frac{kmM}{d(d + l)}.$$

**Figure 10**

→ The force between particles of mass $m$ and $M$ separated by a distance of $d + \frac{l}{2}$ is

$$F = \frac{kmM}{\left(d + \frac{l}{2}\right)^2}.$$

Hence the force between the particle and the rod is *not* that obtained by assuming that all the mass of the rod is concentrated at its center. However, if the rod is replaced by a spherical shell or sphere, the force between the objects can be found by assuming that the mass is at the center. This is shown in the following theorem.

*THEOREM 1* The force between a uniform spherical shell of mass $M$ and a particle of mass $m$ located outside the shell at a distance $R$ from its center is given by $F = \frac{kmM}{R^2}$.

*Proof* Assume the radius of the shell is $r$ and consider the strip shown in Figure 11.

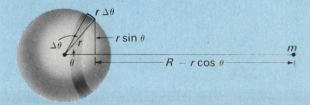

**Figure 11**

The following example assumes some knowledge about the decomposition of a force into components.

---

*Example 9*  The force between a uniform ring of mass $M$ and radius $r$ and a particle of mass $m$ located $R$ units from the center of the ring on a line perpendicular to the ring (Fig. 12) is given by

$$F = \frac{kmMR}{(R^2 + r^2)^{3/2}}$$

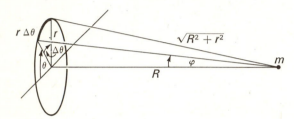

**Figure 12**

*Solution*  Decompose the ring into arcs of length $r\,\Delta\theta$ as shown in Figure 12. The mass $\Delta M$ of one of these arcs must satisfy

$$\frac{\Delta M}{M} = \frac{r\,\Delta\theta}{r2\pi}, \quad \text{so} \quad \Delta M = \frac{M\,\Delta\theta}{2\pi}.$$

Hence, for small $\Delta\theta$ the force $\Delta F$ between the particle and the arc is approximately

$$\Delta F = \frac{km\,\Delta M}{(\sqrt{R^2 + r^2})^2} = \frac{kmM\,\Delta\theta}{2\pi(R^2 + r^2)}.$$

The horizontal component of this force is $\Delta F \cos\phi$ and, because

$$\cos\phi = \frac{R}{\sqrt{R^2 + r^2}},$$

we have

$$\Delta F \cos\phi = \frac{kmMR}{2\pi} \frac{\Delta\theta}{(R^2 + r^2)^{3/2}}.$$

Since the vertical components of the force cancel, the total force $F$ between the ring and the particle is found by taking $\Sigma\,\Delta F \cos\phi$ and then letting $\Delta\theta \to 0$. We obtain the integral

Since the area of the strip is approximately $2\pi r (\sin\theta) r\,\Delta\theta$ (see 6.3C) and the area of the shell is $4\pi r^2$, the mass $\Delta M$ of the strip must satisfy

$$\frac{\Delta M}{M} = \frac{2\pi r (\sin\theta) r\,\Delta\theta}{4\pi r^2}, \quad \text{so} \quad \Delta M = \frac{M\sin\theta\,\Delta\theta}{2}.$$

For $\Delta\theta$ small, we may assume that the strip is almost a ring, as in Example 9. Using the result from this example we then see that the force $\Delta F$ between the strip and the particle is given by

$$\Delta F = \frac{km\,\Delta M(R - r\cos\theta)}{((r\sin\theta)^2 + (R - r\cos\theta)^2)^{3/2}}$$

$$= \frac{kmM}{2} \frac{(R - r\cos\theta)\sin\theta\,\Delta\theta}{(R^2 - 2rR\cos\theta + r^2)^{3/2}}.$$

Now summing these forces and letting $\Delta\theta \to 0$, we obtain

$$F = \frac{kmM}{2} \int_0^\pi \frac{(R - r\cos\theta)\sin\theta\,d\theta}{(R^2 - 2rR\cos\theta + r^2)^{3/2}}.$$

Let $u = R^2 - 2rR\cos\theta + r^2$, so $du = 2rR\sin\theta\,d\theta$. Making these substitutions in the above integral and changing limits we have

$$F = \frac{kmM}{8rR^2} \int_{(R-r)^2}^{(R+r)^2} \frac{R^2 - r^2 + u}{u^{3/2}}\,du$$

$$= \frac{kmM}{8rR^2} \int_{(R-r)^2}^{(R+r)^2} ((R^2 - r^2)u^{-3/2} + u^{-1/2})\,du$$

$$= \frac{kmM}{8rR^2} ((R^2 - r^2)(-2u^{-1/2}) + 2u^{1/2}) \Big|_{(R-r)^2}^{(R+r)^2}$$

$$= \frac{kmM}{8rR^2} (4r - (-4r))$$

$$= \frac{kmM}{R^2} \; \bullet$$

---

*Problem 3*  Assume the particle lies inside the shell and go through the above proof, changing signs where necessary, to show that the force between the particle and the shell is zero.

$$F = \int_0^{2\pi} \frac{kmMR}{2\pi(R^2 + r^2)^{3/2}} \, d\theta = \frac{kmMR}{(R^2 + r^2)^{3/2}}.$$

**Example 10** Place a particle at the point 0 and a rod of length $l$ on the interval $[d, d + l]$. Suppose the particle has mass $m$ and the density of the rod at a point is equal to twice the point's distance from the end of the rod that is nearer the particle. Find the force of attraction between the particle and the rod. (We assume the rod is very thin and has unit cross-sectional area.)

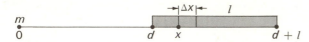

**Figure 13**

> **Corollary** *If R is the distance between the centers of two spheres of masses $M_1$ and $M_2$, then the force between the spheres is*
>
> $$F = \frac{kM_1 M_2}{R^2}.$$
>
> **Proof** This is done in two steps. First consider the case of a sphere and a particle. Decomposing the sphere into concentric shells and applying the theorem gives the result. Hence a sphere acts on a particle as if all its mass were concentrated at its center. Now, given two spheres $A$ and $B$, imagine that $A$ has been decomposed into a large number of point masses. The attraction between $B$ and each of these points is given by treating $B$ as a point mass; hence, summing these attractions, we have that $B$ may be treated as a point mass. This argument may also be applied to reduce $A$ to a point mass ●

**Solution** Referring to Figure 13 we now see that the mass of the subinterval of length $\Delta x$ is approximately $\Delta M = 2(x - d)\,\Delta x$. Hence the force $\Delta F$ between the particle and the subinterval is approximately

$$\Delta F = \frac{km\,\Delta M}{x^2} = \frac{km2(x - d)\,\Delta x}{x^2}.$$

Adding these forces and letting $\Delta x \to 0$, we obtain

$$F = \int_d^{d+l} \frac{2km(x - d)}{x^2} \, dx = 2km \int_d^{d+l} \left(\frac{1}{x} - \frac{d}{x^2}\right) dx$$

$$= 2km \left(\ln x + \frac{d}{x}\right)\Big|_d^{d+l}$$

$$= 2km \left(\ln\left(\frac{d + l}{d}\right) - \frac{l}{d + l}\right).$$

## D. Exercises

**B1.** Assume the force on a particle at the point $x$ is given by $F(x) = \sin x \cos x$. How much work is performed in moving the particle

    a) From $x = 0$ to $x = \frac{1}{2}\pi$?
    b) From $x = \frac{1}{2}\pi$ to $x = \pi$?
    c) From $x = 0$ to $x = \pi$?

**B2.** A cube with side 4 ft. is filled with a fluid of density $\rho$ and is buried with its top two feet below ground. How much work is required in pumping the fluid to the surface?

**B3.** A spring with Hooke's constant 2 lb./ft. is to be stretched $\frac{3}{2}$ feet. How much work will be required?

**B4.** It requires 10 in. lbs. of work to stretch an ideal spring 2 inches. How much work is done in stretching the spring an additional 3 inches?

**B5.** The Grand Trapezoid Dam measures 200 ft. at its base, 600 ft. on its top, and its nonparallel sides are $200\sqrt{5}$ ft. Assuming the density of water is $\rho$, what is the total force exerted on the dam?

**D6.** Find the work done in moving a particle from $a$ to $b$ under the influence of the force $F(x)$.

a) $F(x) = 4x^3$, $a = 0$, $b = 2$
b) $F(x) = 4x^3$, $a = 2$, $b = 0$
c) $F(x) = 4x^3$, $a = 0$, $b = -2$
d) $F(x) = 4x^3$, $a = -2$, $b = 2$

e) $F(x) = \dfrac{x}{\sqrt{25 + x^2}}$, $a = 0$, $b = 12$

f) $F(x) = 2\cos 2x$, $a = 0$, $b = \dfrac{\pi}{4}$

g) $F(x) = \dfrac{2}{x^3}$, $a = 1$, $b = \infty$

h) $F(x) = 2x^3$, $a = 1$, $b = \infty$

a) 16   b) $-16$   c) 16   d) 0   e) 8   f) 1
g) 1   h) $\infty$

**D7.** A cube with side 2 feet is filled with a fluid of density $\rho$. What force does the fluid exert on
a) the bottom   b) the top   c) one side?

a) $8\rho g$   b) 0   c) $4\rho g$

**M8.** In each of the cases below, a force $F$ is given. Compute the work required to move a particle from the point $a$ to the point $b$.

a) $F(x) = \dfrac{1}{\sqrt{x}}$, $a = 0$, $b = 1$

b) $F(x) = \dfrac{1}{1 + x^2}$, $a = 0$, $b = 1$

c) $F(x) = \dfrac{1}{1 + x^2}$, $a = 0$, $b = \infty$

d) $F(x) = \dfrac{1}{x^2}$, $a = 0$, $b = 4$

**M9.** A triangular plate $ABC$ is submerged in water with its plane vertical. The side $AB$, 4 feet long, is 1 foot below the surface, while $C$ is 5 feet below $AB$. Find the total force on one face of the plate.

**M10.** A semicircular plate is submerged in water with its plane vertical and its diameter in the surface. Find the force on one face of the plate if its diameter is 2 feet.

**M11.** An ice cream cone has a height of 5 inches and its radius at the top is 1 inch. Assuming it is filled with melted ice cream having density $\rho$, determine the force of the ice cream on the cone.

**M12.** Show that the force between the two rods shown below is given by the formula

$$F = \frac{kM_1 M_2}{l_1 l_2} \ln \frac{(d + l_1)(d + l_2)}{d(d + l_1 + l_2)}.$$

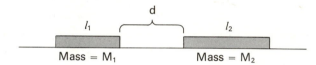

**M13.** The face of a dam has the shape of the region bounded by the curves $y = 4 - x^2$ and $y = 0$. If water, having density $\rho$, is at the level $y = 2$, determine the total force on the dam.

**M14.** Consider the coordinate plane as a plate with constant density $\rho$. Given a particle of mass $m$ not on the plate, prove that the force between the particle and the plate is equal to $2\rho km$ where $k$ is the gravitational constant.

(Hint: Use the formula $F = \dfrac{kmMR}{(R^2 + r^2)^{3/2}}$ for the force between a uniform ring of mass $M$ and radius $r$, and particle of mass $m$ located $R$ units from the center of the ring on a line perpendicular to the ring.)

**M15.** Two electrons repel each other with a force inversely proportional to the square of the distance between them. If one electron is held fixed at the point $(1, 0)$ on the $x$-axis, find the work required to move a second electron along the $x$-axis from the point $(-1, 0)$ to the origin.

**M16.** Find the work done in pumping all the water out of a conical reservoir of radius 10 feet at the top, altitude 8 feet, to a height of 6 feet above the top of the reservoir.

## 6.5  *Moments and Centers of Mass*

→  The material in this section is optional and may be omitted by the non-science major.

The definitions and formulas presented in this section are important for the study of rotational or orbital motion.

**A. Motivation (a System of Particles)**  Suppose we are given a system of $n$ particles of masses

$$m_1, \quad m_2, \quad \ldots, \quad m_n$$

located at the points

$$(x_1, y_1), \quad (x_2, y_2), \quad \ldots, \quad (x_n, y_n)$$

respectively. We define

$$M = m_1 \quad + m_2 \quad + \cdots + m_n,$$
$$M_y = m_1 x_1 \quad + m_2 x_2 \quad + \cdots + m_n x_n,$$
$$M_x = m_1 y_1 \quad + m_2 y_2 \quad + \cdots + m_n y_n,$$
$$I_y = m_1 x_1^2 + m_2 x_2^2 + \cdots + m_n x_n^2,$$
$$I_x = m_1 y_1^2 + m_2 y_2^2 + \cdots + m_n y_n^2,$$

$$\bar{x} = \frac{M_y}{M} \quad \text{and} \quad \bar{y} = \frac{M_x}{M}.$$

$M_x$ and $M_y$ are called the *first moments about the x and y axes* respectively. $I_x$ and $I_y$ are called the *second moments, or moments of inertia, about the x and y axes* respectively. The point $(\bar{x}, \bar{y})$ is defined as the *center of mass* of the system of particles.

We will now generalize these definitions to the case of continuous distributions of matter.

Some motivation for this definition can be seen from the following see-saw example. Two particles of masses $m_1$ and $m_2$ are placed on the ends of the interval $[x_1, x_2]$ (Fig. 1).

→

**Figure 1**

The point $x$ at which a fulcrum must be placed in order that the see-saw balance must satisfy

$$(x - x_1)m_1 = (x_2 - x)m_2, \quad \text{so} \quad x = \frac{m_1 x_1 + m_2 x_2}{m_1 + m_2}$$

is the x-coordinate of the center of mass.

More generally, a planar region composed of n particles $(x_1, y_1), (x_2, y_2), \ldots, (x_n, y_n)$ with masses $m_1, m_2, \ldots, m_n$ respectively would balance on a pin point placed at the center of mass $(\bar{x}, \bar{y})$.

(Vahan)

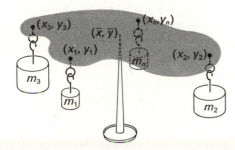

**Figure 2**

### B. Plane Regions (Constant Density)

Consider a region in the plane (Fig. 3) having mass $M$, area $A$, and constant density $\rho$. Then for any portion of $A$ the mass $\Delta M$ of this portion satisfies

$$\Delta M = \rho \, \Delta A.$$

Of course if the region lies in the plane, it has no thickness and, hence, can have no mass. To avoid this difficulty we assume the region has thickness of one unit and that the unit is negligibly small. Such regions, for example a piece of paper, are called *laminas*.

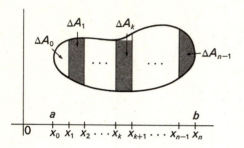

**Figure 3**

Divide the region into strips of width $\Delta x = \frac{b-a}{n}$ (Fig. 3) and consider the expression

$$x_0 \, \Delta M_0 + x_1 \, \Delta M_1 + \cdots + x_{n-1} \, \Delta M_{n-1}$$
$$= \rho(x_0 \, \Delta A_0 + x_1 \, \Delta A_1 + \cdots + x_{n-1} \, \Delta A_{n-1}).$$

The regions we are considering are of a very special type. For a region as shown in Figure 5, we would take $h(x)$ to be the sum $h(x) = h_1(x) + h_2(x)$ of the distances $h_1(x)$ and $h_2(x)$.

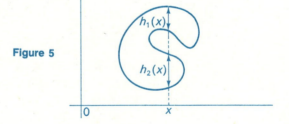

Figure 5

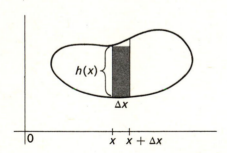

**Figure 4**

Letting $h(x)$ be the distance shown in Figure 4, we see that $\Delta A_k$ is approximately given by

$$\Delta A_k = h(x_k) \, \Delta x;$$

hence the above finite sum is approximately given by

$$\rho(x_0 h(x_0) \, \Delta x + x_1 h(x_1) \, \Delta x + \cdots + x_{n-1} h(x_{n-1}) \, \Delta x)$$

$$= \sum \rho x h(x) \, \Delta x.$$

We define $M_y$ to be the limit of this sum as $\Delta x$

approaches zero, that is,

$$M_y = \int_a^b \rho x h(x)\, dx.$$

Similarly we let

$$I_y = \int_a^b \rho x^2 h(x)\, dx.$$

Referring to Figure 6, we also define

$$M_x = \int_c^d \rho y k(y)\, dy,$$

$$I_x = \int_c^d \rho y^2 k(y)\, dy,$$

and

$$\bar{x} = \frac{M_y}{M} \quad \text{and} \quad \bar{y} = \frac{M_x}{M}.$$

As in 6.5A, we refer to $M_y$ and $M_x$ as the *first moments* about the $y$ and $x$ axes respectively and to $I_y$ and $I_x$ as the *second moments* or moments of inertia about the $y$ and $x$ axes respectively. The point $(\bar{x}, \bar{y})$ is called the *center of mass* of the planar region.

Since $M = \rho A$, we have

$$\bar{x} = \frac{\rho \int_a^b x h(x)\, dx}{\rho A} = \frac{\int_a^b x h(x)\, dx}{A}$$

and

$$\bar{y} = \frac{\rho \int_c^d y k(y)\, dy}{\rho A} = \frac{\int_c^d y k(y)\, dy}{A}.$$

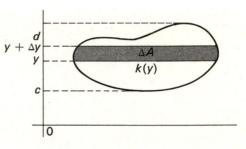

**Figure 6**

→   In this case we consider the expression

$$y_0\, \Delta M_0 + y_1\, \Delta M_1 + \cdots + y_{n-1}\, \Delta M_{n-1}$$
$$= \rho(y_0\, \Delta A_0 + y_1\, \Delta A_1 + \cdots + y_{n-1}\, \Delta A_{n-1})$$
$$= \rho(y_0 k(y_0)\, \Delta y + y_1 k(y_1)\, \Delta y + \cdots + y_{n-1} k(y_{n-1})\, \Delta y)$$
$$= \sum \rho y k(y)\, \Delta y$$

and define $M_x$ to be the limit of this expression as $\Delta y \to 0$.

→   When the density $\rho$ is constant, the center of mass $(\bar{x}, \bar{y})$ is often called the *centroid* and is independent of $\rho$.

→   Notice that $A$ can be computed from either of the formulas

$$A = \int_a^b h(x)\, dx \quad \text{or} \quad A = \int_c^d k(y)\, dy.$$

---

*Example 1*  Consider the region in the plane bounded by the curves $y = x$ and $y = x^2/3$ and having constant density $\rho$. Find $A$, $M_y$, $I_y$, $\bar{x}$, $M_x$, $I_x$, and $\bar{y}$.

*Solution*  Solving the equations $y = x$ and $y = x^2/3$ simultaneously, we see that the curves intersect at the points $(0, 0)$ and $(3, 3)$, and the region is as shown (Fig. 7). Since $h(x) = x - (x^2/3)$, we have

$$A = \int_0^3 h(x)\, dx = \int_0^3 \left( x - \frac{x^2}{3} \right) dx$$

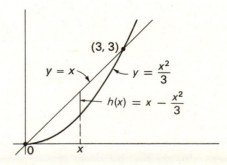

**Figure 7**

$$= \frac{x^2}{2} - \frac{x^3}{9}\Big|_0^3 = \frac{3}{2},$$

$$M_y = \rho \int_0^3 xh(x)\,dx = \rho \int_0^3 \left(x^2 - \frac{x^3}{3}\right)dx$$

$$= \rho\left(\frac{x^3}{3} - \frac{x^4}{12}\right)\Big|_0^3 = \frac{9}{4}\rho,$$

$$I_y = \rho \int_0^3 x^2 h(x)\,dx = \rho \int_0^3 \left(x^3 - \frac{x^4}{3}\right)dx$$

$$= \rho\left(\frac{x^4}{4} - \frac{x^5}{15}\right)\Big|_0^3 = \frac{81}{20}\rho,$$

$$\bar{x} = \frac{M_y}{M} = \frac{M_y}{A\rho} = \frac{(9/4)\rho}{(3/2)\rho} = \frac{3}{2}.$$

To find $k(y)$ we solve the original equations for $x$ to obtain $x = y$ and $x = \sqrt{3y}$. Hence

$$k(y) = \sqrt{3y} - y \quad \text{(Fig. 8)},$$

so that

$$M_x = \rho \int_0^3 yk(y)\,dy = \rho \int_0^3 (\sqrt{3}y^{3/2} - y^2)\,dy$$

$$= \rho\left(\frac{2\sqrt{3}}{5}y^{5/2} - \frac{y^3}{3}\right)\Big|_0^3$$

$$= \rho\left(\frac{2\sqrt{3}}{5}3^{5/2} - 9\right) = \frac{9}{5}\rho,$$

$$I_x = \rho \int_0^3 y^2 k(y)\,dy = \rho \int_0^3 (\sqrt{3}y^{5/2} - y^3)\,dy$$

$$= \rho\left(\frac{2\sqrt{3}}{7}y^{7/2} - \frac{y^4}{4}\right)\Big|_0^3 = \frac{81}{28}\rho,$$

$$\bar{y} = \frac{M_x}{M} = \frac{M_x}{A\rho} = \frac{(9/5)\rho}{(3/2)\rho} = \frac{6}{5}.$$

The center of mass, or centroid in the case of constant density, can be used to find volumes of revolution as shown by the following theorem.

*THEOREM 1* (*Pappus (circa* 320 B.C.)) *Suppose we are given a region in the first quadrant having area A. Let $V_x$ and $V_y$ denote the volumes obtained by revolving this region about the x-axis and y-axis respectively. We then have*

---

*Problem 1* Determine the center of mass of the region in the first quadrant having constant density $\rho$ and bounded by the coordinate axes and the graph of $y = 4 - x^2$.

*Answer:* $(\bar{x}, \bar{y}) = (\frac{3}{4}, \frac{8}{5})$

---

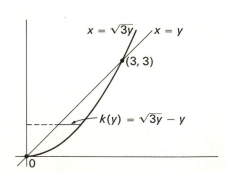

**Figure 8**

$$V_x = 2\pi \bar{y} A$$

*and*

$$V_y = 2\pi \bar{x} A.$$

*Proof* Consider the case of $V_y$ (Fig. 9). Using the cylindrical shell method for computing volumes (6.2C), we have

$$V_y = \int_a^b 2\pi x h(x)\, dx = 2\pi A \frac{\int_a^b x h(x)\, dx}{A}$$

$$= 2\pi A \bar{x}.$$

The proof for $V_x$ is similar and will be omitted ● →

We will now show that the centroid of the region bounded by a circle is the center of the circle.

---

**Example 2** We will show that the centroid of the disc

$$x^2 + y^2 \leq r^2$$

is its center $(0, 0)$.

*Solution* We have (Fig. 11) $h(x) = 2\sqrt{r^2 - x^2}$. Hence

$$\bar{x} = \frac{V_y}{2\pi A} = \frac{\int_{-r}^{r} 2\pi x h(x)\, dx}{2\pi (\pi r^2)}$$

$$= \frac{\int_{-r}^{r} x 2\sqrt{r^2 - x^2}\, dx}{\pi r^2}$$

$$= \frac{-2(r^2 - x^2)^{3/2}}{3\pi r^2} \Big|_{-r}^{r} = 0.$$

Similarly, $\bar{y} = 0$, so $(\bar{x}, \bar{y}) = (0, 0)$ is the centroid of the disc.

---

We may generalize this result to the region bounded by the circle

$$(x - 4)^2 + (y - 3)^2 = 4.$$

Using Pappus' Theorem and the result stated in Problem 2, we have

$$V_x = 2\pi \bar{y} A = 2\pi \cdot 3 \cdot \pi \cdot 2^2 = 24\pi^2$$

and

$$V_y = 2\pi \bar{x} A = 2\pi \cdot 4 \cdot \pi \cdot 2^2 = 32\pi^2.$$

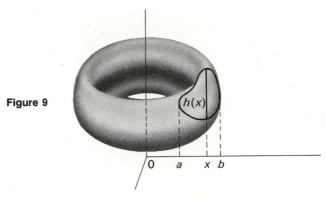

**Figure 9**

This is only a special case of the theorem. The region need not be confined to the first quadrant. More generally, take a region with area $A$ and any line $L$ not passing through the region (Fig. 10). Then the volume $V$ obtained by rotating the region about the line $L$ is given by the formula $V = 2\pi\, dA$ where $d$ is the perpendicular distance from the centroid $(\bar{x}, \bar{y})$ to the line $L$.

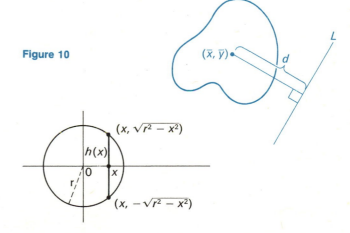

**Figure 10**

**Figure 11**

---

**Problem 2** Show that the centroid of the region bounded by the circle $(x - h)^2 + (y - k)^2 = r^2$ is the point $(h, k)$. (Hint: Use the result in Example 2 and suitably adjust the formulas for $\bar{x}$ and $\bar{y}$.)

When the region in the plane is a curve (Fig. 12) with density $\rho$ (think of a wire), then the mass of an arc of the curve is proportional to its length. Since $\Delta s$ is approximately given by $\sqrt{(\Delta x)^2 + (\Delta y)^2}$, $M_y$ is approximately given by

$$\sum x\rho\, \Delta s = \sum x\rho\, \sqrt{1 + \left(\frac{\Delta y}{\Delta x}\right)^2}\, \Delta x.$$

Letting $\Delta x \to 0$, we have

$$M_y = \int_a^b x\rho\, \sqrt{1 + \left(\frac{dy}{dx}\right)^2}\, dx.$$

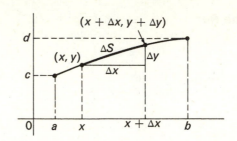

**Figure 12**

**Example 3** Assume that the arc of the circle $x^2 + y^2 = 4$ in the first quadrant has constant density $\rho$ (Fig. 13). We will compute its centroid.

**Solution** We have

$$M_y = \int_0^2 \rho x \sqrt{1 + (y')^2}\, dx,$$

and implicitly differentiating the equation $x^2 + y^2 = 4$ gives

$$2x + 2yy' = 0 \quad \text{or} \quad y' = \frac{-x}{y}.$$

Hence

$$M_y = \int_0^2 \rho x \sqrt{1 + \frac{x^2}{y^2}}\, dx = \int_0^2 \rho \frac{x}{y} \sqrt{x^2 + y^2}\, dx$$

$$= \int_0^2 \rho \frac{x}{y} \cdot 2\, dx = \rho \int_0^2 2x(4 - x^2)^{-1/2}\, dx = 4\rho.$$

Since the length of the wire is $\pi$, its mass is $\pi\rho$, so

$$\bar{x} = \frac{M_y}{M} = \frac{4\rho}{\pi\rho} = \frac{4}{\pi}.$$

By symmetry we have

$$\bar{y} = \frac{4}{\pi},$$

so $(\frac{4}{\pi}, \frac{4}{\pi})$ is the centroid. Notice that the centroid does *not* lie on the wire.

Similarly,

$$M_x = \int_c^d y\rho\, \sqrt{1 + \left(\frac{dx}{dy}\right)^2}\, dy,$$

$$I_y = \int_a^b x^2\rho\, \sqrt{1 + \left(\frac{dy}{dx}\right)^2}\, dx,$$

and

$$I_x = \int_c^d y^2\rho\, \sqrt{1 + \left(\frac{dx}{dy}\right)^2}\, dy.$$

**Figure 13**

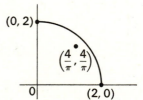

**Problem 3** Suppose a wire has constant density $\rho$ and is in the shape of the curve $y = x^2$ $(0 \le x \le \sqrt{2})$. Show that

$$M_x = \tfrac{1}{2}\rho \int_0^2 \sqrt{y + 4y^2}\, dy.$$

**C. Plane Regions (Variable Density)** When a region with variable density $\rho$ is given, the first and second moments and the center of mass are defined. However, our present techniques only allow us to compute these quantities in special cases.

---

*Example 4* Consider the region in the plane bounded by the parabola $y = x^2/2$, the line $x = 2$, and the $x$-axis (Fig. 15). Suppose the density at the point $(x, y)$ is given by the formula $\rho(x, y) = 3x$. We will compute the mass $M$ and the first moment $M_y$.

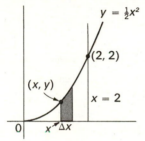

**Figure 15**

*Solution* The mass of the shaded strip is approximately

$$\rho(x, y)y\,\Delta x = 3x\,\frac{x^2}{2}\,\Delta x.$$

Adding the strips and then letting $\Delta x \to 0$, we obtain

$$M = \int_0^2 \frac{3x^3}{2}\,dx = \frac{3x^4}{8}\bigg|_0^2 = 6.$$

In a similar way we have

$$M_y = \int_0^2 \rho(x, y)xy\,dx = \int_0^2 3x \cdot x \cdot \frac{x^2}{2}\,dx$$

$$= \int_0^2 \frac{3x^4}{2}\,dx = \frac{3x^5}{10}\bigg|_0^2 = \frac{48}{5}.$$

---

*Example 5* Suppose the parabola $y = x^2/2$ $(1 \leq x \leq 2)$ has density given by $\rho(x, y) = 2x$. We will compute its mass.

In general, if the density $\rho = \rho(x, y)$ of a region $R$ depends on both $x$ and $y$, we divide the region into squares as shown (Fig. 14). Then the mass of the shaded square is approximately given by

$$\Delta M = \rho\,\Delta A = \rho\,\Delta x\,\Delta y.$$

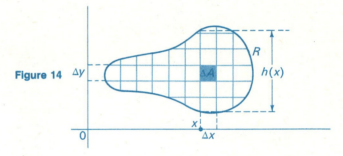

**Figure 14**

Adding these masses, we find that $M$ is approximately given by

$$\sum \rho\,\Delta x\,\Delta y$$

and now, letting $\Delta x = \Delta y \to 0$ we obtain

$$M = \int_R\!\!\int \rho(x, y)\,dx\,dy.$$

Similarly,

$$M_y = \int_R\!\!\int \rho(x, y)x\,dx\,dy$$

These are examples of *double integrals,* and in your next course on calculus, you will learn how to evaluate them.

---

*Problem 4* Suppose that the region described in Example 4 has density given by the formula $\rho(x, y) = y$. Compute $M$, $M_x$, and $I_x$. (Caution: Be sure you integrate with respect to $y$ and not with respect to $x$.)

*Answer:* $M = \frac{4}{5}$, $M_x = \frac{16}{21}$, and $I_x = \frac{8}{9}$.

*Solution*    Since $\Delta s$ is approximately equal to

$$\sqrt{1 + \left(\frac{\Delta y}{\Delta x}\right)^2}\,\Delta x, \text{ we have}$$

$$\Delta M = \rho(x, y)\,\Delta s = 2x\sqrt{1 + \left(\frac{\Delta y}{\Delta x}\right)^2}\,\Delta x.$$

Adding and letting $\Delta x \to 0$, we obtain

$$M = \int_1^2 2x\sqrt{1 + (y')^2}\,dx = \int_1^2 2x\sqrt{1 + x^2}\,dx$$

$$= \tfrac{2}{3}(1 + x^2)^{3/2}\big|_1^2 = \tfrac{2}{3}(5^{3/2} - 2^{3/2}).$$

---

## D. Exercises

**B1.** Consider the region in the plane bounded by the curves $y = x$ and $y = x^2$ and assume it has constant density $\rho$. Find $A$, $M_y$, $I_y$, and $\bar{x}$.

**B2.** Compute $M_x$, $I_x$, $\bar{y}$, and the center of mass for the region described in Exercise B1.

**B3.** Find $\bar{x}$, given the region bounded by $y = \sin x$, $y = \cos x$, $x = 0$, and $x = \tfrac{\pi}{4}$.

**B4.** Find the centroid of the arc in the first quadrant of the curve $x^{2/3} + y^{2/3} = 1$.

**B5.** Find the centroid of the curve described parametrically by the equations

$$x = 3\sin t, \quad y = 3\cos t \quad (0 \le t \le \tfrac{1}{3}\pi).$$

**D6.** In each of the cases below, find $A$, $M$, $M_y$, and $\bar{x}$ assuming the region described has constant density $\rho$.
a) $y = 6x$, $\quad y = 0$, $\quad x = 4$
b) $y = 36x^2$, $\quad y = 0$, $\quad x = 1$
c) $y = x^3$, $\quad y = 0$, $\quad x = 1$

a) $A = 48$, $\quad M = 48\rho$, $\quad M_y = 128\rho$, $\quad \bar{x} = \tfrac{8}{3}$
b) $A = 12$, $\quad M = 12\rho$, $\quad M_y = 9\rho$, $\quad \bar{x} = \tfrac{3}{4}$
c) $A = \tfrac{1}{4}$, $\quad M = \tfrac{1}{4}\rho$, $\quad M_y = \tfrac{1}{5}\rho$, $\quad \bar{x} = \tfrac{4}{5}$

**D7.** In each of the cases below, find $A$, $M$, $M_x$, and $\bar{y}$ assuming the region described has constant density $\rho$.
a) $y = 6x$, $\quad y = 24$, $\quad x = 0$
b) $y = 36x^2$, $\quad y = 1$, $\quad x = 0$
c) $y = x^3$, $\quad y = 1$, $\quad x = 0$
(First solve for $x$ in terms of $y$.)

a) $A = 48$, $\quad M = 48\rho$, $\quad M_x = 768\rho$, $\quad \bar{y} = 16$
b) $A = \tfrac{1}{9}$, $\quad M = \tfrac{\rho}{9}$, $\quad M_x = \tfrac{\rho}{15}$, $\quad \bar{y} = \tfrac{3}{5}$

c) $A = \tfrac{3}{4}$, $\quad M = \tfrac{3}{4}\rho$, $\quad M_x = \tfrac{3}{7}\rho$, $\quad \bar{y} = \tfrac{4}{7}$

**D8.** Determine $M_x$ and $\bar{y}$ for each of the regions described in Exercise D6.

a) $M_x = 384\rho$, $\quad \bar{y} = 8$
b) $M_x = \tfrac{13}{30}\rho$, $\quad \bar{y} = \tfrac{13}{360}$
c) $M_x = \tfrac{4}{7}\rho$, $\quad \bar{y} = \tfrac{16}{7}$

**M9.** Find the centroid of the arc in the first quadrant of the curve $x^{2/3} + y^{2/3} = a^{2/3}$.

**M10.** Consider the region in the plane bounded by the lines $y = 2x + 3$, $y = 0$, $x = 1$, and $x = 3$. Suppose its density is given by $\rho(x, y) = x^{-1}$. Compute $M$ and $M_y$.

**M11.** A wire is in the shape of the parametrically defined curve

$$x = t\ln t \quad \text{and} \quad y = \cos t \quad (3 \le t \le 8).$$

Assuming the density of the wire is given by the formula

$$\rho(t) = (\ln et^2 + \ln^2 t + \sin^2 t)^{-1/2},$$

determine its mass.

**M12.** A wire is in the shape of the parametrically defined curve $x = 6t^2$ and $y = t^4$ ($0 \le t \le 2$) and its density $\rho$ is equal to $x$. Find the mass of the wire.

**M13.** Consider the curve with constant density $\rho$ determined by the polar equation $r = f(\theta)$ ($\theta_1 \le \theta \le \theta_2$). Establish the formula

$$M_y = \rho \int_{\theta_1}^{\theta_2} f(\theta)\cos\theta\sqrt{(f(\theta))^2 + (f'(\theta))^2}\,d\theta,$$

and derive a similar formula for $M_x$.

**M14.** Use the results of Exercise M13 to find $M_x$ and $M_y$ given the curve $r = \sin\theta$ $(0 \leq \theta \leq \frac{1}{2}\pi)$.

**M15.** Consider the region bounded by

$$y = x^2, \quad y = 0, \quad x = 0, \quad \text{and} \quad x = 1.$$

Suppose its density is given by $\rho(x, y) = x$.

a) Show that $M_x = \int_0^1 \frac{1}{2}xy^2\,dx$.
b) Compute $\bar{x}$ and $\bar{l}$.

**M16.** Show that the distance from the centroid of a triangle to a side is one-third of the altitude to that side.

**M17.** Find the center of mass of a solid right circular cone if the density varies as the distance from the base.

---

## 6.6 Approximate Integration

Suppose a function $f$ is continuous on $[a, b]$ and $[a, b]$ is divided into intervals of equal length (Fig. 1). Consider the formulas:

$$A_n = \Delta x(f(x_0) + f(x_1) + \cdots + f(x_{n-1}))$$

$$T_n = \frac{\Delta x}{2}(f(x_0) + 2f(x_1) + 2f(x_2)$$

$$+ \cdots + 2f(x_{n-1}) + f(x_n))$$

and for $n$ an even integer

$$P_n = \frac{\Delta x}{3}(f(x_0) + 4f(x_1) + 2f(x_2) + 4f(x_3)$$

$$+ 2f(x_4) + \cdots + 4f(x_{n-1}) + f(x_n)).$$

We will show that $A_n$, $T_n$, and $P_n$ all approach $\int_a^b f(x)\,dx$ as $n \to \infty$; hence any of these formulas can be used to approximate the definite integral. These ideas will be illustrated with several examples.

**A. Derivation of the Formulas** The formula for $A_n$ has been developed in 5.2A and merely represents the sum of the areas of the rectangles shown in Figure 2. Hence

$$A_n \to \int_a^b f(x)\,dx \quad \text{as} \quad n \to \infty.$$

The formula for $T_n$ is derived by approximating the region by a finite number of trapezoids. Referring to Figure 3, we see that the area of the $k$th trapezoid is

$$\frac{f(x_{k-1}) + f(x_k)}{2}\Delta x,$$

so the sum of the areas of the trapezoids is

⟶ Let $\Delta x = \frac{b-a}{n}$, where $n > 0$ is an integer, and let $x_0 = a$, $x_1 = a + \Delta x$, and in general, $x_k = a + k\,\Delta x$ for $k = 1, 2, \ldots, n$.

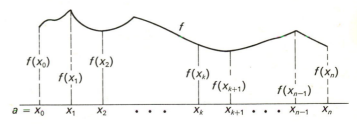

**Figure 1**

⟶ These formulas are especially useful when no antiderivative can be easily found for the given integrand. For very large values of $n$, the approximations are very accurate and computers can do the necessary calculations rapidly and efficiently. (See the comment in 6.6B.)

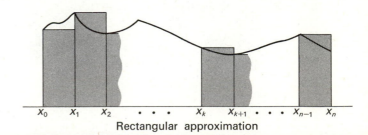

Rectangular approximation

**Figure 2**

$$T_n = \left( \frac{f(x_0) + f(x_1)}{2} + \frac{f(x_1) + f(x_2)}{2} \right.$$
$$\left. + \cdots + \frac{f(x_{n-1}) + f(x_n)}{2} \right) \Delta x$$
$$= \frac{\Delta x}{2} [f(x_0) + 2f(x_1) + 2f(x_2)$$
$$+ \cdots + 2f(x_{n-1}) + f(x_n)].$$

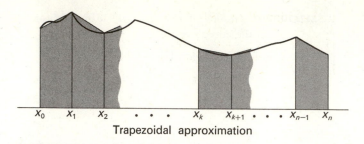

Trapezoidal approximation

**Figure 3**

Intuitively we see that for $f \geq 0$, $T_n$ approaches the area under $f$ as $n \to \infty$. Thus

$$T_n \to \int_a^b f(x)\, dx \quad \text{as } n \to \infty.$$

To derive the formula for $P_n$ we will use the following lemma:

**LEMMA** *Suppose $p$ is the parabola that passes through the points $(a, A)$, $(a + h, B)$ and $(a + 2h, C)$. Then*

$$\int_a^{a+2h} p(x)\, dx = \frac{h}{3}(A + 4B + C).$$

*Proof* The argument is given in the comments.

Suppose now that $n$ is even and consider the three points $(x_0, f(x_0))$, $(x_1, f(x_1))$, $(x_2, f(x_2))$ (Fig. 4). Letting $p$ denote the parabola passing through these points, we have, using the lemma,

$$\int_{x_0}^{x_2} p(x)\, dx = \frac{\Delta x}{3}(f(x_0) + 4f(x_1) + f(x_2)).$$

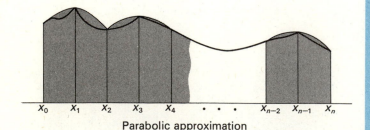

Parabolic approximation

**Figure 4**

The parabola whose equation is

$$p(x) = B + \left( \frac{C - A}{2h} \right)(x - a - h)$$
$$+ \left( \frac{A + C - 2B}{2h^2} \right)(x - a - h)^2$$

does pass through the three points since $p(a) = A$, $p(a + h) = B$, and $p(a + 2h) = C$. In fact it is the only parabola that does so. (See Exercises M11 and M12.)

$$\int_a^{a+2h} p(x)\, dx = Bx + \left( \frac{C - A}{2h} \right) \frac{(x - a - h)^2}{2}$$
$$+ \left( \frac{A + C - 2B}{2h^2} \right) \frac{(x - a - h)^3}{3} \Big|_a^{a+2h}$$
$$= \left[ B(a + 2h) + \left( \frac{C - A}{2h} \right) \frac{h^2}{2} \right.$$
$$\left. + \left( \frac{A + C - 2B}{2h^2} \right) \frac{h^3}{3} \right]$$
$$- \left[ Ba + \left( \frac{C - A}{2h} \right) \frac{h^2}{2} - \left( \frac{A + C - 2B}{2h^2} \right) \frac{h^3}{3} \right]$$
$$= \left[ \frac{h}{3}(A + 4B + C) \right] \bullet$$

Doing the same for the points

$$(x_2, f(x_2)), \quad (x_3, f(x_3)), \quad (x_4, f(x_4))$$

and continuing to the last triple,

$$(x_{n-2}, f(x_{n-2})), \quad (x_{n-1}, f(x_{n-1})), \quad (x_n, f(x_n)),$$

we find that the sum of the areas of the parabolic regions is

$$P_n = \frac{\Delta x}{3}(f(x_0) + 4f(x_1) + f(x_2))$$

$$+ \frac{\Delta x}{3}(f(x_2) + 4f(x_3) + f(x_4))$$

$$+ \cdots + \frac{\Delta x}{3}(f(x_{n-2}) + 4f(x_{n-1}) + f(x_n))$$

$$= \frac{\Delta x}{3}(f(x_0) + 4f(x_1) + 2f(x_2) + 4f(x_3)$$

$$+ \cdots + 4f(x_{n-1}) + f(x_n)).$$

Here again for $f \geq 0$, it is intuitively clear that $P_n$ approximates the area under the graph of $f$, and as $n \to \infty$ we have

$$P_n \to \int_a^b f(x)\,dx.$$

**B. Examples** We will now use our approximation formulas to estimate

$$\int_2^{10} \frac{dx}{1 + x}$$

(The value of this integral to 11 places is 1.29928302765.)

---

**Example 1** ($n = 4$) In this case $\Delta x = \frac{8}{4} = 2$, $x_0 = 2$, $x_1 = 4$, $x_2 = 6$, $x_3 = 8$, and $x_4 = 10$. Hence (to five places)

$$A_4 = 2(f(2) + f(4) + f(6) + f(8))$$

$$= 2(\tfrac{1}{3} + \tfrac{1}{5} + \tfrac{1}{7} + \tfrac{1}{9}) = \tfrac{1488}{945} = 1.57460,$$

$$T_4 = 1(f(2) + 2f(4) + 2f(6) + 2f(8) + f(10))$$

$$= 1 \cdot (\tfrac{1}{3} + \tfrac{2}{5} + \tfrac{2}{7} + \tfrac{2}{9} + \tfrac{1}{11})$$

$$= \tfrac{13848}{10395} = 1.33218,$$

and

$$P_4 = \tfrac{2}{3}(f(2) + 4f(4) + 2f(6) + 4f(8) + f(10))$$

$$= \tfrac{2}{3}(\tfrac{1}{3} + \tfrac{4}{5} + \tfrac{2}{7} + \tfrac{4}{9} + \tfrac{1}{11})$$

$$= \tfrac{13544}{10395} = 1.30293.$$

---

→ There are many other approximation techniques possible. In the parabolic one we found a second degree equation passing through three points on the graph of f. Analogously we could find a cubic passing through four points, or a fourth degree equation through five points. However, especially when using a computer, the parabolic approximation usually suffices.

→ As previously stated, these approximation techniques are especially useful when combined with the efficiency of the computer. The following comment discusses briefly the history and uses of the computer.

*"Life is like a sewer: what you get out of it depends upon what you put into it." And so it is with a computer. The digital computer basically adds, subtracts, multiplies, divides, and stores results. It is the knowledge and cleverness of the operator which makes the machine such a powerful tool.*

*Communicating knowledge to an early computer was an arduous and extremely tedious task, for the machine had to be told at each step which elementary arithmetic operation to perform and where to store the result. In fact, the program for the simplest of calculations could be fifty steps long. However, the development of program "languages" greatly simplified the task.*

*These languages are used in conjunction with a compiler which is in a sense a "pre-program," since it translates statements from the more sophisticated language into elementary machine language.*

*These developments have made the computer a powerful instrument in performing certain mathematical operations. Matrix multiplication, integration, differentiation, determining roots of a function and solutions to differential equations are a few examples of problems regularly solved by computers.*

*We now put something in the previously mentioned sewer*

**Example 2** ($n = 8$)   In this case $\Delta x = \frac{10-2}{8} = 1$, and $x_0 = 2$, $x_1 = 3$, $x_2 = 4$, $x_3 = 5$, $x_4 = 6$, $x_5 = 7$, $x_6 = 8$, $x_7 = 9$, $x_8 = 10$. Hence, to five places

$$A_8 = 1[\tfrac{1}{3} + \tfrac{1}{4} + \tfrac{1}{5} + \tfrac{1}{6} + \tfrac{1}{7} + \tfrac{1}{8} + \tfrac{1}{9} + \tfrac{1}{10}]$$
$$= 1.42897$$
$$T_8 = \tfrac{1}{2}[\tfrac{1}{3} + \tfrac{2}{4} + \tfrac{2}{5} + \tfrac{2}{6} + \tfrac{2}{7} + \tfrac{2}{8} + \tfrac{2}{9} + \tfrac{2}{10} + \tfrac{1}{11}]$$
$$= 1.30779$$
$$P_8 = \tfrac{1}{3}[\tfrac{1}{3} + \tfrac{4}{4} + \tfrac{2}{5} + \tfrac{4}{6} + \tfrac{2}{7} + \tfrac{4}{8} + \tfrac{2}{9} + \tfrac{4}{10} + \tfrac{1}{11}]$$
$$= 1.29961$$

As Examples 1 and 2 indicate, the parabolic approximation is usually the most accurate method to use. Notice, however, that when $n$ is large, even $A_n$ closely approximates the exact value.

The integral in the next example cannot be evaluated by any of the methods of Chapter 5. We will estimate it by using parabolic approximation and then give additional estimates which were determined by a computer.

**Example 3**   Estimate $E = \int_0^1 e^{-x^2}\,dx$.

*Solution*   Letting $n = 4$ we have $\Delta x = \tfrac{1}{4}$ and $x_0 = 0$, $x_1 = \tfrac{1}{4}$, $x_2 = \tfrac{1}{2}$, $x_3 = \tfrac{3}{4}$, $x_4 = 1$. Rounding off to two places, we have

$$P_4 = \tfrac{1}{12}[f(0) + 4f(\tfrac{1}{4}) + 2f(\tfrac{1}{2}) + 4f(\tfrac{3}{4}) + f(1)]$$
$$= \tfrac{1}{12}[1 + 3.76 + 1.56 + 2.29 + .37]$$
$$= .75$$

The results using the computer were (letting $n = N$):

| | |
|---|---|
| $N = 10$ | $E = .7462100386619567$ |
| $N = 50$ | $E = .7467978596687316$ |
| $N = 100$ | $E = .7468148469924926$ |

## C. Exercises

In Exercises B1–B3 evaluate the definite integral by first using antiderivatives. Then approximate the integral by finding $T_4$ and $P_4$ and compare the results.

**B1.** $\int_0^4 x^2\,dx$   **B2.** $\int_0^\pi \sin x\,dx$

*and see what comes out. The explanation given will illustrate how a computer evaluates integrals by the approximation techniques developed in this section.*

*Using the trapezoidal rule and the following program (approximately in FORTRAN language), we will evaluate*

$$\int_0^1 e^{-x^2}\,dx.$$

```
READ, N
DELTX = 1/N
SUM = (EXP(0) + EXP(-1))/2
N1 = N - 1
DØ 3 I = 1,N1
X = 1/N
Y = EXP(-(X*X))
3   SUM = SUM + Y
SUM = SUM*DELTX
PRINT, SUM
STØP
```

*Here $z = \int_0^1 y(x)\,dx = \int_0^1 e^{-x^2}\,dx$ and N represents the number of pieces into which the interval [0, 1] is divided. The program steps are quite simple. At the heart of the program is a DØ loop. Specifically, the statement DØ 3 I = 1, NI tells the computer to perform the steps following this statement up to program step number 3 for every value of I from 1 to NI.*

*The calculated values of the integral are given in the text.*

*(Marshall Sylvan and Leonard Evens)*

**B3.** $\displaystyle\int_1^5 \frac{2x + 1}{x^2 + x}\,dx$

**B4.** Estimate $\displaystyle\int_1^5 x^x\,dx$ by computing $P_4$.

**B5.** Approximate $\int_0^{10} x^2\,dx$ by computing $A_{10}$ and $T_{10}$. Now compute $P_2$ and explain why $P_2$ is the exact value of the integral.

**D6.** Suppose $f(x) = 3 + 2x$. Evaluate the following.

a) $f(2) + f(3) + f(4) + f(5)$

b) $\displaystyle\sum_{k=1}^{6} f(k)$

c) $\displaystyle\sum_{k=1}^{3} f(2k)$

d) $2f(2) + 3f(3) + 4f(4)$

e) $f(1) + f(\frac{1}{2}) + f(\frac{1}{3}) + f(\frac{1}{4})$

f) $\displaystyle\sum_{k=0}^{3} f\!\left(\frac{k}{k+1}\right)$

g) $\displaystyle\sum_{k=1}^{4} k^2 f(k)$

h) $\displaystyle\sum_{k=0}^{3} f^2(k)$

a) 40    b) 60    c) 33    d) 85    e) $\frac{97}{6}$    f) $\frac{95}{6}$
g) 290    h) 164

**D7.** Suppose $f(x) = 3 + 2x$ over the interval $[0, 8]$. Find

a) $T_2$     c) $T_4$     e) $T_8$     g) $T_{100}$
b) $P_2$     d) $P_4$     f) $P_8$     h) $P_{1000}$

a) 88    b) 88    c) 88    d) 88    e) 88    f) 88
g) 88    h) 88

**M8.** Approximate $\displaystyle\int_1^5 \sqrt{35 + x}\,dx$ by finding $T_4$.

**M9.** Approximate $\displaystyle\int_1^3 \ln x\,dx$ by finding $T_4$.

**M10.** Approximate $\displaystyle\int_0^4 \frac{dx}{1 + x^2}$ by finding $A_4$, $T_4$, and $P_4$.

**M11.** Assume the graphs of the parabolas

$$p_1(x) = a_1 + b_1 x + c_1 x^2$$

and

$$p_2(x) = a_2 + b_2 x + c_2 x^2$$

intersect at three distinct points. Show that $a_1 = a_2$, $b_1 = b_2$, and $c_1 = c_2$.

**M12.** Show that the parabola whose equation is

$$p(x) = B + \left(\frac{C - A}{2h}\right)(x - a - h)$$
$$+ \left(\frac{A + C - 2B}{2h^2}\right)(x - a - h)^2$$

is the only parabola passing through the points

$$(a, A), \quad (a + h, B), \quad \text{and} \quad (a + 2h, C).$$

(See Exercise M11.)

**M13.** Given integers $m$, $N$ with $1 \le m \le N$, show that

$$\cos x + \cos(x + \theta) + \cdots + \cos(x + N\theta) = 0$$

where $\theta = \dfrac{2\pi m}{N + 1}$.

**M14.** Show that for any integer $N \ge 2$

$$A_N = \int_0^{2\pi} \cos x\,dx.$$

(See Exercise M13.)

**M15.** Evaluate $P_{26}$ given $\displaystyle\int_0^5 (1 - 2x + 7x^2)\,dx$.

**M16.** Show that $\displaystyle\lim_{n \to \infty} \left(\sum_{k=0}^{n-1} \frac{(2k)^3}{n^4}\right) = 2$.

---

## 6.7   *Sample Exams*

*Sample Exam 1 (45–60 minutes)*

1. Find the area of the region in the plane that lies above the x-axis and below the graph of $y = 4x - x^2$.

2. Consider the region bounded by the graphs of $y = \sqrt{1 - x^2}$ and the line $y = 0$. Find the volume of the solid obtained when this region is rotated about the x-axis.

3. Find the length of the arc of $y = \frac{2}{3}x^{3/2}$ between $x = 3$ and $x = 8$.

4. The graph of $y = \frac{1}{3}x^3$ is rotated about the $x$-axis. Find the area of the surface that lies between $x = 0$ and $x = 3^{1/4}$.

5. A wire is placed on the interval $[0, 4]$ and its density at the point $x$ is given by the formula $\rho(x) = 3x^2$. Determine the mass of the wire. (Do not bother about units.)

## Sample Exam 2 (45–60 minutes)

1. Find the area of the region bounded by the graphs of $y^2 = x$ and $y = x - 2$.

2. Consider the region in the first quadrant bounded by the graph of $y^2 = x$ and the lines $y = 0$ and $x = 4$.
   a) What is the volume obtained if this region is rotated about the $x$-axis?
   b) What is the volume obtained if this region is rotated about the $y$-axis?

3. A plate of metal in the first quadrant is in the shape of the region
$$x^2 + y^2 = 4 \quad (x, y \geq 0),$$
and its density is given by the formula $\rho(x, y) = 2x$. Find the mass of the region. (Do not bother about units.)

4. Consider the arc that lies on the graph of $y = x^2$ between the points $(0, 0)$ and $(2, 4)$. Set up but *do not* evaluate integrals that give:
   a) The length of the arc
   b) The area of the surface obtained by rotating this arc about the $x$-axis
   c) The area of the surface obtained by rotating this arc about the $y$-axis

5. Determine the center of mass of the region having constant density that lies below the $x$-axis and above the graph of $x^2 = y + 4$.

## Sample Exam 3 (45–60 minutes)

1. A four-leaf clover is in the shape of the curve $r = \sin 4\theta$. Find its area.

2. A particle moves along the curve
$$x = e^t \cos t, \quad y = e^t \sin t.$$
How far does it travel during a 2-second interval if it starts at $t = 3$?

3. Find the area of the surface generated by rotating about the $y$-axis the arc of $x = y^3$ ($0 \leq y \leq 1$).

4. A slice of pizza is in the shape of the region bounded by the graph of $y = x^3$, the line $x = 1$, and the $x$-axis. Suppose the density at a point $(x, y)$ is given by the formula $\rho(x, y) = x$. Find
   a) the mass of the slice of pizza, and
   b) its first moment $M_y$ about the $y$-axis.

5. A cylindrical drum of height 4 feet and radius 1 foot is filled with an inhomogeneous fluid. Assume the density at a point of the fluid is the square of the distance between the point and the top of the drum. Find the work required to remove the fluid from the drum.

## Sample Exam 4 (45–60 minutes)

1. The area of the region above the graph of $y = x^2$ and below some line through the origin is equal to 36. Find the equations of two lines with this property.

2. A particle moves along the path defined by the equations
$$x = 3t^2, \quad y = \frac{2}{3}t^3.$$
Starting at $(0, 0)$ how many units of time must elapse before the particle travels a distance of $\frac{196}{3}$ units?

3. Suppose a tank is in the shape of the surface obtained by rotating the arc $y = x^4$ ($0 \leq x \leq 1$) about the $y$-axis. If the tank is filled with a fluid of constant density $\rho$, how much work must be done to remove the fluid from the tank?

4. A sea shell is in the shape of the surface generated by rotating the curve $x = \dfrac{y^5}{5}$ ($0 \leq y \leq 1$) about the $y$-axis.
   If the density at a point on the shell is equal to the square of the height of the point, determine the mass of the shell. (Do not bother about units.)

5. Find the first moments $M_x$ and $M_y$ of the region of constant density $\rho$ which consists of all points $(x, y)$ that satisfy
$$0 \leq y \leq \cos x, \quad 0 \leq x \leq \tfrac{1}{2}\pi.$$

## Sample Exam 5 (Chapters 1–6, approximately 2 hours)

Throughout this exam you will be concerned with the function
$$f(x) = \frac{e^x + e^{-x}}{2}.$$

1. Graph the function $f$ and discuss the features described by the $M_2I_2$ ACIDS.

2. Compute
   a) $f'(\ln 2)$  c) $f^{(122)}(\ln 3)$
   b) $f''(\ln \frac{1}{3})$  d) $f^{(246)}(\ln 4)$

3. Given $g(x) = x^2$, find
   a) $(f \circ g)'((\ln 6)^{1/2})$  b) $(g \circ f)'(0)$

4. Suppose $h(x) = \int_0^{f(x)} t^2 \, dt$. Find $h'(\ln 2)$.

5. Find the area of the region bounded by the graph of $f$ and the lines $y = 0$, $x = 0$, and $x = \ln 5$.

6. The region in the first quadrant bounded by the graph of $f$, $y = 0$, $x = 0$, and $x = \ln 2$ is rotated about the $x$-axis. Find the volume of the solid that is generated.

7. The region in the first quadrant that lies below the graph of $f$ between $x = 0$ and $x = \ln 2$ is rotated about the $y$-axis. Find the volume of the solid that is generated.

8. Find the length of the arc of $f$ that lies between the lines $x = \ln 4$ and $x = \ln 6$.

9. The arc of $f$ between the lines $x = 0$ and $x = \ln 2$ is rotated about the $x$-axis. Find the area of the surface generated.

10. A thin plate of constant density $\rho$ is in the shape of the region below the graph of $f$ that lies above the $x$-axis and between the lines $x = 0$ and $x = \ln 3$. Find the first moment $M_y$.

## Sample Exam 6 (Chapters 1–6, approximately 3 hours)

1. A line passes through the point $(1, 3)$ and intersects the positive $x$- and $y$-axes to form a triangle. What is the slope of the line that minimizes the area of the triangle?

2. A rectangle has length 10 and width 4. At a given instant, its length and width begin to increase at the rate of 3 units per second and 4 units per second respectively. At what rate is its area changing when $t$ seconds have elapsed?

3. Graph the function $g(x) = \dfrac{x}{(x - 2)^2}$ and discuss the $M_2I_2$ ACIDS features.

4. Evaluate the limits
   a) $\lim\limits_{x \to \infty} \dfrac{x^2 + 4x + 4}{x^2 + 3x + 2}$

b) $\lim\limits_{x \to 0} \dfrac{\sin 4x}{\sin x}$

c) $\lim\limits_{x \to 2} \dfrac{x^3 - 8}{x - 2}$

5. Find the maximum possible area of a rectangle which has one side on the positive $x$-axis, one side on the positive $y$-axis, and one vertex on the curve $y = e^{-x}$.

6. Find $f'(x)$ given
   a) $f(x) = \sqrt{1 + \sqrt{x}}$
   b) $e^{f(x)} = \ln(1 + x^2)$
   c) $f''(x) = e^x + 2xe^{x^2}$

7. A glass in the shape of a cylinder with radius $\frac{3}{2}$ inches and height 6 inches is filled with milk of density $\rho$. A child drinks the milk using a 10 inch straw. How much work does the child do in emptying the glass?

8. Find the indefinite integral
$$\int \frac{x^2 + 4x + 1}{(x^2 - 1)(x + 2)} \, dx.$$

9. Find the length of the arc described by the equations
$$x = \tfrac{1}{2}\ln(1 + t^2), \qquad y = \operatorname{Tan}^{-1} t$$
where $0 \le t \le 1$.

10. A wire is in the shape of the graph of
$$y = e^x \quad (\ln \sqrt{3} \le x \le \ln \sqrt{8})$$
and the density of the wire at a point $(x, y)$ is given by the formula $\rho(x, y) = e^{2x}$. Find the mass of the wire.

11. A hemispherical bowl of radius 3 is filled with a fluid whose density at a point is equal to the square of the depth of the point. What is the mass of the contents of the bowl?

12. Use parabolic approximation with $n = 4$ to estimate the value of $\int_2^6 x^3 \, dx$.

In the first two sections of this chapter we study the problem of adding an infinite collection of numbers. As will be seen, the addition can often be carried out, and the sum is a real number. These notions are generalized in the next two sections, where we add an infinite number of functions. It is shown that most of the important functions studied earlier can be written in the form of a power series

$$a_0 + a_1x + a_2x^2 + \cdots.$$

The fifth section deals with complex numbers, and the previous results in the chapter are used to define and discuss certain important complex valued functions. The concepts developed in this chapter will be applied in Chapter 8, and they form the basis for several advanced branches of mathematics.

# Series

## 7.1 Infinite Sums

Adding two numbers or a finite number of numbers is a simple process. It is also possible to find the sum of an infinite collection of numbers, but in this case the situation is more complicated. The basic definitions concerning infinite sums will be given in this section.

**A. Motivation** We will begin our discussion with an intuitively simple example.

→ *You already know something about infinite sums. For example, you know that $\frac{1}{3}$ can be written as 0.333 . . . . This means that the infinite sum 0.3 + 0.03 + 0.003 + · · · (which is another way of writing 0.333 . . .) is equal to the number $\frac{1}{3}$. So whether $\frac{1}{3}$ looks like an infinite sum or like an ordinary rational number depends on the way you look at it.*

*(Bob)*

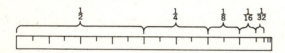

**Figure 1**

Imagine a ruler (Fig. 1). Now successively divide it in half as shown in Figure 2. There are then infinitely many pieces of lengths

$$\tfrac{1}{2}, \tfrac{1}{4}, \tfrac{1}{8}, \tfrac{1}{16}, \tfrac{1}{32}, \cdots$$

that make up the original ruler. Hence the sum of the lengths of the pieces must equal the length of the ruler or

$$1 = \tfrac{1}{2} + \tfrac{1}{4} + \tfrac{1}{8} + \tfrac{1}{16} + \tfrac{1}{32} + \cdots.$$

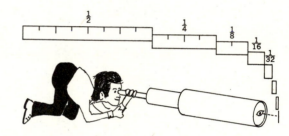

**Figure 2**

**B. The Geometric Series** We will now generalize the example given in 7.1A.

Suppose $x$ is any real number not equal to 1 and let

$$S_n = x + x^2 + x^3 + \cdots + x^n.$$

We will find a formula for $S_n$. Multiplying $S_n$ by $x$ we obtain

$$xS_n = x^2 + x^3 + x^4 + \cdots + x^{n+1},$$

and now subtracting $xS_n$ from $S_n$, we obtain

$$
\begin{aligned}
S_n - xS_n &= (x + x^2 + x^3 + \cdots + x^n) \\
&\quad -(x^2 + x^3 + \cdots + x^n + x^{n+1}) \\
&= x - x^{n+1} \quad \text{or} \\
(1 - x)S_n &= x - x^{n+1}.
\end{aligned}
$$

→ In order to make this statement we must ignore the fact that the ruler is composed of a finite number of "indivisible" atoms.

→ *The reason for excepting 1 will be clear in a moment.*

*(Ed)*

→ For example, if $x = \frac{1}{10}$ and $n = 8$ then

$$
\begin{aligned}
S_8 &= \tfrac{1}{10} + (\tfrac{1}{10})^2 + (\tfrac{1}{10})^3 + \cdots + (\tfrac{1}{10})^8 \\
&= .1 + .01 + .001 + \cdots + .00000001 \\
&= .11111111
\end{aligned}
$$

Dividing by $1 - x$ then gives

$$S_n = \frac{x - x^{n+1}}{1 - x} = \frac{x}{1 - x} - \frac{x^{n+1}}{1 - x}.$$

We assume now that $-1 < x < 1$ and examine the behavior of this formula as $n \to \infty$. Since $x^{n+1} \to 0$ as $n \to \infty$, we have

$$\frac{x^{n+1}}{1 - x} \to 0 \quad \text{as} \quad n \to \infty.$$

It follows that

$$S_n = \frac{x}{1 - x} - \frac{x^{n+1}}{1 - x} \to \frac{x}{1 - x} \quad \text{as} \quad n \to \infty.$$

Denoting $\frac{x}{1-x}$ by $S$, we then have that $S$ is the sum of all the powers of $x$. In other words, for $-1 < x < 1$, we have

$$(*) \qquad \frac{x}{1 - x} = x + x^2 + x^3 + \cdots = \sum_{n=1}^{\infty} x^n.$$

In the example in 7.1A we had $x = \frac{1}{2}$. Substituting this value in equation (*) gives

$$1 = \frac{1/2}{1 - (1/2)} = \frac{1}{2} + \frac{1}{4} + \frac{1}{8} + \frac{1}{16} + \cdots,$$

which was our previous observation.

Adding 1 to each side of the equation $\dfrac{x}{1 - x} = \displaystyle\sum_{n=1}^{\infty} x^n$ gives

$$1 + \frac{x}{1 - x} = 1 + \sum_{n=1}^{\infty} x^n$$

or

$$(\#) \qquad \frac{1}{1 - x} = 1 + x + x^2 + x^3 + \cdots = \sum_{n=0}^{\infty} x^n.$$

Here it is also assumed that $-1 < x < 1$. The expressions $\sum_{n=1}^{\infty} x^n$ and $\sum_{n=0}^{\infty} x^n$ in the formulas (*) and (#) above are examples of geometric series. A geometric series is a sum of numbers where the ratio of any term to the term before it is constant.

*We don't have to worry about the denominator ever being zero because we so foresightedly forbade x to equal 1. Even if it did, all the powers of x would be 1, so $S_n$ would equal n simply enough.*

*(Ed)*

*The reason we restrict to $x \in (-1, 1)$ is to insure that $S_n$ tends to a finite limit as $n \to \infty$. For example, if $x = 2$ then*

$$S_n = \frac{2 - 2^{n+1}}{1 - 2} = 2^{n+1} - 2$$

*and clearly $S_n \to \infty$ as $n \to \infty$.*

---

**Problem 1** Show that

$$\tfrac{12}{3} + \tfrac{24}{9} + \tfrac{48}{27} + \tfrac{96}{81} + \cdots = 12.$$

(Hint: Factor out a 6.)

---

*A geometric series is but one example of an infinite collection of numbers whose sum exists. The series*

$$1 + \frac{1}{2^2} + \frac{1}{3^2} + \frac{1}{4^2} + \cdots$$

*also has a finite sum but is not a geometric series.*

**C. Definitions**  Suppose we are given an infinite collection of real numbers

$$a_1, a_2, a_3, \ldots .$$

The expression

$$a_1 + a_2 + a_3 + \cdots \quad \text{or} \quad \sum_{k=1}^{\infty} a_k$$

→ The first subscript in the series need not be 1, nor do the indices have to be the positive integers. For example,

$$a_6 + a_7 + a_8 + \cdots = \sum_{k=6}^{\infty} a_k$$

is then called a *series*. For each positive integer $n$, $a_n$ is called the $n$th *term* and

$$S_n = a_1 + a_2 + \cdots + a_n = \sum_{k=1}^{n} a_k$$

$$b_{-2} + b_{-1} + b_0 + b_1 + b_2 + \cdots = \sum_{k=-2}^{\infty} b_k$$

is called the $n$th *partial sum*.

If the partial sums $S_n$ approach some real number $S$ as $n \to \infty$, that is, if

$$S_n \to S \quad \text{as} \quad n \to \infty,$$

$$c_0 + c_2 + c_4 + c_6 + \cdots = \sum_{k=0}^{\infty} c_{2k}$$

are also instances of series.

then the series is called *convergent*. The number $S$ is called the *sum* of the series and we write

→ It is important to realize that if the series $\sum_{k=1}^{\infty} a_k$ converges, then a series obtained from it by adding or changing a finite number of terms also converges (not necessarily to the same sum, of course). In particular, $\sum_{k=k_0}^{\infty} a_k$ is convergent, where $k_0 > 1$.

$$S = a_1 + a_2 + \cdots = \sum_{k=1}^{\infty} a_k.$$

A series that is not convergent is said to be *divergent*.

We will now give some examples of convergent and divergent series.

These notions of convergence and divergence are related to those given in the Section (5.6) on improper integrals, as → will be seen in Theorem 6 (7.2D).

---

*Example 1*  Using the definitions above, show that the series

$$x + x^2 + x^3 + \cdots = \sum_{n=1}^{\infty} x^n$$

converges to the sum $\frac{x}{1-x}$ when $-1 < x < 1$.

*Solution*  From 7.1B we have

$$S_n = x + x^2 + \cdots + x^n = \frac{x - x^{n+1}}{1-x} \quad \text{for} \quad x \neq 1.$$

When $-1 < x < 1$ it follows that the partial sums $S_n$ satisfy

$$S_n \to S = \frac{x}{1-x} \quad \text{as} \quad n \to \infty.$$

*Problem 2*  Two men on unicycles are 40 miles apart at noon and traveling toward each other at the rate of ten miles an hour. A fly traveling at the rate of 20 miles an hour begins at noon and flies back and forth between the noses of the unicyclists. How far does the fly fly before being squashed? (Hint: The problem can be solved by finding the sum of a certain series, but there is a more elegant method of solution.)

(John von Neumann (1903–1957), one of the most phenomenal mathematicians of this century, was renowned for the incredible speed in which he solved problems. He also made significant contributions to physics, economics, and the theory of automata. One day (so the story goes) a mathematician friend of his decided to determine if von Neumann thought more like a mathematician or a physicist. He devised the above problem, reasoning that if von Neumann was a

Hence the series

$$x + x^2 + x^3 + \cdots$$

is convergent and we have

$$\sum_{n=1}^{\infty} x^n = \frac{x}{1 - x}.$$

mathematician he would use the elegant method and answer quickly. If he thought like a physicist he would solve the problem by the more straightforward but lengthy method of series. The question was asked and von Neumann immediately gave the answer as 40 miles. The friend replied "Ah ha, so you are a mathematician," and explained. Von Neumann replied that he had not noticed the elegant method and merely summed the series.)

---

**Example 2**  (*The harmonic series*) Show that the series

$$1 + \tfrac{1}{2} + \tfrac{1}{3} + \tfrac{1}{4} + \cdots,$$

called the *harmonic series,* is divergent.

*Solution*   We write out the series with parentheses around the terms as follows:

$$1 + \tfrac{1}{2} + (\tfrac{1}{3} + \tfrac{1}{4}) + (\tfrac{1}{5} + \tfrac{1}{6} + \tfrac{1}{7} + \tfrac{1}{8})$$
$$+ (\tfrac{1}{9} + \tfrac{1}{10} + \tfrac{1}{11} + \tfrac{1}{12} + \tfrac{1}{13} + \tfrac{1}{14} + \tfrac{1}{15} + \tfrac{1}{16})$$
$$+ \cdots.$$

Now $\tfrac{1}{3} + \tfrac{1}{4} > \tfrac{1}{4} + \tfrac{1}{4} = \tfrac{1}{2}$, and $\tfrac{1}{5} + \tfrac{1}{6} + \tfrac{1}{7} + \tfrac{1}{8} > \tfrac{1}{8} + \tfrac{1}{8} + \tfrac{1}{8} + \tfrac{1}{8} = \tfrac{1}{2}$, and $\tfrac{1}{9} + \cdots + \tfrac{1}{16} > \tfrac{1}{16} + \cdots + \tfrac{1}{16} = \tfrac{1}{2}$. The next 16 terms from $\tfrac{1}{17}$ to $\tfrac{1}{32}$ add up to more than $\tfrac{1}{2}$, and the following 32 terms similarly add up to more than $\tfrac{1}{2}$. Continuing in this manner, we find that by taking enough terms of the series, the partial sums $S_n$ can be made arbitrarily large. Hence the $S_n$'s cannot converge to a real number and the series diverges. This case is often described by writing $S_n \to \infty$ or

$$1 + \tfrac{1}{2} + \tfrac{1}{3} + \cdots = \infty.$$

→   *The harmonic series diverges, but it does so very slowly. For example, the first hundred terms add up to approximately 5, and the first thousand terms add up to approximately $7\tfrac{1}{2}$. For a partial sum to exceed 100 it would have to consist of at least $10^{50}$ terms.*

*(Vahan)*

---

**THEOREM 1**   *If the series* $a_1 + a_2 + a_3 + \cdots$ *is convergent, then* $a_n \to 0$ *as* $n \to \infty$.

It is important to realize that the converse of this theorem is not true. For example, the harmonic series is divergent even though its terms, $a_n = \tfrac{1}{n}$, approach zero as $n \to \infty$.

→   *Proof*   Let $S = \Sigma_{k=1}^{\infty} a_k$ and $S_n = \Sigma_{k=1}^{n} a_k$. We then have that $S_n \to S$ as $n \to \infty$. Consider $S_n - S_{n-1}$. Since $S_n \to S$ as $n \to \infty$, it also follows that $S_{n-1} \to S$ as $n \to \infty$. Therefore the difference $S_n - S_{n-1}$ must approach zero as $n \to \infty$. In other words, $a_n = S_n - S_{n-1} \to 0$ as $n \to \infty$ ●

## Resolution of a Paradox

The most famous paradox of Zeno concerns Achilles, who runs to overtake a tortoise. Although Achilles is the faster runner, he can never pass the tortoise, for when he reaches the place at which the tortoise began, the tortoise will have moved ahead to a new position. When Achilles reaches the new position, the tortoise will have again moved ahead, and so on ad infinitum. The argument is faulty, for the increments into which Achilles' path has been divided form a convergent series, and so do the time intervals.

Suppose that Achilles travels at 10 miles/hour and a speedy tortoise chugs along at 5 miles/hour and is given a 1 mile head start. Examine the accompanying table.

| Time (hours) | 0 | $\frac{1}{10}$ | $\frac{3}{20}$ | $\frac{7}{40}$ | $\frac{15}{80}$ | $\cdots$ |
|---|---|---|---|---|---|---|
| Position of Achilles | 0 | 1 | $\frac{3}{2}$ | $\frac{7}{4}$ | $\frac{15}{8}$ | $\cdots$ |
| Position of tortoise | 1 | $\frac{3}{2}$ | $\frac{7}{4}$ | $\frac{15}{8}$ | $\frac{31}{16}$ | $\cdots$ |

Notice that Achilles successively takes the positions that the tortoise has just occupied. The sum of the time increments is given by

$$0 + \frac{1}{10} + \frac{1}{20} + \frac{1}{40} + \frac{1}{80} + \cdots = \frac{1}{10} \sum_{n=0}^{\infty} \left(\frac{1}{2}\right)^n$$

$$= \frac{1}{10}\left(\frac{1}{1-(1/2)}\right) = \frac{1}{5}.$$

The sum of the intervals of distance Achilles covers is

$$0 + 1 + \frac{1}{2} + \frac{1}{4} + \frac{1}{8} + \cdots = \sum_{n=0}^{\infty} \left(\frac{1}{2}\right)^n = \frac{1}{1-\frac{1}{2}} = 2,$$

and the sum of the intervals of distance that the tortoise traverses is

$$1 + \frac{1}{2} + \frac{1}{4} + \frac{1}{8} + \cdots = \frac{1}{1-(1/2)} = 2.$$

Achilles and the tortoise are then both 2 miles from the starting point at time $t = \frac{1}{5}$, and Achilles passes the tortoise immediately thereafter.

### D. Exercises

**B1.** Find the sum of each of the following series.

a) $\sum_{k=1}^{\infty} \frac{1}{3^k}$    c) $\sum_{k=0}^{\infty} \frac{4}{5^{2k}}$

b) $\sum_{k=0}^{\infty} \frac{(-1)^k}{4^{k+1}}$    d) $\sum_{k=0}^{\infty} \frac{2^{k+3}}{3^{k+2}}$

**B2.** Consider the series $1 + 2 + 3 + 4 + 5 + \cdots$.
a) Evaluate $S_1, S_2, \ldots, S_7$.

b) Show that $S_n = \frac{n(n+1)}{2}$ for any integer $n > 0$.
Hint: $2S_n$ equals

$$1 + \quad 2 \quad + \cdots + (n-1) + n$$
$$+ n + (n-1) + \cdots + \quad 2 \quad + 1$$

**B3.** Given the series, $\sum_{n=0}^{\infty} (-1)^n$:
a) Evaluate $S_{2000}$ and $S_{2001}$.
b) Find a general formula for $S_n$ and explain why the series is divergent.

**B4.** Given the series

$$\ln \tfrac{1}{2} + \ln \tfrac{2}{3} + \ln \tfrac{3}{4} + \ln \tfrac{4}{5} + \cdots,$$

show that $S_n = -\ln(n+1)$, and hence that the series is divergent.

**B5.** Find

a) $\displaystyle\sum_{k=0}^{\infty} \frac{6}{10^{2k+1}},$  b) $\displaystyle\sum_{k=1}^{\infty} \frac{3}{10^{2k}},$

and then show that

c) $\dfrac{7}{11} = \dfrac{6}{10} + \dfrac{3}{10^2} + \dfrac{6}{10^3} + \dfrac{3}{10^4} + \dfrac{6}{10^5}$

$\qquad\qquad + \dfrac{3}{10^6} + \cdots.$

**B6.** Express the following rational numbers in terms of an infinite series as shown in Exercise B5c for the number $\frac{7}{11}$.

a) $\frac{1}{3}$   b) $\frac{4}{15}$   c) $\frac{23}{99}$   d) $\frac{19}{33}$

**D7.** Write each of the following series using the summation notation.

a) $a_3 + a_4 + a_5 + \cdots$
b) $a_2 + a_4 + a_6 + a_8 + \cdots$
c) $a_1 + a_3 + a_5 + a_7 + \cdots$
d) $-a_1 + a_2 - a_3 + a_4 - \cdots$
e) $a_1 - a_2 + a_3 - a_4 + \cdots$
f) $a_1 + 2a_2 + 3a_3 + 4a_4 + \cdots$
g) $a_1 + 2a_2 + 4a_3 + 8a_4 + \cdots$
h) $a_1 + a_4 + a_9 + a_{16} + \cdots$

a) $\displaystyle\sum_{k=3}^{\infty} a_k$   b) $\displaystyle\sum_{k=1}^{\infty} a_{2k}$   c) $\displaystyle\sum_{k=1}^{\infty} a_{2k-1}$

d) $\displaystyle\sum_{k=1}^{\infty} (-1)^k a_k$   e) $\displaystyle\sum_{k=1}^{\infty} (-1)^{k+1} a_k$   f) $\displaystyle\sum_{k=1}^{\infty} k a_k$

g) $\displaystyle\sum_{k=1}^{\infty} 2^{k-1} a_k$   h) $\displaystyle\sum_{k=1}^{\infty} a_{k^2}$

**D8.** Find the sum of each of the following series.

a) $\displaystyle\sum_{k=0}^{\infty} \left(\tfrac{2}{3}\right)^k$   c) $\displaystyle\sum_{k=0}^{\infty} \frac{2^k}{5^k}$   e) $\displaystyle\sum_{k=1}^{\infty} \left(\tfrac{5}{6}\right)^k$

b) $\displaystyle\sum_{k=0}^{\infty} \frac{3^k}{4^k}$   d) $\displaystyle\sum_{k=0}^{\infty} \frac{9^k}{10^k}$   f) $\displaystyle\sum_{k=1}^{\infty} \left(\tfrac{2}{7}\right)^k$

g) $\displaystyle\sum_{k=2}^{\infty} \frac{1}{4^k}$   h) $\displaystyle\sum_{k=0}^{\infty} \left(\frac{-2}{3}\right)^k$

a) 3   b) 4   c) $\frac{5}{3}$   d) 10   e) 5   f) $\frac{2}{5}$
g) $\frac{1}{12}$   h) $\frac{3}{5}$

**D9.** Give a formula for the $n$th term of each of the following series.

a) $1 + \dfrac{1}{2!} + \dfrac{1}{3!} + \dfrac{1}{4!} + \cdots$

b) $-1 + \tfrac{1}{2} - \tfrac{1}{3} + \tfrac{1}{4} - \tfrac{1}{5} + \cdots$
c) $1 - \tfrac{1}{3} + \tfrac{1}{5} - \tfrac{1}{7} + \tfrac{1}{9} - \cdots$
d) $\tfrac{2}{1} + \tfrac{4}{4} + \tfrac{8}{9} + \tfrac{16}{16} + \tfrac{32}{25} + \cdots$
e) $\tfrac{1}{3} + \tfrac{5}{9} + \tfrac{9}{27} + \tfrac{13}{81} + \tfrac{17}{243} + \cdots$

a) $\dfrac{1}{n!}$   b) $\dfrac{(-1)^n}{n}$   c) $\dfrac{(-1)^{n+1}}{2n-1}$   d) $\dfrac{2^n}{n^2}$

e) $\dfrac{4n-3}{3^n}$

**M10.** Find the ratio of the $(n+1)$th term to the $n$th term in the series

$$\frac{x}{1\cdot 2} + \frac{x^2}{2\cdot 3} + \frac{x^3}{3\cdot 4} + \cdots.$$

**M11.** Give a formula for the $n$th term of the series

$$\frac{\sqrt{x}}{2} + \frac{x}{2\cdot 4} + \frac{x\sqrt{x}}{2\cdot 4\cdot 6} + \frac{x^2}{2\cdot 4\cdot 6\cdot 8} + \cdots.$$

**M12.** Express the sum of each of the following series in the form $\frac{m}{n}$ where $m$ and $n$ are positive integers.

a) $\dfrac{3}{10} + \dfrac{8}{10^2} + \dfrac{3}{10^3} + \dfrac{8}{10^4} + \dfrac{3}{10^5} + \dfrac{8}{10^6} + \cdots$

b) Let $a = \dfrac{1}{10} + \dfrac{2}{10^2} + \dfrac{3}{10^3}$  and find

$$a + \frac{a}{10^3} + \frac{a}{10^6} + \cdots.$$

**M13.** What rational numbers are represented by the following infinite decimals?
a) $.123123123\ldots$
b) $.56565656\ldots$
c) $.78787878\ldots$

**M14.** Use the fact that if

$$P_n = r_1 \cdot r_2 \cdot \cdots \cdot r_n \quad \text{then} \quad \ln P_n = \sum_{k=1}^{n} \ln r_k$$

to show that

$$\ln \frac{3}{2} = \sum_{n=3}^{\infty} \ln \left( \frac{n^2}{(n+1)(n-1)} \right).$$

**M15.** a) Check that the formula

$$1^2 + 2^2 + \cdots + n^2 = \frac{n(n+1)(2n+1)}{6}$$

is valid for $n = 1, 2, 3, 4,$ and $5$.

    b) Assume the formula is valid for an integer $k$ and then show that it is valid for the integer $k + 1$.

    c) Think about parts (a) and (b) until you realize that you have proven the formula for all positive integers $n$. (This is an example of a proof by mathematical induction. See Example 5, 9.1C.)

**M16.** a) Show that $\displaystyle\sum_{k=1}^{\infty} \left( \frac{1}{k} - \frac{1}{k+1} \right) = 1$ and

    b) using this fact, now prove that

$$\sum_{k=1}^{\infty} \frac{1}{k(k+1)} = 1.$$

**M17.** Using the method of Exercise M16 prove that

    a) $\displaystyle\sum_{k=1}^{\infty} \frac{1}{k(k+2)} = \frac{3}{4}$

    b) $\displaystyle\sum_{k=1}^{\infty} \frac{1}{k(k+3)} = \frac{11}{18}.$

**M18.** Generalize the results of Exercises M16 and M17 by showing that for any integer $r > 0$

$$\sum_{k=1}^{\infty} \frac{1}{k(k+r)} = \frac{1}{r}\left( 1 + \frac{1}{2} + \cdots + \frac{1}{r} \right).$$

**M19.** Show that

    a) $\displaystyle\sum_{k=1}^{\infty} \frac{1}{k(k+1)(k+2)} = \frac{1}{4}$

    b) $\displaystyle\sum_{k=1}^{\infty} \frac{1}{(2k-1)(2k+1)} = \frac{1}{2}.$

**M20.** Show that

    a) $\displaystyle\sum_{k=1}^{\infty} \frac{1}{(4k-3)(4k+1)} = \frac{1}{4}$

    b) $\displaystyle\sum_{k=1}^{\infty} \left( \frac{1}{k^p} - \frac{1}{(k+1)^p} \right) = 1 \quad \text{for } p > 0.$

**M21.** Show that

    a) $\displaystyle\sum_{k=1}^{\infty} \frac{k}{(k+1)!} = 1$

    b) $\displaystyle\sum_{k=1}^{\infty} \frac{k^2 - k - 1}{k!} = 1.$

**M22.** A ball is dropped from a height of 10 feet. Each time it bounces it rises to a height of $\frac{3h}{4}$, where $h$ is the height of the previous bounce. Find the total distance the ball travels before coming to rest.

---

## 7.2   *Tests for Convergence*

We now give some tests for the convergence and divergence of series. Using a particular test we may find that a particular series converges, but determining what number the series converges to is more difficult and requires other methods.

**A. The Comparison Test**   We will state this test in the form of a theorem.

$\rightarrow$   We will postpone the proof until 9.2B. In the meantime the reader should consider the result as being almost evident. To see how it is related to decimal expansions for real numbers, look at Example 2.

*THEOREM 1*   *Suppose that* $0 \le a_n \le b_n$ *for* $n = 1, 2, 3, \ldots$ .

a) If $\sum\limits_{n=1}^{\infty} b_n$ is convergent, then $\sum\limits_{n=1}^{\infty} a_n$ is convergent.

b) If $\sum\limits_{n=1}^{\infty} a_n$ is divergent, then $\sum\limits_{n=1}^{\infty} b_n$ is divergent.

→ *In words this says that if each term of a series (of non-negative terms) is less than the corresponding term of a convergent series then the first series must be convergent.*

*(Vahan)*

Note also that (a) and (b) are equivalent, for (a) is equivalent to its contrapositive "If $\Sigma_{n=1}^{\infty} a_n$ is not convergent, then $\Sigma_{n=1}^{\infty} b_n$ is not convergent," which is statement (b).

**Example 1** Show that the series

$$\frac{1}{1\cdot 2} + \frac{1}{2\cdot 2^2} + \frac{1}{3\cdot 2^3} + \cdots + \frac{1}{n\cdot 2^n} + \cdots$$

converges.

*Solution* Since $0 \leq \dfrac{1}{n\cdot 2^n} \leq \dfrac{1}{2^n}$ and the series $\sum\limits_{n=1}^{\infty} \dfrac{1}{2^n}$ is convergent, it follows by (a) of Theorem 1 that the given series converges.

The comparison test can also be used to show that certain series diverge.

**Example 3** Show that the series

$$1 + \frac{1}{\sqrt{2}} + \frac{1}{\sqrt{3}} + \frac{1}{\sqrt{4}} + \cdots \quad \text{diverges.}$$

*Solution* In this case we apply (b) of Theorem 1. Since

$$\frac{1}{n} \leq \frac{1}{\sqrt{n}} \quad (n = 1, 2, 3, \ldots)$$

and $\sum\limits_{n=1}^{\infty} \dfrac{1}{n}$ diverges, it follows that $\sum\limits_{n=1}^{\infty} \dfrac{1}{\sqrt{n}}$ also diverges.

**Example 2** (*Decimal representation*) Let $\Sigma_{n=1}^{\infty} b_n$ be the series $\dfrac{9}{10} + \dfrac{9}{10^2} + \dfrac{9}{10^3} + \cdots$. It is convergent since

$$\sum_{n=1}^{\infty} b_n = 9\left(\frac{1}{10} + \frac{1}{10^2} + \frac{1}{10^3} + \cdots\right)$$

$$= 9\left(\frac{1/10}{1 - (1/10)}\right) = 9\left(\frac{1/10}{9/10}\right) = 1.$$

Observe that this shows that $0.9999\ldots$ equals 1.

Next take the series $\sum\limits_{n=1}^{\infty} a_n = \dfrac{d_1}{10} + \dfrac{d_2}{10^2} + \dfrac{d_3}{10^3} + \cdots$ where each $d_n$ is a digit (that is, $d_n = 0, 1, 2, 3, 4, 5, 6, 7, 8,$ or 9). Since $a_n = \dfrac{d_n}{10^n} \leq \dfrac{9}{10^n}$, it follows from the comparison test (Theorem 1) that the series $\Sigma_{n=1}^{\infty} a_n$ converges to a real number. This should come as no surprise to the reader since

$$\frac{d_1}{10} + \frac{d_2}{10^2} + \frac{d_3}{10^3} + \cdots$$

is just the real number $.d_1 d_2 d_3 \ldots$ . The series merely clarifies the meaning of the decimal representation.

**B. Absolute Convergence** A series

$$a_1 + a_2 + a_3 + \cdots$$

is said to be *absolutely convergent* (or to converge absolutely) if

$$|a_1| + |a_2| + |a_3| + \cdots$$

is convergent.

→ If $a_n \geq 0$ for all $n$, then convergence and absolute convergence are equivalent.

More generally, if $\Sigma_{n=1}^{\infty} a_n$ is a series with finitely many negative terms (or finitely many positive terms), then convergence and absolute convergence are the same.

An important fact about absolute convergence is given in the next theorem.

**THEOREM 2**   *If the series*

$$a_1 + a_2 + a_3 + \cdots$$

*converges absolutely, then it converges.*

We may now generalize the applicability of the comparison test.

**COROLLARY**   *Suppose $b_1 + b_2 + \cdots$ is a convergent series and*

$$0 \leq |a_n| \leq b_n \quad \text{for all } n.$$

*Then*

$$a_1 + a_2 + a_3 + \cdots$$

*is convergent.*

*Proof*   By the comparison test (Theorem 1)

$$|a_1| + |a_2| + |a_3| + \cdots$$

is convergent. Hence the series $\sum_{n=1}^{\infty} a_n$ is absolutely convergent and by the theorem it is convergent ●

It is important to realize that the converse of the above theorem is false. Indeed it is possible for a series to converge without converging absolutely. Such a series is called *conditionally convergent*.

A class of conditionally convergent series is discussed below.

**THEOREM 3**   *Suppose*

$$a_1 \geq a_2 \geq a_3 \geq \cdots \geq 0$$

*and*

$$a_n \to 0 \quad \text{as} \quad n \to \infty.$$

*Then*

$$a_1 - a_2 + a_3 - a_4 + a_5 - \cdots$$

*is convergent.*

*Proof*   We will plot several of the partial sums.

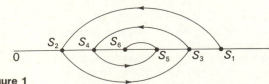

**Figure 1**

*Proof*   Let $A_n = a_1 + \cdots + a_n$, $B_n = |a_1| + \cdots + |a_n|$, and

$$C_n = \frac{a_1 + |a_1|}{2} + \cdots + \frac{a_n + |a_n|}{2}.$$

Since $\sum_{n=1}^{\infty} |a_n|$ is convergent, $B_n \to B$ for some real number $B$. Now, for each $n$,

$$0 \leq \frac{a_n + |a_n|}{2} \leq |a_n|,$$

so it follows by the comparison test that the series $\sum_{n=1}^{\infty} \frac{a_n + |a_n|}{2}$

is convergent. Hence $C_n \to C$ for some real number $C$. Therefore

$$2\left(C_n - \frac{B_n}{2}\right) \to 2\left(C - \frac{B}{2}\right)$$

and since $2(C_n - \frac{B_n}{2}) = A_n$, we have $A_n \to 2(C - \frac{B}{2})$. Hence $\sum_{n=1}^{\infty} a_n$ converges ●

---

*Example 4*   Show that the series

$$1 - 1 + \tfrac{1}{2} - \tfrac{1}{2} + \tfrac{1}{3} - \tfrac{1}{3} + \tfrac{1}{4} - \tfrac{1}{4} + \cdots$$

is conditionally convergent.

*Solution*   By comparing this series with the harmonic series (7.1C) we see that it is not absolutely convergent. However, its partial sums satisfy $S_{2n} = 0$ and $S_{2n-1} = \frac{1}{n}$, and thus converge to zero. Hence the series is convergent. Since it is not absolutely convergent it is therefore conditionally convergent.

---

A rigorous proof of this result is as follows. Consider the partial sums with even indices:

$$S_{2n} = (a_1 - a_2) + (a_3 - a_4) + \cdots + (a_{2n-1} - a_{2n}).$$

Since each factor in parentheses is nonnegative, we have that $S_{2n} \leq S_{2n+2} \leq \cdots$. We also have

$$S_{2n} = a_1 - (a_2 - a_3) - \cdots - (a_{2n-2} - a_{2n-1}) - a_{2n} \leq a_1.$$

Hence $0 \leq \cdots \leq S_{2n} \leq S_{2n+2} \leq \cdots \leq a_1$ for all $n$. Since the partial sums are increasing and bounded, it follows (Theorem 4, 9.2B) that $S_{2n} \to S$ for some real number $S$. Furthermore, the odd partial sums satisfy $S_{2n+1} = S_{2n} + a_{2n+1}$ and $a_{2n+1} \to 0$; hence, $S_{2n+1} \to S$. We conclude that $S_n \to S$, and the series converges ●

Referring to Figure 1, we see that the partial sums oscillate back and forth in such a way that each $S_n$ lies between $S_{n-2}$ and $S_{n-1}$. Since $|S_n - S_{n-1}| = a_n \to 0$ and $|S_n| \le a_1$ for all $n$, then $S_n$ must converge to some real number ●

A series of alternating positive and negative terms is called, naturally enough, an *alternating series*.

## C. The Ratio and Root Tests

**THEOREM 4** (Ratio test) *Suppose $\Sigma_{n=1}^{\infty} a_n$ is a series and $a_n > 0$ for all $n$. Then*

a) *if* $\lim_{n \to \infty} \dfrac{a_{n+1}}{a_n} < 1,$

If $\lim_{n \to \infty} \dfrac{a_{n+1}}{a_n}$ does not exist, this test does not apply.

the series converges; and

b) *if* $\lim_{n \to \infty} \dfrac{a_{n+1}}{a_n} > 1$ *(including $+\infty$), the series diverges; but*

c) *if* $\lim_{n \to \infty} \dfrac{a_{n+1}}{a_n} = 1,$

*no conclusion can be drawn.*

*Proof*  a) Let $\lim_{n \to \infty} \dfrac{a_{n+1}}{a_n} = \theta < 1$ and choose $r$ satisfying $\theta < r < 1$. Since $\dfrac{a_{n+1}}{a_n} \to \theta < r$ as $n \to \infty$, it follows that $\dfrac{a_{n+1}}{a_n} < r$ for all $n$ greater than or equal to some integer $k$. Hence we have

To see this more clearly, consider Figure 2. Since $\dfrac{a_{n+1}}{a_n} \to \theta$ as $n \to \infty$, we see that when the subscript is large enough $(n \ge k)$ the number $\dfrac{a_{n+1}}{a_n}$ lies in the shaded interval.

$$a_{k+1} < ra_k, \quad a_{k+2} < ra_{k+1}, \quad a_{k+3} < ra_{k+2}, \quad \text{etc.}$$

From this it follows that

$$a_{k+1} < ra_k,$$
$$a_{k+2} < ra_{k+1} < r(ra_k) = r^2 a_k,$$
$$a_{k+3} < ra_{k+2} < r(ra_{k+1}) < r(r^2 a_k) = r^3 a_k,$$

and in general, for all $m \ge 1$,

$$a_{k+m} < r^m a_k.$$

**Figure 2**

Since $\Sigma_{m=1}^{\infty} a_k r^m = \dfrac{a_k r}{1-r}$ we have, by the comparison test, that $\Sigma_{m=1}^{\infty} a_{k+m}$ is convergent. We conclude that $\Sigma_{n=1}^{\infty} a_n$ is also convergent.

b) The proof here is similar to that in (a). Suppose

$$\lim_{n \to \infty} \frac{a_{n+1}}{a_n} = \phi > 1$$

Since $\dfrac{a_{n+1}}{a_n} \to \phi > s$, the numbers $\dfrac{a_{n+1}}{a_n}$ lie in the shaded interval (Figure 3) for all $n$ sufficiently large $(n \ge k)$.

and choose $s$ satisfying $1 < s < \phi$ (Fig. 3). In this case we have

$$\frac{a_{n+1}}{a_n} > s$$

**Figure 3**

for all $n$ greater than or equal to some $k$. As in part (a), it can be shown that, for all $m \geq 1$,

$$a_{k+m} > s^m a_k.$$

Since the terms $s^m$ do not approach zero, the series $\sum_{m=1}^{\infty} a_k s^m$ is divergent (Theorem 1, 7.1C). By the comparison test $\sum_{m=1}^{\infty} a_{k+m}$ is divergent. It follows that $\sum_{n=1}^{\infty} a_n$ is also divergent.

c) Finally, to show that no conclusion can be drawn when $\lim_{n \to \infty} \frac{a_{n+1}}{a_n} = 1$, we consider the two series

$$\text{A) } \sum_{n=1}^{\infty} \frac{1}{n} \quad \text{and} \quad \text{B) } \sum_{n=1}^{\infty} \frac{1}{n^2}.$$

Both series have the property that $\lim_{n \to \infty} \frac{a_{n+1}}{a_n} = 1$. However, as we know, the series (A) diverges (harmonic series), and as we will see in Example 8(a), series (B) converges. ●

COROLLARY  *Suppose $\sum_{n=1}^{\infty} a_n$ is a series with $a_n \neq 0$ for all n. Then if*

$$\lim_{n \to \infty} \left| \frac{a_{n+1}}{a_n} \right| < 1,$$

*the series $\sum_{n=1}^{\infty} a_n$ is absolutely convergent.*

THEOREM 5  (Root test) *Suppose $\sum_{n=1}^{\infty} a_n$ is a series and $a_n \geq 0$ for all n. Then*

a) *if*  $\lim_{n \to \infty} a_n^{1/n} < 1,$

  *the series is convergent, and*

b) *if*  $\lim_{n \to \infty} a_n^{1/n} > 1,$

**Example 5**  Show that the series $\displaystyle\sum_{n=1}^{\infty} \frac{1}{n!}$ is convergent.

*Solution*  Since  $a_n = \dfrac{1}{n!}$,

$$\frac{a_{n+1}}{a_n} = \frac{1/(n+1)!}{1/n!} = \frac{1}{(n+1)!} \frac{n!}{1}$$

$$= \frac{n(n-1)(n-2) \cdots 2 \cdot 1}{(n+1)(n)(n-1)(n-2) \cdots 2 \cdot 1} = \frac{1}{n+1}$$

As $n \to \infty$, $\frac{1}{n+1} \to 0 < 1$; hence, by the ratio test, the series is convergent.

**Example 6**  Show that the series $\displaystyle\sum_{n=1}^{\infty} \frac{n!}{100^n}$ is divergent.

*Solution*  We have $\dfrac{a_{n+1}}{a_n} = \dfrac{(n+1)!/100^{n+1}}{n!/100^n}$

$$= \frac{(n+1)!}{100^{n+1}} \frac{100^n}{n!}$$

$$= \frac{n+1}{100}.$$

As $n \to \infty$, $\frac{n+1}{100} \to \infty$; hence, by the ratio test, the series is divergent.

→  Proof of corollary:  From part (a) of the above theorem, the series $\sum_{n=1}^{\infty} |a_n|$ is convergent. This means that the series $\sum_{n=1}^{\infty} a_n$ is absolutely convergent.

→  Proof of (a):  Choose s such that $\lim_{n \to \infty} a_n^{1/n} < s < 1$. Then, as in Theorem 4, there is a positive integer k such that $a_m^{1/m} < s$ for all $m \geq k$. Hence $a_m < s^m$ for all $m \geq k$. Since $\sum_{n=k}^{\infty} s^n$ converges, by the comparison test it follows that $\sum_{n=k}^{\infty} a_n$ is convergent. It follows that $\sum_{n=1}^{\infty} a_n$ is also convergent.

*the series is divergent, but*

c) *if* $\lim\limits_{n\to\infty} a_n^{1/n} = 1$,

   *no conclusion can be drawn.*

The proof of this result is similar to that for Theorem 4. We give the argument for (a) in the comments. The following corollary will be used in the next section.

**COROLLARY** *Suppose* $\sum_{n=1}^{\infty} a_n$ *is a series. Then if*

$$\lim_{n\to\infty} |a_n|^{1/n} < 1,$$

*the series is absolutely convergent.*

The proof is immediate since the theorem gives that $\sum\limits_{n=1}^{\infty} |a_n|$

is convergent, and this, by definition, means that $\sum\limits_{n=1}^{\infty} a_n$ is absolutely convergent ●

---

**Example 7** Determine whether the series $\sum_{n=1}^{\infty} \left(\frac{1}{n}\right)^n$ is convergent or divergent.

*Solution*   Since  $\lim_{n\to\infty} a_n^{1/n} = \lim_{n\to\infty} \left(\left(\frac{1}{n}\right)^n\right)^{1/n} = \lim_{n\to\infty} \frac{1}{n} = 0$, the series is convergent by the root test.

**Problem 1**  Determine whether the series $\sum\limits_{n=1}^{\infty} \dfrac{3^{2n}}{n^n 2^{3n}}$

is convergent.

*Answer:*  It is convergent.

---

**D. The Integral Test**  We will use improper integrals in the following useful test for convergence.

**THEOREM 6**  (Integral test) *Suppose f is continuous on* $[1, \infty)$, *satisfies* $f(x) \geq 0$ *for all x, and is decreasing. Consider the integral*

$$\int_1^{\infty} f(x)\, dx$$

*and the series*

$$\sum_{n=1}^{\infty} f(n).$$

*Then*

a) *if the integral is convergent the series is convergent, and*

b) *if the integral is divergent the series is divergent.*

*Proof*  a) Referring to Figure 4a we see that

$$f(2) + f(3) + \cdots < \int_1^{\infty} f(x)\, dx.$$

Hence $\int_1^\infty f(x)\,dx < \infty$ implies

$$\sum_{n=1}^\infty f(n) = f(1) + f(2) + f(3) + \cdots$$

$$< f(1) + \int_1^\infty f(x)\,dx < \infty.$$

b) From Figure 4b we have

$$\sum_{n=1}^\infty f(n) = f(1) + f(2) + f(3) + \cdots$$

$$> \int_1^\infty f(x)\,dx,$$

so if $\int_1^\infty f(x)\,dx = \infty$ then $\sum_{n=1}^\infty f(n) = \infty$ ●

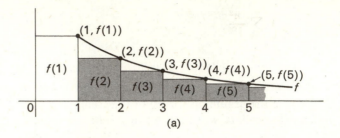

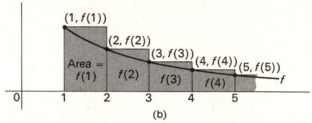

**Figure 4**

---

*Example 8* Test the convergence of the following series:

a) $\displaystyle\sum_{n=1}^\infty \frac{1}{n^2}$    b) $\displaystyle\sum_{n=1}^\infty \frac{1}{\sqrt{n}}$

*Solution* a) Since $\displaystyle\int_1^\infty \frac{1}{x^2}\,dx = \lim_{r\to\infty}\int_1^r \frac{1}{x^2}\,dx =$

$\displaystyle\lim_{r\to\infty}\left(-\frac{1}{r} + 1\right) = 1$, it follows from Theorem 6(a) that

$\displaystyle\sum_{n=1}^\infty \frac{1}{n^2}$ is convergent.

b) Since $\displaystyle\int_1^\infty \frac{1}{\sqrt{x}}\,dx = \lim_{r\to\infty}\int_1^r \frac{1}{\sqrt{x}}\,dx =$

$\displaystyle\lim_{r\to\infty}(2\sqrt{r} - 2) = \infty$, it follows from Theorem 6(b) that $\displaystyle\sum_{n=1}^\infty \frac{1}{\sqrt{n}}$ is divergent.

*Problem 2* Show that $\sum_{n=1}^\infty 1/n^r$ is convergent for $r > 1$ and divergent for $0 \le r \le 1$.

---

### E. Exercises

**B1.** Use the comparison test to determine which of the following series converge.

a) $\displaystyle\sum_{n=1}^\infty \frac{\sin^2 n}{2^n}$    b) $\displaystyle\sum_{n=1}^\infty \frac{2 + \cos n}{n}$

c) $\displaystyle\sum_{n=1}^\infty \frac{1}{\sqrt{1 + 2^{2n}}}$    d) $\displaystyle\sum_{n=1}^\infty \frac{1}{\sqrt{2n^2 - 1}}$

**B2.** Use the ratio test to determine which of the following series converge.

a) $\sum_{n=1}^{\infty} \frac{n^2}{10^n}$

c) $\sum_{n=1}^{\infty} \frac{n^n}{n!}$

b) $\sum_{n=1}^{\infty} \frac{n}{2^n}$

d) $\sum_{n=1}^{\infty} \frac{(n!)^2}{(2n)!}$

**B3.** Use the root test to determine which of the following series converge. (Hint: $n^{1/n} \to 1$ as $n \to \infty$.)

a) $\sum_{n=1}^{\infty} \frac{n^n}{(2n)^n}$

c) $\sum_{n=1}^{\infty} \frac{3^n}{n^3}$

b) $\sum_{n=1}^{\infty} \frac{n}{2^n}$

d) $\sum_{n=1}^{\infty} \left( \frac{3+2n}{2+3n} \right)^n$

**B4.** Use the integral test to determine which of the following series converge.

a) $\sum_{n=1}^{\infty} \frac{e^{1/n}}{n^2}$

b) $\sum_{n=1}^{\infty} \frac{\text{Tan}^{-1} n}{1+n^2}$

c) $\sum_{n=2}^{\infty} \frac{1}{n \ln n}$

d) $\sum_{n=1}^{\infty} \frac{2n}{\sqrt{1+n^2}}$

**B5.** Determine the convergence and absolute convergence of the following series.

a) $\sum_{n=1}^{\infty} \frac{(-1)^n}{\sqrt{n}}$

c) $\sum_{n=1}^{\infty} \frac{\cos n\pi}{n}$

b) $\sum_{n=1}^{\infty} \frac{(-1)^n}{n^2}$

d) $\sum_{n=1}^{\infty} \frac{(-n)^3}{1+n^4}$

**D6.** Compare each of the following series with either the convergent $\sum_{n=1}^{\infty} \frac{1}{n^2}$ or the divergent $\sum_{n=1}^{\infty} \frac{1}{n}$ to determine if it converges.

a) $\sum_{n=1}^{\infty} \frac{3}{n}$

e) $\sum_{n=1}^{\infty} \frac{1}{3n^2}$

b) $\sum_{n=2}^{\infty} \frac{1}{n-1}$

f) $\sum_{n=1}^{\infty} \frac{1}{n^2+1}$

c) $\sum_{n=2}^{\infty} \frac{1}{\sqrt{n^2-1}}$

g) $\sum_{n=1}^{\infty} \frac{1}{\sqrt{n^4+1}}$

d) $\sum_{n=1}^{\infty} \frac{1}{\sqrt{n}}$

h) $\sum_{n=1}^{\infty} \frac{1}{n^3}$

(a) to (d) divergent    (e) to (h) convergent.

**D7.** Each of the following series has the form

$$a_1 + a_2 + a_3 + a_4 + \cdots.$$

Find a formula for $\frac{a_{n+1}}{a_n}$ and compute $\lim_{n \to \infty} \frac{a_{n+1}}{a_n}$.

a) $\frac{1}{2} + \frac{2}{3} + \frac{3}{4} + \frac{4}{5} + \cdots$

b) $\frac{1}{1 \cdot 2} + \frac{1}{2 \cdot 2^2} + \frac{1}{3 \cdot 2^3} + \frac{1}{4 \cdot 2^4} + \cdots$

c) $\frac{1}{2 \cdot 3} + \frac{2}{3 \cdot 4} + \frac{3}{4 \cdot 5} + \frac{4}{5 \cdot 6} + \cdots$

d) $1 + \frac{2^2}{2!} + \frac{3^3}{3!} + \frac{4^4}{4!} + \cdots$

e) $\frac{1}{5} + \frac{1}{7} + \frac{1}{9} + \frac{1}{11} + \cdots$

f) $\frac{3}{2} + \frac{9}{4} + \frac{27}{8} + \frac{81}{16} + \cdots$

g) $7 + 11 + 15 + 19 + 23 + \cdots$

h) $1 + \frac{1}{8} + \frac{1}{27} + \frac{1}{64} + \frac{1}{125} + \cdots$

a) $\frac{(n+1)^2}{n(n+2)} \to 1$    b) $\frac{n}{2n+2} \to \frac{1}{2}$

c) $\frac{(n+1)^2}{n(n+3)} \to 1$    d) $\left(1 + \frac{1}{n}\right)^n \to e$

e) $\frac{3+2n}{5+2n} \to 1$    f) $\frac{3}{2} \to \frac{3}{2}$

g) $\frac{7+4n}{3+4n} \to 1$    h) $\left(\frac{n}{n+1}\right)^3 \to 1$

**M8.** Show that (a) is divergent and (b) is convergent.

a) $\sum_{n=1}^{\infty} \frac{1}{\sqrt{n^2+3n+2}}$    b) $\sum_{n=1}^{\infty} \frac{1}{\sqrt{n^3+2n+3}}$

**M9.** Show that (a) is convergent and (b) is divergent.

a) $\sum_{n=2}^{\infty} \frac{1}{n(\ln n)^2}$    b) $\sum_{n=2}^{\infty} \frac{1}{n \ln n}$

**M10.** Use the integral test to determine the convergence or divergence of the following.

a) $\sum_{n=1}^{\infty} \frac{n}{1+n^2}$    b) $\sum_{n=1}^{\infty} \frac{n}{(1+n^2)^2}$

**M11.** Discuss the convergence of the following series.

a) $\sum_{n=1}^{\infty} (-1)^n \frac{\ln n}{n}$    b) $\sum_{n=1}^{\infty} (-1)^n \frac{\ln n}{n^2}$

**M12.** Show that each of the following series converges.

a) $\sum_{n=2}^{\infty} \dfrac{1}{n^2 - 1}$

b) $\sum_{n=3}^{\infty} \dfrac{1}{n^3 - 3n^2 + 2n}$

c) $\sum_{n=4}^{\infty} \dfrac{1}{n^4 - 81}$        (See also Exercise M13.)

**M13.** Suppose $p$ is a polynomial whose degree is greater than one, and $p(n) \neq 0$ for $n \geq k$. Show that

$$\sum_{n=k}^{\infty} \frac{1}{p(n)} \quad \text{is convergent.}$$

**M14.** Determine the convergence of the following series.

a) $\sum_{n=1}^{\infty} (-1)^n \dfrac{n^2}{1 + n^3}$

b) $\sum_{n=1}^{\infty} (-1)^n \dfrac{2n}{1 + n^2}$

c) $\sum_{n=1}^{\infty} \dfrac{1 \cdot 3 \cdot 5 \cdot \cdots \cdot (2n + 1)}{n!}$

d) $\sum_{n=1}^{\infty} \dfrac{n}{3n^2}$

e) $\sum_{n=1}^{\infty} \dfrac{n^2}{3^n}$    f) $\sum_{n=1}^{\infty} \dfrac{1}{n}\left(1 + \dfrac{1}{n}\right)^n$

**M15.** Determine the convergence of the following series.

a) $\sum_{n=1}^{\infty} \dfrac{n}{1 + n^3}$        d) $\sum_{n=1}^{\infty} \dfrac{1}{10n}$

b) $\sum_{n=1}^{\infty} \dfrac{n^2 - 1}{n^2 + 1}$        e) $\sum_{n=1}^{\infty} \dfrac{n!}{n^n}$

c) $\sum_{n=1}^{\infty} \dfrac{1}{(2n + 1)!}$        f) $\sum_{n=1}^{\infty} \dfrac{2^n}{n^3 + 1}$

**M16.** Show that the series $\sum_{n=1}^{\infty} na^n$ converges for $|a| < 1$.

**M17.** Show that the series $\sum_{n=1}^{\infty} n^2 a^n$ converges for $|a| < 1$.

**M18.** Assume $a_n > 0$ and $\sum_{n=1}^{\infty} \frac{1}{a_n}$ is convergent. Show that

$$\sum_{n=1}^{\infty} \frac{n}{a_1 + a_2 + \cdots + a_n}$$

is convergent. (This exercise is very difficult despite the fact that its solution can be found by elementary means. If you can solve the problem in hours instead of days, you are doing very well.)

**M19.** Suppose $\sum_{n=1}^{\infty} a_n$ is convergent and let

$$S_n = a_1 + a_2 + \cdots + a_n.$$

Prove that

$$\frac{S_1 + \cdots + S_n}{n} \to \sum_{n=1}^{\infty} a_n.$$

---

## 7.3 *Power Series*

We will now study an important type of series of functions. These series, called power series, extend the notion of a polynomial and have many practical and theoretical applications.

→  The results given in this section for power series of real numbers extend immediately to power series of complex numbers and are extensively studied in a course on complex analysis.

**A. Definitions**  Recall that a polynomial is a function that can be written in the form

$$p(x) = a_0 + a_1 x + a_2 x^2 + \cdots + a_n x^n.$$

If we do not stop with the term $a_n x^n$ but continue writing, we obtain a *power series*

$$a_0 + a_1 x + a_2 x^2 + \cdots + a_n x^n + \cdots.$$

The real numbers $a_n$ are called the *coefficients,* and the functions $a_n x^n$ are called *terms.* Notice that for each real number $x$, the power series generates a series of real numbers. The power series is said to converge for those values of $x$ which yield a convergent series of real numbers. For example, the geometric series

$$1 + x + x^2 + \cdots$$

converges to $\frac{1}{1-x}$ for $-1 < x < 1$. We will now investigate the convergence properties of a general power series.

**B. The Convergence of a Power Series**  The following result contains the primary fact about the convergence of a power series.

*THEOREM 1   Suppose*

$$a_0 + a_1 x + a_2 x^2 + \cdots$$

*is a power series and*

$$\sum_{n=0}^{\infty} a_n s^n$$

*is convergent for some $s \neq 0$. Then the series*

$$\sum_{n=0}^{\infty} a_n x^n$$

*is absolutely convergent whenever $|x| < |s|$.*

*Proof*  Since $\Sigma_{n=0}^{\infty} a_n s^n$ is convergent, it follows from Theorem 1 of 7.1C that $a_n s^n \to 0$ (as $n \to \infty$). Hence there must be a number $M$ such that

$$|a_n s^n| \leq M \quad \text{for all} \quad n.$$

Now if $|x| < |s|$ we have that $\theta = \left|\frac{x}{s}\right| < 1$. Hence

$$|a_n x^n| = \left| a_n \left(\frac{x}{s}\right)^n s^n \right| = \left|\frac{x}{s}\right|^n |a_n s^n| < \theta^n M.$$

Since $\Sigma_{n=0}^{\infty} \theta^n M$ is convergent and $|a_n x^n| < \theta^n M$, it follows from the comparison test that $\Sigma_{n=0}^{\infty} |a_n x^n|$ is convergent  ●

→ A series of the form $a_0 + a_1(x - c) + a_2(x - c)^2 + \cdots$ is also called a power series. Moreover, the first nonzero coefficient need not be $a_0$. For example, $a_2 x^3 + a_3 x^4 + \cdots + a_n x^{n+1} + \cdots$ is again an example of a power series. However, neither the series

$$x^{-1} + x^{-2} + x^{-3} + \cdots$$

nor the series

$$x^\pi + x^{\pi^2} + x^{\pi^3} + \cdots$$

is a power series. To be a power series the exponents of $x$ must be nonnegative integers.

→ *I have coined and donated to the English language the word "infinomial" to suggest a polynomial of infinite length.*
*(Ed)*

→ In the case where the series has the form $\Sigma_{n=0}^{\infty} a_n(x - c)^n$, the convergence of $\Sigma_{n=0}^{\infty} a_n s^n$ implies the absolute convergence of $\Sigma_{n=0}^{\infty} a_n(x - c)^n$ for all $|x - c| < |s|$. This follows from Theorem 1 with $x$ replaced by $x - c$.

→ Since $a_n s^n \to 0$ it follows that $|a_n s^n| \leq 1$ for all $n$ greater than some integer $k$. Simply let $M$ equal the maximum of

$$\{|a_0|, |a_1 s|, |a_2 s^2|, \ldots, |a_k s^k|, 1\}.$$

→ $$\sum_{n=0}^{\infty} \theta^n M = M \sum_{n=0}^{\infty} \theta^n = \frac{M}{1 - \theta}.$$

We will now define the *radius of convergence R* of a power series. If the series $\Sigma_{n=0}^{\infty} a_n x^n$ diverges whenever $x \neq 0$, its radius of convergence is defined as zero; and if it converges for all $x$, its radius of convergence is said to be infinite. The remaining case is when the series converges for some but not all $x \neq 0$. Then the largest number $R$ having the property that

$$\sum_{n=0}^{\infty} a_n x^n$$

is convergent whenever $|x| < R$ is called the *radius of convergence*. The interval $(-R, R)$ is known as the *interval of convergence*.

A power series $\Sigma_{n=0}^{\infty} a_n x^n$ with interval of convergence $(-r, r)$ satisfies the property that for each $x \in (-r, r)$, $\Sigma_{n=0}^{\infty} a_n x^n$ converges and hence represents a real number.

Denoting the sum by $f(x) = \displaystyle\sum_{n=0}^{\infty} a_n x^n,$

we have that $f$ is a function defined on the open interval $(-r, r)$.

## C. Computing the Radius of Convergence
The tests used previously to determine the convergence of an ordinary series can often be applied to find the radius of convergence of a power series.

---

*Example 1*  Show that the series

$$1 - x + \frac{x^2}{2} - \frac{x^3}{3} + \cdots$$

has radius of convergence equal to 1.

*Solution*  We will show that this series is absolutely convergent for $x \in (-1, 1)$. Since the series can be written as

$$1 + \sum_{n=1}^{\infty} \frac{(-1)^n x^n}{n},$$

the series of corresponding absolute values is

$$1 + \sum_{n=1}^{\infty} \frac{|x|^n}{n}.$$

→ In 10.2C a formula for computing the radius of convergence of a power series will be given in terms of its coefficients.

→ The word "largest" is made precise in the following way: Let

$$S = \left\{ b > 0: \sum_{n=0}^{\infty} a_n x^n \text{ converges for } |x| < b \right\}.$$

It follows from Theorem 1 and the completeness property of the real numbers (Axiom 8, Chapter 9.1) that $S$ is one of the intervals $(0, R)$ or $(0, R]$ where $R$ is a real number.

→ You will often use the simple fact that the series $\Sigma_{n=k}^{\infty} a_n x^n$, for any positive integer $k$, has the same interval of convergence as the series $\Sigma_{n=0}^{\infty} a_n x^n$.

(Pat)

→ For example, the series $\Sigma_{n=0}^{\infty} x^n$ has radius of convergence 1 and is equal to the function $f(x) = \frac{1}{1-x}$ for $-1 < x < 1$.

---

*Problem 1*  Find the radius of convergence of

a) $\displaystyle\sum_{n=0}^{\infty} \frac{n! x^n}{3^n}$  b) $\displaystyle\sum_{n=1}^{\infty} n x^n$  c) $\displaystyle\sum_{n=1}^{\infty} \frac{n^n}{(n+1)!} x^n$

*Answer:*  a) 0   b) 1   c) 1/e

The ratio $\dfrac{a_{n+1}}{a_n}$ of the terms is equal to $\dfrac{|x|^{n+1}/(n+1)}{|x|^n/n}$,

or $\frac{n}{n+1}|x|$. As $n \to \infty$ the ratio approaches $|x|$. Hence the series converges for $|x| < 1$ or $x \in (-1, 1)$, and the radius of convergence is 1.

**Example 2**   Show that the power series

$$1 + x + \frac{x^2}{2!} + \frac{x^3}{3!} + \cdots$$

converges for all $x$; that is, its radius of convergence is infinite.

*Solution*   The ratio test will again be applied to determine the radius of convergence.

Since $\left|\dfrac{a_{n+1}}{a_n}\right| = \dfrac{|x|^{n+1}/(n+1)!}{|x|^n/n!} = \dfrac{|x|}{n+1}$,

$$\lim_{n\to\infty} \frac{|x|}{n+1} = |x| \lim_{n\to\infty} \frac{1}{n+1} = |x|0 = 0.$$

Hence the series converges absolutely for all $x$.

If a power series converges on a finite interval $(-r, r)$, it may converge on both, one, or neither of the endpoints $r$ and $-r$. These cases must usually be checked separately. As an illustration, the series given in Example 1 converges for $x \in (-1, 1)$. If we let $x = 1$, the series becomes

$$1 - 1 + \frac{1}{2} - \frac{1}{3} + \cdots = \sum_{n=2}^{\infty} \frac{(-1)^n}{n},$$

and this is convergent by Theorem 3 of 7.2B. However, if we let $x = -1$, the series becomes

$$1 + 1 + \frac{1}{2} + \frac{1}{3} + \cdots = 1 + \sum_{n=1}^{\infty} \frac{1}{n} \quad \text{(the harmonic series)},$$

which is divergent (Example 2, 7.1C). Hence the series converges for $-1 < x \le 1$.

## D. Exercises

**B1.** Show that the following power series converge only when $x = 0$.

a) $\displaystyle\sum_{n=0}^{\infty} n! x^n$        b) $\displaystyle\sum_{n=1}^{\infty} (nx)^n$

**B2.** Show that the radius of convergence is 1 for each of the following power series.

a) $\displaystyle\sum_{n=0}^{\infty} n^2 x^n$      c) $\displaystyle\sum_{n=1}^{\infty} \frac{x^n}{n(n+1)}$

b) $\displaystyle\sum_{n=1}^{\infty} \frac{(-1)^n}{n} x^n$      d) $\displaystyle\sum_{n=1}^{\infty} \frac{x^{2n}}{n^2}$

**B3.** Show that each of the following power series converges for all $x$.

a) $\displaystyle\sum_{n=0}^{\infty} \frac{x^{2n}}{n!}$        b) $\displaystyle\sum_{n=1}^{\infty} \frac{n! x^n}{n^n}$

c) $\displaystyle\sum_{n=1}^{\infty} \left(\frac{x}{n}\right)^n$      d) $\displaystyle\sum_{n=2}^{\infty} \frac{x^n}{(n-2)!}$

**B4.** Determine the radius of convergence for the following power series.

a) $\displaystyle\sum_{n=0}^{\infty} \left(\frac{-x}{2}\right)^n$      c) $\displaystyle\sum_{n=0}^{\infty} nx^n$

b) $\displaystyle\sum_{n=0}^{\infty} \frac{n}{n+1} \frac{x^n}{4^n}$      d) $\displaystyle\sum_{n=0}^{\infty} \frac{nx^n}{5^n}$

**B5.** If the radius of convergence of $\sum_{n=0}^{\infty} a_n x^n$ is $R$, show that for $d > 0$ the radius of convergence of $\sum_{n=0}^{\infty} a_n d^n x^n$ is $\frac{R}{d}$.

**D6.** The following series are all variations of the series $\sum_{n=0}^{\infty} x^n$ whose radius of convergence is 1. Determine the radius of convergence of each.

a) $\displaystyle\sum_{n=0}^{\infty} (-1)^n x^n$    b) $\displaystyle\sum_{n=0}^{\infty} x^{2n}$    c) $\displaystyle\sum_{n=0}^{\infty} nx^n$

d) $\sum_{n=1}^{\infty} \dfrac{x^n}{n}$          f) $\sum_{n=0}^{\infty} \dfrac{x^n}{3^n}$          h) $\sum_{n=0}^{1000} x^n$

e) $\sum_{n=0}^{\infty} 2^n x^n$          g) $\sum_{n=0}^{\infty} n4^n x^n$

a) 1     b) 1     c) 1     d) 1     e) $\frac{1}{2}$     f) 3
g) $\frac{1}{4}$     h) $\infty$

**D7.** Show that each of the following series converges for all values of $x$.

a) $\sum_{n=0}^{\infty} \dfrac{x^n}{n!}$          e) $\sum_{n=0}^{\infty} \dfrac{x^{2n}}{(2n)!}$

b) $\sum_{n=0}^{\infty} \dfrac{x^n}{(n+1)!}$          f) $\sum_{n=0}^{\infty} \dfrac{(-1)^n x^{2n}}{(2n)!}$

c) $\sum_{n=1}^{\infty} \dfrac{x^n}{(n-1)!}$          g) $\sum_{n=0}^{\infty} (-1)^n \dfrac{x^{2n+1}}{(2n+1)!}$

d) $\sum_{n=0}^{\infty} \dfrac{x^{3n}}{n!}$          h) $\sum_{n=1000}^{\infty} \dfrac{x^n}{n!}$

Determine the radius of convergence for each of the following series.

**M8.** a) $\sum_{n=0}^{\infty} \dfrac{x^n}{1+n^2}$          b) $\sum_{n=0}^{\infty} (1+n^2) x^n$

**M9.** a) $\sum_{n=0}^{\infty} (-1)^n \sqrt{n}\, x^n$     b) $\sum_{n=1}^{\infty} \dfrac{x^n}{n^{3/2}}$

**M10.** a) $\sum_{n=0}^{\infty} \dfrac{n}{n+3} \dfrac{x^n}{2^n}$          b) $\sum_{n=1}^{\infty} \dfrac{3^n x^n}{n}$

**M11.** a) $\sum_{n=0}^{\infty} x^{3n+7}$          b) $\sum_{n=0}^{\infty} x^{7n+3}$

**M12.** a) $\sum_{n=0}^{\infty} x^{n!}$          b) $\sum_{n=0}^{\infty} \dfrac{x^{n!}}{n!}$

**M13.** a) $\sum_{n=0}^{\infty} \left(\dfrac{2n+1}{n+1}\right) x^n$     b) $\sum_{n=0}^{\infty} \left(\dfrac{3n+2}{2n+3}\right) \dfrac{x^n}{n!}$

**M14.** a) $\sum_{n=0}^{\infty} \dfrac{x^n}{\sqrt{1+n^2}}$     b) $\sum_{n=0}^{\infty} \dfrac{x^n}{(n^2+1)!}$

**M15.** a) $\sum_{n=0}^{\infty} \pi^n x^n$          b) $\sum_{n=0}^{\infty} \dfrac{x^n}{e^n}$

**M16.** Determine the interval of convergence of the following series.

a) $\sum_{n=0}^{\infty} \dfrac{(3x)^n}{2^{n+1}}$          b) $\sum_{n=1}^{\infty} \dfrac{(n!)^3 x^n}{2n!}$

**M17.** Determine the radius of convergence of the following series.

a) $\sum_{n=1}^{\infty} \dfrac{n^n}{n!} x^n$

b) $\sum_{n=1}^{\infty} \left(1 + \dfrac{a}{n} + \dfrac{b}{n^2}\right)^{n^2} x^n$

**M18.** Assume $\sum_{n=0}^{\infty} a_n x^n$ has radius of convergence $r$. Show that if there exists a constant $M$ such that $|a_n| \le M$ for every integer $n$, then $r \ge 1$.

## 7.4   *Some Important Expansions*

A number of interesting applications follow from the fact that many familiar functions can be written in the form of a power series. We will discuss some of the most common cases.

→   Not every function can be written in the form of a power series which converges to the function. The function $f(x) = e^{-1/x^2}$ is one such example and is commented on in 7.4B.

### A. Coefficients of Power Series   Given a power series, its coefficients can be found in terms of suitable derivatives. We have the following theorem:

→   The question of determining when a function $f$ has a power series expansion is more difficult and will be dealt with in Chapter 10.

**THEOREM 1** If $f(x) = \displaystyle\sum_{n=0}^{\infty} a_n x^n$,

then $\qquad a_n = \dfrac{f^{(n)}(0)}{n!}$.

*Proof* We will find several successive derivatives of $f$. →

$f(x)$
$= a_0 + a_1 x + a_2 x^2 + a_3 x^3 + a_4 x^4 + \cdots$

$f'(x)$
$= \qquad a_1 + 2a_2 x + 3a_3 x^2 + 4a_4 x^3 + \cdots$

$f^{(2)}(x)$
$= \qquad\qquad 2a_2 + 3 \cdot 2a_3 x + 4 \cdot 3a_4 x^2 + \cdots$

$f^{(3)}(x)$
$= \qquad\qquad\qquad 3 \cdot 2a_3 + 4 \cdot 3 \cdot 2a_4 x + \cdots$

$f^{(4)}(x)$
$= \qquad\qquad\qquad\qquad 4 \cdot 3 \cdot 2a_4 + \cdots$

Now letting $x = 0$ in the above formulas we have

$$f(0) = a_0$$
$$f'(0) = a_1$$
$$f^{(2)}(0) = 2a_2$$
$$f^{(3)}(0) = 3 \cdot 2a_3$$
$$f^{(4)}(0) = 4 \cdot 3 \cdot 2a_4.$$

The pattern is clear and we have, in general, →

$$f^{(n)}(0) = n! a_n$$

or

$$a_n = \frac{f^{(n)}(0)}{n!} \quad \bullet$$

**COROLLARY** If $g(x) = \sum_{n=0}^{\infty} a_n (x - c)^n$,

then $\qquad a_n = \dfrac{g^{(n)}(c)}{n!}$.

*Proof* Let $u = x - c$, so $x = u + c$. Letting

$$h(u) = g(u + c) = g(x),$$

we then have

$$h(u) = \sum_{n=0}^{\infty} a_n u^n.$$

By the above theorem

Of course we must assume that the power series $\Sigma_{n=0}^{\infty} a_n x^n$ has a nonzero radius of convergence. Otherwise the function $f$ is defined only at $x = 0$, and it makes no sense to speak of its derivatives.

Also, the argument given here is based on the fact that a power series can be differentiated (and integrated) just like a polynomial. This will be shown in 10.2C and is based on the notion of the uniform convergence of a power series, which will be given later. Until then the reader should have no difficulty in accepting the argument in the text.

The reader familiar with induction should have no difficulty in proving this result using the induction technique.

*Problem 1* Prove the result as stated above.

*Problem 2* Suppose $f(x) = a_0 + a_1 x + a_2 x^2 + a_3 x^3$ and $f(0) = 2$, $f'(0) = 6$, $f''(0) = 8$, $f'''(0) = 12$. Find $a_0$, $a_1$, $a_2$, and $a_3$.

*Answer:* $a_0 = 2$, $a_1 = 6$, $a_2 = 4$, $a_3 = 2$.

$$a_n = \frac{h^{(n)}(0)}{n!}.$$

Since $h^{(n)}(0) = g^{(n)}(c)$, this becomes $a_n = \frac{g^{(n)}(c)}{n!}$ ●

Indeed $h(u) = g(u + c)$, so $h'(u) = g'(u + c)$. Repeating this we get $h''(u) = g''(u + c)$, and more generally $h^{(n)}(u) = g^{(n)}(u + c)$. Now, letting $u = 0$, we obtain

$$h^{(n)}(0) = g^{(n)}(c).$$

**B. Three Important Examples** The functions $e^x$, $\sin x$, and $\cos x$ all have power series expansions that converge for all values of $x$. We will now find them, using Theorem 1.

It will be shown in 10.3 that these functions indeed have power series representations.

*THEOREM 2* *We have*

a)  $\quad e^x = 1 + x + \frac{x^2}{2!} + \frac{x^3}{3!} + \cdots$

b)  $\quad \cos x = 1 - \frac{x^2}{2!} + \frac{x^4}{4!} - \frac{x^6}{6!} + \cdots$

c)  $\quad \sin x = x - \frac{x^3}{3!} + \frac{x^5}{5!} - \frac{x^7}{7!} + \cdots$.

*Proof* a) Letting $f(x) = e^x$, we have

$$f^{(n)}(x) = e^x$$

for all $n$. Hence $f^{(n)}(0) = 1$ for all $n$, so

$$a_n = \frac{f^{(n)}(0)}{n!} = \frac{1}{n!}$$

and the formula for $e^x$ is verified.

b) Now let $g(x) = \cos x$, so $g(0) = 1$. Then

$$\begin{aligned}
g'(x) &= -\sin x, &&\text{so} && g'(0) = 0 \\
g^{(2)}(x) &= -\cos x, &&\text{so} && g^{(2)}(0) = -1 \\
g^{(3)}(x) &= \sin x, &&\text{so} && g^{(3)}(0) = 0 \\
g^{(4)}(x) &= \cos x, &&\text{so} && g^{(4)}(0) = 1
\end{aligned}$$

and from this point on, the derivatives begin to repeat. Now $a_n = \frac{g^{(n)}(0)}{n!}$, but we clearly have $g^{(n)}(0) = 0$ whenever $n$ is odd. Moreover, since $g(0) = g^{(4)}(0) = g^{(8)}(0) = \cdots = 1$ and $g^{(2)}(0) = g^{(6)}(0) = g^{(10)}(0) = \cdots = -1$, we have

$$\cos x = 1 - \frac{x^2}{2!} + \frac{x^4}{4!} - \frac{x^6}{6!} + \frac{x^8}{8!} - \cdots.$$

c) The proof for $\sin x$ is similar and will be left as an exercise ●

There are some subtleties connected with power series. For example, the function defined by

$$f(x) = \begin{cases} e^{-1/x^2} & (x \neq 0) \\ 0 & (x = 0) \end{cases}$$

is infinitely differentiable and

$$f^{(n)}(0) = 0 \quad \text{for} \quad n = 0, 1, 2, \ldots.$$

Hence the function

$$g(x) = \sum_{n=0}^{\infty} \frac{f^{(n)}(0)}{n!} x^n = \sum_{n=0}^{\infty} 0 x^n = 0$$

has little to do with $f$ since $f(x) = g(x)$ only for $x = 0$. In general, the series $\sum_{n=0}^{\infty} \frac{f^{(n)}(0)}{n!} x^n$ is called the power series expansion of $f$ at 0, and it requires some effort to determine if and where

$$f(x) = \sum_{n=0}^{\infty} \frac{f^{(n)}(0)}{n!} x^n.$$

Many useful and curious formulas can be derived using these results. For example, letting $x = 1$ in the expansion for $e^x$ gives

$$e = 1 + 1 + \frac{1}{2!} + \frac{1}{3!} + \frac{1}{4!} + \frac{1}{5!} + \cdots,$$

Also, since $\sin \pi = 0$, letting $x = \pi$ in the expansion for $\sin x$ gives

$$0 = \pi - \frac{\pi^3}{3!} + \frac{\pi^5}{5!} - \frac{\pi^7}{7!} + \cdots$$

or

$$\pi = \frac{\pi^3}{3!} - \frac{\pi^5}{5!} + \frac{\pi^7}{7!} - \cdots$$

The three formulas above will be used in Section 7.5 to define the complex exponential functions.

## C. Manipulating the Geometric Series

In the last section we differentiated a power series term by term. Now we will integrate one. Start with the geometric series

$$\frac{1}{1-x} = 1 + x + x^2 + x^3 + \cdots \quad (-1 < x < 1).$$

Replacing $x$ by $-x$ gives the series

$$\frac{1}{1+x} = 1 - x + x^2 - x^3 + \cdots \quad (-1 < x < 1).$$

Now integrating we have

$$\ln(1+x) = \int \frac{dx}{1+x}$$

$$= \left(x - \frac{x^2}{2} + \frac{x^3}{3} - \cdots\right) + C.$$

The constant $C$ must equal zero because if we let $x = 0$ we obtain

$$0 = \ln 1 = C.$$

Hence we have that

$$\ln(1+x)$$

$$= x - \frac{x^2}{2} + \frac{x^3}{3} - \frac{x^4}{4} + \cdots \quad (-1 < x < 1).$$

We will now use the series

$$(*) \qquad \frac{1}{1+x} = \sum_{n=0}^{\infty} (-1)^n x^n$$

to derive another important series. Replacing $x$ by $x^2$ in (*) gives

$$\frac{1}{1+x^2} = 1 - x^2 + x^4 - x^6 + \cdots \quad (-1 < x < 1).$$

Integrating this we obtain

$$\operatorname{Tan}^{-1} x = \int \frac{dx}{1+x^2}$$

$$= \left(x - \frac{x^3}{3} + \frac{x^5}{5} - \frac{x^7}{7} + \cdots\right) + C$$

$$(-1 < x < 1).$$

If $f(x) = \sum_{n=0}^{\infty} a_n x^n$ has radius of convergence $r$, then each of the series

$$f'(x) = \sum_{n=1}^{\infty} n a_n x^{n-1}$$

and

$$\int f(x)\, dx = C + \sum_{n=0}^{\infty} \frac{a_n x^{n+1}}{n+1}$$

also has radius of convergence $r$. This result will be proven using the formula for the radius of convergence (10.2C) and the following limit:

$$\lim_{n \to \infty} n^{1/n} = 1.$$

The series $\frac{1}{1+x} = 1 - x + x^2 - x^3 + \ldots$ does not converge at either of the endpoints $x = \pm 1$, but its integrated series

$$\ln(1+x) = x - \frac{x^2}{2} + \frac{x^3}{3} - \frac{x^4}{4} + \cdots$$

does converge at $x = 1$. We note that by letting $x = 1$ in the above series we obtain

$$\ln 2 = 1 - \tfrac{1}{2} + \tfrac{1}{3} - \tfrac{1}{4} + \tfrac{1}{5} - \cdots.$$

The constant $C$ is again zero because letting $x = 0$ gives

$$0 = \text{Tan}^{-1} 0 = C.$$

Therefore

Tan$^{-1} x$

$$= x - \frac{x^3}{3} + \frac{x^5}{5} - \frac{x^7}{7} + \cdots \quad (-1 < x < 1).$$

It can be shown that this series converges to the value of Tan$^{-1} x$ at $x = 1$ but not at $x = -1$.

Since $\tan \frac{\pi}{4} = 1$, it follows that Tan$^{-1} 1 = \frac{\pi}{4}$, so letting $x = 1$ in the power series gives

$$\frac{\pi}{4} = \text{Tan}^{-1} 1 = 1 - \frac{1}{3} + \frac{1}{5} - \frac{1}{7} + \frac{1}{9} - \cdots;$$

hence

$$\pi = 4(1 - \frac{1}{3} + \frac{1}{5} - \frac{1}{7} + \frac{1}{9} - \cdots).$$

## D. Exercises

**B1.** Find the power series expansions for the following.

a) $f(x) = \dfrac{1}{x + 2}$    b) $g(x) = \dfrac{2}{3x - 4}$

c) $h(x) = \dfrac{1}{1 - x^2}$

**B2.** Find the power series expansions for
a) $y = \sin 2x$   and   b) $y = \cos(x^2)$.

**B3.** Use the formula $\cos^2 x = \dfrac{1 + \cos 2x}{2}$ to establish the equation

$$\cos^2 x = 1 + \sum_{n=1}^{\infty} \frac{(-1)^n 2^{2n-1} x^{2n}}{(2n)!}.$$

**B4.** Given   Tan$^{-1} x = x - \dfrac{x^3}{3} + \dfrac{x^5}{5} - \dfrac{x^7}{7} + \cdots$

a) Find a power series expansion for Tan$^{-1} x^2$.

b) Evaluate   $\dfrac{d^{10}}{dx^{10}} (\text{Tan}^{-1} x^2)\Big|_{x=0}$,

$\dfrac{d^{17}}{dx^{17}} (\text{Tan}^{-1} x^2)\Big|_{x=0}$,   and

$\dfrac{d^{80}}{dx^{80}} (\text{Tan}^{-1} x^2)\Big|_{x=0}$.

**B5.** Show that for $-1 < x < 1$

$$\frac{\cos x}{1 - \sin x} = 1 + x + \frac{x^2}{2} + \frac{x^3}{3} + \cdots.$$

**D6.** Given that $e^x = \sum_{n=0}^{\infty} \dfrac{x^n}{n!}$ determine what functions are represented by the following power series.

a) $\sum_{n=0}^{\infty} \dfrac{(-x)^n}{n!}$    b) $\sum_{n=0}^{\infty} \dfrac{(-1)^n x^n}{n!}$

c) $\sum_{n=0}^{\infty} \dfrac{2^n x^n}{n!}$    e) $\sum_{n=0}^{\infty} \dfrac{(-1)^n 4^n x^n}{n!}$

d) $\sum_{n=0}^{\infty} \dfrac{x^n}{3^n n!}$    f) $\sum_{n=0}^{\infty} \dfrac{x^{2n}}{n!}$

a) $e^{-x}$   b) $e^{-x}$   c) $e^{2x}$   d) $e^{x/3}$   e) $e^{-4x}$
f) $e^{x^2}$

**D7.** Given that $\frac{1}{1-x} = \sum_{n=0}^{\infty} x^n$ determine what functions are represented by the following power series.

a) $\sum_{n=0}^{\infty} 2^n x^n$    d) $\sum_{n=0}^{\infty} (-x)^{2n}$

b) $\sum_{n=0}^{\infty} (-1)^n 3^n x^n$    e) $\sum_{n=0}^{\infty} (-1)^n x^{5n}$

c) $\sum_{n=0}^{\infty} x^{3n}$    f) $\sum_{n=0}^{\infty} 3 x^{2n}$

a) $\dfrac{1}{1 - 2x}$   b) $\dfrac{1}{1 + 3x}$   c) $\dfrac{1}{1 - x^3}$   d) $\dfrac{1}{1 - x^2}$

e) $\dfrac{1}{1 + x^5}$   f) $\dfrac{3}{1 - x^2}$

**D8.** Knowing the power series representation for $e^x$, it is easy to find the power series representation of related functions. For example, since $e^x = \sum_{n=0}^{\infty} \dfrac{x^n}{n!}$ it follows that

$$e^{-3x} = \sum_{n=0}^{\infty} \frac{(-3x)^n}{n!} = \sum_{n=0}^{\infty} \frac{(-3)^n x^n}{n!}.$$

Given the power series representations of $e^x$, $\frac{1}{1-x}$, and $\ln(1 - x)$, determine the series that represent the following functions.

a) $f(x) = \dfrac{1}{1 - 3x}$    e) $f(x) = \ln(1 - x^3)$

b) $f(x) = \dfrac{2}{1 + 3x}$    f) $f(x) = \ln\sqrt{1 - x}$

c) $f(x) = \dfrac{1}{1 - x^2}$    g) $f(x) = \dfrac{2}{1 - x^3}$

d) $f(x) = 3e^{2x}$    h) $f(x) = \dfrac{1}{2 - x}$

a) $\displaystyle\sum_{n=0}^{\infty} 3^n x^n$    b) $\displaystyle\sum_{n=0}^{\infty} 2(-3)^n x^n$

c) $\displaystyle\sum_{n=0}^{\infty} x^{2n}$    d) $\displaystyle\sum_{n=0}^{\infty} \dfrac{3 \cdot 2^n \cdot x^n}{n!}$

e) $\displaystyle\sum_{n=1}^{\infty} \dfrac{x^{3n}}{n}$    f) $\displaystyle\sum_{n=1}^{\infty} \dfrac{x^n}{2n}$

g) $\displaystyle\sum_{n=0}^{\infty} 2x^{3n}$    h) $\displaystyle\sum_{n=0}^{\infty} \dfrac{x^n}{2^{n+1}}$.

**M9.** Find power series expansions for the following functions.

a) $y = xe^x$    b) $y = x^2 \sin x$    c) $y = x^3 \cos x$

**M10.** Find the first five terms of the power series expansions for the following functions.

a) $y = e^x \sin x$

b) $y = e^x \cos x$

**M11.** Find the first four terms of the power series expansions for the following.

a) $y = \sec x$

b) $y = \tan x$

**M12.** Find the first four terms of the power series expansions for the following.

a) $y = \sqrt{1 + \sin x}$

b) $y = e^{\cos x}$

**M13.** a) Show that

$$\frac{1}{\sqrt{1 - x^2}} = 1 + \frac{1}{2}x^2 + \frac{1 \cdot 3}{2 \cdot 4}x^4 + \frac{1 \cdot 3 \cdot 5}{2 \cdot 4 \cdot 6}x^6 + \cdots.$$

b) Use the result of (a) to establish the identity

$$\mathrm{Sin}^{-1} x = x + \frac{1}{2 \cdot 3}x^3 + \frac{1 \cdot 3}{2 \cdot 4 \cdot 5}x^5$$

$$+ \frac{1 \cdot 3 \cdot 5}{2 \cdot 4 \cdot 6 \cdot 7}x^7 + \cdots.$$

**M14.** Show that

$$\sin^2 x = \sum_{n=1}^{\infty} \frac{(-1)^{n+1} 2^{2n-1} x^{2n}}{(2n)!}.$$

**M15.** a) Show that

$$\int_0^x e^{-t^2}\, dt = \sum_{n=0}^{\infty} \frac{(-1)^n x^{2n+1}}{(2n + 1)n!}.$$

b) Estimate $\int_0^1 e^{-t^2}\, dt$ by using the first 4 nonzero terms of the series in (a) with $x = 1$.

**M16.** Use power series to find the value of the following expressions at $x = 0$.

a) $\dfrac{d^{24}}{dx^{24}} \sin(x^2)$    b) $\dfrac{d^{25}}{dx^{25}} e^{(x^3)}$    c) $\dfrac{d^{26}}{dx^{26}}\left(\dfrac{1}{1 - x^4}\right)$

## 7.5  *Complex Numbers*

The numbers we have dealt with heretofore have all been real numbers. In this section we will study a larger collection of numbers called the complex numbers which will prove to be exceptionally useful in Chapters 8, 9, and 10.

→   Since complex numbers do not occur in an obvious way as the measure of ordinary physical quantities, students often find them a source of confusion. Historically, mathematicians had the same difficulty. (See the historical insert at the end of this section.)

**A. Points in the Plane as Numbers**   The complex numbers are merely points in the plane. When we consider a point $(x, y)$ in the plane as a complex number, we will use a new notation and write

$$x + iy = (x, y)$$

as illustrated in Figures 1 and 2. Also, we will often use a single letter $z$ to denote a complex number and write

$$z = x + iy \quad (x, y \text{ real}).$$

In this case $x$ is called the *real part* of $z$ (written Re $z = x$) and $y$ is called the *imaginary part* of $z$ (written Im $z = y$). Notice that when $y = 0$, $z$ is simply a real number, lying on the $x$-axis of the plane.

Given complex numbers $a + ib$ and $c + id$ where $a$, $b$, $c$, and $d$ are real numbers, we have

$$(a + ib) + (c + id) = (a + c) + i(b + d) \quad \text{(Fig. 3)}.$$

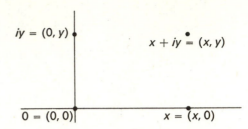

**Figure 1**

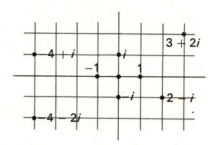

**Figure 2**

**Figure 3**

This is very natural, but the rule for multiplication is not quite as obvious.

We define

$$(a + ib)(c + id) = (ac - bd) + i(ad + bc).$$

In particular, it follows that $ii = -1$. This definition can easily be remembered by simply multiplying in the usual way for real numbers and replacing $ii = i^2$ by $-1$ whenever it appears.

With these definitions of addition and multiplication, the usual rules of arithmetic apply as well to complex numbers as to real numbers.

**B. Properties of Complex Numbers** The *absolute value* of a complex number $z = x + iy$ is defined to be

$$|z| = (x^2 + y^2)^{1/2}$$

and is merely the distance between $z$ and the origin

The reader might think that it would be more natural to define

$$(a, b)(c, d) = (ac, bd).$$

We could do this, but the resulting system is not very interesting and there would be no number $z$ satisfying $z^2 = -1$.

Indeed,

$$(a + ib)(c + id) = (a + ib)c + (a + ib)(id)$$
$$= ac + ibc + iad + iibd$$
$$= ac + ibc + iad - bd$$
$$= ac - bd + i(ad + bc).$$

For example, the commutative law for addition $z_1 + z_2 = z_2 + z_1$, associative law for multiplication $(z_1 z_2)z_3 = z_1(z_2 z_3)$, and distributive law $z_1(z_2 + z_3) = z_1 z_2 + z_1 z_3$ all hold.

Notice that when $z = x$ is a real number, $|z| = \sqrt{x^2} = |x|$ is the usual absolute value of $x$.

(Fig. 4). Observe that $|z| = 0$ if and only if $z = 0$. The *argument* of $z$, arg $z$, is the angle shown. As in polar coordinates we adopt the convention that if $z = 0$, the argument of $z$ is any real number.

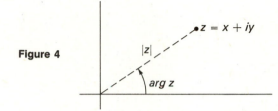

Figure 4

Two frequently used properties of absolute value are given in the following:

**THEOREM 1** *If $u$ and $v$ are complex numbers, we have*
    a) $|uv| = |u|\,|v|$,     *and*
    b) $|u + v| \leq |u| + |v|$    (*Triangle inequality*).

*Geometric proof of the triangle inequality* The complex numbers $u$, $v$, and $u + v$ are indicated in Figure 5, and the nonnegative numbers $|u|$, $|v|$, and $|u + v|$ are as shown in Figure 6.

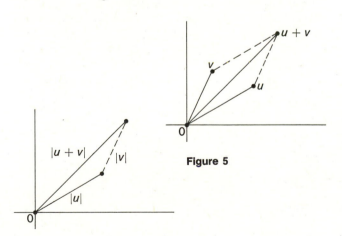

Figure 5

Figure 6

It is clear that $|u + v| \leq |u| + |v|$ since the length $|u + v|$ of one side of a triangle must be less than or equal to the sum of the lengths $(|u| + |v|)$ of the other two sides. (Compare with the algebraic proof on the right.) ●

*Algebraic proof of Theorem 1* Let $u = a + ib$ and $v = c + id$

a) Since $uv = (ac - bd) + i(ad + bc)$ we have

$$
\begin{aligned}
|uv| &= \sqrt{(ac - bd)^2 + (ad + bc)^2} \\
&= \sqrt{a^2c^2 + b^2d^2 - 2acbd + a^2d^2 + b^2c^2 + 2adbc} \\
&= \sqrt{a^2c^2 + b^2d^2 + a^2d^2 + b^2c^2} \\
&= \sqrt{(a^2 + b^2)(c^2 + d^2)} \\
&= \sqrt{a^2 + b^2}\sqrt{c^2 + d^2} = |u|\,|v|.
\end{aligned}
$$

b) We will first derive an auxiliary inequality. Since

$$0 \leq (ad - bc)^2 = (ad)^2 + (bc)^2 - 2abcd$$

we have

$$2abcd \leq (ad)^2 + (bc)^2.$$

Hence

$$
\begin{aligned}
(ac + bd)^2 &= (ac)^2 + (bd)^2 + 2abcd \\
&\leq (ac)^2 + (bd)^2 + (ad)^2 + (bc)^2 \\
&= (a^2 + b^2)(c^2 + d^2).
\end{aligned}
$$

Taking positive square roots we have

$$(*) \quad ac + bd \leq \sqrt{a^2 + b^2}\sqrt{c^2 + d^2} = |u|\,|v|.$$

Now

$$
\begin{aligned}
|u + v|^2 &= |(a + c) + i(b + d)|^2 \\
&= (a + c)^2 + (b + d)^2 \\
&= a^2 + 2ac + c^2 + b^2 + 2bd + d^2 \\
&= (a^2 + b^2) + (c^2 + d^2) + 2(ac + bd) \\
&= |u|^2 + |v|^2 + 2(ac + bd) \\
&\leq |u|^2 + |v|^2 + 2|u|\,|v| \quad \text{(by formula (*))} \\
&= (|u| + |v|)^2.
\end{aligned}
$$

Taking positive square roots now gives the desired inequality ●

The *complex conjugate* $\bar{z}$ of a complex number $z = x + iy$ is defined by the formula

$$\bar{z} = x - iy$$

and is merely the reflection of $z$ about the $x$-axis (Fig. 7).

Some simple properties of complex conjugates are listed below:

a) $\overline{u + v} = \bar{u} + \bar{v}$

b) $\overline{(\bar{u})} = u$

c) $u\bar{u} = |u|^2$

d) $\overline{uv} = \bar{u}\,\bar{v}$

e) $u = \bar{u}$ if and only if $u$ is real.

*Proof of (d)*  Let $u = a + ib$ and $v = c + id$. Then

$$uv = ac - bd + i(ad + bc), \quad \text{so}$$

$\overline{uv} = ac - bd - i(ad + bc)$. Since $\bar{u} = a - ib$ and $\bar{v} = c - id$, we have

$$\bar{u}\,\bar{v} = ac - (-b)(-d) + i(a(-d) + (-b)c)$$
$$= ac - bd - i(ad + bc) = \overline{uv} \quad \bullet$$

The expression $z = x + iy$ is called the *standard form* of a complex number. The inverse of a complex number $z$ can easily be written in standard form:

$$z^{-1} = \frac{1}{z} = \frac{1}{z}\frac{\bar{z}}{\bar{z}} = \frac{\bar{z}}{|z|^2} = \frac{x}{|z|^2} - i\frac{y}{|z|^2}.$$

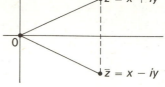

**Figure 7**

*Problem 1*  Verify the properties (a), (b), (c), and (e) listed for complex conjugation.

→  We can take the inverse of a complex number $z = a + ib$ whenever $z \neq 0$, and this occurs whenever $a$ or $b$ is not zero.

*Example 1*

a)  $(3 - 4i)^{-1} = \dfrac{1}{3 - 4i} = \dfrac{1}{3 - 4i}\dfrac{3 + 4i}{3 + 4i}$

$$= \frac{3 + 4i}{3^2 + 4^2} = \frac{3}{25} + \frac{4}{25}i.$$

b) Given $(2 - 4i)z = 3 + i$, we will solve for $z$ by writing

$$z = \frac{3 + i}{2 - 4i} = \frac{3 + i}{2 - 4i}\left(\frac{2 + 4i}{2 + 4i}\right)$$

$$= \frac{6 - 4 + i(2 + 12)}{4 + 16}$$

$$= \frac{2 + 14i}{20} = \frac{1}{10} + \frac{7}{10}i.$$

*Problem 2*  Write the given complex numbers in the form $a + ib$.

a) $(2 - 3i)(3 + 5i)$     b) $1 + i + i^2 + i^3$

c) $(\frac{1}{2} - \frac{1}{2}i)^{-1}$

*Answer:*  a) $21 + i$  b) $0$  c) $1 + i$

## C. The Complex Exponential Functions

We have already defined $a^x$ where $a > 0$ and $x$ is any real number. Now we define $a^z$ for any complex number $z$. This will be done in several steps. First of all, for $x$ and $y$ real, and $e = 2.71828\ldots$ we define

$$e^{iy} = \cos y + i \sin y.$$

Then, given $z = x + iy$, we define

$$e^z = e^{(x+iy)} = e^x e^{iy}.$$

Finally, for $a > 0$ and any complex number $z$ we define

$$a^z = e^{z \ln a}.$$

Returning now to the original equation

$$e^{iy} = \cos y + i \sin y,$$

we have

$$|e^{iy}| = (\cos^2 y + \sin^2 y)^{1/2} = 1,$$

so the point $e^{iy}$ is one unit from the origin $O$ (Fig. 8).

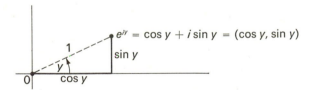

**Figure 8**

Several other points are plotted in Figure 9.

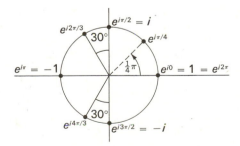

**Figure 9**

We know that every complex number of the form $e^{iy} = \cos y + i \sin y$ has absolute value 1. We will now

With this definition the laws of exponents for real exponents (front cover) hold for complex exponents as well.

We will show, for example, that $a^{u+v} = a^u a^v$ when $u$ and $v$ are complex. Notice first that

$$e^{ix+iy} = e^{i(x+y)} = \cos(x+y) + i \sin(x+y)$$
$$= (\cos x \cos y - \sin x \sin y)$$
$$+ i(\sin x \cos y + \sin y \cos x)$$
$$= (\cos x + i \sin x)(\cos y + i \sin y)$$
$$= e^{ix} e^{iy}.$$

Now, given $u = a + ib$ and $v = c + id$, we have

$$e^{u+v} = e^{(a+c)+i(b+d)} = e^{a+c} e^{ib+id}$$
$$= e^a e^c e^{ib} e^{id} = e^{a+ib} e^{c+id} = e^u e^v.$$

Finally,

$$a^{u+v} = e^{(u+v)\ln a} = e^{u \ln a + v \ln a}$$
$$= e^{u \ln a} e^{v \ln a} = a^u a^v.$$

The remaining laws of exponents are proven in a similar way.

The reader might wonder why $e^{ix}$ was defined in the way it was. One reason stems from the laws of exponents. As seen above, the reason that $e^{ix} e^{iy}$ is equal to $e^{ix+iy}$ is that the formulas for $\cos(x + y)$ and $\sin(x + y)$ work out just right. But there is an even better reason for this definition of $e^{ix}$ and that involves series.

Recall that for $t$ real,

$$e^t = 1 + t + \frac{t^2}{2!} + \frac{t^3}{3!} + \frac{t^4}{4!} + \cdots,$$

$$\cos t = 1 - \frac{t^2}{2!} + \frac{t^4}{4!} - \frac{t^6}{6!} + \cdots,$$

and

$$\sin t = t - \frac{t^3}{3!} + \frac{t^5}{5!} - \frac{t^7}{7!} + \cdots.$$

Now, formally replacing $t$ by $ix$ in the first equation, we obtain

$$e^{ix} = 1 + ix + \frac{(ix)^2}{2!} + \frac{(ix)^3}{3!} + \frac{(ix)^4}{4!} + \cdots$$

$$= 1 + ix - \frac{x^2}{2!} - i\frac{x^3}{3!} + \frac{x^4}{4!} + \cdots$$

$$= \left(1 - \frac{x^2}{2!} + \frac{x^4}{4!} - \frac{x^6}{6!} + \cdots\right)$$

$$+ i\left(x - \frac{x^3}{3!} + \frac{x^5}{5!} - \cdots\right)$$

$$= \cos x + i \sin x.$$

Hence, in a formal sense we have $e^{ix} = \cos x + i \sin x$.

To make the above argument precise we must first discuss series of complex numbers. This will be done in Chapters 9 and 10, and the above equation will be a theorem, not a definition.

show that conversely every complex number with absolute value 1 has the form $\cos\theta + i\sin\theta$. Let $w = a + ib$ have absolute value 1. Then $w = (a, b)$ and it must lie on the unit circle (Fig. 10). It follows that $a = \cos\theta$ and $b = \sin\theta$ where $\theta = \arg w$. Thus

$$w = \cos\theta + i\sin\theta \quad \text{or} \quad w = e^{i\theta} = e^{i\,\arg w}.$$

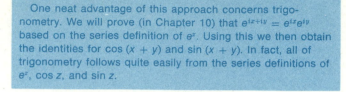

One neat advantage of this approach concerns trigonometry. We will prove (in Chapter 10) that $e^{ix+iy} = e^{ix}e^{iy}$ based on the series definition of $e^z$. Using this we then obtain the identities for $\cos(x + y)$ and $\sin(x + y)$. In fact, all of trigonometry follows quite easily from the series definitions of $e^z$, $\cos z$, and $\sin z$.

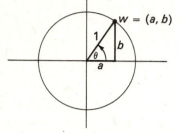

**Figure 10**

Now let $z = a + ib$ be any complex number. See Figure 11. Observe that $a = |z|\cos\theta$ and $b = |z|\sin\theta$, where $\theta = \arg z$. Thus

$$z = a + ib = |z|\cos\theta + i|z|\sin\theta$$
$$= |z|(\cos\theta + i\sin\theta)$$

or

**Figure 11**

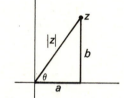

(*) $$z = |z|e^{i\,\arg z}.$$

The latter expression is called the *polar form* for a complex number $z$.

---

*Example 2*  Find the polar form of:
a) $2i$       c) $2 + 2i$
b) $-3$       d) $-1 + i\sqrt{3}$

*Solution*  We first plot each of the points (Fig. 12).

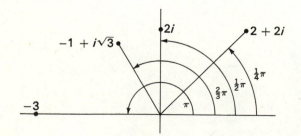

**Figure 12**

---

*Problem 3*  Find the polar form of:
a) $14i$       c) $-1 - i$
b) $-8$        d) $2 - i2\sqrt{3}$

*Answer:*  a) $14e^{i(\pi/2)}$    b) $8e^{i\pi}$    c) $\sqrt{2}e^{i(5\pi/4)}$
d) $4e^{i(5\pi/3)}$

a) Since $|2i| = 2$ and $\arg 2i = \frac{\pi}{2}$, we have

$$2i = |2i|e^{i \arg 2i} = 2e^{i(\pi/2)}.$$

b) Since $|-3| = 3$ and $\arg -3 = \pi$, we have

$$-3 = |-3|e^{i \arg -3} = 3e^{i\pi}.$$

c) Since $|2 + 2i| = \sqrt{2^2 + 2^2} = 2\sqrt{2}$ and $\arg (2 + 2i) = \frac{\pi}{4}$, we have (using formula (*))

$$2 + 2i = 2\sqrt{2}e^{i(\pi/4)}.$$

d) Since $|-1 + i\sqrt{3}| = \sqrt{(-1)^2 + (\sqrt{3})^2} = 2$ and $\arg (-1 + i\sqrt{3}) = \frac{2\pi}{3}$, we have (using formula (*))

$$-1 + i\sqrt{3} = 2e^{i(2\pi/3)}.$$

---

**D. Some Formulas**   Since $e^{ix} = \cos x + i \sin x$, we have

$$e^{-ix} = \cos x - i \sin x.$$

Adding these equations gives

$$e^{ix} + e^{-ix} = 2\cos x,$$

so

$$\cos x = \frac{e^{ix} + e^{-ix}}{2}.$$

If instead we subtract the equations, we get

$$e^{ix} - e^{-ix} = 2i \sin x,$$

or

$$\sin x = \frac{e^{ix} - e^{-ix}}{2i}.$$

Now suppose $f$ is a function whose domain is the real numbers but which takes complex values. Then, denoting the real and imaginary parts of $f(x)$ by $u(x)$ and $v(x)$ respectively, we have

$$f(x) = u(x) + iv(x).$$

We define the derivative of $f$ with respect to $x$ by the formula

$$\frac{df}{dx} = \frac{du}{dx} + i\frac{dv}{dx}.$$

This definition has been introduced simply to prove the following easily remembered theorem.

---

*Example 3*   Find all complex numbers $z$ satisfying

$$z^3 = 1.$$

*Solution*   First note that $e^{i\theta} = \cos \theta + i \sin \theta$ is equal to 1 if and only if $\theta = 2n\pi$ where $n$ is an integer.

Since $z^3 = 1$, we have $|z|^3 = 1$ and it follows that $|z| = 1$. Hence $z$ has the form $z = e^{ix}$. Because $1 = z^3 = (e^{ix})^3 = e^{3ix}$ we then have that $x$ satisfies $3x = 2n\pi$. Letting $n = 0$, 1, and 2 respectively gives $x = 0$, $\frac{2\pi}{3}$, and $\frac{4\pi}{3}$, so that

$$z_1 = 1, \quad z_2 = e^{2\pi i/3}, \quad \text{and} \quad z_3 = e^{4\pi i/3}$$

are the roots of $z^3 - 1 = 0$ ($e^{6\pi i/3} = e^{2\pi i} = 1 = z_1$ is not a new root). Writing $z_2$ and $z_3$ in the form $a + ib$, we have $z_2 = -1 + i\sqrt{3}$ and $z_3 = -1 - i\sqrt{3}$.

---

$\longrightarrow$   This follows from the elementary trigonometric identity:

$$e^{-ix} = e^{i(-x)} = \cos(-x) + i \sin(-x)$$
$$= \cos x - i \sin x.$$

$\longrightarrow$   Notice the analogy between these equations and the equations

$$\cosh x = \frac{e^x + e^{-x}}{2}$$

and

$$\sinh x = \frac{e^x - e^{-x}}{2}.$$

In fact, we have that

$$\cosh x = \cos ix$$

and

$$\sinh x = i \sin(-ix).$$

$\longrightarrow$   For example, if $f(x) = (ix)^2 - (ix)^3$, we then have

$$f(x) = i^2x^2 - i^3x^3 = -x^2 + ix^3,$$

so

$$u(x) = -x^2 \quad \text{and} \quad v(x) = x^3.$$

$\longrightarrow$   Notice the analogy between this and the derivative of a point described parametrically. If $P = (x(t), y(t))$, then we defined the velocity $v$ by the equation

$$v = \frac{dP}{dt} = \left(\frac{dx}{dt}, \frac{dy}{dt}\right).$$

**THEOREM 2**  *Given any complex number m,*

$$\frac{d}{dx} e^{mx} = m e^{mx}.$$

*Proof*  Suppose $m = a + ib$. Then

$$\frac{d}{dx} e^{mx} = \frac{d}{dx} e^{(a+ib)x} = \frac{d}{dx}(e^{ax}e^{ibx})$$

$$= \frac{d}{dx}(e^{ax}\cos bx + ie^{ax}\sin bx)$$

$$= \frac{d}{dx} e^{ax}\cos bx + i\frac{d}{dx} e^{ax}\sin bx$$

$$= ae^{(a+ib)x} + ibe^{(a+ib)x}$$

$$= (a + ib)e^{(a+ib)x}$$

$$= me^{mx}.$$

In this step we used the Product Rule as follows:

$$\frac{d}{dx} e^{ax}\cos bx + i\frac{d}{dx} e^{ax}\sin bx$$

$$= e^{ax}(-b\sin bx) + ae^{ax}\cos bx$$
$$\quad + i(e^{ax}b\cos bx + ae^{ax}\sin bx)$$

$$= ae^{ax}(\cos bx + i\sin bx) + be^{ax}(i\cos bx - \sin bx)$$

$$= ae^{ax}e^{ibx} + ibe^{ax}(\cos bx + i\sin bx)$$

$$= ae^{ax+ibx} + ibe^{ax}e^{ibx} = ae^{(a+ib)x} + ibe^{(a+ib)x}$$

## *"The True Metaphysics of $\sqrt{-1}$ Is Hard."* (Gauss)

Just as it took many centuries for negative and irrational numbers to gain acceptance among mathematicians, so it was many years before complex or "imaginary" numbers were recognized.

Before the sixteenth century, complex numbers were simply considered nonexistent. The earliest mention of the root of a negative number was by Heron of Alexandria in the first century. Near the year A.D. 275, Diophantus acknowledged the unsolvability of certain quadratic equations, and in 850 Mahavira clearly stated that a negative number has no square root. By 1494, L. Pacioli had calculated that $x^2 + c = bx$ is insoluble unless $b^2 \geq 4c$.

With the coming of the sixteenth century, complex numbers entered a phase of pseudo existence. The equation $x^2 + 1 = 0$ was no longer said to be without roots. Its roots were called $\pm i = \pm\sqrt{-1}$, but were called imaginary numbers. Complex numbers existed only as fictitious symbols and were not considered to have any real meaning. But there is nothing that prevents a mathematician from manipulating meaningless symbols and, starting with Cardan in 1545, the formal theory of complex numbers was developed over a 150 year period. Through the work of DeMoivre, Cotes, and Euler the relations between the complex exponential functions and the trigonometric functions were established. There were many formulas relating complex numbers and real numbers, for example $e^{i\pi} = -1$, but complex numbers were still not considered quite legitimate. What was needed was an application. Just as mathematicians would not accept negative numbers until the thirteenth century when Fibonacci interpreted them in problems dealing with money as a loss instead of a gain, so it was that mathematicians required an interpretation of complex numbers before they would acknowledge their true existence.

The interpretation was finally given independently by two amateur mathematicians. The Norwegian C. Wessel (1745–1818) in 1797 and the Swiss J. R. Argand (1768–1822) in 1806 both represented the complex number $a + bi$ geometrically as the point $(a, b)$ in the plane. Unfortunately, neither Argand nor Wessel had much contact with the scientific community and their work remained virtually unnoticed.

It was ultimately through the great German mathematician Karl Friedrich Gauss (1777–1855) that complex numbers became universally accepted. Between 1799, when Gauss began his study of complex numbers in his doctor's dissertation, and 1831, he independently arrived at the discoveries of Wessel and Argand, and proved the fundamental theorem of algebra concerning the existence of complex roots of polynomial equations. After Gauss, mathematicians wholeheartedly embraced complex numbers, and the theory of functions of a complex variable has become a significant branch of mathematics.

## E. Exercises

**B1.** Given $u = 2 + i$ and $v = 2 + 5i$, find the following.

     a) $uv$    c) $\bar{u}v$    e) $\dfrac{u}{v}$

     b) $u\bar{v}$    d) $\overline{uv}$    f) $|uv|$

**B2.** Write each of the following complex numbers $z$ in its polar form $z = |z|e^{i\,\arg z}$.

     a) $4i$    b) $-1 - i\sqrt{3}$    c) $2 - 2i$    d) $\sqrt{3} + i$

**B3.** Write each of the following complex numbers in the form $a + ib$ ($a, b \in R$).

     a) $e^{i\pi/2}$    b) $3e^{i\pi/3}$    c) $-4e^{i\pi/6}$    d) $2e^{i20\pi/3}$

**B4.** Find three complex numbers $z$ satisfying the equation $z^3 = 27$.

**B5.** Write $f(t)$ in the form $f(t) = x(t) + iy(t)$ ($x(t)$, $y(t) \in R$) given that

     a) $f(t) = (1 + it)^3$    b) $f(t) = (i + t)^2 e^{it}$.

**B6.** Given $f(x) = e^{4ix}$ find

     a) $\dfrac{df}{dx}\left(\dfrac{\pi}{2}\right)$    c) $\dfrac{d^3f}{dx^3}\left(\dfrac{\pi}{8}\right)$

     b) $\dfrac{d^2f}{dx^2}\left(\dfrac{\pi}{4}\right)$    d) $\dfrac{d^4f}{dx^4}\left(\dfrac{\pi}{24}\right)$.

**D7.** Write the complex number $\frac{1}{z}$ in the form $a + ib$ ($a, b \in R$) given that $z$ is equal to

     a) $\dfrac{1}{5} - \dfrac{2i}{5}$    c) $\dfrac{3}{25} + \dfrac{4i}{25}$    e) $2 - 3i$

     b) $\dfrac{2}{13} - \dfrac{3i}{13}$    d) $1 + 2i$    f) $-1 - i$

     a) $1 + 2i$    b) $2 + 3i$    c) $3 - 4i$

     d) $\dfrac{1}{5} - \dfrac{2}{5}i$    e) $\dfrac{2}{13} + \dfrac{3}{13}i$    f) $\dfrac{-1}{2} + \dfrac{i}{2}$

**D8.** Plot the following complex numbers $z$ and compute $|z|$.

     a) $3i$    d) $1 + i\sqrt{3}$    g) $-2 - 2i\sqrt{3}$

     b) $1 + i$    e) $\sqrt{3} - i$    h) $\sqrt{3} + i\sqrt{3}$

     c) $-3 + 3i$    f) $6$

     a) $3$   b) $\sqrt{2}$   c) $3\sqrt{2}$   d) $2$   e) $2$   f) $6$

     g) $4$   h) $\sqrt{6}$

**D9.** Each of the complex numbers given in Exercise D8 can be written in the form

$$z = |z|e^{i\,\arg z}$$

where $0 \le \arg z < 2\pi$. Determine $\arg z$ in each case.

     a) $\dfrac{\pi}{2}$    b) $\dfrac{\pi}{4}$    c) $\dfrac{3\pi}{4}$    d) $\dfrac{\pi}{3}$    e) $\dfrac{11\pi}{6}$

     f) $0$    g) $\dfrac{4\pi}{3}$    h) $\dfrac{\pi}{4}$

**D10.** Write the following complex numbers in the form $a + ib$ ($a, b \in R$).

     a) $i^4$                  e) $\left(\dfrac{\sqrt{3}}{2} + \dfrac{i}{2}\right)^3$

     b) $\left(\dfrac{-1}{2} + i\dfrac{\sqrt{3}}{2}\right)^3$    f) $\left(\dfrac{-\sqrt{3}}{2} + \dfrac{i}{2}\right)^3$

     c) $\left(\dfrac{-1}{2} - i\dfrac{\sqrt{3}}{2}\right)^3$    g) $\left(\dfrac{1 + i}{\sqrt{2}}\right)^2$

     d) $\left(\dfrac{1}{2} + i\dfrac{\sqrt{3}}{2}\right)^6$    h) $(-i)^3$

     a) $1$    b) $1$    c) $1$    d) $1$    e) $i$    f) $i$

     g) $i$    h) $i$

**M11.** Solve the following equations for real numbers $x$ and $y$.

     a) $\left(\dfrac{1 + i}{1 - i}\right)^2 + \dfrac{1}{x + iy} = 1 + i$

     b) $(3 - 2i)(x + iy) = 2(x - 2iy) + 2i - 1$

**M12.** Show that $|\bar{z}| = |z|$.

**M13.** Verify the following identities.
 a) $\sinh ix = i \sin x$
 b) $\cosh ix = \cos x$
 c) $\tanh ix = i \tan x$

**M14.** Find all complex numbers $z$ satisfying
 a) $z^3 = 1$     c) $z^3 = i$
 b) $z^3 = -1$    d) $z^3 = -i$

**M15.** Find twelve complex numbers $z$ satisfying the equation $z^{12} = 1$. (Hint: See Exercise M14.)

**M16.** Find all roots of the equation
$$z^3 + 3z^2 + 3z + 9 = 0.$$

**M17.** Find all roots of the equation
$$x^4 + 4x^2 + 16 = 0.$$

**M18.** Suppose $p(z) = a_0 + a_1 z + a_2 z^2 + \cdots + a_n z^n$ is a polynomial whose coefficients $a_0, a_1, \ldots, a_n$ are real numbers. Show that $p(z) = 0$ if and only if $p(\bar{z}) = 0$.

**M19.** Suppose $z^n = 1$ where $n > 1$ is a positive integer and $z \neq 1$. Show that
$$1 + z + z^2 + \cdots + z^{n-1} = 0.$$

**M20.** Verify the following power series expansions.
 a) $\sinh x = x + \dfrac{x^3}{3!} + \dfrac{x^5}{5!} + \cdots$
 b) $\cosh x = 1 + \dfrac{x^2}{2!} + \dfrac{x^4}{4!} + \cdots$

**M21.** Use the identity $2 \cos x = e^{ix} + e^{-ix}$ to establish the identity
$$1 + 2\cos x + 2\cos 2x + \cdots + 2\cos nx$$
$$= \frac{\sin (n + (1/2))x}{\sin (x/2)}.$$

**M22.** Show that $\int_0^{2\pi} e^{inx}\, dx = 0$ whenever $n$ is a nonzero integer.

**M23.** Use the formula $\cos bx = \dfrac{e^{ibx} + e^{-ibx}}{2}$ to derive the formula
$$\int e^{ax} \cos bx\, dx = \frac{e^{ax}(a \cos bx + b \sin bx)}{a^2 + b^2} + C.$$

**M24.** Replace the sine and cosine functions by exponential functions and derive the following formulas where $m$ and $n$ are integers.
 a) $\displaystyle\int_0^{2\pi} \sin mx \cos nx\, dx = 0$
 b) $\displaystyle\int_0^{2\pi} \sin mx \sin nx\, dx = 0 \quad (m \neq n)$
 c) $\displaystyle\int_0^{2\pi} \cos mx \cos nx\, dx = 0 \quad (m \neq n)$
 d) $\displaystyle\int_0^{2\pi} \sin^2 mx\, dx = \int_0^{2\pi} \cos^2 mx\, dx = \pi \quad (m \neq 0)$

## 7.6 Sample Exams

### Sample Exam 1 (45–60 minutes)

**1.** Assume that a dropped ball always rebounds to 60% of the height from which it is dropped. If a ball is dropped from 3 ft., what is the total distance it eventually travels?

**2.** Use the comparison test to determine which of the following series converge.
 a) $\displaystyle\sum_{n=2}^{\infty} \frac{1}{\sqrt{n^2 - 1}}$    b) $\displaystyle\sum_{n=1}^{\infty} \frac{3}{2n + 1}$    c) $\displaystyle\sum_{n=1}^{\infty} \frac{1}{n^2 \sec^2 n}$

**3.** Find the radius of convergence of
 a) $\displaystyle\sum_{n=0}^{\infty} (-1)^n x^n$   and   b) $\displaystyle\sum_{n=0}^{\infty} \frac{(n + 1)x^n}{(2n + 1)n!}$.

**4.** Given the series $\frac{1}{1-x} = 1 + x + x^2 + x^3 + \cdots$, determine what functions are represented by the following series.
 a) $1 - x^2 + x^4 - x^6 + \cdots$
 b) $x^3 + x^6 + x^9 + x^{12} + \cdots$
 c) $1 - 2x + 4x^2 - 8x^3 + 16x^4 - \cdots$

5. Write the solution of the equation

$$z(2 + 3i) = 4 - 5i$$

in the form $z = a + ib$ where $a$ and $b$ are real numbers.

## Sample Exam 2 (45–60 minutes)

1. Find the sum of each of the following series.

   a) $\frac{2}{3} + \frac{4}{9} + \frac{8}{27} + \frac{16}{81} + \frac{32}{243} + \cdots$

   b) $\displaystyle\sum_{n=3}^{\infty} \left(\frac{4}{5}\right)^n$    c) $\displaystyle\sum_{n=0}^{\infty} \frac{3^n}{4^{n+1}}$

2. Given $e^x = 1 + x + \frac{x^2}{2!} + \frac{x^3}{3!} + \cdots$

   a) write down the first 6 terms of the power series expansion of $e^{(x^2)}$, and

   b) evaluate $\left.\dfrac{d^6}{dx^6} e^{(x^2)}\right|_{x=0}$.

3. Compare the following series with either

$$\sum_{n=1}^{\infty} \frac{1}{n} \quad \text{or} \quad \sum_{n=1}^{\infty} \frac{1}{n^2}$$

   to determine whether they converge or diverge.

   a) $\displaystyle\sum_{n=1}^{\infty} \frac{1}{(n+1)(n+2)}$    b) $\displaystyle\sum_{n=1}^{\infty} \frac{\sin^2 n}{n^2}$

   c) $\displaystyle\sum_{n=1}^{\infty} \frac{1}{\sqrt{n}}$

4. For what values of $x$ do these power series converge?

   a) $\displaystyle\sum_{n=1}^{\infty} \frac{x^n}{n}$    b) $\displaystyle\sum_{n=1}^{\infty} \frac{nx^n}{n!}$    c) $\displaystyle\sum_{n=1}^{\infty} \frac{x^n}{\sqrt{n}}$

5. If $x = 3 + 4i$ and $y = 2 - i$, find a) $|x|$, b) $|x + y|$, c) $xy$, and d) $x/y$ in the form $a + ib$.

## Sample Exam 3 (45–60 minutes)

1. Find the sum of each of the following convergent series.

   a) $\displaystyle\sum_{n=0}^{\infty} \left(\frac{4}{9}\right)^n$    c) $\displaystyle\sum_{n=0}^{\infty} \left(\frac{4}{9}\right)^{2n+1}$

   b) $\displaystyle\sum_{n=0}^{\infty} \left(\frac{4}{9}\right)^{2n}$    d) $\displaystyle\sum_{n=0}^{\infty} (-1)^n \left(\frac{4}{9}\right)^{n/2}$

2. Use the integral test to determine whether the following converge.

   a) $\displaystyle\sum_{n=1}^{\infty} \frac{1}{(2n + 1)^{1/2}}$    b) $\displaystyle\sum_{n=2}^{\infty} \frac{1}{n \ln^2 n}$

3. Find the radius of convergence of

   a) $f(x) = \displaystyle\sum_{n=1}^{\infty} \frac{x^{n+1}}{n}$    and    b) $g(x) = \displaystyle\sum_{n=1}^{\infty} n^n x^n$.

4. The series

$$\ln \tfrac{2}{1} - \ln \tfrac{3}{2} + \ln \tfrac{4}{3} - \ln \tfrac{5}{4} + \ln \tfrac{6}{5} - \cdots$$

   a) is convergent, but    b) not absolutely convergent. Prove either (a) or (b).

5. Given $\sin x = \dfrac{x}{1} - \dfrac{x^3}{3!} + \dfrac{x^5}{5!} - \dfrac{x^7}{7!} + \cdots$    find

   a) the power series expansion of $f(x) = \sin 2x^2$,
   b) $f^{(16)}(0)$, and
   c) $f^{(23)}(0)$.

## Sample Exam 4 (45–60 minutes)

1. Assume that $\sum_{n=0}^{\infty} a_n < \infty$ $(a_n > 0)$ and a series $\sum_{n=0}^{\infty} b_n$ $(b_n > 0)$ is given. Prove that if $\lim_{n\to\infty} \dfrac{b_n}{a_n} = c < 1$ then $\sum_{n=0}^{\infty} b_n$ is convergent.

2. Graphically represent all 5 solutions of the equation $z^5 + 32i = 0$.

3. a) Find, by choosing coefficients properly, a power series whose radius of convergence is zero.
   b) Find a power series whose radius of convergence is 78.

4. Prove that $\displaystyle\sum_{n=1}^{\infty} \frac{1}{n(n+1)} = 1$ by first showing that

$$\sum_{n=1}^{k} \frac{1}{n(n+1)} = \frac{k}{k+1}. \quad \text{(Hint: Use the identity}$$

$$\frac{1}{n(n+1)} = \frac{n}{n+1} - \frac{n-1}{n}$$

   or mathematical induction.)

5. a) Show that $\displaystyle\sum_{n=1}^{\infty} \left(\frac{e}{n}\right)^n$ is convergent.

   b) Using (a), show that $\displaystyle\int_1^{\infty} \frac{e^y}{y^y}\, dy$ is convergent.

   c) Using (b), show that $\displaystyle\sum_{n=2}^{\infty} \frac{1}{(\ln n)^{\ln n}}$ is convergent.

## Sample Exam 5 (approximately 2 hours)

1. Write the sum of each of the following series in the form $\frac{m}{n}$ where $m$ and $n$ are integers.

   a) $\dfrac{1}{10} + \dfrac{1}{10^2} + \dfrac{1}{10^3} + \dfrac{1}{10^4} + \cdots$

   b) $\dfrac{2}{10} + \dfrac{3}{10^2} + \dfrac{2}{10^3} + \dfrac{3}{10^4} + \dfrac{2}{10^5} + \cdots$

2. Determine which of the following series converge, and justify your answer.

   a) $\displaystyle\sum_{n=1}^{\infty} \dfrac{2n+3}{3n+2}$   b) $\displaystyle\sum_{n=1}^{\infty} \dfrac{3^n}{\pi^n}$   c) $\displaystyle\sum_{n=1}^{\infty} \dfrac{3^n}{e^n}$

3. Write the first three terms of the power series for
$$f(x) = \cos(\ln(x+1)).$$

4. Given $e^x = \displaystyle\sum_{n=0}^{\infty} \dfrac{x^n}{n!}$ determine what functions are represented by the following power series.

   a) $\displaystyle\sum_{n=0}^{\infty} \dfrac{(-2)^n x^n}{n!}$   b) $\displaystyle\sum_{n=0}^{\infty} \dfrac{(x-1)^{2n}}{n!}$

5. Use the ratio test to determine the convergence or divergence of the following series.

   a) $\dfrac{1}{5} + \dfrac{2}{5^2} + \dfrac{3}{5^3} + \dfrac{4}{5^4} + \cdots$

   b) $\dfrac{1!}{5} + \dfrac{2!}{5^2} + \dfrac{3!}{5^3} + \dfrac{4!}{5^4} + \cdots$

6. Determine the radius of convergence of the following power series.

   a) $\displaystyle\sum_{n=1}^{\infty} \dfrac{x^n}{\sqrt{n}}$   b) $\displaystyle\sum_{n=1}^{\infty} \dfrac{x^n}{n3^n}$

7. Write down the power series representations.

   a) $f(x) = \dfrac{1}{1-x}$   c) $h(x) = \dfrac{1}{(1-x)^2}$

   b) $g(x) = \ln(1-x)$

8. Show that $\displaystyle\lim_{n\to\infty} \left( \sum_{k=1}^{n} \dfrac{k}{n^2} \right) = \dfrac{1}{2}.$

9. Differentiate the following complex valued functions.
   a) $f(x) = e^{2ix} + e^{-2ix}$
   b) $g(x) = (i \ln ix)^3$
   c) $h(x) = (ix)^4 + e^{(2+3i)x}$

## Sample Exam 6 (approximately 3 hours)

1. Use the integral test to show that each of the following series is convergent.

   a) $\displaystyle\sum_{n=1}^{\infty} \dfrac{n}{e^n}$

   b) $\sin \pi + \dfrac{1}{4} \sin \dfrac{\pi}{2} + \dfrac{1}{9} \sin \dfrac{\pi}{3} + \dfrac{1}{16} \sin \dfrac{\pi}{4} + \cdots$

2. Find the sum of $\displaystyle\sum_{n=1}^{\infty} \dfrac{2n+1}{n^2(n+1)^2}.$

   $\left(\text{Hint: Write } \dfrac{2n+1}{n^2(n+1)^2} \text{ in the form } \dfrac{a}{n^2} - \dfrac{b}{(n+1)^2}.\right)$

3. Show that

   a) $\displaystyle\sum_{n=1}^{\infty} \dfrac{1}{n} \sin n\pi$   is convergent,

   b) $\displaystyle\sum_{n=1}^{\infty} \dfrac{1}{n} \sin n\dfrac{\pi}{2}$   is conditionally convergent, and

   c) $\displaystyle\sum_{n=1}^{\infty} \dfrac{1}{n} \sin(2n+1)\dfrac{\pi}{2}$   is divergent.

4. Determine which of the following series converge. If they converge, tell whether they converge absolutely or conditionally.

   a) $\displaystyle\sum_{n=2}^{\infty} (-1)^n \dfrac{n}{n^3+1}$

   b) $\displaystyle\sum_{n=23}^{\infty} \left( \dfrac{24754298}{24754299} \right)^n$

   c) $1 - \dfrac{1}{4} + \dfrac{2}{8} - \dfrac{1}{16} + \dfrac{2}{32} - \dfrac{1}{64} + \dfrac{2}{128} - \cdots$

5. Find the radius of convergence of the following power series.

   a) $\displaystyle\sum_{n=1}^{\infty} \dfrac{x^n}{1+(n-1)^n}$

   b) $\dfrac{x}{1\cdot 2} + \dfrac{x^2}{2\cdot 3} + \dfrac{x^3}{3\cdot 4} + \dfrac{x^4}{4\cdot 5} + \cdots$

   c) $\displaystyle\sum_{n=1}^{\infty} (-1)^n \dfrac{n!(2x)^n}{2\cdot 4\cdot \ldots \cdot 2n}$

6. Determine the power series for the following functions.
   a) $y = \sin 3x$   b) $y = \sinh 2x$

7. Write the following complex numbers in polar form.
   a) $\sqrt{3} + i$     c) $15i$
   b) $-10$         d) $-\sqrt{3} - 3i$

8. If $u = 3 + i$ and $v = 6 + i$ find
   a) $u + v$      d) $|u|\,\bar{v}$
   b) $\overline{u + v}$     e) $|u + v|$
   c) $\overline{uv}$       f) $|u| + |v|$

9. Find four solutions of the equation $z^4 + 1 = 0$. Write each solution in the form $a + ib$ ($a, b$ real) and locate them in the complex plane.

10. a) Determine the power series expansion for the function

$$f(x) = \frac{1}{\sqrt{1 - x}}$$

   b) Compute its radius of convergence.

11. Show that $\displaystyle\sum_{n=2}^{\infty} \ln \frac{n^2}{(n - 1)(n + 1)} = \ln 2.$

Equations involving an unknown function and its derivatives are called differential equations and arise frequently in applications. They come in a great variety of forms, and many different procedures exist for finding their solutions. In the first three sections of this chapter we learn some elementary techniques that can be used to solve many of the most common differential equations. Where these methods fail it is often possible to find the solution in terms of a power series, and this method is described in 8.4. In Section 8.5 we use differential equations in deriving Kepler's three laws of planetary motion.

# Differential Equations

## 8.1  *First Order Equations*

An equation that involves only $x$, a function $y$ of $x$, and the derivative $\frac{dy}{dx}$ is called a *first order differential equation*. The equation

$$x\frac{dy}{dx} + y\left(\frac{dy}{dx}\right)^2 = \sin x - e^y$$

is an example. Many of these equations can be solved using integration and a few additional tricks.

**A. Separation of Variables**   We will start with a familiar example. Suppose

$$\frac{dy}{dx} = f(x).$$

This is a first order differential equation and its solution is simply

$$y = \int f(x)\,dx.$$

Suppose now we are given an equation of the form

$$\frac{dy}{dx} = \frac{f(x)}{g(y)}.$$

We say that the *variables* in this equation are *separated* because the equation may be rewritten as

$$g(y)\,dy = f(x)\,dx,$$

in which only expressions containing $y$ appear on one side of the equation and only expressions containing $x$ appear on the other. We solve by integrating:

$$\int g(y)\,dy = \int f(x)\,dx.$$

---

*Example 1*   Solve the differential equation

$$\frac{dy}{dx} = \frac{2x}{3y^2}.$$

*Solution*   Writing this equation in the form

$$3y^2\,dy = 2x\,dx$$

and integrating, we obtain

More precisely, given a function $F$ of three variables, the equation

$$F\left(x, y, \frac{dy}{dx}\right) = 0$$

is a first order differential equation. For example if

$$F(x_1, x_2, x_3) = x_1 x_3 + x_2 x_3{}^2 - \sin x_1 + e^{x_2},$$

then $F(x, y, \frac{dy}{dx}) = 0$ is the equation given in the text.

This is the solution to the differential equation, subject of course to actually evaluating the integral $\int f(x)\,dx$. We see from this that the variety of techniques for solving differential equations is at least as great as the number of techniques of integration.

A proof that this is indeed the solution of the original differential equation simply involves implicit differentiation. Suppose $F$ and $G$ are antiderivatives of $f$ and $g$ respectively. If $F$ and $G$ satisfy $F(x) = G(y) + C$, then, differentiating with respect to $x$, we obtain

$$\frac{dF(x)}{dx} = \frac{dG(y)}{dy}\frac{dy}{dx}$$

or

$$f(x) = g(y)\frac{dy}{dx}, \quad \text{so} \quad \frac{dy}{dx} = \frac{f(x)}{g(y)}.$$

$$y^3 + C_1 = x^2 + C_2.$$

This may be written in the form

$$y^3 = x^2 + C,$$

where $C = C_2 - C_1$.

---

**Example 2**   Solve the differential equation

$$2x^2y\frac{dy}{dx} - y^2 - 1 = 0.$$

*Solution*   Rewriting the equation as

$$2x^2y\frac{dy}{dx} = y^2 + 1$$

and then as

$$\frac{2y\,dy}{1 + y^2} = \frac{dx}{x^2},$$

we see that the variables are separated. Integrating we obtain

$$\int \frac{2y\,dy}{1 + y^2} = \int \frac{dx}{x^2}$$

or

$$\ln(1 + y^2) = \frac{-1}{x} + C.$$

Rewriting this in terms of exponentials, we have

$$1 + y^2 = e^{-1/x+C} = e^{-1/x}e^C$$

or

$$y^2 = Ae^{-1/x} - 1$$

where $A = e^C$.

---

**Example 3**   We will solve

$$xy' + y = y^2.$$

*Solution*   Rewriting the equation as

$$\frac{dy}{y^2 - y} = \frac{dx}{x}$$

---

We can also write this solution in the form

$$y = (x^2 + C)^{1/3}.$$

In what follows we will not attempt to write the solutions in terms of $y$ as a function of $x$, because this is often difficult or impossible to do. For example, the solution of

$$\frac{dy}{dx} = \frac{1}{1 + 2y + 3y^2 + 4y^3 + 5y^4}$$

is

$$y + y^2 + y^3 + y^4 + y^5 = x + C,$$

and this is the most convenient form of the solution.

---

**Problem 1**   Solve the equation $yy' + 3y' - x - 1 = 0$ by separating the variables.

*Answer:*   $(y + 3)^2 = (x + 1)^2 + C$

---

*Some confusion arose when we first began solving differential equations, and arrived at different answers, depending on our method of solution. For example, in solving Problem 1, $yy' + 3y' - x - 1 = 0$, I wrote the answer as*

$$y^2 + 6y = x^2 + 2x + C$$

*which is correct when checked by differentiation. However, the given answer*

$$(y + 3)^2 = (x + 1)^2 + C$$

*is also valid, because only the constant has been altered. The value of this form is that it suggests a family of hyperbolas as the graphic solution to this problem.*

*(Beth)*

---

**Problem 2**   Separate variables to solve

$$y' + e^{-y}\sin x = 0.$$

*Answer:*   $e^y = \cos x + C$

and integrating, we obtain

$$\int \frac{dy}{y^2 - y} = \int \frac{dx}{x}.$$

This may be written as

$$\int \left( \frac{1}{y - 1} - \frac{1}{y} \right) dy = \int \frac{dx}{x},$$

and we have

$$\ln|y - 1| - \ln|y| = \ln|x| + C$$

or

$$\ln \left| \frac{y - 1}{xy} \right| = C.$$

Taking exponentials now gives

$$\left| \frac{y - 1}{xy} \right| = e^C = A.$$

We now will write the solution as

$$\frac{y - 1}{xy} = A$$

and, solving for $y$, we obtain

$$y = \frac{1}{1 - Ax}$$

where $A$ is an arbitrary constant.

---

**B. Linear Differential Equations** A differential equation of the form

$$a_1(x)y' + a_0(x)y = b(x)$$

where $a_0(x)$, $a_1(x)$, and $b(x)$ $(a_1(x) \neq 0)$ are functions of $x$ is said to be a *linear differential equation*. We will write the equation in a more convenient form. Dividing by $a_1(x)$ we obtain

$$y' + P(x)y = Q(x)$$

where $P = \dfrac{a_0}{a_1}$ and $Q = \dfrac{b}{a_1}$. The solution to this differential equation is given in the following.

*THEOREM 1* *Given*

$$y' + P(x)y = Q(x),$$

In the text no justification was given for removing the absolute value sign in passing from

$$\left| \frac{y - 1}{xy} \right| = A \quad \text{to} \quad \frac{y - 1}{xy} = A.$$

The reader can easily check, however, that $y = \frac{1}{1 - Ax}$ is a solution to the differential equation. Moreover, there are many more solutions than those indicated above. For example, for $B < C$ the function

$$f(x) = \begin{cases} \dfrac{1}{1 - Bx} & (x < B) \\ \dfrac{1}{1 - Cx} & (x > C) \end{cases}$$

is also a solution of the given differential equation.

The important feature of linear equations is that $y$ and $y'$ appear to the first power. The equation $y' = y^2$ is *not* a linear differential equation, but the equation $y' = x^2$ is.

*the solution is*

$$y = e^{-\int P} \int e^{\int P} Q + Ce^{-\int P}$$

*where C is an arbitrary constant.*

*Proof*  It can easily be checked that the indicated solution satisfies the equation. One will need to recall that

$$\frac{d}{dx} e^{\int P} = e^{\int P} \frac{d}{dx} \int P = e^{\int P} P.$$

(We have written $\int P$ in place of the more cumbersome $\int P(x)\, dx$.)  ●

---

*Example 4*  We will solve the equation

$$\frac{dy}{dx} + 2xy = xe^{x^2}$$

by applying the formula given in Theorem 1.

*Solution*  Observe that $P = 2x$ and $Q = xe^{x^2}$. It follows that $\int P = \int 2x = x^2$; hence $e^{\int P} = e^{x^2}$ and $e^{-\int P} = e^{-x^2}$. Using Theorem 1, we have

$$y = e^{-x^2} \int e^{x^2} xe^{x^2}\, dx + Ce^{-x^2}$$

$$= \frac{e^{-x^2}}{4} \int e^{2x^2} 4x\, dx + Ce^{-x^2}$$

$$= \frac{e^{-x^2}}{4} e^{2x^2} + Ce^{-x^2} = \frac{e^{x^2}}{4} + Ce^{-x^2}.$$

---

*Example 5*  We will illustrate the derivation of the formula in Theorem 1 by finding the solution of

$$xy' + (1 + x)y = x$$

by the same method.

*Solution*  We write the equation as

$$(*) \qquad y' + \frac{1 + x}{x} y = 1.$$

Since $P(x) = \frac{1+x}{x}$, we have $\int P(x)\, dx = x + \ln x$. Hence $e^{\int P} = e^{x + \ln x} = e^x e^{\ln x} = xe^x$. Multiplying both sides of equation (*) by $e^{\int P} = xe^x$ we have

---

*The reader may wonder where this solution came from. Here is a derivation (which may also be unsatisfying).*
*Multiply the differential equation by $e^{\int P}$ to obtain*

$$(*) \qquad e^{\int P} y' + e^{\int P} Py = e^{\int P} Q,$$

*and notice that*

$$\frac{d}{dx} (e^{\int P} y) = e^{\int P} y' + e^{\int P} Py.$$

*Hence equation (*) can be written as*

$$\frac{d}{dx} (e^{\int P} y) = e^{\int P} Q.$$

*Integrating, we have*

$$e^{\int P} y = \int e^{\int P} Q + C,$$

*and then multiplying by $e^{-\int P}$ we obtain*

$$y = e^{-\int P} \int e^{\int P} Q + Ce^{-\int P}.$$

*In evaluating $\int P(x)\, dx$ we may let the constant of integration be any convenient value, usually zero. Indeed, if we replace $\int P$ by $A + \int P$ in the formula for y given above, we obtain*

$$y = e^{-A - \int P} \int e^{A + \int P} Q\, dx + Ce^{-A - \int P}$$

$$= e^{-\int P} e^{-A} \int e^A e^{\int P} Q\, dx + Ce^{-A} e^{-\int P}$$

$$= e^{-\int P} e^{-A} e^A \int e^{\int P} Q\, dx + (Ce^{-A})e^{-\int P}$$

$$= e^{-\int P} \int e^{\int P} Q\, dx + De^{-\int P} \quad \text{where} \quad D = Ce^{-A}.$$

*Since D is an arbitrary constant, this solution is the same as the original one.*

$$xe^x\left(y' + \frac{1+x}{x}y\right) = xe^x$$

or

$$xe^x y' + (1+x)e^x y = xe^x.$$

This equation may be written in the form

$$\frac{d}{dx}(yxe^x) = xe^x.$$

Integrating it we obtain

$$\int \frac{d}{dx}(yxe^x)\,dx = \int xe^x\,dx$$

or

$$yxe^x = xe^x - e^x + C.$$

Now solving for $y$ we have $y = 1 - \dfrac{1}{x} + C\dfrac{1}{xe^x}$.

---

**C. Changing Variables**   Often a suitable change of variable will reduce a particular equation to a linear equation or one in which the variables are separated. We will illustrate this method with several examples.

---

**Example 6**   The equation

$$y' + \frac{1}{x}y = 3x^2 y^2$$

is not linear because the $y^2$ factor appears. We will find the solution by first transforming the equation to a linear one with the substitution $y = \frac{1}{u}$.

**Solution**   Letting $y = \dfrac{1}{u}$ we have $y' = \dfrac{-u'}{u^2}$. Substituting these equalities into the original differential equation gives

$$\frac{-u'}{u^2} + \frac{1}{x}\left(\frac{1}{u}\right) = \frac{3x^2}{u^2}$$

or

(*) $$u' - \frac{u}{x} = -3x^2,$$

which is a linear equation. Since $P = \frac{-1}{x}$ we have

$$\int P = -\ln x = \ln\frac{1}{x} \quad (x > 0)$$

**Problem 3**   Solve the differential equation $y' - 2y = 0$.

*Answer:*   $y = Ce^{2x}$

---

**Problem 4**   Solve the differential equation $y' + 3x^2 y = x^2$ using Theorem 1, and then check by finding the solution with the separation of variable technique.

*Answer:*   $y = \frac{1}{3} + Ce^{-x^3}$

---

This is a special case of *Bernoulli's equation*, which has the form

$$y' + P(x)y = Q(x)y^n.$$

We will show that the substitution $y = u^{1/(1-n)}$ transforms the equation to one that is linear in $u$. We have $u = y^{1-n}$, so $u' = (1-n)y^{-n}y'$. Solving for $y'$ gives

$$y' = \frac{y^n}{1-n}u',$$

and substituting this expression in the original equation yields

$$\frac{y^n}{1-n}u' + P(x)y = Q(x)y^n.$$

Dividing by $y$, we have

$$\frac{y^{n-1}}{1-n}u' + P(x) = Q(x)y^{n-1}$$

or

$$\frac{u^{-1}}{1-n}u' + P(x) = Q(x)u^{-1}$$

or

$$u' + (1-n)uP(x) = (1-n)Q(x),$$

and this is a linear equation.

so $e^{\int P} = e^{\ln(1/x)} = \frac{1}{x}$ and $e^{-\int P} = x$. Using Theorem 1 we have

$$u = x \int \frac{1}{x}(-3x^2)\,dx + Cx$$

$$= x\left(\frac{-3}{2}x^2\right) + Cx = \frac{-3x^3 + 2Cx}{2}.$$

Resubstituting $\frac{1}{y}$ for $u$, we obtain $y = \dfrac{2}{2Cx - 3x^3}$,

which may be put in the simpler form $y = \dfrac{2}{Ax - 3x^3}$

by letting $2C = A$.

---

**Example 7** We will solve the equation

$$\frac{dy}{dx} = \frac{x^2 + y^2}{2xy}$$

by transforming it to one in which the variables are separated.

*Solution* Let

$$y = ux,$$

so that

$$\frac{dy}{dx} = \frac{du}{dx}x + u.$$

The original differential equation then becomes

$$\frac{du}{dx}x + u = \frac{x^2 + u^2x^2}{2xux} = \frac{1 + u^2}{2u}.$$

Hence

$$\frac{du}{dx}x = \frac{1 + u^2}{2u} - u = \frac{1 - u^2}{2u}.$$

Writing this as $\dfrac{-2u\,du}{1 - u^2} = \dfrac{-dx}{x}$, we see that the variables are now separated. Integration yields

$$\ln(1 - u^2) = -\ln x + C = \ln\left(\tfrac{A}{x}\right)$$

where $\ln A = C$.

Taking exponentials, we have

**Problem 5** Use the substitution $y = u^2$ to reduce

$$xy' - 2y + 6x^3\sqrt{y} = 0$$

to a linear equation. Then solve it using formula (*) of Example 6.

*Answer:* $y = \left(\dfrac{Ax - 3x^3}{2}\right)^2$

---

→ The situation in which the substitution $y = ux$ transforms the equation to one in which the variables are separated is described in the following.

*Theorem 2* Suppose $\frac{dy}{dx} = F(x, y)$ and $F(tx, ty) = F(x, y)$ for all $t$, $x$, and $y$. Then, letting $y = ux$ transforms the equation to

$$\frac{du}{F(1, u) - u} = \frac{dx}{x}.$$

*Proof* Since $y = ux$, we have $\frac{dy}{dx} = \frac{du}{dx}x + u$. Hence

$$\frac{du}{dx}x + u = F(x, ux) = F(1, u) \quad \text{or} \quad \frac{du}{dx}x = F(1, u) - u.$$

Algebraic manipulation of this equality then yields

$$\frac{du}{F(1, u) - u} = \frac{dx}{x} \quad \bullet$$

$$1 - u^2 = \frac{A}{x},$$

and finally, resubstituting $\frac{y}{x}$ for $u$, we obtain

$$1 - \frac{y^2}{x^2} = \frac{A}{x} \quad \text{or} \quad x^2 - y^2 = Ax.$$

**Problem 6**   Solve $y' = \frac{y}{x} + \sec \frac{y}{x}$ using the substitution $y = ux$.

---

**Example 8**   We will solve the equation

$$\frac{dy}{dx} = \frac{x + y}{x - y}$$

using the substitution $y = ux$.

*Solution*   Let $y = ux$, so $\frac{dy}{dx} = \frac{du}{dx} x + u$.
With these substitutions the equation becomes

$$\frac{du}{dx} x + u = \frac{x + ux}{x - ux} = \frac{1 + u}{1 - u}.$$

Separating the variables, we have

$$\frac{dx}{x} = \frac{1 - u}{1 + u^2} du.$$

Integrating, we obtain

$$\int \frac{dx}{x} = \int \frac{(1 - u)\, du}{1 + u^2} = \int \frac{du}{1 + u^2} - \frac{1}{2} \int \frac{2u\, du}{1 + u^2}$$

so

$$\ln x = \text{Tan}^{-1} u - \tfrac{1}{2} \ln (1 + u^2) + C.$$

This gives

$$\ln x + \ln \sqrt{1 + u^2} = \text{Tan}^{-1} u + C$$

or

$$\ln x \sqrt{1 + u^2} = \text{Tan}^{-1} u + C.$$

Resubstituting $\frac{y}{x}$ for $u$, we obtain

$$\text{Tan}^{-1} \frac{y}{x} + C = \ln \sqrt{x^2 + y^2} = \tfrac{1}{2} \ln (x^2 + y^2)$$

or

$$2 \, \text{Tan}^{-1} \frac{y}{x} + K = \ln (x^2 + y^2)$$

where $K = 2C$.

*Answer:*   $\sin \frac{y}{x} = \ln Ax$

---

**Problem 7**   Given $\dfrac{dy}{dx} = \dfrac{x + y + 2}{x - y - 4}$, determine $h$ and $k$ so that the substitutions $x = v + h, \, y = u + k$ will transform the equation into

$$\frac{du}{dv} = \frac{v + u}{v - u}.$$

Then use Example 8 to write down the solution.

*Answer:*

$$2 \, \text{Tan}^{-1} \left( \tfrac{y+3}{x-1} \right) = \ln ((x - 1)^2 + (y + 3)^2) + K$$

and $h = 1, \, k = -3$.

**D. Some Applications**   Consider the differential equation

$$\frac{dy}{dx} = ky$$

where $k$ is a constant. Separating the variables, we obtain

$$\frac{dy}{y} = k\,dx,$$

so

$$\ln y = kx + C.$$

Taking exponentials then gives

$$y = e^{kx+C} = e^{kx}e^{C}$$

or

$$y = Be^{kx}$$

where $B = e^{C}$.

---

**Example 9**   A certain bacterium is assumed to reproduce at a rate proportional to the amount of bacteria present. We will show that the equation that describes the amount $A$ of bacteria at time $t$ has the form

$$A(t) = A_0 e^{kt}$$

where $A_0$ is the amount of bacteria present at $t = 0$ and $k$ is a constant that depends on the specific type of bacteria present.

*Solution*   Our sole problem is to translate the phrase "... to reproduce at a rate proportional to the amount..." into mathematical terms. Then calculus will take over.

The rate at which the bacteria reproduce is $\frac{dA}{dt}$. Since it is given that this is proportional to the amount of bacteria, we see that

$$\frac{dA}{dt} \bigg/ A \quad \text{is constant.}$$

Denoting this constant of proportionality by $k$, we have

$$\frac{dA}{dt} \bigg/ A = k \quad \text{or} \quad \frac{dA}{dt} = kA.$$

This differential equation has the solution

$$A(t) = Be^{kt} \quad \text{for some constant } B.$$

Since this equation is particularly important, we will prove that every real valued solution of $y' = ky$ indeed has the form $y = Be^{kx}$ where $B$ is an arbitrary real number.

**Theorem 3**   *Suppose f is differentiable on $(-\infty, \infty)$ and satisfies $f' = kf$. Then $f(x) = Be^{kx}$ for some constant B.*

*Proof*   We have that

$$\frac{d}{dx}\left(\frac{f(x)}{e^{kx}}\right) = \frac{e^{kx}f'(x) - f(x)ke^{kx}}{(e^{kx})^2} = \frac{e^{kx}}{e^{2kx}}(f'(x) - f(x)k)$$

$$= \frac{1}{e^{kx}}(f'(x) - kf(x)) = 0$$

since $(f' - kf)(x) = 0$ for all $x$ and $e^{kx} \neq 0$. Since $\frac{f(x)}{e^{kx}}$ is defined on $(-\infty, \infty)$ and has zero derivative, it must be a constant (Corollary 1, 4.1C). Hence $\frac{f(x)}{e^{kx}} = B$ or $f(x) = Be^{kx}$ ●

---

**Example 10**   We refer to the solution given in Example 9. Assume now that 10 thousand bacteria are present at $t = 0$ and their population doubles every 3 hours. We will determine the constants $A_0$ and $k$.

*Solution*   Since $A(t) = A_0 e^{kt}$ and $A(0) = 10^4$, we have $A_0 = 10^4$ so $A(t) = 10^4 e^{kt}$. By assumption we have that when $t = 3$, $A(3) = 2A_0 = 2 \cdot 10^4$. Hence $2 \cdot 10^4 = A(3) = 10^4 e^{k3}$ or $e^{3k} = 2$. This yields $3k = \ln 2$ or $k = \frac{\ln 2}{3}$. Replacing $k$ by this value in $A(t) = 10^4 e^{kt}$, we then have

$$A(t) = 10^4 (e^{\ln 2})^{t/3} = 10^4 \cdot 2^{t/3}.$$

Hence the equation is $A(t) = 10^4 \cdot 2^{t/3}$.

---

Since $A(0) = Be^{k0} = B$, the constant $B$ represents the amount of bacteria present at $t = 0$. We therefore write $B = A_0$ and have $A(t) = A_0 e^{kt}$.

Notice that because we assumed that the amount of bacteria increases, the constant $k$ must be positive ($k = 0$ signifies sterility, and $k$ negative means decay).

Example 11, which concerns radioactive decay, is mathematically similar to the bacteria problems given above.

If you can find the solution to the problem presented in Example 12 you are doing very well.

(Estelle)

**Example 11** A radioactive substance decays at a rate proportional to the amount present. Assuming the "half life" (time for half the substance to decay) is 5 years, how long will it be until only 1% of the original substance remains?

*Solution* The derivation of the equation is the same as that in Example 9. Letting $A(t)$ equal the amount of the substance present at time $t$, we have

$$A(t) = A_0 e^{kt}$$

where this time $k < 0$. We are also given that

$$A(5) = \frac{A_0}{2}$$

so

$$\frac{A_0}{2} = A(5) = A_0 e^{5k} \quad \text{or} \quad \frac{1}{2} = e^{5k}.$$

Hence, $5k = \ln\frac{1}{2}$; that is, $k = \frac{1}{5}\ln\frac{1}{2}$. Therefore

$$A(t) = A_0 e^{(1/5)(\ln(1/2))t} = A_0 e^{(t/5)(\ln(1/2))}$$

$$= A_0 e^{\ln(1/2)^{t/5}} = A_0 \left(\tfrac{1}{2}\right)^{t/5} = A_0 2^{-t/5}.$$

Our problem is to find the $t$ that satisfies $A(t) = \frac{A_0}{100}$.

Setting $\frac{A_0}{100} = A(t) = A_0 2^{-t/5}$, we have

$$\tfrac{1}{100} = 2^{-t/5}$$

and, taking the logs of both sides, we obtain

$$-\ln 100 = \frac{-t}{5}\ln 2.$$

This gives $t = \dfrac{5\ln 100}{\ln 2}$, which is approximately 33 years.

**Example 12** It starts snowing steadily sometime before noon. At noon a man begins to shovel snow from the sidewalk at a constant rate. He clears 2 blocks the first 2 hours and 1 block the next 2 hours. When did it start snowing?

*Solution* Let $t = 0$ correspond to noon, $h(t)$ equal the height of the snow at time $t$, and $x(t)$ equal the number of blocks the man has shoveled at time $t$. Since it is snowing at a constant rate we have $\frac{dh}{dt} = a$ for some constant $a$. Integrating this gives $h = at + b$ where $b$ is a constant. Now let $c$ be the number of cubic feet of snow the man shovels per hour ($c$ is constant). Since $\frac{dx}{dt}$ is inversely proportional to $h = at + b$ we have

$$\frac{dx}{dt} = \frac{\lambda}{h} = \frac{\lambda}{at + b}$$

where $\lambda$ is a constant of proportionality. Integrating this equation we obtain

$$\frac{a}{\lambda}x(t) = \ln(at + b) + C.$$

Because $x(0) = 0$ we have $\frac{a}{\lambda} \cdot 0 = \ln b + C$, so $C = -\ln b$. Therefore $\frac{a}{\lambda}x(t) = \ln(at + b) - \ln b = \ln\left(\frac{a}{b}t + 1\right)$. From the hypothesis that $x(2) = 2$ and $x(4) = 3$, we have

$$\frac{2a}{\lambda} = \ln\left(\frac{2a}{b} + 1\right) \quad \text{and} \quad \frac{3a}{\lambda} = \ln\left(\frac{4a}{b} + 1\right)$$

respectively. Dividing these two equations gives

$$\frac{2}{3} = \frac{\ln\left(\dfrac{2a}{b} + 1\right)}{\ln\left(\dfrac{4a}{b} + 1\right)}$$

or

$$2\ln\left(\frac{4a}{b} + 1\right) = 3\ln\left(\frac{2a}{b} + 1\right);$$

that is,     $\ln\left(\dfrac{4a}{b} + 1\right)^2 = \ln\left(\dfrac{2a}{b} + 1\right)^3$.

Hence       $\left(\dfrac{4a}{b} + 1\right)^2 = \left(\dfrac{2a}{b} + 1\right)^3$.

Now, it clearly began snowing when $h = at + b$ was equal to zero, and this occurs when $t = -\frac{b}{a}$. Substituting $-\frac{1}{t}$ for $\frac{a}{b}$ in the last equation then gives

$$\left(\dfrac{-4}{t} + 1\right)^2 = \left(\dfrac{-2}{t} + 1\right)^3$$

or, after some algebra,

$$t^2 - 2t - 4 = 0.$$

Hence

$$t = 1 - \sqrt{5} \text{ or approximately } 1 - \frac{56}{25} = \frac{-31}{25} \text{ hours.}$$

It therefore began to snow around 10:46 A.M.

---

## E. Exercises

**B1.** Use the method of separation of variables to solve the following differential equations.

   a) $\dfrac{dy}{dx} = \dfrac{xy}{1 + x^2}$      b) $\dfrac{dy}{dx} = \dfrac{1 + x^2}{xy}$

**B2.** Consider the differential equation

$$(1 - x^2)\dfrac{dy}{dx} + 2xy = 2x.$$

   a) Solve it by the method of separation of variables.

   b) Write it in the form $y' + Py = Q$ and evaluate

$$y = e^{-\int P} \int e^{\int P} Q + Ce^{-\int P}.$$

**B3.** a) Reduce the Bernoulli equation

$$-2y' + y = xy^3$$

     to a linear equation with the substitution $u = \dfrac{1}{y^2}$.

   b) Solve the linear equation obtained in (a).

**B4.** Use the substitution $y = ux$ to solve the differential equation

$$\dfrac{dy}{dx} = \dfrac{x^2 + 3y^2}{2xy}.$$

**B5.** Brand X mothballs evaporate, losing half their volume every 3 weeks. If the volume of each mothball is initially 1 in.$^3$ and the mothball is ineffective when its volume reaches 0.1 in.$^3$, how long will these mothballs be effective?

**D6.** Separate the variables to solve each of the following differential equations.

   a) $\dfrac{dy}{dx} = \dfrac{x}{y}$      e) $\dfrac{dy}{dx} = y$

   b) $\dfrac{dy}{dx} = \dfrac{y}{x}$      f) $\dfrac{dy}{dx} = \dfrac{1}{y}$

   c) $\dfrac{dy}{dx} = xy$      g) $\dfrac{dy}{dx} = x$

   d) $\dfrac{dy}{dx} = \dfrac{-x}{y}$      h) $\dfrac{dy}{dx} = \dfrac{1}{x}$

   a) $x^2 - y^2 = C$     b) $y = Ax$     c) $y = Ae^{x^2/2}$
   d) $x^2 + y^2 = C$     e) $y = Ae^x$     f) $y^2 = 2x + C$

   g) $y = \dfrac{x^2}{2} + C$     h) $y = \ln|x| + C$

**D7.** Write each of the following differential equations in the form $y' + P(x)y = Q(x)$ and evaluate $\int P(x)\,dx$, ignoring the constant of integration.

a) $xy' = x + y$    d) $y' = \dfrac{x + 2xy}{1 - x^2}$

b) $x^2y' = y - x$    e) $y' + x^2y' = x^3 - xy$

c) $y' = 3x^2$    f) $y' + y^2 = (x + y)^2$

a) $-\ln x$    b) $\frac{1}{x}$    c) $0$    d) $\ln(1 - x^2)$

e) $\ln\sqrt{1 + x^2}$    f) $-x^2$

Find the solutions to the differential equations given in Exercises M8–M15.

**M8.** $\dfrac{dy}{dx} = \dfrac{1 + x^2}{1 - y^2}$

**M9.** $y' - 3xy = x$

**M10.** $xy' + y = \sin 2x$

**M11.** $xy' + 12y + 4 = 0$

**M12.** a) $y' + y\cos x = 2xe^{\sin x}$

b) $y' - y\cot x = \sin x$

**M13.** $(1 + e^{-y/x})\dfrac{dy}{dx} = \dfrac{y - x}{x}$

**M14.** $(2x - y)y' = 2x + 5y$

**M15.** $y' = \frac{1}{2}\tan^2(x + 2y)$   (Hint: Let $v = x + 2y$.)

**M16.** The current $I$ in an electric circuit consisting of a coil of inductance $L$, a resistance $R$, and a generator that produces a voltage $E\cos\omega t$ as shown in the figure below satisfies the differential equation

$$L\frac{dI}{dt} + RI = E\cos\omega t \quad (L, R, E = \text{const.} > 0).$$

Find the solution $I$ as a function of $t$.

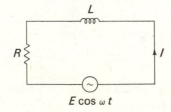

$E\cos\omega t$

**M17.** The rate of growth of an amount of money deposited in a bank that gives an interest rate compounded *continuously* is proportional to the amount of money in the bank, where the constant of proportionality is the interest rate. If \$1000 is deposited in a bank that gives a 5% annual interest rate compounded *daily*, estimate the amount of money in the bank in ten years using continuous compounding. (It can be shown that the estimate will differ from the true amount by less than ten cents.)

**M18.** Assume that the population of the United States increases at a rate proportional to the population, and that the population was 150 million in 1950 and 200 million in 1970.

a) Estimate the population for the year 2000.

b) Approximately when will the population be one billion?

**M19.** A certain substance in solution decomposes at a rate proportional to its weight. If 9 gm. of the substance are initially present and 1 gm. decomposes in the first hour, how long will it take before 5 gm. have decomposed?

**M20.** Show that the equation $y' + P(x)y = Q(x)y\ln y$ can be solved by letting $\ln y = v$.

**M21.** Find $y = f(x)$ given that

a) $\displaystyle\int_0^x f(t)\,dt = \int_0^x 2\pi f(t)\sqrt{1 + f'(t)^2}\,dt$

b) $\displaystyle\int_0^x f(t)\,dt = \int_0^x (1 + f'(t))\,dt$

**M22.** At low speeds an object falling in the atmosphere experiences a resistance proportional to the velocity, while at high speeds the resistance is proportional to the square of the velocity. In each of the above cases the velocity of the object does not exceed a certain limiting velocity. Solve each of the following differential equations and determine the limiting velocity in each case.

a) $m\dfrac{dv}{dt} = mg - kv$   $(m, g, k > 0)$

b) $m\dfrac{dv}{dt} = mg - kv^2$   $(m, g, k > 0)$

<table>
<tr><td>

8.2
</td><td>

### *Special Second Order Equations*
</td></tr>
</table>

An equation that contains only $x, y, y'$, and $y''$ is called a differential equation of *second order*. We will find the solutions to a few special types below.

    We will use the term *general solution* to mean the expression containing two arbitrary constants which represents the family of solutions to a differential equation.

*In other words, a differential equation is of second order if the highest derivative that occurs is the second.*

(Mark)

It is usually true that the general solution encompasses all possible solutions of the differential equation, but this is not always the case, and the situation can become complicated.

**A. One of Our Variables Is Missing**  Although a general second order equation contains the four factors

$$x, y, \frac{dy}{dx}, \quad \text{and} \quad \frac{d^2y}{dx^2},$$

if either $x$ or $y$ does not appear then a suitable substitution can be made that transforms the equation to one of first order.

---

**Example 1**  Find the solution of

$$xy'' = y'.$$

*Solution*  We let $y' = p$, so $y'' = \frac{dp}{dx}$ and the equation becomes

$$x\frac{dp}{dx} = p$$

or

$$\frac{dp}{p} = \frac{dx}{x}.$$

Integration yields

$$\ln p = \ln x + C = \ln Ax$$

where $C = \ln A$, so $p = Ax$. We must still solve for $y$. Since $p = \frac{dy}{dx}$, we have

$$\frac{dy}{dx} = Ax,$$

which gives, when integrated,

$$y = \frac{Ax^2}{2} + B = Dx^2 + B$$

where $D = \frac{A}{2}$.

---

*Case 1 (y missing)*  We are given an equation involving

$$x, \frac{dy}{dx}, \frac{d^2y}{dx^2}.$$

Let $\frac{dy}{dx} = p$, so $\frac{d^2y}{dx^2} = \frac{dp}{dx}$. We then obtain an equation involving

$$x, p, \frac{dp}{dx}$$

which is of first order.

*Case 2 (x missing)*  In this case we have an equation involving

$$y, \frac{dy}{dx}, \frac{d^2y}{dx^2}.$$

Let $\frac{dy}{dx} = p$, so $\frac{d^2y}{dx^2} = \frac{dp}{dx} = \frac{dp}{dy}\frac{dy}{dx} = \frac{dp}{dy}p$, and we obtain a first order equation containing

$$y, p, \frac{dp}{dy}p.$$

    In each of the cases above, the solution is obtained in two steps. We first solve the first order equation to find $p$. Then we integrate the equation

$$\frac{dy}{dx} = p \quad \text{to obtain} \quad y = \int p(x)\, dx.$$

The reader should notice that two arbitrary constants appear: one is obtained in solving for $p$ and the other occurs when we solve for $y$.

**Example 2**  Find the solution of

$$yy'' + y'^2 = y'.$$

*Solution*  We let $y' = p$ so that $y'' = \frac{dp}{dy}p$. The differential equation then becomes

$$y\frac{dp}{dy}p + p^2 = p$$

or

$$y\frac{dp}{dy} = 1 - p.$$

Hence $\frac{dp}{1-p} = \frac{dy}{y}$ and, integrating, we obtain

$$-\ln(1 - p) = \ln Ay, \quad \text{so} \quad (1 - p)^{-1} = Ay.$$

Solving for $p$ we have $p = \frac{Ay-1}{Ay}$, and therefore the equation

$$\frac{dy}{dx} = p = \frac{Ay - 1}{Ay}.$$

Hence

$$\frac{Ay}{Ay - 1}dy = dx \quad \text{or} \quad \left(1 + \frac{1}{Ay - 1}\right)dy = dx.$$

Integrating we obtain

$$y + \frac{1}{A}\ln(Ay - 1) = x + C$$

or

$$\frac{1}{A}\ln(Ay - 1) = x - y + C.$$

This gives

$$\ln(Ay - 1) = A(x - y) + AC.$$

Taking exponentials we then have

$$Ay - 1 = e^{A(x-y)}E$$

where $E = e^{AC}$, and we write this answer in the form

$$(Ay - 1)e^{Ay} = Ee^{Ax}.$$

**B. Constant Coefficients**  A second order equation of the form $ay'' + by' + cy = 0$ with $a, b, c$ $(a \neq 0)$

**Problem 1**  Solve each of the following equations:
a) $y'' + y' = e^{2x}$
b) $y'' - y = 0$

*Answer:*  a) $y = \dfrac{e^{2x}}{6} + Ae^{-x} + B$

b) $y = Ae^x + Be^{-x}$

It was necessary in this example to make a substitution and then integrate twice successively. It would not be realistic to say that this is often necessary or not often necessary, for differential equations are so varied that those who solve them use any method that works, which means just about anything. The methods here cover only a small fraction of all differential equations. Most are unsolvable by elementary means.

(Steve)

real constants is particularly easy to solve. Associated with each such equation is the quadratic equation

$$am^2 + bm + c = 0,$$

called the *auxiliary equation*. Let $m_1$ and $m_2$ denote the roots of this equation.

**THEOREM 1** *Suppose $ay'' + by' + cy = 0$ is a differential equation with constant coefficients. If $m_1$ and $m_2$ are the roots of the auxiliary equation and $m_1 \neq m_2$, then*

$$y = Ae^{m_1 x} + Be^{m_2 x}$$

*is the general solution of the differential equation.*

*Proof* Substituting the indicated solution into the differential equation, we obtain

$$a(Ae^{m_1 x} + Be^{m_2 x})'' + b(Ae^{m_1 x} + Be^{m_2 x})'$$
$$+ c(Ae^{m_1 x} + Be^{m_2 x})$$
$$= aAm_1{}^2 e^{m_1 x} + aBm_2{}^2 e^{m_2 x} + bAm_1 e^{m_1 x}$$
$$+ bBm_2 e^{m_2 x} + cAe^{m_1 x} + cBe^{m_2 x}$$
$$= Ae^{m_1 x}(am_1{}^2 + bm_1 + c) + Be^{m_2 x}(am_2{}^2 + bm_2 + c),$$

and this equals 0 because $m_1$ and $m_2$ are roots of the auxiliary equation ●

---

*Example 3* Find the solution of

$$y'' + y' - 2y = 0.$$

*Solution* The auxiliary equation for this differential equation is

$$m^2 + m - 2 = 0.$$

Its roots are $m_1 = 1$ and $m_2 = -2$. Using Theorem 1 we find that the general solution of the differential equation is

$$y = Ae^x + Be^{-2x}.$$

---

We have already seen examples of this type of equation. For example

$$y'' + y = 0 \quad (\text{here } a = 1, b = 0, c = 1)$$

has $y = A \sin x + B \cos x$ as its solution and

$$y'' - y = 0 \quad (\text{here } a = 1, b = 0, c = -1)$$

has $y = Ae^x + Be^{-x}$ as its solution.

In the proof we only show that $y = Ae^{m_1 x} + Be^{m_2 x}$ is a solution for arbitrary constants $A$ and $B$. It can also be shown that any solution of the given differential equation must have this form.

When the roots $m_1$ and $m_2$ are complex, the formula

$$y = Ae^{m_1 x} + Be^{m_2 x}$$

can be written in the form

$$y = e^{rx}(C \cos sx + D \sin sx).$$

Indeed, in this case we have

$$m_1 = r + is \quad \text{and} \quad m_2 = r - is$$

where $r$ and $s$ are real. This follows from the quadratic formula since

$$m_1 = \frac{-b + \sqrt{b^2 - 4ac}}{2a} \quad \text{and} \quad m_2 = \frac{-b - \sqrt{b^2 - 4ac}}{2a}$$

and the roots are complex conjugates whenever $b^2 - 4ac < 0$. We then have

$$Ae^{m_1 x} + Be^{m_2 x}$$
$$= Ae^{(r+is)x} + Be^{(r-is)x} = e^{rx}(Ae^{isx} + Be^{-isx})$$
$$= e^{rx}(A(\cos sx + i \sin sx) + B(\cos sx - i \sin sx))$$
$$= e^{rx}((A + B) \cos sx + (iA - iB) \sin sx)$$
$$= e^{rx}(C \cos sx + D \sin sx)$$

with $C = A + B$ and $D = iA - iB$.

**Example 4** Find the solution of

$$y'' - 2y' + 2y = 0.$$

*Solution* Its auxiliary equation is

$$m^2 - 2m + 2 = 0.$$

The roots of the auxiliary equation are $m_1 = 1 + i$ and $m_2 = 1 - i$. Hence the general solution is

$$y = Ae^{(1+i)x} + Be^{(1-i)x} = e^x(Ae^{ix} + Be^{-ix}).$$

**THEOREM 2** *Suppose we are given the differential equation $ay'' + by' + cy = 0$ with constant coefficients. If the roots of the auxiliary equation are equal, $m_1 = m_2$, then*

$$y = Ae^{m_1 x} + Bxe^{m_1 x}$$

*is the general solution to the differential equation.*

*Proof* We simply verify by substituting the indicated expression into the differential equation. Since

$$y' = (Ae^{m_1 x} + Bxe^{m_1 x})'$$
$$= Am_1 e^{m_1 x} + Bxm_1 e^{m_1 x} + Be^{m_1 x}$$

and

$$y'' = (Ae^{m_1 x} + Bxe^{m_1 x})''$$
$$= Am_1{}^2 e^{m_1 x} + Bxm_1{}^2 e^{m_1 x} + Bm_1 e^{m_1 x} + Bm_1 e^{m_1 x},$$

we have

$$a(Ae^{m_1 x} + Bxe^{m_1 x})'' + b(Ae^{m_1 x} + Bxe^{m_1 x})'$$
$$+ c(Ae^{m_1 x} + Bxe^{m_1 x})$$
$$= a(Am_1{}^2 e^{m_1 x} + Bxm_1{}^2 e^{m_1 x} + Bm_1 e^{m_1 x} + Bm_1 e^{m_1 x})$$
$$+ b(Am_1 e^{m_1 x} + Bxm_1 e^{m_1 x} + Be^{m_1 x})$$
$$+ c(Ae^{m_1 x} + Bxe^{m_1 x})$$
$$= Ae^{m_1 x}(am_1{}^2 + bm_1 + c) + Bxe^{m_1 x}(am_1{}^2 + bm_1 + c)$$
$$+ Be^{m_1 x}(2am_1 + b).$$

Since $am_1{}^2 + bm_1 + c = 0$, the first two terms of this expression are zero, and because $m_1 = \frac{-b}{2a}$ or $2am_1 + b = 0$, the third term is also zero. Hence the indicated $y$ is the solution  ●

**Problem 2** Solve the equation

$$y'' - 5y' + 6y = 0.$$

*Answer:* $y = Ae^{2x} + Be^{3x}$

Since the solutions of the auxiliary equation $am^2 + bm + c = 0$ are

$$\frac{-b \pm \sqrt{b^2 - 4ac}}{2a},$$

the case of equal roots occurs when $b^2 - 4ac = 0$ and so

$$m_1 = m_2 = \frac{-b}{2a}.$$

**Problem 3** Write the solution of $y'' - 2y' + 2y = 0$ in terms of real functions. (See Example 4.)

*Answer:* $y = e^x(C \cos x + D \sin x)$

**Example 5** Solve

a) $y'' - 4y' + 4y = 0$, and more generally
b) $y'' - 2ay' + a^2 y = 0$.

*Solution* a) The auxiliary equation for $y'' - 4y' + 4y = 0$ is $m^2 - 4m + 4 = (m - 2)^2 = 0$. Hence $m_1 = m_2 = 2$, so

$$y = Ae^{2x} + Bxe^{2x}$$

is the general solution of the differential equation.
b) In this case the auxiliary equation is

$$m^2 - 2am + a^2 = 0$$

or $(m - a)^2 = 0$. Hence $m_1 = m_2 = a$ and the general solution of the differential equation is

$$y = e^{ax}(A + Bx).$$

*Example 6* (Cauchy's equation) The equation

$$ax^2y'' + bxy' + cy = 0$$

can be transformed to an equation with constant coefficients by using the substitution $u = \ln x$.

*Solution* Let $u = \ln x$. Then

$$\frac{dy}{dx} = \frac{dy}{du}\frac{du}{dx} = \frac{dy}{du}\frac{1}{x}$$

and

$$\frac{d^2y}{dx^2} = \frac{d}{dx}\left(\frac{dy}{du}\frac{1}{x}\right) = \frac{-1}{x^2}\frac{dy}{du} + \frac{1}{x}\frac{d}{du}\left(\frac{dy}{du}\right)\frac{du}{dx}$$

$$= \frac{-1}{x^2}\frac{dy}{du} + \frac{1}{x}\frac{d}{du}\left(\frac{dy}{du}\right)\frac{1}{x} = \frac{-1}{x^2}\frac{dy}{du} + \frac{1}{x^2}\frac{d^2y}{du^2}.$$

Substituting these expressions for $y'$ and $y''$ into the original equation, we obtain

$$0 = ax^2\left(\frac{-1}{x^2}\frac{dy}{du} + \frac{1}{x^2}\frac{d^2y}{du^2}\right) + bx\left(\frac{dy}{du}\frac{1}{x}\right) + cy$$

$$= a\frac{d^2y}{du^2} + (b-a)\frac{dy}{du} + cy,$$

which is an equation with constant coefficients.

**C. Applications** Imagine a spring hanging with one end fixed and a mass $M$ attached to its free end (Fig. 1). Assume that the spring satisfies Hooke's Law; that

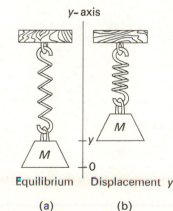

**Figure 1**

y- axis

Equilibrium   Displacement $y$

(a)    (b)

→ The slightly more general equation

$$ax^2y'' + bxy' + cy = r(x)$$

is often referred to as Euler's equation. The subject of differential equations is full of equations named after the mathematicians who first studied them and recognized their importance.

---

**Problem 4** Solve the equation $4y'' - 12y' + 9y = 0$.

*Answer:* $y = e^{3x/2}(A + Bx)$

---

*Example 7* We will solve $2x^2y'' + xy' - 3y = 0$.

*Solution* Letting $u = \ln x$ and using the formula from Example 6, we have

$$2\frac{d^2y}{du^2} - \frac{dy}{du} - 3y = 0.$$

The auxiliary equation

$$2m^2 - m - 3 = 0$$

has roots $m_1 = \frac{3}{2}$ and $m_2 = -1$. Hence

$$y = Ae^{3u/2} + Be^{-u}$$

and, replacing $u$ by $\ln x$, we obtain

$$y = Ae^{3(\ln x)/2} + Be^{-\ln x}$$
$$= Ae^{\ln (x^{3/2})} + Be^{\ln (x^{-1})} = Ax^{3/2} + Bx^{-1}.$$

is, the force exerted by the spring is proportional to the distance it is displaced. Orient the $y$-axis so that its positive direction is up and the equilibrium position of the end of the spring with the mass attached is at the origin. If $F$ represents force, we have

$$F = -ky \quad (k > 0)$$

because the force is downward (negative) when the displacement is upward (positive) and vice versa. We now take Newton's Second Law, $F = Ma$, and replace $F$ by $-ky$ and $a$ (the acceleration) by $d^2y/dt^2$ to obtain

$$M\frac{d^2y}{dt^2} = -ky.$$

This is a second order, linear differential equation whose solution is

$$y = A \sin\left(t\sqrt{\frac{k}{M}}\right) + B \cos\left(t\sqrt{\frac{k}{M}}\right).$$

We rewrite this as

$$y = \sqrt{A^2 + B^2}\left(\frac{A}{\sqrt{A^2 + B^2}}\sin t\sqrt{\frac{k}{M}}\right.$$

$$\left. + \frac{B}{\sqrt{A^2 + B^2}}\cos t\sqrt{\frac{k}{M}}\right)$$

$$= \sqrt{A^2 + B^2}\left(\cos\theta\sin t\sqrt{\frac{k}{M}} + \sin\theta\cos t\sqrt{\frac{k}{M}}\right)$$

where

$$\sin\theta = \frac{B}{\sqrt{A^2 + B^2}} \quad \text{and} \quad \cos\theta = \frac{A}{\sqrt{A^2 + B^2}}.$$

Using the identity $\sin(\alpha + \beta) = \sin\alpha\cos\beta + \cos\alpha\sin\beta$, we have

$$y = \sqrt{A^2 + B^2}\sin\left(t\sqrt{\tfrac{k}{M}} + \theta\right)$$

or, more commonly,

$$y = T\sin(\omega t + \theta)$$

where

$$T = \sqrt{A^2 + B^2}$$

and

$$\omega = \sqrt{\tfrac{k}{M}}.$$

The number $T$ is called the *amplitude* of the motion of the spring (because $-T \le y \le T$) and the number $\frac{2\pi}{\omega}$ is termed the *period* since the motion repeats itself

→ To solve the equation, write it as $My'' + ky = 0$. The auxiliary equation is $Mm^2 + k = 0$, so $m = \pm i\sqrt{\frac{k}{M}}$ and

$$y = Ce^{it\sqrt{k/M}} + De^{-it\sqrt{k/M}}.$$

This may be written in the form

$$y = A \sin\left(t\sqrt{\frac{k}{M}}\right) + B \cos\left(t\sqrt{\frac{k}{M}}\right)$$

by Theorem 1 (8.2B).

→ The angle $\theta$ is shown in Figure 2. As is indicated there, the signs of $A$ and $B$ determine which quadrant $\theta$ lies in. (When $A < 0$ and $B < 0$, then $\theta$ lies in the third quadrant, and when $A > 0$ and $B < 0$, then $\theta$ lies in the fourth quadrant.)

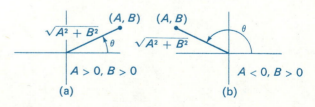

**Figure 2**

every $\frac{2\pi}{\omega}$ seconds. The quantity $\theta$ is called the *phase* and indicates whether the motion begins at equilibrium or at some other position. The motion is called *simple harmonic motion*.

When initial conditions about the velocity, position, phase, or amplitude of the motion are given, $T$ and $\theta$ can be determined.

---

**Example 8**　Assume that a mass of 4 units is attached to the free end of a spring with Hooke's constant 1. If at $t = 0$, $y = 0$ and the spring is given an initial velocity of 3 ft./sec., describe the motion.

*Solution*　We have that $y(0) = 0$ and $\left.\dfrac{dy}{dt}\right|_{t=0} = 3$. Putting these values into the equations

$$y = T \sin (\omega t + \theta)$$

and

$$\frac{dy}{dt} = \omega T \cos (\omega t + \theta),$$

we have

$$0 = T \sin \theta \quad \text{and} \quad 3 = \omega T \cos \theta.$$

From the first equation we obtain $\theta = 0$, so the second equation becomes

$$3 = \omega T \cos 0 = \omega T, \quad \text{or} \quad T = \frac{3}{\omega}.$$

Since $\omega = \sqrt{\dfrac{k}{M}} = \dfrac{1}{2}$, $T = \dfrac{3}{\omega} = \dfrac{3}{1/2} = 6$ so the

motion is described by the formula

$$y(t) = 6 \sin \frac{t}{2}.$$

---

The original equation for the spring

$$M \frac{d^2y}{dt^2} = -ky$$

was derived with the implicit assumption that there was no resistance of any kind against the spring. In many applications there is often a resistance that is proportional to the velocity (air resistance, for

→ The spring's periodic motion is clear from the graph (Fig. 3).

*Do not be misled by this graph. It is not a picture of the path that the spring travels (the path is only a segment) but it shows the displacement of the spring with respect to time.*

*(Pat)*

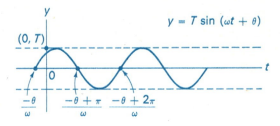

**Figure 3**

*The reader will not be shocked to learn that the formula describing simple harmonic motion says that the spring moves up and down periodically. We certainly did not need mathematics to deduce this. Notice, however, that the formula shows that the period*

$$\frac{2\pi}{\omega} = 2\pi \sqrt{\frac{M}{k}}$$

*depends only on the spring constant k and the mass M. This may be a surprise to the intuition because it means that the number of oscillations per second is independent of both the initial displacement of the spring and its initial velocity.*

*(Vahan)*

*The surprise will be dispelled, however, if the reader recalls that the strings of a guitar vibrate with simple harmonic motion. No matter how hard one plucks the lowest string or how far he displaces it when he plucks it, the instrument will still produce the note E.*

*(Steve)*

---

**Problem 5**　Assume that the spring in Example 8 now satisfies $y(0) = 8$ and $y'(0) = 0$. Determine a formula for the motion of the spring.

*Answer:*　$y = 8 \sin \frac{1}{2}(t + \pi)$

example). In such cases, the force $F$ is given by

$$F = -ky - c\frac{dy}{dt} \quad (k > 0, c > 0)$$

where $c$ is the constant relating the force of resistance to the velocity. (The resistance opposes the motion. Hence, when $\frac{dy}{dt} > 0$ the resistance must be $-c\frac{dy}{dt} < 0$ where $c > 0$.) Using Newton's Law again, we replace $F$ by $M\,d^2y/dt^2$ to obtain the differential equation

$$M\frac{d^2y}{dt^2} + c\frac{dy}{dt} + ky = 0$$

whose auxiliary equation is $Mm^2 + cm + k = 0$.

We will now look at two cases.

*Case 1*  (Small resistance, or $c^2 < 4Mk$)

In this case the roots of the auxiliary equation are

$$\frac{-c \pm \sqrt{c^2 - 4Mk}}{2M} = -\alpha \pm \omega i$$

where $\alpha = \dfrac{c}{2M}$ and $\omega = \dfrac{\sqrt{4Mk - c^2}}{2M}$. The solution may be written in the form

$$y = e^{-\alpha t}(A \cos \omega t + B \sin \omega t)$$

or as

$$y = e^{-\alpha t} T \sin(\omega t + \theta)$$

where, as before,

$$T = \sqrt{A^2 + B^2} \quad \text{and} \quad \sin\theta = \frac{B}{\sqrt{A^2 + B^2}}.$$

It is similar to the motion without resistance except that it is "damped out"; that is, the amplitude decreases with time because $e^{-\alpha t} \to 0$ as $t \to \infty$. The situation is illustrated in Figure 4.

Notice that for $y > 0$ and $\frac{dy}{dt} > 0$, both the forces $-ky$ (due to the spring) and $-c\frac{dy}{dt}$ (due to the resistance) act to restore the rest position and hence must have the same sign.

There is a third case between these two when $c^2 = 4Mk$. The analysis of this case is quite similar to that of Case 2, and is left as an exercise for the reader.

*Case 2*  (Large resistance, or $c^2 > 4Mk$)

In this case the roots of the auxiliary equation are

$$m_1 = \frac{-c + \sqrt{c^2 - 4Mk}}{2M}$$

and

$$m_2 = \frac{-c - \sqrt{c^2 - 4Mk}}{2M}.$$

Since $\sqrt{c^2 - 4Mk} < c$, each of these roots is negative and the solution has the form

$$y = Ae^{m_1 t} + Be^{m_2 t}.$$

Notice that $y$ can assume the value 0 at most once. Setting $y = 0$ we see that this can occur only when

$$t = \frac{1}{m_2 - m_1} \ln\left(-\frac{A}{B}\right)$$

which is possible if $AB < 0$. In this case the resistance is so great that the spring can oscillate at most once. The situation is illustrated in Figure 5.

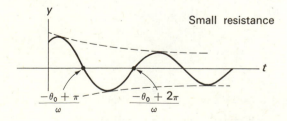

Small resistance

Figure 4

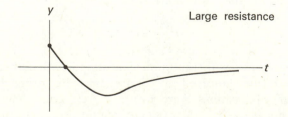

Large resistance

Figure 5

## D. Exercises

**B1.** Use the substitutions $\dfrac{dy}{dx} = p$ and $\dfrac{d^2y}{dx^2} = \dfrac{dp}{dx}$ in solving the following differential equations.

a) $xy'' - 2y' + 6x = 0$

b) $xy'' - y' - 3x^2 = 0$

**B2.** Solve the following differential equations using the substitutions given in Exercise B1.

a) $y'' + 3(y')^2 = 0$     b) $y'' + 3x(y')^2 = 0$

**B3.** Use the substitutions $\dfrac{dy}{dx} = p$ and $\dfrac{d^2y}{dx^2} = \dfrac{dp}{dy}\,p$ in solving the following differential equations.

a) $yy'' = -3(y')^2$     b) $yy'' = 3(y')^2$

**B4.** a) Find the general solution to the differential equation

$$y'' - 3y' + 2y = 0.$$

b) Find the solution $y = y(x)$ satisfying $y(0) = 4$ and $y'(0) = 6$.

**B5.** a) Write the solution of $y'' - 6y' + 13y = 0$ in terms of exponential functions.

b) Put the above solution in terms of real functions.

**D6.** Use the substitutions $\dfrac{dy}{dx} = p$ and $\dfrac{d^2y}{dx^2} = \dfrac{dp}{dx}$ in solving the following differential equations.

a) $xy'' = y'$     d) $(1 + x^2)y'' = 2xy'$

b) $xy'' = 2y'$     e) $y'' + x^3y'' = 3x^2y'$

c) $xy'' = 3y'$     f) $y'' \sin x = y' \cos x$

a) $y = Ax^2 + B$     b) $y = Ax^3 + B$

c) $y = Ax^4 + B$     d) $y = 3Ax + Ax^3 + B$

e) $y = 4Ax + Ax^4 + B$     f) $y = A \cos x + B$

**D7.** Use the substitutions $\dfrac{dy}{dx} = p$ and $\dfrac{d^2y}{dx^2} = \dfrac{dp}{dy}\,p$ in solving the following.

a) $y'' = 0$     d) $yy'' = -(y')^2$

b) $y'' = 2$     e) $yy'' = -2(y')^2$

c) $yy'' = (y')^2$     f) $(1 + y^2)y'' = 2y(y')^2$

a) $y = Ax + B$     b) $y = x^2 + Ax + B$

c) $y = Ae^{Bx}$     d) $y^2 = Ax + B$

e) $y^3 = Ax + B$     f) $\mathrm{Tan}^{-1} y = Ax + B$

**D8.** Write the solutions of the following differential equations in terms of exponential functions.

a) $y'' - 9y = 0$     d) $y'' + 4y = 0$

b) $y'' - 3y' = 0$     e) $y'' - 8y' + 16y = 0$

c) $y'' - 5y' + 6y = 0$     f) $y'' - 2y' + 5y = 0$

a) $y = Ae^{3x} + Be^{-3x}$     b) $y = Ae^{3x} + B$

c) $y = Ae^{3x} + Be^{2x}$     d) $y = Ae^{2ix} + Be^{-2ix}$

e) $y = Ae^{4x} + Bxe^{4x}$     f) $y = e^x(Ae^{2ix} + Be^{-2ix})$

Find the solutions of the differential equations given in Exercises M9–M17.

**M9.** a) $y'' - (y')^2 = 4$     b) $y'' = (1 + (y')^2)^3$

**M10.** $yy'' - (y')^2 = y'$

**M11.** $xy'' = 1 + y'$

**M12.** a) $y'' - 4y = 0$     b) $y'' + 4y = 0$

**M13.** a) $y'' + 4y = 0$     b) $y'' - 4y = 0$

**M14.** a) $y'' + 4y' - 4y = 0$     b) $y'' + 4y' + 4y = 0$

**M15.** $yy'' + (y')^2 + 1 = 0$

**M16.** $2yy'' + (y')^2 + 1 = 0$

**M17.** $y'' = (1 + (y')^2)^{3/2}$

**M18.** Use the substitution $u = \ln x$ in solving the following differential equations.

a) $x^2y'' - xy' + y = 0$     b) $x^2y'' + xy' + y = 0$

**M19.** a) Show that the function

$$g(x) = \int_0^x Af(t)\,dt$$

satisfies the differential equation $f(x)y'' = f'(x)y'$. (See Exercise D6.)

b) Show the function

$$h(y) = \int_0^y \frac{1}{Af(t)}\,dt$$

satisfies the differential equation $f(y)y'' = f'(y)(y')^2$. (See Exercise D7.)

**M20.** A mass of 16 units is attached to the end of a 4 ft. spring having Hooke's constant 2. Describe the motion of the weight if it is raised 1 ft. and let go.

**M21.** A mass of 8 units is attached to the end of a spring having Hooke's constant 12. Describe the motion of the weight if it is raised 5 inches and then thrust upward with velocity 5 ft./sec.

**M22.** A certain spring operates without friction, has Hooke's constant 10, and is known to oscillate with period $\frac{\pi}{3}$. What mass must be suspended on its free end?

**M23.** When a mass $x$ at the end of a vertical spring is set into vibration, the period is 1.5 sec. When 8 units are added, the period becomes 2.5 sec. Find $x$.

**M24.** The angle $\theta$ that a pendulum of length $L$ in feet makes with the vertical as it swings back and forth satisfies the differential equation

$$\frac{d^2\theta}{dt^2} = -\frac{g}{L}\sin\theta$$

where $g = 32$ ft/sec$^2$ is the acceleration due to gravity. Assuming that $\theta$ is small enough so that $\sin\theta$ may be replaced by $\theta$ (recall that $\frac{\sin\theta}{\theta} \to 1$ as $\theta \to 0$) find the solution $\theta$ for a pendulum of length 64 ft.

## 8.3  Linear Second Order Equations

The differential equation

$$a(x)y'' + b(x)y' + c(x)y = r(x)$$

with $a$, $b$, $c$, $r$ functions of $x$ ($a \neq 0$) is clearly second order. Equations of this form are called *linear* because $y$ and its derivatives do not appear in products or in other functions. If $r(x)$ is zero for all $x$, that is, if $r$ is the zero function, then the equation is said to be *homogeneous*.

**A. Two Theorems**  The following two theorems are related and are vital to the solution of many common differential equations.

*THEOREM 1*  *Let $u = u(x)$ and $v = v(x)$ be solutions of the linear, homogeneous equation*

$$a(x)y'' + b(x)y' + c(x)y = 0.$$

*Then, for all constants $A$ and $B$, $Au + Bv$ is also a solution of the differential equation.*

*Example 1*  (Illustration of Theorem 1) It is easily checked that the functions

$$u(x) = x \quad \text{and} \quad v(x) = x^3$$

are solutions of the differential equation

$$x^2y'' - 3xy' + 3y = 0.$$

The theorem says that the functions $2x + 7x^3$, $x + x^3$, $-653x - 8357x^3$, and in general $Ax + Bx^3$ (for arbitrary constants $A$ and $B$) are solutions of the above differential equation.

When $r$ is not the zero function the equation is called *nonhomogeneous* (or *inhomogeneous*). For example,

$$x^5y'' - 3x^4y' + (x^2 + 1)y = 0$$

is homogeneous while

$$y'' = x + 1$$

is nonhomogeneous.

The proof is fairly easy, for it relies only on basic calculus.

*Proof*  Since $u$ and $v$ are solutions we have

$$a(x)u'' + b(x)u' + c(x)u = 0$$

and

$$a(x)v'' + b(x)v' + c(x)v = 0.$$

Multiplying the first equation by $A$, the second equation by $B$, and adding gives

$$a(x)(Au'' + Bv'') + b(x)(Au' + Bv') + c(x)(Au + Bv) = 0.$$

Since $A$ and $B$ are constants, we have

$$Au'' + Bv'' = (Au + Bv)''$$

and

$$Au' + Bv' = (Au + Bv)'.$$

The equation then becomes

$$a(x)(Au + Bv)'' + b(x)(Au + Bv)' + c(x)(Au + Bv) = 0,$$

and this means that $Au + Bv$ is indeed a solution ●

**THEOREM 2** *Suppose that*

$$a(x)y'' + b(x)y' + c(x)y = r(x)$$

*is a second order differential equation. Let $p = p(x)$ be a solution of this equation (called a particular solution) and let $u = u(x)$ and $v = v(x)$ be solutions to the corresponding homogeneous equation*

$$a(x)y'' + b(x)y' + c(x)y = 0.$$

*Then any function of the form*

$$y = p(x) + Au(x) + Bv(x)$$

*where A and B are real constants is a solution of the original differential equation.*

→   Proof   Since $u$ and $v$ are solutions of the homogeneous equation, we have by Theorem 1 that $Au + Bv$ is also a solution. Hence

$$a(x)(Au + Bv)'' + b(x)(Au + Bv)' + c(x)(Au + Bv) = 0.$$

Also, $p(x)$ is a solution of the nonhomogeneous equation, so

$$a(x)p'' + b(x)p' + c(x)p = r(x).$$

Adding these two equalities and collecting the terms yields

$$a(x)(p + Au + Bv)'' + b(x)(p + Au + Bv)' + c(x)(p + Au + Bv) = r(x).$$

We conclude, therefore, that

$$y = p(x) + Au(x) + Bv(x)$$

is a solution of the original differential equation ●

---

**Example 2** (Illustration of Theorem 2) It is easily verified that the function

$$p(x) = x^2$$

is a solution of the differential equation

$$x^2y'' - 3xy' + 3y = -x^2.$$

We know from Example 1 that $u(x) = x$ and $v(x) = x^3$ are solutions of the homogeneous equation. Theorem 2 states that every function of the form

$$y = x^2 + Ax + Bx^3$$

where $A$ and $B$ are arbitrary constants is a solution.

---

**Problem 1** Observing that $y = \cos x$ is a solution to the differential equation $y'' - 4y = -5 \cos x$, find the most general solution to this differential equation.

*Answer:* $y = \cos x + Ae^{2x} + Be^{-2x}$.

---

→   You cannot *put a constant in front of the $x^2$ term. For example, $4x^2 + Ax + Bx^3$ is never a solution of the original equation. Be careful because this is an easy mistake to make.*

(Jean)

---

**B. Finding a Second Solution**   There are times when only one solution of the equation

$$a(x)y'' + b(x)y' + c(x)y = 0$$

is known. The following theorem then provides a method of finding a second solution.

**THEOREM 3** *Suppose u is a solution of*

$$a(x)y'' + b(x)y' + c(x)y = 0.$$

*Then*

$$v = u \int \frac{e^{-\int b/a}}{u^2}$$

→   In what follows, we will often use the shorthand notation of replacing $a(x)$ by $a$ and $b(x)$ by $b$.

*is also a solution. (In evaluating $\int_a^b$, the constant of integration can be taken to be any convenient value, and in particular zero since it only changes $v$ by a constant factor.)*

---

**Example 3** We know that

$$u = \cos x$$

is a solution of

$$y'' + y = 0$$

and we will now see what Theorem 3 specifies as a second solution. Since $b = 0$, $\int \frac{b}{a} = C$ and we may take the constant $C$ equal to zero. Therefore $e^{-\int b/a} = 1$ and we have

$$v = u \int \frac{e^{-\int b/a}}{u^2} = \cos x \int \frac{dx}{\cos^2 x} = \cos x \int \sec^2 x \, dx$$

$$= \cos x \tan x = \sin x.$$

---

**Example 4** The constant function $u = 1$ is a solution of $x^2 y'' + 2xy' = 0$. We will find a second solution.

*Solution* Since $\dfrac{b}{a} = \dfrac{2x}{x^2} = \dfrac{2}{x}$,

$$e^{-\int b/a} = e^{-\int 2/x} = e^{-2 \ln x} = e^{\ln x^{-2}} = \frac{1}{x^2}.$$

Hence

$$v = 1 \int \frac{e^{-\int b/a}}{1^2} = \int \frac{dx}{x^2} = \frac{-1}{x}$$

is a second solution.

---

**C. Finding a Particular Solution** We will now show how to find a particular solution to an equation of the form

$$y'' + P(x)y' + Q(x)y = R(x).$$

The first method, called *variation of parameters,* is based on knowing two solutions $u$ and $v$ of the homogeneous equation

$$y'' + P(x)y' + Q(x)y = 0.$$

---

**Proof** To simplify notation, let $R = e^{-\int b/a}$ $\left(\text{so } R' = \dfrac{-bR}{a}\right)$.

Then

$$v = u \int \frac{R}{u^2},$$

$$v' = \frac{R}{u} + u' \int \frac{R}{u^2}, \quad \text{and}$$

$$v'' = \frac{uR' - u'R}{u^2} + \frac{u'R}{u^2} + u'' \int \frac{R}{u^2}$$

$$= \frac{R'}{u} + u'' \int \frac{R}{u^2} = \frac{-bR}{au} + u'' \int \frac{R}{u^2}.$$

Hence

$$av'' + bv' + cv$$

$$= a\left(\frac{-bR}{au} + u'' \int \frac{R}{u^2}\right) + b\left(\frac{R}{u} + u' \int \frac{R}{u^2}\right) + cu \int \frac{R}{u^2}$$

$$= \frac{-bR}{u} + au'' \int \frac{R}{u^2} + \frac{bR}{u} + bu' \int \frac{R}{u^2} + cu \int \frac{R}{u^2}$$

$$= (au'' + bu' + cu) \int \frac{R}{u^2} = 0 \quad \bullet$$

---

**Problem 2** Given that $u = x$ is a solution of

$$x^3 y'' + xy' - y = 0.$$

Find a second solution $v$.

*Answer:* $v = -xe^{1/x}$

---

We have not yet clarified what we mean by "two solutions" of a differential equation. For example, $\sin x$ and $2 \sin x$ are solutions of $y'' + y = 0$, but they are not essentially different. When we speak of two solutions $u$ and $v$ we will assume that neither is a constant multiple of the other. They are called *linearly independent* solutions.

**THEOREM 4** *Suppose u and v are linearly independent solutions of the homogeneous equation*

$$y'' + P(x)y' + Q(x)y = 0.$$

*Then*

$$y = M(x)u(x) + N(x)v(x)$$

*is a solution of the nonhomogeneous equation*

$$y'' + P(x)y' + Q(x)y = R(x)$$

*where*

$$M = \int \frac{-Rv}{uv' - u'v} \quad and \quad N = \int \frac{Ru}{uv' - u'v}.$$

*Proof* Letting

$$y = Mu + Nv = u \int \frac{-Rv}{uv' - u'v} + v \int \frac{Ru}{uv' - u'v}$$

yields

$$y' = u' \int \frac{-Rv}{uv' - u'v} + v' \int \frac{Ru}{uv' - u'v}$$

and

$$y'' = R + u'' \int \frac{-Rv}{uv' - u'v} + v'' \int \frac{Ru}{uv' - u'v}.$$

Substituting these expressions in the differential equation yields

$$y'' + P(x)y' + Q(x)y$$

$$= \left( R + u'' \int \frac{-Rv}{uv' - u'v} + v'' \int \frac{Ru}{uv' - u'v} \right)$$

$$+ P(x) \left( u' \int \frac{-Rv}{uv' - u'v} + v' \int \frac{Ru}{uv' - u'v} \right)$$

$$+ Q(x) \left( u \int \frac{-Rv}{uv' - u'v} + v \int \frac{Ru}{uv' - u'v} \right)$$

$$= (u'' + P(x)u' + Q(x)u) \int \frac{-Rv}{uv' - u'v}$$

$$+ (v'' + P(x)v' + Q(x)v) \int \frac{Ru}{uv' - u'v} + R$$

$$= R.$$

Hence $Mu + Nv$ is a solution ●

---

**Example 6** The equation

$$x^2 y'' - (x^2 + 2x)y' + (x + 2)y = 0$$

has solutions

---

→ The expression $uv' - u'v$ is known as the *Wronskian*. "It is named after Hoëné Wronski (1778–1853), who is said to have begun life as an army officer, redeemed himself by becoming a mathematician, fallen from grace by becoming a philosopher, and completed his downfall by going insane."

The above quote of Ralph Palmer Agnew is taken from his delightful book, *Differential Equations* (McGraw-Hill, 1960), which we heartily recommend to the ambitious student.

---

→ Here are the details concerning the computations of $y'$ and $y''$.

$$y' = u \left( \frac{-Rv}{uv' - u'v} \right) + u' \int \frac{-Rv}{uv' - u'v} + v \left( \frac{Ru}{uv' - u'v} \right)$$

$$+ v' \int \frac{Ru}{uv' - u'v} = u' \int \frac{-Rv}{uv' - u'v} + v' \int \frac{Ru}{uv' - u'v}$$

$$y'' = u' \left( \frac{-Rv}{uv' - u'v} \right) + u'' \int \frac{-Rv}{uv' - u'v} + v' \left( \frac{Ru}{uv' - u'v} \right)$$

$$+ v'' \int \frac{Ru}{uv' - u'v}$$

$$= \frac{Rv'u - Ru'v}{uv' - u'v} + u'' \int \frac{-Rv}{uv' - u'v} + v'' \int \frac{Ru}{uv' - u'v}$$

---

**Example 5** We will solve the equation $y'' + y = \sec x$.

*Solution* We know that $u = \sin x$ and $v = \cos x$ are solutions to the homogeneous equation $y'' + y = 0$. Applying Theorem 4, $u' = \cos x$, $v' = -\sin x$, and $R = \sec x$, so

$$M = \int \frac{-\sec x \cos x \, dx}{-\sin^2 x - \cos^2 x} = \int dx = x$$

and

$$N = \int \frac{\sec x \sin x \, dx}{-\sin^2 x - \cos^2 x}$$

$$= -\int \tan x \, dx = \ln \cos x.$$

$$u = x \quad \text{and} \quad v = xe^x.$$

We will determine a particular solution and then the general solution of

$$x^2 y'' - (x^2 + 2x)y' + (x + 2)y = x^3.$$

*Solution*  Dividing the nonhomogeneous equation through by $x^2$, we have that $R = x$. Also

$$uv' - u'v = x\frac{d}{dx}(xe^x) - \left(\frac{d}{dx}x\right)xe^x$$

$$= x(xe^x + e^x) - xe^x = x^2 e^x.$$

Hence, by Theorem 4

$$M = \int \frac{-Rv}{uv' - u'v} = \int \frac{-xxe^x\, dx}{x^2 e^x} = -x$$

and

$$N = \int \frac{Ru}{uv' - u'v} = \int \frac{xx\, dx}{x^2 e^x} = \int e^{-x}\, dx = -e^{-x}.$$

(We have taken the constants of integration to be zero.) It follows that

$$y = Mu + Nv = -xx + (-e^{-x})xe^x = -x^2 - x$$

is a particular solution, so by Theorem 2 (8.3A) the general solution is

$$Ax + Bxe^x - x^2 - x = Cx + Bxe^x - x^2.$$

The procedure described in the example below is known as the method of *undetermined coefficients* and can be effectively used if the coefficients involved are not too complicated.

**Example 7**  We will find particular solutions for
a) $y'' + 2y' + 3y = 9x - 6$   and
b) $y'' + 2y' + 3y = \sin x$.

*Solution*  a) We simply try a solution of the form

$$y = ax + b.$$

Then $y' = a$ and $y'' = 0$. Substituting these expressions into the original equation gives

$$0 + 2a + 3(ax + b) = 9x - 6.$$

We therefore must have

$$3a = 9 \quad \text{and} \quad 2a + 3b = -6.$$

Therefore $y = Mu + Nv = x \sin x + (\ln \cos x) \cos x$ is a particular solution of the differential equation and

$$x \sin x + (\ln \cos x) \cos x + A \sin x + B \cos x$$

is the general solution.

---

**Problem 3**  Determine a particular solution for each of the following equations.
a) $y'' + 2y' - 3y = 12$
b) $y'' - y = x$

*Answer*  a) $-4$  b) $-x$

---

→  The reason we have taken the constants of integration to be zero is mainly one of convenience. If we had left them in, the result would be something like

$$Ax + Bxe^x - x^2 + (c_1 - 1)x + c_2 xe^x$$

and, after combining terms, we would have had to replace the multiple constants by a single constant anyway.

(Pat)

→  You might redo Problem 3 using this method. The advantage in this method lies in the fact that you do not need to know the solution of the homogeneous equation to find a particular solution of the nonhomogeneous equation.

(Greg)

→  It is not always obvious what the form of the particular solution should be. Compare (a) and (b) in Problem 4.

---

**Problem 4**  a) Find a particular solution to $y'' + y = \sin 2x$ by the method of undetermined coefficients.
b) Now find a particular solution for $y'' + y = \sin x$ by variation of parameters.

Solving these equations, we have $a = 3$ and $b = -4$.
Hence

$$y = 3x - 4$$

is a particular solution.

b) In this case we try

$$y = a \sin x + b \cos x$$

in the equation to obtain

$$-a \sin x - b \cos x + 2(a \cos x - b \sin x)$$
$$+3(a \sin x + b \cos x) = \sin x.$$

We therefore must have

$$-a - 2b + 3a = 1 \quad \text{and} \quad -b + 2a + 3b = 0.$$

Solving, we obtain $a = \frac{1}{4}$ and $b = -\frac{1}{4}$, so

$$y = \tfrac{1}{4}(\sin x - \cos x)$$

is a particular solution.

*Answer:* a) $\dfrac{-1}{3} \sin 2x$

b) $\dfrac{\sin^3 x}{2} - \dfrac{x \cos x}{2} + \dfrac{\sin 2x \cos x}{4}$

---

### D. Exercises

**B1.** Find the general solution of the differential equation

$$x^2 y'' - (x^2 + 2x)y' + (x + 2)y = 0$$

given that $u = x$ is one solution.

**B2.** Find a particular solution of

$$y'' + 3y' + 2y = 3e^{2x}$$

and then determine the general solution.

**B3.** Use the method of variation of parameters to find the general solution of the differential equation

$$y'' + y = \tan x.$$

**B4.** Given that $u = x + 1$ and $v = e^x$ are solutions of

$$xy'' - (x + 1)y' + y = 0,$$

find the general solution of

$$xy'' - (x + 1)y' + y = x^2.$$

**B5.** Use the method of undetermined coefficients to find the general solutions of the following equations.
a) $y'' + 2y' = 3x^2$   b) $y'' + 2y' = \cos x$

**D6.** Find particular solutions of the form $y = Ax + B$ for the following differential equations.
a) $3y'' + 2y' + y = 2x + 3$

b) $7y'' + 2y' + y = 2x + 3$
c) $3y'' - 5y' + 2y = 4x - 12$
d) $y'' + 3xy' + 2y = 10x - 2$
e) $2y'' - 3x^3 y' + 3x^2 y = -3x^2$
f) $x^2 y'' + y' \sin x + 2y \sin x = 4x \sin x$

In each case the answer is $y = 2x - 1$.

**D7.** Find particular solutions of the form $y = A \sin x + B \cos x$ for the following differential equations.
a) $y'' = \cos x$
b) $y'' - y = 2 \cos x$
c) $y'' + y' + y = \cos x$
d) $y'' + y' - y = 5 \cos x$
e) $y'' + y' - y = -3 \sin x - \cos x$
f) $2y'' + 3y' = \sin x + 5 \cos x$

a) $y = -\cos x$        b) $y = -\cos x$
c) $y = \sin x$         d) $y = \sin x - 2 \cos x$
e) $y = \sin x + \cos x$   f) $y = \sin x - \cos x.$

**M8.** Given that $u = x$ is a solution to

$$(1 - x)y'' + xy' - y = 0,$$

determine a second solution not of the form $Ax$ where $A$ is a constant.

**M9.** Given that $u = x$ is a solution to

$$(x - x^2)y'' - y' + yx^{-1} = 0,$$

find a second solution not of the form $Ax$ where $A$ is a constant.

Find the general solutions for the equations given in Exercises M10–M20.

**M10.** a) $y'' - 3y' + 2y = 2x - 1$
b) $y'' - 3y' + 2y = 6e^{2x}$

**M11.** a) $y'' - 3y' - 4y = x$  b) $y'' - 3y' - 4y = e^{-x}$

**M12.** $y'' + 2y' + y = e^x \cos x$

**M13.** a) $y'' + y = 2e^x$  b) $y'' + y = -8 \cos 3x$
c) $y'' + y = 2e^x - 8 \cos 3x$

**M14.** $y'' - 3y' + 2y = 2x - 1 + 6e^{2x}$ (See Exercise M10.)

**M15.** $y'' - 2y' = e^x \sin x$

**M16.** $y'' - 4y' + 4y = x^3 e^{2x} + xe^{2x}$

**M17.** $y'' + y = \csc x$

**M18.** $y'' - 4y' + 3y = \dfrac{e^x}{e^x + 1}$

**M19.** $y'' - y = e^{-x} \sin e^{-x} + \cos e^{-x}$

**M20.** $y'' + 2y' + y = 2 \cos 2x + 3x + 2 + 3e^x$

**M21.** Find particular solutions of the form $y = Ax^n e^x$ for each of the following.
a) $y'' + 3y' + 4y = e^x$  b) $y'' + 2y' - 3y = e^x$
c) $y'' - 2y' + y = e^x$

In each of the following exercises two solutions of the homogeneous equation are given. Use the method of variation of parameters to find a particular solution of the nonhomogenous equation.

**M22.** $y'' + y = \sec^3 x$; $u = \sin x$, $v = \cos x$

**M23.** $y'' - y = xe^x$; $u = e^x$, $v = e^{-x}$

**M24.** $y'' - 5y' + 6y = 2e^x$; $u = e^{3x}$, $v = e^{2x}$

**M25.** $4x^2 y'' + 4xy' + (4x^2 - 1)y = 3x^{3/2} \sin x$;

$$u = \frac{\sin x}{\sqrt{x}}, v = \frac{\cos x}{\sqrt{x}}.$$

**M26.** The current $I$ of an electrical circuit that contains a capacitor of capacity $C$, a coil of inductance $L$, and a generator that produces a voltage $E(t)$, as shown in the figure below, satisfies the differential equation

$$L \frac{d^2 I}{dt^2} + \frac{1}{C} I = \frac{dE}{dt} \quad (L, C = \text{const.} > 0).$$

Find the solution $I$ as a function of $t$ if

a) $E = \sin \alpha t$, $\alpha = \text{constant} \neq \sqrt{1/LC}$ and

b) $E = \sin \alpha t$, $\alpha = \sqrt{1/LC}$.

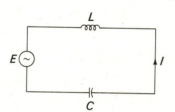

---

## 8.4  *Series Solutions*

The solution to a differential equation can often be found in terms of a series. The basic idea is to let

$$y = a_0 + a_1 x + a_2 x^2 + \cdots + a_n x^n + \cdots,$$
$$y' = a_1 + 2a_2 x + 3a_3 x^2 + \cdots$$
$$+ (n + 1)a_{n+1} x^n + \cdots,$$

and

→  One of the more difficult aspects of this section is our "juggling" of summands. For example, we will frequently use identities such as

$$\sum_{n=2}^{\infty} n(n - 1)a_n x^{n-2} = \sum_{n=0}^{\infty} (n + 2)(n + 1)a_{n+2} x^n$$

and

$$y'' = 2a_2 + 6a_3x + 12a_4x^2 + \cdots$$
$$+ (n + 1)(n + 2)a_{n+2}x^n + \cdots,$$

which we may write more compactly in the form

$$y = \sum_{n=0}^{\infty} a_n x^n,$$

$$y' = \sum_{n=1}^{\infty} na_n x^{n-1},$$

and

$$y'' = \sum_{n=2}^{\infty} n(n - 1)a_n x^{n-2}.$$

We substitute these series into the differential equation and obtain an infinite number of equations involving the coefficients

$$a_0, a_1, a_2, \ldots.$$

When these equations can be solved, they lead to the solution of the differential equation.

## A. Some Easy Examples

*Example 1*　We already know that the solution to

$$y' = x + 3$$

is $y = \dfrac{x^2}{2} + 3x + C$. We will rederive the solution using the series method.

*Solution*　Replacing $y'$ by its series representation we have

$$a_1 + 2a_2x + 3a_3x^2 + \cdots = x + 3.$$

Equating coefficients we obtain

$$a_1 = 3, \quad 2a_2 = 1, \quad \text{and} \quad na_n = 0 \quad \text{for} \quad n \geq 3.$$

Hence

$$a_1 = 3, \quad a_2 = \tfrac{1}{2}, \quad \text{and} \quad a_n = 0 \quad \text{for} \quad n \geq 3.$$

Since $y = a_0 + a_1x + a_2x^2 + \cdots$, we have

$$(1 + x^2) \sum_{n=1}^{\infty} na_n x^{n-1}$$

$$= a_1 + 2a_2x + \sum_{n=2}^{\infty} ((n + 1)a_{n+1} + (n - 1)a_{n-1})x^n.$$

*You should not feel embarrassed if the only way you can check these identities is by writing out several terms to "see what's going on." For example, to prove the second identity above, write things out as follows:*

$$(1 + x^2) \sum_{n=1}^{\infty} na_n x^{n-1}$$

$$= (1 + x^2)a_1 + (1 + x^2)2a_2x + (1 + x^2)\sum_{n=3}^{\infty} na_n x^{n-1}$$

$$= a_1 + 2a_2x + a_1x^2 + 2a_2x^3 + \sum_{n=3}^{\infty} na_n(x^{n-1} + x^{n+1})$$

$$= a_1 + 2a_2x + a_1x^2 + 2a_2x^3$$

$$+ \sum_{n=2}^{\infty} (n + 1)a_{n+1}x^n + \sum_{n=4}^{\infty} (n - 1)a_{n-1}x^n$$

$$= a_1 + 2a_2x + a_1x^2 + 2a_2x^3 + 3a_3x^2 + 4a_4x^3$$

$$+ \sum_{n=4}^{\infty} ((n + 1)a_{n+1} + (n - 1)a_{n-1})x^n$$

$$= a_1 + 2a_2x + \sum_{n=2}^{\infty} ((n + 1)a_{n+1} + (n - 1)a_{n-1})x^n.$$

(Bob)

*In "equating coefficients" we are using the fact that if*

$$a_0 + a_1x + a_2x^2 + \cdots = b_0 + b_1x + b_2x^2 + \cdots$$

*then $a_n = b_n$ for all $n$. The proof of this result is as follows: Letting*

$$f(x) = \sum_{n=0}^{\infty} a_n x^n = \sum_{n=0}^{\infty} b_n x^n,$$

*we have $\dfrac{f^{(n)}(0)}{n!} = a_n$ by Theorem 1 (7.4A). Similarly we have $\dfrac{f^{(n)}(0)}{n!} = b_n$, so*

$$a_n = \frac{f^{(n)}(0)}{n!} = b_n.$$

$$y = a_0 + 3x + \tfrac{1}{2}x^2 + 0x^3 + 0x^4 + \cdots$$

$$= a_0 + 3x + \frac{x^2}{2}.$$

➞ Notice that when we equated coefficients, no condition on $a_0$ was obtained. This coefficient plays the role of the constant of integration.

*Example 2* We will solve the equation $y' + y = 0$ by the series method.

*Solution* Replacing $y$ and $y'$ by their series representations we have

$$\sum_{n=1}^{\infty} na_n x^{n-1} + \sum_{n=0}^{\infty} a_n x^n = 0.$$

Rewriting $\sum_{n=1}^{\infty} na_n x^{n-1}$ as $\sum_{n=0}^{\infty} (n+1)a_{n+1}x^n$ in the above equation, we obtain

$$\sum_{n=0}^{\infty} (n+1)a_{n+1}x^n + \sum_{n=0}^{\infty} a_n x^n = 0$$

or

$$\sum_{n=0}^{\infty} ((n+1)a_{n+1} + a_n)x^n = 0.$$

*It is easier to see how this formula was arrived at if you write out a few of the equations as follows:*

$$a_1 + a_0 = 0,$$
$$2a_2 + a_1 = 0,$$
$$3a_3 + a_2 = 0,$$
$$4a_4 + a_3 = 0, \text{ etc.}$$

Since the right-hand side of the equation is zero, it must be that all the coefficients are zero. Hence

$$(n+1)a_{n+1} + a_n = 0 \quad \text{for all } n.$$

This leads to the formula

$$a_n = \frac{(-1)^n a_0}{n!}.$$

*Hence*

$$a_1 = -a_0$$

$$a_2 = \frac{-a_1}{2} = \frac{-(-a_0)}{2} = \frac{a_0}{2}$$

$$a_3 = \frac{-a_2}{3} = \frac{-1}{3}\frac{a_0}{2} = \frac{-a_0}{3!}$$

$$a_4 = \frac{-a_3}{4} = \frac{-1}{4}\frac{-a_0}{3!} = \frac{a_0}{4!}$$

➞ *The pattern $a_n = (-1)^n \dfrac{a_0}{n!}$ is now clear.*

(Bob)

Hence

$$y = \sum_{n=0}^{\infty} a_n x^n = \sum_{n=0}^{\infty} \frac{(-1)^n a_0}{n!} x^n$$

$$= a_0 \left(1 - x + \frac{x^2}{2!} - \frac{x^3}{3!} + \cdots\right)$$

$$= a_0 e^{-x}$$

➞ In this case it was easy to recognize the series expansion of $e^{-x}$, but quite often the series does not represent the expansion of a familiar function.

We will now take up the case of a second order equation.

*Example 3* Solve the equation

$$y'' + y = 0$$

using the series method.

*Solution*  Replacing $y''$ and $y$ by their series expansions we have

$$y'' + y = \sum_{n=0}^{\infty} (n + 2)(n + 1)a_{n+2}x^n + \sum_{n=0}^{\infty} a_n x^n = 0.$$

Hence $\sum_{n=0}^{\infty} ((n + 2)(n + 1)a_{n+2} + a_n)x^n = 0,$

so all the coefficients must be zero and we have

$$(n + 2)(n + 1)a_{n+2} + a_n = 0 \quad \text{for all } n.$$

This leads to two sets of equations involving the even and odd indices, respectively. Their solutions are

$$a_{2n} = (-1)^n \frac{a_0}{(2n)!}$$

and

$$a_{2n+1} = (-1)^n \frac{a_1}{(2n + 1)!}.$$

Hence

$$y = \sum_{n=0}^{\infty} a_n x^n = \sum_{n=0}^{\infty} a_{2n} x^{2n} + \sum_{n=0}^{\infty} a_{2n+1} x^{2n+1}$$

$$= a_0 \sum_{n=0}^{\infty} (-1)^n \frac{x^{2n}}{(2n)!} + a_1 \sum_{n=0}^{\infty} (-1)^n \frac{x^{2n+1}}{(2n + 1)!}$$

$$= a_0 \cos x + a_1 \sin x.$$

We have

| | |
|---|---|
| $2 \cdot 1 a_2 + a_0 = 0$ | $3 \cdot 2 a_3 + a_1 = 0$ |
| $4 \cdot 3 a_4 + a_2 = 0$ | $5 \cdot 4 a_5 + a_3 = 0$ |
| $6 \cdot 5 a_6 + a_4 = 0$ | $7 \cdot 6 a_7 + a_5 = 0$ |
| $8 \cdot 7 a_8 + a_6 = 0$ | $9 \cdot 8 a_9 + a_7 = 0$ |

and

$\cdots$ $\cdots$

This gives

$$a_2 = \frac{-a_0}{2}, \qquad a_3 = \frac{-a_1}{3!},$$

$$a_4 = \frac{-a_2}{4 \cdot 3} = \frac{a_0}{4!}, \qquad a_5 = \frac{-a_3}{5 \cdot 4} = \frac{a_1}{5!},$$

and

$$a_6 = \frac{-a_4}{6 \cdot 5} = \frac{-a_0}{6!}, \qquad a_7 = \frac{-a_5}{7 \cdot 6} = \frac{-a_1}{7!},$$

$$a_8 = \frac{-a_6}{8 \cdot 7} = \frac{a_0}{8!}, \qquad a_9 = \frac{-a_7}{9 \cdot 8} = \frac{a_1}{9!},$$

and, in general,

$$a_{2n} = \frac{-a_{2n-2}}{2n(2n - 1)} = \frac{(-1)^n a_0}{(2n)!}$$

and

$$a_{2n+1} = \frac{a_{2n-1}}{(2n + 1)2n} = \frac{(-1)^n a_1}{(2n + 1)!}.$$

---

**Problem 1**  Use series to solve the equation

$$y' - y + x = 0.$$

*Answer:* $(a_0 - 1)e^x + x + 1$

---

## B. More Complicated Examples

---

*Example 4*  We will solve the equation

$$0 = (x^2 + 1)y'' + 6xy' + 6y$$

by the series method.

*Solution*  Substituting the series expansions for $y$, $y'$, $y''$ into the differential equation, we have

$$0 = (x^2 + 1) \sum_{n=0}^{\infty} (n + 1)(n + 2)a_{n+2}x^n$$

$$+ 6x \sum_{n=0}^{\infty} (n + 1)a_{n+1}x^n + 6 \sum_{n=0}^{\infty} a_n x^n$$

$$= \sum_{n=0}^{\infty} (n + 1)(n + 2)a_{n+2}x^{n+2}$$

$$+ \sum_{n=0}^{\infty} (n + 1)(n + 2)a_{n+2}x^n$$

$$+ \sum_{n=0}^{\infty} 6(n + 1)a_{n+1}x^{n+1} + \sum_{n=0}^{\infty} 6a_n x^n$$

$$= \sum_{n=2}^{\infty} (n - 1)na_n x^n + \sum_{n=0}^{\infty} (n + 1)(n + 2)a_{n+2}x^n$$

It is convenient now to write all summands in the form which displays the coefficients of $x^n$. We use the identities

$$\sum_{n=0}^{\infty} (n + 1)(n + 2)a_{n+2}x^{n+2} = \sum_{n=2}^{\infty} (n - 1)na_n x^n$$

and

$$\sum_{n=0}^{\infty} 6(n + 1)a_{n+1}x^{n+1} = \sum_{n=1}^{\infty} 6na_n x^n.$$

$$+ \sum_{n=1}^{\infty} 6na_n x^n + \sum_{n=0}^{\infty} 6a_n x^n$$

$$= 2a_2 + 6a_3 x + 6a_1 x + 6a_0 + 6a_1 x$$

$$+ \sum_{n=2}^{\infty} ((n - 1)na_n + (n + 1)(n + 2)a_{n+2}$$

$$+ 6na_n + 6a_n)x^n$$

$$= (2a_2 + 6a_0) + (12a_1 + 6a_3)x$$

$$+ \sum_{n=2}^{\infty} ((n + 2)(n + 3)a_n$$

$$+ (n + 1)(n + 2)a_{n+2})x^n.$$

This is obtained by writing

$$\sum_{n=0}^{\infty} (n + 1)(n + 2)a_{n+2}x^n$$

as

$$2a_2 + 6a_3 x + \sum_{n=2}^{\infty} (n + 1)(n + 2)a_{n+2}x^n,$$

$$\sum_{n=1}^{\infty} 6na_n x^n \quad \text{as} \quad 6a_1 x + \sum_{n=2}^{\infty} 6na_n x^n,$$

and

$$\sum_{n=0}^{\infty} 6a_n x^n \quad \text{as} \quad 6a_0 + 6a_1 x + \sum_{n=2}^{\infty} 6a_n x^n.$$

Now setting all coefficients equal to zero, we obtain

$$2a_2 + 6a_0 = 0,$$
$$12a_1 + 6a_3 = 0,$$

and

$$(n + 2)(n + 3)a_n + (n + 1)(n + 2)a_{n+2} = 0 \quad (n \geq 2).$$

Hence

$$a_2 = -3a_0, \quad a_3 = -2a_1,$$

and

$$a_{n+2} = \frac{-(n + 3)}{n + 1} a_n \quad (n \geq 2).$$

Notice, however, that this equation is also valid when $n = 0$ or $1$; so we may write

$$a_{n+2} = \frac{-(n+3)}{n+1} a_n \quad (n \geq 0).$$

This yields the equations

$$a_{2n} = (-1)^n(2n+1)a_0$$

and

$$a_{2n+1} = (-1)^n(n+1)a_1.$$

Hence

$$y = \sum_{n=0}^{\infty} a_n x^n = \sum_{n=0}^{\infty} a_{2n} x^{2n} + \sum_{n=0}^{\infty} a_{2n+1} x^{2n+1}$$

$$= a_0 \sum_{n=0}^{\infty} (-1)^n(2n+1)x^{2n}$$

$$+ a_1 \sum_{n=0}^{\infty} (-1)^n(n+1)x^{2n+1}.$$

---

**Example 5** The equation

$$(1 - x^2)y'' - 2xy' + k(k+1)y = 0$$

is called *Legendre's Equation*. We will solve the equation for the case that $k$ is a nonnegative integer.

*Solution* Substituting the series expansions for $y$, $y'$, and $y''$ in the equation we have

$$0 = (1 - x^2) \sum_{n=0}^{\infty} (n+2)(n+1)a_{n+2}x^n$$

$$- 2x \sum_{n=0}^{\infty} (n+1)a_{n+1}x^n + k(k+1) \sum_{n=0}^{\infty} a_n x^n$$

$$= \sum_{n=0}^{\infty} (n+2)(n+1)a_{n+2}x^n$$

$$- \sum_{n=0}^{\infty} (n+2)(n+1)a_{n+2}x^{n+2}$$

$$- \sum_{n=0}^{\infty} 2(n+1)a_{n+1}x^{n+1} + \sum_{n=0}^{\infty} k(k+1)a_n x^n$$

$$= \sum_{n=0}^{\infty} (n+2)(n+1)a_{n+2}x^n - \sum_{n=2}^{\infty} n(n-1)a_n x^n$$

This set of equations divides itself into the sets of equations with even and odd indices. We have

$$a_4 = \frac{-5}{3} a_2 = \frac{-5}{3}(-3a_0) = 5a_0,$$

$$a_6 = \frac{-7}{5} a_4 = \frac{-7}{5}(5a_0) = -7a_0,$$

and in general

$$a_{2n} = (-1)^n(2n+1)a_0.$$

Also $\quad a_5 = \dfrac{-6}{4} a_3 = \dfrac{-3}{2}(-2a_1) = 3a_1,$

$$a_7 = \frac{-8}{6} a_5 = \frac{-4}{3}(3a_1) = -4a_1,$$

and in general

$$a_{2n+1} = (-1)^n(n+1)a_1.$$

---

**Problem 2** This solution may also be written in the form

$$\frac{a_0(1 - x^2) + a_1 x}{(1 + x^2)^2}.$$

Prove this by establishing the following identities.

a) $\displaystyle \sum_{n=0}^{\infty} (-1)^n(2n+1)x^{2n} = \frac{1 - x^2}{(1 + x^2)^2}$

b) $\displaystyle \sum_{n=0}^{\infty} (-1)^n(n+1)x^{2n+1} = \frac{x}{(1 + x^2)^2}$

$$-\sum_{n=1}^{\infty} 2na_n x^n + \sum_{n=0}^{\infty} k(k+1)a_n x^n$$

$$= 2a_2 + 6a_3 x - 2a_1 x + k(k+1)a_0$$

$$+ k(k+1)a_1 x + \sum_{n=2}^{\infty} ((n+2)(n+1)a_{n+2}$$

$$- n(n-1)a_n - 2na_n + k(k+1)a_n)x^n$$

$$= 2a_2 + k(k+1)a_0 + (6a_3 - 2a_1$$

$$+ k(k+1)a_1)x + \sum_{n=2}^{\infty} ((n+1)(n+2)a_{n+2}$$

$$+ (k(k+1) - n(n+1))a_n)x^n$$

$$= \sum_{n=0}^{\infty} ((n+1)(n+2)a_{n+2}$$

$$+ (k(k+1) - n(n+1))a_n)x^n.$$

Setting all coefficients equal to zero yields

$$(n+1)(n+2)a_{n+2} + (k(k+1) - n(n+1))a_n = 0$$

for all integers $n \geq 0$. Hence

$$a_{n+2} = \frac{n(n+1) - k(k+1)}{(n+1)(n+2)} a_n$$

or

$$a_{n+2} = \frac{(n-k)(n+k+1)}{(n+1)(n+2)} a_n.$$

We will now complete the derivation for $k = 0$. In this case we have

$$a_{n+2} = \frac{n(n+1)}{(n+1)(n+2)} a_n = \frac{n}{n+2} a_n.$$

For $n = 0$ we have $a_2 = \frac{0}{2} a_0$, so $a_2 = 0$, $a_4 = \frac{2}{4} a_2 = 0$, and in general $a_{2n} = 0$. For the coefficients with odd indices we have

$$a_3 = \frac{1}{3} a_1, \quad a_5 = \frac{3}{5} a_3 = \frac{3}{5}\left(\frac{1}{3} a_1\right) = \frac{a_1}{5},$$

$$a_7 = \frac{5}{7} a_5 = \frac{5}{7}\left(\frac{a_1}{5}\right) = \frac{a_1}{7},$$

and in general

$$a_{2n+1} = \frac{a_1}{2n+1}.$$

In obtaining this equation we used the identities

$$\sum_{n=0}^{\infty} (n+2)(n+1)a_{n+2}x^n$$

$$= 2a_2 + 6a_3 x + \sum_{n=2}^{\infty} (n+2)(n+1)a_{n+2}x^n,$$

$$\sum_{n=1}^{\infty} 2na_n x^n = 2a_1 x + \sum_{n=2}^{\infty} 2na_n x^n,$$

and

$$\sum_{n=0}^{\infty} k(k+1)a_n x^n$$

$$= k(k+1)a_0 + k(k+1)a_1 x + \sum_{n=2}^{\infty} k(k+1)a_n x^n.$$

This follows since

$$n(n+1) - k(k+1) = n^2 + n - k^2 - k$$
$$= n^2 - k^2 + n - k$$
$$= (n-k)(n+k) + (n-k)$$
$$= (n-k)(n+k+1).$$

**Problem 3** Show that for $k = 1$ the solution is

$$y = a_1 x - a_0 \sum_{n=0}^{\infty} \frac{x^{2n}}{2n-1}.$$

Hence the solution is

$$y = \sum_{n=0}^{\infty} a_n x^n = a_0 - a_1 \sum_{n=0}^{\infty} \frac{x^{2n+1}}{2n+1}.$$

---

## C. Exercises

**B1.** The differential equation

$$y'' = 12x^2 + 2$$

has $y = a_0 + a_1 x + x^2 + x^4$ as its general solution. Arrive at this solution by substituting the series

$$y = \sum_{k=0}^{\infty} a_k x^k$$

in the equation and solving for the coefficients.

**B2.** Verify that

$$u = \sum_{k=0}^{\infty} \frac{(-1)^k x^{2k}}{1 \cdot 3 \dots (2k-1)}$$

and

$$v = \sum_{k=0}^{\infty} \frac{(-1)^k x^{2k+1}}{2^k k!}$$

are solutions of the differential equation

$$y'' + xy' + 2y = 0.$$

**B3.** Use series to find the general solution of the differential equation

$$y'' + xy' + y = 0.$$

**B4.** Given $(k+1)(k+2)a_{k+1} = a_{k-1}$ for $k \geq 1$ and $a_0 = a_1 = 1$, evaluate the $a_k$'s.

**D5.** Determine a general formula for $a_n$ given that the equations below hold for all integers $k \geq 0$.
   a) $a_{k+1} = (k+1)a_k$   b) $a_{k+1} = 2a_k$

c) $a_{k+1} = ka_k$   f) $a_{k+1} = \dfrac{3a_k}{k+1}$

d) $a_{k+1} = -a_k$   g) $a_{k+1} - a_k = 0$

e) $a_{k+1} = -(k+1)a_k$   h) $a_{k+1} = (k-1)a_k$

a) $a_n = n!a_0$   b) $a_n = 2^n a_0$   c) $a_n = 0$

d) $a_n = (-1)^n a_0$   e) $a_n = (-1)^n n!a_0$

f) $a_n = \dfrac{3^n}{n!} a_0$   g) $a_n = a_0$

h) $a_1 = -a_0$ and $a_n = 0$ for $n \geq 2$.

**M6.** a) Given

$$a_{k+2} = \frac{2k-2}{(k+2)(k+1)} a_k \quad (k \geq 0),$$

find a general formula for $a_k$.
b) Solve the differential equation

$$y'' - 2xy' + 2y = 0.$$

**M7.** Given the differential equation

$$y'' - xy' - 2y = 0,$$

find the solution which satisfies the conditions

$$y(0) = 1 \quad \text{and} \quad y'(0) = 0.$$

**M8.** By using series solve the equation

$$(x^2 - 2x)y'' + (2 - 2x)y' + 2y = 0$$

given that $y(0) = y(1) = 1$.

**M9.** By using series solve the equation $y'' = y$ given that $y(0) = 1$ and $y'(0) = 0$.

---

**8.5  Kepler's Laws**

Kepler arrived at his three laws of planetary motion after prodigious calculations using vast amounts of experimental data. Now using Newton's Laws we will derive Kepler's three laws which state:

1) The radius vector (Fig. 1) from the sun to any planet sweeps out area at a constant rate.

2) The planets describe elliptical orbits about the sun.

3) The square of the period of revolution of a planet is proportional to the cube of the length of the major axis of the ellipse it describes.

→ Historically, Kepler's Laws came first and Newton used them to make an educated guess at his universal law of gravitation.

→ Given the ellipse whose equation is $\dfrac{x^2}{a^2} + \dfrac{y^2}{b^2} = 1$ where $a > b > 0$, the segment joining $(-a, 0)$ to $(a, 0)$ is called the *major axis*. Its length is clearly $2a$.

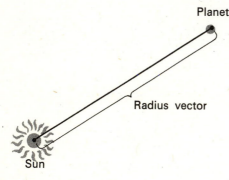

Planet

Radius vector

Sun

**Figure 1**

**A. Kepler's First Law** We will use our knowledge of complex exponentials. Let

$$z(t) = r(t)e^{i\theta(t)}$$

so

$$v(t) = \frac{dz}{dt} = re^{i\theta}i\frac{d\theta}{dt} + \frac{dr}{dt}e^{i\theta}$$

$$= \left(\frac{dr}{dt} + ir\frac{d\theta}{dt}\right)e^{i\theta}.$$

Also

$$a(t) = \frac{d^2z}{dt^2} = \frac{dv}{dt} = \frac{d}{dt}\left(\left(\frac{dr}{dt} + ir\frac{d\theta}{dt}\right)e^{i\theta}\right)$$

$$= \left(\frac{dr}{dt} + ir\frac{d\theta}{dt}\right)e^{i\theta}i\frac{d\theta}{dt}$$

$$+ \left(\frac{d^2r}{dt^2} + ir\frac{d^2\theta}{dt^2} + i\frac{dr}{dt}\frac{d\theta}{dt}\right)e^{i\theta}$$

or

$$(*) \quad a(t)$$

$$= \left(\frac{d^2r}{dt^2} - r\left(\frac{d\theta}{dt}\right)^2 + i\left(2\frac{d\theta}{dt}\frac{dr}{dt} + r\frac{d^2\theta}{dt^2}\right)\right)e^{i\theta}.$$

→ We will simplify our notation by suppressing the letter $t$ and writing $z$, $r$, and $\theta$ in place of $z(t)$, $r(t)$, and $\theta(t)$. Also, observe that $z = re^{i\theta}$ is simply the polar form of the complex number $z$ (Fig. 2).

**Figure 2**

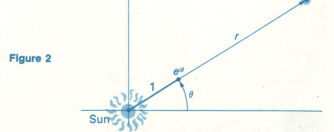

$z = re^{i\theta}$

$r$

$e^{i\theta}$

1

$\theta$

Sun

Now we will suppose that $z$ is the position of a planet, so that $v$ is its velocity and $a$ its acceleration. By Newton's Law of Gravitation we know that the force exerted by the sun on the planet is given by the formula

$$F = \frac{-kMe^{i\theta}}{r^2}.$$

→ Here, $k$ is the gravitational constant. Also, the minus sign is taken because the force $F$ tends to pull the planet toward the sun; hence it is in the direction of $-z$, not in the direction of $z$.

In Newton's Second Law, $ma = F$, we replace $F$ by $-\frac{kMe^{i\theta}}{r^2}$ and $a$ by the expression (*) derived above to obtain

$$m\left(\frac{d^2r}{dt^2} - r\left(\frac{d\theta}{dt}\right)^2 + i\left(2\frac{d\theta}{dt}\frac{dr}{dt} + r\frac{d^2\theta}{dt^2}\right)\right)e^{i\theta}$$

$$= \frac{-kM}{r^2}e^{i\theta}$$

or, dividing out $e^{i\theta}$,

$$\frac{d^2r}{dt^2} - r\left(\frac{d\theta}{dt}\right)^2 + i\left(2\frac{d\theta}{dt}\frac{dr}{dt} + r\frac{d^2\theta}{dt^2}\right) = \frac{-C}{r^2}$$

→ This is a differential equation, but one we are not familiar with because it contains two unknown functions of $t$.

where

$$C = \frac{-kM}{m}.$$

We can equate the real and imaginary parts of this equation. Since the right side of the equation is real, we must have

→ Recall that $a + ib = c + id$ if and only if $a = c$ and $b = d$.

$$2\frac{d\theta}{dt}\frac{dr}{dt} + r\frac{d^2\theta}{dt^2} = 0$$

or

$$r^2\frac{d^2\theta}{dt^2} + 2r\frac{d\theta}{dt}\frac{dr}{dt} = 0.$$

We obtained this equation by multiplying the previous one by $r$. The effect of this is to put the equation in the form $\frac{d}{dx}uv$.

Integrating, we obtain

$$r^2\frac{d\theta}{dt} = K \quad \text{for some constant } K.$$

Now look at Figure 3, which depicts the area that $r$ sweeps out in an increment of time $\Delta t$. For $\Delta\theta$ sufficiently small, $\alpha$ is approximately a right angle and the area swept out is approximated by a right triangular region. Hence we have that $\Delta A$ is approximately equal to $\frac{1}{2}r^2\Delta\theta$. Dividing by $\Delta t$ and then letting $\Delta t \to 0$, we obtain

**Figure 3**

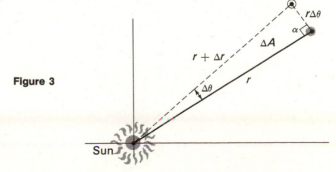

$$\frac{dA}{dt} = \frac{1}{2} r^2 \frac{d\theta}{dt}.$$

But we just found from our differential equation that $r^2 \frac{d\theta}{dt}$ is a constant. Hence $\frac{dA}{dt}$ is constant, and this is the mathematical statement of Kepler's First Law.

→   Before this was proven, many astronomers thought the motion of the planets about the sun must be circular. As we shall soon see, the condition we have derived is not strong enough to limit motion to a circle.

### B. Kepler's Second Law   We return to the equation

$$m \frac{d^2 z}{dt^2} = ma = \frac{-kM}{r^2} e^{i\theta}.$$

Rewriting it we have

$$\frac{d^2 z}{dt^2} = \frac{-kM e^{i\theta}}{mr^2} = \frac{C e^{i\theta}}{r^2} \quad \text{where} \quad C = -\frac{kM}{m},$$

which can be written in the form

$$\frac{d^2 z}{dt^2} = \frac{C}{K} e^{i\theta} \frac{d\theta}{dt}.$$

→   We use here the fact that $r^2 \frac{d\theta}{dt} = K$ is constant as shown in 8.5A. Also, we may assume $K > 0$, for this merely stipulates the direction, counterclockwise, the planet is traveling.

Integrating both sides of this equation gives

$$\frac{dz}{dt} = -i \frac{C}{K} e^{i\theta} + R$$

where $R$ is a constant of integration, and we rewrite this equation as

$$\frac{dz}{dt} = -i \frac{C}{K} e^{i\theta} + i c_0 e^{i\theta_0}.$$

→   For convenience, we write $R$ in the form $R = i c_0 e^{i\theta_0}$ where $c_0 > 0$.

We now have two expressions for $\frac{dz}{dt}$: this one, and the one derived at the beginning of 8.5A. Equating them gives

$$-i \frac{C}{K} e^{i\theta} + i c_0 e^{i\theta_0} = \left( \frac{dr}{dt} + ir \frac{d\theta}{dt} \right) e^{i\theta}$$

or

$$-i \frac{C}{K} + i c_0 e^{-i(\theta - \theta_0)} = \frac{dr}{dt} + ir \frac{d\theta}{dt}.$$

→   This can be written as

$$-i \frac{C}{K} + i c_0 (\cos(\theta - \theta_0) - i \sin(\theta - \theta_0)) = \frac{dr}{dt} + ir \frac{d\theta}{dt}$$

or

$$c_0 \sin(\theta - \theta_0) + i \left( c_0 \cos(\theta - \theta_0) - \frac{C}{K} \right) = \frac{dr}{dt} + ir \frac{d\theta}{dt}.$$

Equating real parts and imaginary parts yields

$$c_0 \sin(\theta - \theta_0) = \frac{dr}{dt}$$

and

$$c_0 \cos(\theta - \theta_0) - \frac{C}{K} = r \frac{d\theta}{dt}.$$

Since $\frac{d\theta}{dt} = \frac{K}{r^2}$, we can write the second equation as $\longrightarrow$

$$r = \frac{K^2}{-C + Kc_0 \cos(\theta - \theta_0)}.$$

For convenience, suppose that the axes are positioned so that $\theta_0 = 0$. Then we can write

(1) $\qquad r = \frac{K^2}{-C + Kc_0 \cos\theta} = \frac{-G}{1 - H\cos\theta}.$ $\longrightarrow$

We will now write this equation in terms of rectangular coordinates. Replacing $r$ by $\sqrt{x^2 + y^2}$ and $\cos\theta$ by $\frac{x}{r}$ in (1), we obtain the equation

(2) $\qquad x^2 + y^2 = (Hx - G)^2.$ $\longrightarrow$

When $H = 1$ this equation becomes

$$y^2 = G^2 - 2Gx,$$

which is the equation of a parabola and clearly cannot represent the orbit of a planet about the sun.

For $H \neq 1$ we may write equation (2) in the form

(3) $\quad (1 - H^2)\left(x + \frac{GH}{1 - H^2}\right)^2 + y^2 = \frac{G^2}{1 - H^2}.$ $\longrightarrow$

When $H^2 > 1$ this becomes the equation of a hyperbola, which also cannot represent the orbit of a planet about the sun. This leaves us with $0 \leq H^2 < 1$, and equation (3) is then the equation of an ellipse, as stated by Kepler's Second Law.

**C. Kepler's Third Law** We will make use of Kepler's First and Second Laws in deriving his final one. First of all we simplify equation (3) by letting

$$a^2 = \frac{G^2}{(1 - H^2)^2} \quad \text{and} \quad b^2 = \frac{G^2}{1 - H^2} \quad (a, b > 0). \quad \longrightarrow$$

Equation (3) then becomes

(4) $\qquad \frac{(x - h)^2}{a^2} + \frac{y^2}{b^2} = 1$

where $h = aH = -\frac{GH}{1 - H^2}$ and $2a$ is the length of the major axis. $\longrightarrow$

---

This follows from the fact that $r^2 \frac{d\theta}{dt} = K$, and hence

$$\frac{d\theta}{dt} = \frac{K}{r^2} \quad \text{and} \quad c_0 \cos(\theta - \theta_0) - \frac{C}{K} = \frac{K}{r}.$$

Therefore

$$\frac{1}{r} = \frac{Kc_0 \cos(\theta - \theta_0) - C}{K^2}$$

or

$$r = \frac{K^2}{-C + Kc_0 \cos(\theta - \theta_0)}.$$

Here $G = K^2/C$ and $H = Kc_0/C$ are constants which we introduce to simplify some calculations below.

To see this, write equation (1) in the form

$$r - Hr\cos\theta = -G$$

and replace $\cos\theta$ by $\frac{x}{r}$ to obtain

$$r - Hx = -G.$$

Hence

$$\sqrt{x^2 + y^2} = r = Hx - G,$$

and squaring this gives equation (2).

To derive this formula we multiply out equation (2) to obtain

$$x^2 + y^2 = H^2x^2 - 2GHx + G^2$$

or

$$(1 - H^2)x^2 + 2GHx + y^2 = G^2.$$

Completing the square, we obtain

$$(1 - H^2)\left(x^2 + \frac{2GH}{1 - H^2}x + \frac{G^2H^2}{(1 - H^2)^2}\right) + y^2 = G^2 + \frac{G^2H^2}{1 - H^2}$$

or

$$(1 - H^2)\left(x + \frac{GH}{1 - H^2}\right)^2 + y^2 = \frac{G^2}{1 - H^2}.$$

Since $0 \leq H^2 < 1$ we have $1 - H^2 > 0$ and this insures that $a^2$ and $b^2$ are positive real numbers.

Because $0 \leq H^2 < 1$, it follows that $0 < 1 - H^2 \leq 1$, so $(1 - H^2)^2 \leq 1 - H^2$. Hence

$$a^2 = \frac{G^2}{(1 - H^2)^2} \geq \frac{G^2}{1 - H^2} \geq b^2,$$

so $a \geq b$ and $2a$ is indeed the length of the major axis.

By Kepler's First Law we know that the rate $\frac{dA}{dt}$ at which the area is swept out is constant. In fact, we have $\frac{dA}{dt} = \frac{K}{2}$ where $K^2 = GC$ (see Equation (1)). Hence, if $T$ denotes the time for one orbit, we have

$$A = \frac{dA}{dt} T = KT \quad \text{or} \quad T = \frac{A}{K}.$$

By Kepler's Second Law the orbit is an ellipse whose area is given by $A = \pi ab$. Hence

$$T^2 = \frac{A^2}{K^2} = \frac{\pi^2 a^2 b^2}{GC} = \frac{\pi^2 G^4}{GC(1 - H^2)^3}$$

$$= \frac{\pi^2}{C} \left( \frac{G}{1 - H^2} \right)^3 = \frac{\pi^2}{|C|} a^3.$$

→ See 5.5B for the derivation of the formula $A = \pi ab$ for the area of an ellipse whose equation is given by equation (4).

In other words,

$$T^2 = \frac{\pi^2}{8|C|} (2a)^3,$$

which is the mathematical expression of Kepler's Third Law, and where $\pi^2/8|C|$ is the constant of proportionality.

## 8.6 Sample Exams

### Sample Exam 1 (45–60 minutes)

1. Solve $\dfrac{dy}{dx} = \dfrac{x - xy^2}{x^2 y - y}$ by separating the variables.

2. Find the general solutions to the following differential equations.
   a) $y'' + 2y' - 35y = 0$      b) $y'' - 2y' + 13y = 0$

3. Guess at a particular solution to
$$y'' + 6y = \cos 2x$$
and then determine the general solution.

4. Solve the differential equation $xy'' = 4y'$ with the aid of an appropriate substitution.

### Sample Exam 2 (45–60 minutes)

1. Find the solution of
$$(3x^2 + 1)y' = 6xy - 6x.$$

2. Find the general solution to the differential equation

$$yy'' + (1 + y)(y')^2 = 0$$

using an appropriate substitution.

3. Find the general solutions to the following differential equations.
   a) $y'' + 4y = 0$      b) $y'' + 4y' + 4y = 0$.

4. a) Use the method of undetermined coefficients to find a particular solution to
$$y'' + y' + y = 2x + 6.$$

   b) Now find the general solution of this equation.

### Sample Exam 3 (45–60 minutes)

1. Find the general solution of the following differential equations.
   a) $y'' + 4y' + 2y = e^x$      b) $y'' + y' - 2y = e^x$

2. Use appropriate substitutions to solve the following differential equation:
$$(1 + x^2)y'' + xy' = 4x.$$

**3.** Show that the equation

$$x^2 y'' + xy' + (x^2 - 3)y = 0$$

has no nonzero solution of the form $y = \sum_{n=0}^{\infty} a_n x^n$.

## Sample Exam 4 (45–60 minutes)

**1.** Use the method of variation of parameters to find a particular solution of

$$y'' + y = \sin x \cos x.$$

**2.** Given that $u = x^2$ is one solution of the differential equation $xy'' - y' = 0$, find the general solution.

**3.** Given that $u = x^3$ and $v = x$ are solutions of the differential equation

$$x^2 y'' - 3xy' + 3y = 0,$$

find a particular solution of

$$x^2 y'' - 3xy' + 3y = -x^2$$

using the method of variation of parameters.

We now begin again. Starting with notions from set theory, we assume the existence of a set having certain properties. The set is called the real numbers, and our description of it is referred to as an axiomatic one. We then review the definitions of limits and continuous functions and prove a number of important theorems. The final result is the proof of the existence of the integral for a continuous function.

# Continuity

343

## 9.1 Set Theoretical Notions

The axiomatic description of the real numbers which is given at the end of this section is most easily phrased using notions and notations from set theory. We will now take up these matters.

→ Although the real numbers are not defined until the end of this section, we will use them to illustrate some of the notions. This will not interfere with the logical development of the material.

**A. General Remarks About Sets** We will use the term *set* to refer to any collection of objects provided it is clear which objects belong to the collection and which do not. The objects are called *elements* or *members* of the set.

Given a set $A$ and elements $a$ and $b$, we write $a \in A$ to signify that $a$ is an element of $A$ and $b \notin A$ to signify that $b$ is not an element of $A$.

Some sets can be described by simply listing their elements. For example, if a set $A$ consists of the elements 1 and 2, we write

$$A = \{1, 2\}.$$

The set $B$ containing the integers $1, 2, \ldots, n$ is written as $B = \{1, 2, \ldots, n\}$ and the set of all natural numbers $N$ is written as

$$N = \{1, 2, 3, \ldots\}.$$

There is a very special set that contains no elements at all. It is well defined because it is clear which objects are members, namely none of them. The set is called the *null set* or *empty set* and is denoted by the letter $\phi$.

Now suppose that $A$ and $B$ are sets. We define the following:

*Subset* If every element of $A$ is also an element of $B$, we call $A$ a *subset* of $B$. We write

$$A \subset B \quad \text{or} \quad B \supset A.$$

When both $A \subset B$ and $B \supset A$ we say the sets are *equal* and write $A = B$.

*Intersection* The set that consists of all elements belonging to both $A$ and $B$ is called the *intersection* of $A$ and $B$ and is denoted by

$$A \cap B.$$

→ As shown in Examples 1 and 2, a grammatically correct statement referring to a collection of objects does not necessarily define a set. In fact, there is no general agreement on what constitutes a precise definition of a set. In texts on set theory the concepts of set, element, and the relation of membership are usually undefined.

**Example 1** The collection of states in the United States is a set containing 50 elements. However, the collection of beautiful states does not constitute a set because there is no general agreement as to what makes a state beautiful.

**Example 2** (Russell's Paradox) Let us assume we do know exactly when a collection of objects is a set. This leads to trouble as the following argument shows.

Let $A$ denote the collection of all sets. (With the above assumption, $A$ is a set.) Now let $B$ denote the set of all sets with the property that they do not contain themselves as elements. In symbols,

$$A = \{x: x \text{ is a set}\}$$
$$B = \{x \in A: x \notin x\}.$$

Clearly $B$ is a set. We will determine whether $B \in B$ or $B \notin B$.

If $B \in B$ then $B$ does contain itself as an element, but since $B$ consists of only those sets that do not contain themselves as elements, we must have $B \notin B$. Hence $B \in B$ implies $B \notin B$, which is absurd. We must therefore have $B \notin B$. This means that $B$ does not contain itself as an element; but by definition of $B$, $B$ is an element of $B$. Hence $B \notin B$ implies $B \in B$, and this again is impossible.

*Union* The set whose elements belong to either $A$ or $B$ is called the *union* of $A$ and $B$ and is denoted by

$$A \cup B.$$

*Product* The set consisting of all pairs $(a, b)$ where $a \in A$ and $b \in B$ is called the *cartesian product* of $A$ and $B$ and is denoted by

$$A \times B.$$

Two elements $(a_1, b_1)$ and $(a_2, b_2)$ of $A \times B$ are considered equal if

$$a_1 = a_2 \quad \text{and} \quad b_1 = b_2.$$

*Difference* If $A \subset B$ we write $B - A$ to denote the set of all members of $B$ that do not belong to $A$.

There are a number of simple relations that exist between these definitions, but we will not need to use them extensively. The following example and problem are given merely so that the reader can test his understanding of the new notions.

---

**Example 3** We will prove that

$$A \times (B \cup C) = (A \times B) \cup (A \times C).$$

*Proof* Suppose $(x, y) \in A \times (B \cup C)$. Then $x \in A$, and $y \in B$ or $y \in C$. Hence $(x, y) \in A \times B$ or $(x, y) \in A \times C$, and this means that $(x, y) \in (A \times B) \cup (A \times C)$. Therefore all elements of $A \times (B \cup C)$ are in $(A \times B) \cup (A \times C)$, so

$$A \times (B \cup C) \subset (A \times B) \cup (A \times C).$$

Conversely, if $(x, y) \in (A \times B) \cup (A \times C)$, then $(x, y) \in A \times B$ or $(x, y) \in A \times C$. Hence $x \in A$ and $y \in B$, or $x \in A$ and $y \in C$, which implies that $y \in B \cup C$. Therefore

$$(x, y) \in A \times (B \cup C).$$

We have shown that all elements of $(A \times B) \cup (A \times C)$ are elements of $A \times (B \cup C)$, so $(A \times B) \cup (A \times C) \subset A \times (B \cup C)$. Combining this with the first inclusion yields

$$A \times (B \cup C) = (A \times B) \cup (A \times C).$$

→ Mathematically the word "or" is used in the sense of "and/or." For example, the phrase

"*x* or *y* has property *P*"

means that either *x* has property *P*, *y* has property *P*, or possibly both *x* and *y* have property *P*. The common usage of the word "or" is slightly different. Consider the statement

"Mike or Eve has a dollar."

The usual meaning attached to this statement implies that only one dollar exists and either Mike or Eve has it. By contrast, the mathematical sense of this statement would allow the possibility that Mike and Eve each have a dollar.

→ It is also possible to extend these definitions to an infinite number of sets. For example, if for each $n \in N$, $A_n$ is a set we write

$$x \in \bigcup_{n=1}^{\infty} A_n \quad \text{if} \quad x \in A_n \quad \text{for some } n \in N,$$

$$x \in \bigcap_{n=1}^{\infty} A_n \quad \text{if} \quad x \in A_n \quad \text{for every } n \in N.$$

---

**Problem 1** Prove that $A \cap (B \cup C) = (A \cap B) \cup (A \cap C)$.

A notation that will frequently be useful in describing subsets is

$$\{x \in A: \text{"such and such"}\}.$$

This stands for the subset of $A$ consisting of all elements $x$ satisfying the property stated in "such and such."

Here are some examples.

a) $A \cap B = \{x \in A \cup B: x \in A \text{ and } x \in B\}$

b) If $A \subset B$, then $B - A = \{x \in B: x \notin A\}$

→ c) $\{x \in R: x = y^2, y \in R\}$ is the set of nonnegative real numbers.

d) $\{(x, y) \in R \times R: x < y\}$ is the set of all points in the plane that lie above the line $y = x$.

## Georg Cantor and Set Theory

Set theory has been a subject of fierce controversy ever since its creation by Georg Cantor nearly a century ago. The great mathematician David Hilbert described it as

One of the most beautiful realizations of human activity in the domain of the purely intelligible. . . . No one shall expel us from the paradise which Cantor has created for us.

On the other hand, speaking of mathematics built upon the foundation of Cantor's set theory, the equally famous mathematician Herman Weyl said

We must learn a new modesty. We have stormed the heavens, but have succeeded only in building fog upon fog, a mist which will not support anybody who earnestly desires to stand upon it.

Georg Ferdinand Ludwig Phillip Cantor (1845–1918) was born in Russia and his family moved to Germany when he was nine. He took an early interest in theology and philosophy, and in the manner in which the concepts of continuity and infinity were treated in these disciplines. By the age of fifteen he was devoted to mathematics, but reluctantly began the study of engineering because of the strong wishes of his father. This pursuit depressed him to such an extent that when he entered the university his father allowed him to study mathematics. Five years later, at the age of twenty-two, he received his doctorate.

Cantor's thesis was in number theory, but his early fascination with the notions of continuity and infinity remained. He began a serious study of analysis and found the treatment of these concepts, from that point of view, wholly inadequate. These investigations led, in 1874, to his first revolutionary publication. In this paper Cantor defines two sets to have the same magnitude (cardinality) if a bijection exists between them, and then proves the existence of a hierarchy of infinite sets of ever increasing magnitudes. It was the beginning of a new theory which Cantor developed in additional publications.

The bold ideas and powerful methods of Cantor's set theory immediately came under severe attack. Cantor took these assaults on his work so personally that at times he doubted his own theory, and when he was thirty-nine suffered a nervous breakdown. The remainder of his life was marked by spells of irrationality and depression, but between these periods were times of exceptional creativity, and it was then that Cantor did some of his best work. His theory gradually attracted followers, but Cantor was not temperamentally or emotionally suited for the controversy that surrounded it, and he died in a mental institution.

Today, set theory is generally accepted by the mathematical world, and aspects of it have even found their way into the elementary school curriculum. However, the debate over it has never died out, and currently its opponents appear to be on the rise. We will not attempt to give opinions on either side of the question, but will indicate a few places where differences of opinion exist.

As Russell's paradox (Example 2, 9.1A) shows, it is dangerous to speak too loosely of sets, and this was dramatically illustrated in 1903 by the mathematical logician F. L. G. Frege (1845–1925). He first read of the paradox in a letter from Russell received just after he had completed a comprehensive work on set theory. The effect of the example as expressed by Frege himself was as follows:

Hardly anything more unwelcome can befall a scientific writer than that one of the foundations of the edifice be shaken after the work is finished. . . . It is not just a matter of my particular method of laying the foundations, but of whether a logical foundation for arithmetic is possible at all.

Various methods were found to circumvent Russell's paradox (and other similar ones) but the foundations of set theory are still not firm. There is considerable difference of opinion as to the meaning that is supposed to be attached to the phrase "there exists." This concerns the method of proof, rather than the definition of a concept, and applies to such concrete sets as the natural numbers or the real numbers. For example, using methods of set theory, it is often quite easy to prove that a particular entity exists, but it can happen that no formula or procedure is available for finding or computing this entity. An example of such a result is the Intermediate Value Theorem (9.3B) which will be discussed further in the history section at the end of Chapter 10. Such examples leave us with the serious problem of interpreting a quantity that in one sense exists, but in another sense can never be found.

We do not want to leave the reader with the impression that the mathematical world is currently divided in heated debate on the above points. The majority of mathematicians are concerned with pushing ahead the frontiers of mathematics, and it is a time of great activity and many new theories. Foundational questions seem to belong more to philosophy than to mathematics, and the attitude of many mathematicians is similar to that expressed by Henri Lebesgue a half century ago.

In my opinion a mathematician, in so far as he is a mathematician, need not concern himself with philosophy—an opinion, moreover, which has been expressed by many philosophers.

On the other hand, if the problems become interesting enough, a large number of mathematicians may again return to a study of the foundations. Even if set theory were to be almost totally rejected in the future it would not diminish Cantor's stature as one of the most original and important mathematicians of all time, for the mark of a truly excellent theory is that it provides more questions than answers.

**B. Functions**　　Given sets $A$ and $B$, a subset $G \subset A \times B$ is called the *graph of a function* if for each $x \in A$ there is a unique $y \in B$ such that $(x, y) \in G$. Denoting $y$ by $f(x)$ whenever $(x, y) \in G$, $f$ is then called the *function* (or mapping) defined by the graph $G$. We will denote this by any of the following equivalent notations:

→　In Chapter 1 we defined a function as a rule and then obtained its graph. The advantage of the present definition is that the notion of function is seen to depend only on that of a set.

$$f: A \rightarrow B$$
$$A \xrightarrow{f} B$$

or

$$x \rightarrow f(x): A \rightarrow B$$

(read "$f$ is a function from $A$ to $B$")

Suppose now that

$$A \xrightarrow{f} B \quad \text{and} \quad C \xrightarrow{g} D \quad \text{and} \quad B \subset C.$$

Then the *composition* of $f$ and $g$, $g \circ f$, is defined by the formula

$$(g \circ f)(x) = g(f(x)) \quad (x \in A)$$

and is a function from $A$ to $D$.

An important fact about the composition of functions is given in the following theorem.

**THEOREM 1** *Suppose*

$$A \xrightarrow{f} B, C \xrightarrow{g} D, E \xrightarrow{h} F$$

*are functions, $B \subset C$, and $D \subset E$. Then*

$$(h \circ g) \circ f \quad \text{and} \quad h \circ (g \circ f)$$

*are functions from $A$ to $F$, and*

$$(h \circ g) \circ f = h \circ (g \circ f).$$

*Proof* By definition of composition we have

$$((h \circ g) \circ f)(x) = (h \circ g)(f(x))$$
$$= h(g(f(x))) = h((g \circ f)(x))$$
$$= (h \circ (g \circ f))(x).$$

The two functions are equal for all $x$ in their domain and therefore are equal ●

A particularly simple function defined for each set $A$ is called the *identity function* $I_A$ and is defined by the formula

$$I_A(x) = x \quad \text{for all} \quad x \in A.$$

Suppose now that $f: A \rightarrow B$ is a bijective function. Then for each $y \in B$ there is a unique $x \in A$ with the property that $f(x) = y$. Denoting $x$ by $f^{-1}(y)$ it follows that

$$f^{-1}: B \rightarrow A$$

**Some terminology** Given $f: A \rightarrow B$, the set $A$ is called the *domain* of the function $f$ and $B$ is called the *codomain*. The set

$$f(A) = \{f(a) \in B: a \in A\}$$

is called the *image* of $A$ under $f$. When $f(A) = B$ the function $f$ is said to be *surjective*, and if $x = y$ whenever $f(x) = f(y)$ the function is said to be *injective*. A function that is both injective and surjective is called *bijective* (or a *bijection*).

Suppose that $A_1$, $A_2$, and $B$ are sets with $A_1 \subset A_2$. Let $f_1: A_1 \rightarrow B$ and $f_2: A_2 \rightarrow B$. If

$$f_1(x) = f_2(x) \quad \text{for all } x \in A_1,$$

then $f_1$ is said to be *the restriction* of $f_2$ to $A_1$, and $f_2$ is said to be *an extension* of $f_1$ to $A_2$.

Two functions $f: A \rightarrow B$ and $g: A \rightarrow B$ are said to be *equal* if $f(x) = g(x)$ for all elements in their domain $A$.

The definition of composition is quite obvious when we think of a function as a "rule," but since the definition here is in terms of graphs, we should define composition in the same terms. Indeed, letting $G_f \subset A \times B$ and $G_g \subset C \times D$ denote the graphs of $f$ and $g$ respectively, we then define the subset $G_{g \circ f} \subset A \times D$ to be $G_{g \circ f} = \{(x, w): \text{there is a } y \in B$ with the property that $(x, y) \in G_f$ and $(y, w) \in G_g\}$. It follows that $G_{g \circ f}$ is, in fact, the graph of a function. As an exercise the reader might prove Theorem 1 in terms of graphs.

It follows from this theorem that the notation

$$h \circ g \circ f$$

is unambiguously defined.

If $f: A \rightarrow B$ it follows immediately that

$$f \circ I_A = f \quad \text{and} \quad I_B \circ f = f.$$

---

**Example 4** Let $R$ denote the reals and $R^+ = \{x \in R: x > 0\}$. Then the function

$$f: R \rightarrow R^+$$

defined by the formula $f(x) = e^x$ is bijective and its inverse $g: R^+ \rightarrow R$ is the function $g(x) = \ln x$.

---

is a bijective function with domain $B$ and image $A$. It is called the inverse of $f$.

**THEOREM 2**  *If $f: A \to B$ is bijective and $f^{-1}$ denotes its inverse, then*

$$f^{-1} \circ f = I_A$$

*and*

$$f \circ f^{-1} = I_B.$$

*Proof*  Suppose $x \in A$ and $f(x) = y$. Then by definition $f^{-1}(y) = x$ and we have

$$(f^{-1} \circ f)(x) = f^{-1}(f(x)) = f^{-1}(y) = x = I_A(x).$$

Therefore $(f^{-1} \circ f)(x) = I_A(x)$ for all $x \in A$, and it follows that $f^{-1} \circ f = I_A$. The proof that $f \circ f^{-1} = I_B$ is similar  ●

Another theorem with a simple proof is:

**THEOREM 3**  *Suppose $f: A \to B$ is bijective. Then*

$$(f^{-1})^{-1} = f.$$

*Proof*  We have

$$(f^{-1})^{-1} = (f^{-1})^{-1} \circ I_A = (f^{-1})^{-1} \circ (f^{-1} \circ f)$$

$$= ((f^{-1})^{-1} \circ f^{-1}) \circ f = I_B \circ f = f  ●$$

→ There is a converse to this theorem that we give in the form of a problem.

**Problem 2**  Suppose $f: A \to B$, $g: B \to A$, $g \circ f = I_A$, and $f \circ g = I_B$. Show that $f$ is bijective and $g = f^{-1}$.

**Problem 3**  Suppose that $f: A \to B$ and $g: B \to C$ are bijective. Prove that

$$(g \circ f)^{-1} = f^{-1} \circ g^{-1}.$$

**C. The Set of Real Numbers**  We will assume the existence of a set $R$ and mappings

$$A: R \times R \to R$$

and

$$M: R \times R \to R$$

that satisfy the properties below. The set $R$ is called the *real numbers*, and the mappings $A$ and $M$ are called *addition* and *multiplication* respectively.

*Algebraic properties*  For all $x, y, z \in R$,

*Axiom 1  Commutative laws*

$$x + y = y + x \quad \text{and} \quad xy = yx$$

→ We will not enter into a metaphysical discussion as to whether the set $R$ actually exists. Much of mathematics is based on the assumption that it does.

→ It will be convenient and suggestive to use the more familiar notation

$$A(x, y) = x + y \quad \text{and} \quad M(x, y) = xy$$

to denote addition and multiplication respectively.

*Axiom 2  Associative laws*

$$(x + y) + z = x + (y + z)$$

and

$$(xy)z = x(yz)$$

*Axiom 3  Distributive law*

$$x(y + z) = xy + xz$$

*Axiom 4  Identities*  There exist distinct elements $0 \in R$ and $1 \in R$ that satisfy $0 + x = x$ and $1x = x$. The elements $0$ and $1$ are called respectively the *additive* and *multiplicative identities*.

*Axiom 5  Inverses*  For each $x \in R$ there is an element $-x \in R$ satisfying

$$-x + x = 0,$$

and for each $y \in R$ with $y \neq 0$ there is an element $y^{-1}$ (also written as $\frac{1}{y}$) satisfying

$$y^{-1}y = 1.$$

The element $-x$ is called the *additive inverse* of $x$, and $y^{-1}$ is called the *multiplicative inverse* of $y$.

We will give several of the elementary consequences of these five axioms in the following theorem and problem.

### THEOREM 4

a) *If $x = y$ and $a = b$, then $x + a = y + b$.*
b) *If $x + y = x$, then $y = 0$.*
c) *$0x = 0$ for all $x \in R$.*
d) *$-(-x) = x$.*
e) *If $xy = 0$, then either $x = 0$ or $y = 0$.*

*Proof*  a) Since $x = y$ and $a = b$, we have

$$(x, a) = (y, b) \in R \times R.$$

Hence

$$A(x, a) = A(y, b) \quad \text{or} \quad x + a = y + b.$$

b) Since $x + y = x$ and $-x = -x$, we have

$$-x + (x + y) = -x + x$$

or

$$(-x + x) + y = -x + x.$$

We also have $(y + z)x = yx + zx$. Indeed,

$$\begin{aligned}
(y + z)x &= x(y + z) \quad \text{(by Axiom 1)} \\
&= xy + xz \quad \text{(by Axiom 3)} \\
&= yx + zx \quad \text{(by Axiom 1).}
\end{aligned}$$

Note, however, that we do *not* assume that

$$x + (yz) = (x + y)(x + z),$$

→ which would be the form of the distributive law if the roles of addition and multiplication were reversed.

→ Each of these elements is unique. For example, to show 0 is unique, suppose that $0' \in R$ also satisfies $0' + x = x$ for all $x$. We then have

$$\begin{aligned}
0 &= 0' + 0 \quad \text{(by assumption letting } x = 0) \\
&= 0 + 0' \quad \text{(by Axiom 1)} \\
&= 0' \quad \text{(by Axiom 4).}
\end{aligned}$$

The proof that 1 is unique is similar.

→ Each of these elements is unique. For example, suppose $z$ also satisfies $zy = 1$. Then

$$\begin{aligned}
y^{-1} &= 1y^{-1} \quad \text{(by Axiom 4)} \\
&= (zy)y^{-1} \quad \text{(by assumption)} \\
&= z(yy^{-1}) \quad \text{(by Axiom 2)} \\
&= z(y^{-1}y) \quad \text{(by Axiom 1)} \\
&= z1 \quad \text{(by Axiom 5)} \\
&= 1z = z \quad \text{(by Axiom 1).}
\end{aligned}$$

The proof that $-x$ is unique is similar.

---

*Problem 4*  Prove that

a) If $x = y$ and $a = b$, then $xa = yb$
b) If $ax = ay$ and $a \neq 0$, then $x = y$
c) If $y \neq 0$ ($y \in R$), then $(y^{-1})^{-1} = y$
d) $(-x)(-y) = xy$
e) $x^2 - y^2 = (x + y)(x - y)$

---

Therefore

$$0 + y = 0 \quad \text{and} \quad y = 0.$$

c) Since $xx = (x + 0)x = xx + 0x$, it follows from (b) that $0x = 0$.

d) $-(-x) = -(-x) + 0 = -(-x) + (-x + x)$
$$= (-(-x) + (-x)) + x = 0 + x = x.$$

e) Assuming $x \neq 0$ we will show that $y = 0$. We have

$$y = 1y = (x^{-1}x)y = x^{-1}(xy) = x^{-1}0 = 0 \quad \bullet$$

The operations of addition and multiplication that are defined for the set of real numbers permit us to define similar operations for functions. Suppose $f$, $g$: $A \to R$ are functions. We then define the *sum* $f + g$: $A \to R$ and *product* $fg$: $A \to R$ by the formulas

$$(f + g)(x) = f(x) + g(x) \quad \text{and} \quad (fg)(x) = f(x)g(x).$$

Also when $f$: $A \to R - \{0\}$ we may define $\frac{1}{f}(x) = \frac{1}{f(x)}$. (Do not confuse the multiplicative inverse $\frac{1}{f}$ with the composite inverse $f^{-1}$.)

*Order properties*

**Axiom 6** There is a subset $R^+ \subset R$ with the property that if $x \in R$, then exactly one of the statements

$$x \in R^+, \ x = 0, \ -x \in R^+$$

holds. Moreover, if $x, y \in R^+$, then

$$x + y \in R^+ \quad \text{and} \quad xy \in R^+.$$

The elements of $R^+$ are called *positive real numbers* and the set $R^- = \{x \in R: -x \in R^+\}$ is called the *negative real numbers*. In view of Axiom 6 we have that

$$R^+ \cup R^- \cup \{0\} = R.$$

Some elementary consequences of these notions are given in Theorem 5 and Problem 5 that follow.

**THEOREM 5**

a) *If $x > 0$ and $y < 0$, then $xy < 0$.*

b) *If $x \neq 0$, then $xx > 0$. In particular,*

$$1 = 1 \cdot 1 = (1)(1) > 0.$$

Axioms 1–5 for the real numbers have analogs for functions. For example, we will show that if

$$f, g, h: A \to R \text{ are functions}$$
$$\text{then } f(g + h) = fg + fh.$$

Indeed, we have

$(f(g + h))(x)$
$\quad = f(x)(g + h)(x)$ (by the definition of the product of functions)
$\quad = f(x)(g(x) + h(x))$ (by the definition of the sum of functions)
$\quad = f(x)g(x) + f(x)h(x)$ (by Axiom 3)
$\quad = (fg)(x) + (fh)(x)$ (by the definition of product of functions)
$\quad = (fg + fh)(x)$ (by the definition of sum of functions)

Therefore we have

$$(f(g + h))(x) = (fg + fh)(x)$$

for all $x \in A$, and it follows that

$$f(g + h) = fg + fh$$

by the definition of equality of functions.

When $x \in R^+$ we will write $x > 0$ or $0 < x$, and for $x \in R^-$ we write $x < 0$ or $0 > x$. More generally, given $x, y \in R$ we write $x < y$ or $y > x$ whenever $y - x \in R^+$.
We will use the notation $x \leq y$ or $y \geq x$ to mean

"$x < y$ or $x = y$."

**Problem 5** Prove the following assertions:

a) If $x < y$ and $y < z$ then $x < z$.
b) If $x < 0$ and $y < 0$ then $xy > 0$.

c) *If $x < y$ and $a > 0$, then $ax < ay$.*

*Proof* a) Since $x > 0$ and $y < 0$ we have $x \in R^+$ and $-y \in R^+$. Hence by Axiom 6 $x(-y) = -xy \in R^+$, so $xy \in R^-$, or $xy < 0$.

b) Since $x \neq 0$ we know that either $x \in R^+$ or $x \in R^-$. If $x \in R^+$ then $xx \in R^+$ by Axiom 6. If $x \in R^-$, then $-x \in R^+$ and we have $xx = (-x)(-x) \in R^+$.

c) Since $x < y$ and $a > 0$ we have that $y - x$, $a \in R^+$. Hence $a(y - x) \in R^+$ or $ay - ax \in R^+$. Therefore

$$ay - ax > 0,$$

so $ay > ax$, or $ax < ay$ ●

*Inductive property* A subset $A \subset R$ is said to be *inductive* if $1 \in A$ and whenever $x \in A$ then $x + 1 \in A$.

*Axiom 7* There is a unique inductive subset $N \subset R$ with the property that if $A$ is any other inductive subset of $R$ then $N \subset A$. The set $N$ is called the *natural numbers* or *positive integers*.

We can now define *integers* and *rational* numbers. A real number $x$ is called an *integer* if $x = 0$, $x \in N$, or $-x \in N$. A real number $y$ is called a *rational number* if it can be written in the form $y = \frac{m}{n}$ where $m$ and $n$ are integers and $n \neq 0$.

The *principle of mathematical induction* is based on Axiom 7 and may be described as follows.

Suppose that for each natural number $n$ we are given a statement $S(n)$. Suppose that

$$S(1) \text{ is a true statement}$$

and that

$$S(k + 1) \text{ is true whenever } S(k) \text{ is true.}$$

It then follows that $S(n)$ is true for all natural numbers $n$. To see this, let

$$N_0 = \{n \in N : S(n) \text{ is true}\}.$$

Then we have $1 \in N_0$ and $k + 1 \in N_0$ whenever $k \in N_0$. Hence $N_0$ is an inductive set. Since $N$ is contained in every inductive set we have $N \subset N_0$. But $N_0$ is obviously contained in $N$, so $N = N_0$. Hence $S(n)$ is true for all natural numbers $n$. (An instance of a

---

c) If $x < y$ and $a < 0$ then $ax > ay$.

----

→ For example, the sets $R$ and $R^+$ are inductive sets.

→ In other words, the natural numbers form the smallest inductive subset of $R$.

The reader may wonder why we did not simply define the natural numbers (as in Chapter 1) to be the set containing

$$1, \quad 1 + 1 = 2, \quad 2 + 1 = 3, \quad 3 + 1 = 4, \quad \text{and so on.}$$

The reason is that the phrase "and so on" is too vague, and the purpose of Axiom 7 is to make it precise.

proof by mathematical induction is given in Example 5.)

---

**Example 5** Letting $S(n)$ denote the statement

$$1^2 + 2^2 + \cdots + n^2 = \frac{2n^3 + 3n^2 + n}{6},$$

we have

$$1^2 = \frac{2 \cdot 1^3 + 3 \cdot 1^2 + 1}{6},$$

so the statement $S(1)$ is true. Supposing now that $S(k)$ is true, we have

$$1^2 + 2^2 + \cdots + k^2 = \frac{2k^3 + 3k^2 + k}{6}.$$

Adding $(k + 1)^2$ to each side of this equation gives

$$1^2 + 2^2 + \cdots + k^2 + (k + 1)^2$$

$$= \frac{2k^3 + 3k^2 + k}{6} + (k + 1)^2$$

$$= \frac{2k^3 + 3k^2 + k + 6(k + 1)^2}{6}$$

$$= \frac{2k^3 + 3k^2 + k + 6k^2 + 12k + 6}{6}$$

$$= \frac{(2k^3 + 6k^2 + 6k + 2) + (3k^2 + 6k + 3) + (k + 1)}{6}$$

$$= \frac{2(k + 1)^3 + 3(k + 1)^2 + (k + 1)}{6}.$$

Since this is merely the statement $S(k + 1)$, we have shown that $S(k)$ implies $S(k + 1)$ for all $k$, hence $S(n)$ is true for all $n \in N$.

---

**THEOREM 6**   *Any finite set A of real numbers has both a least element and a greatest element.*

*Proof*   The argument will be by induction on the number of elements in $A$. If $A$ has one element then it is clearly the least element. Suppose now that the result is true for sets with $k$ elements, and let $A = \{a_1, \ldots, a_{k+1}\}$ be a set with $k + 1$ elements. By the induction assumption the set $\{a_1, \ldots, a_k\}$ has

A set $A$ is said to have $n$ elements ($n \in N$) if there is a bijection

$$a: \{1, 2, \ldots, n\} \to A$$

between the sets $\{1, 2, \ldots, n\}$ and $A$. In this case the set $A$ must consist of the elements $\{a(1), a(2), \ldots, a(n)\}$ and we often denote this by simply writing

$$A = \{a_1, a_2, \ldots, a_n\} \quad (a_k = a(k)).$$

A set is said to be *finite* if it is the null set or contains $n$ elements for some $n \in N$. A set is called *infinite* if it is not finite.

a least element. Call it $b$. Hence $b \leq a_i$ for $1 \leq i \leq k$. If

$$b \leq a_{k+1},$$

then $b$ is the least element of $A$. If $a_{k+1} \leq b$, then $a_{k+1} \leq a_i$ for $1 \leq i \leq k$ and $a_{k+1}$ is the least element. Hence in either case $A$ has a least element. By induction it now follows that any finite subset of $R$ has a least element. The argument for finding a greatest element is similar and we omit it ●

**COROLLARY** *Any nonempty subset $A$ of $N$ has a least element.*

To properly understand the technique of mathematical induction, it is important to recognize situations where it applies, but it is also useful to see instances where it does not apply. One such instance is shown in Example 7, which should be read after Example 6.

___

**Example 6** A man has 9 balls that are equal in weight except for one which is lighter (defective) than the rest. He also has a balance scale (without weights), and is required to find the defective ball in two balancings. How does he proceed?

*Solution* He places three balls on each side of the balance.

*Case 1* If they balance, the defective ball must be among the other three. Discarding the first six, he places two of the remaining three balls on the balance, one on each side. If they balance, the defective ball is the remaining one. If they do not balance, the defective ball is the one in the higher tray.

*Case 2* If the six balls do not balance, he discards the heavier three. Choosing two of the lighter three balls, he places them on either side of the scale. If they balance, the remaining ball is the defective one. If they do not, the ball in the higher tray is the lighter one.

___

Returning to more serious matters, we now give an example of how induction can be used to define an infinite number of quantities.

Given any $x \in R$, we define

$$x^1 = x$$

*Proof* Since $A$ is nonempty it contains some element $n$. Let $A' = A \cap \{1, 2, \ldots, n\}$. Then $A'$ is finite, so by the Theorem $A'$ contains a least element $m$. Hence if $a \in A'$ we have $m \leq a$. If $a \in A$ and $a \notin A'$ then $n \leq a$, and since $m \leq n$ it follows that $m \leq a$. Hence $m$ is the least element in $A$ ●

___

**Example 7** Given any finite number of balls, suppose that they are all equal in weight except for one which is lighter than the rest. Then the lighter ball can be found in two weighings.

*Proof* We will use induction on the number of balls. If there is one ball, the result is obvious. Assume that for any $k$ balls the lighter one can be found in two balancings. Now take $k + 1$ balls. We place one in our pocket and are then left with $k$ balls. If the lighter ball is amongst these $k$, our induction assumption provides a method of finding it in two weighings. If we do not find the defective ball among these $k$ balls after two balancings, then the defective ball must be the one in our pocket. Hence the result is true for $k + 1$ balls and, by induction, true for any finite number of balls.

___

**Problem 6** Find the error in the argument given in Example 7.

___

A classic problem related to the one in Example 6 is given below.

___

**Problem 7** The man in Example 6 has 12 balls and one is defective (either heavy or light). Without knowing whether it is heavy or light, find it in three weighings.

(Hint: Do not use induction, and do not attempt to solve the problem in the space provided here.)

___

and, assuming $x^k$ has been defined, we define $x^{k+1} = x^k x$. This, then, defines $x^n$ for all $n \in N$.

To see this, let

$$N_0 = \{n \in N: x^n \text{ is defined}\}.$$

Then $1 \in N_0$ and $k + 1 \in N_0$ whenever $k \in N_0$. Hence $N_0$ is an inductive set and $N_0 \supset N$. But $N \supset N_0$, so $N_0 = N$ and $x^n$ is defined for all $n \in N$.

*Axiom 8 Supremum* Suppose $A$ is a subset of $R$ that is bounded from above. Then there is an $m \in R$ that satisfies

a) $a \leq m$ for all $a \in A$, and
b) if $a \leq x$ for all $a \in A$, then $m \leq x$.

A set $A \subset R$ having the property that there is a number $v$ satisfying $a \leq v$ for all $a \in A$ is said to be *bounded from above*. If a set $B$ has the property that $u \leq b$ for all $b \in B$ and some $u \in R$, then $B$ is said to be *bounded from below*. A set is said to be *bounded* if it is bounded from above and from below.

The number $m$ is unique, is called the *supremum* of $A$, and is denoted by

$$m = \sup A.$$

To prove $m$ is unique, suppose $m'$ also satisfies the conditions (a) and (b). Since $a \leq m'$ for all $a \in A$, we have $m \leq m'$. Reversing the roles of $m$ and $m'$, we then obtain $m' \leq m$. Hence $m = m'$.

A corresponding property for sets bounded from below is given in the following theorem:

**THEOREM 7** *If $B$ is a subset of $R$ that is bounded from below, then there is a unique element $n \in R$ that satisfies*

    a) *$n \leq b$ for all $b \in B$, and*
    b) *if $x \leq b$ for all $b \in B$, then $x \leq n$.*

The number $n$ given in this theorem is called the *infimum* of $B$ and is denoted by "inf $B$."

We also introduce here the related notions of the maximum and minimum of a set. If $\sup A$ exists and belongs to $A$, then $\sup A$ is called the *maximum* of $A$ and is denoted by "max $A$." Similarly, if $\inf B$ exists and belongs to $B$, then $\inf B$ is called the *minimum* of $B$ and is denoted by "min $B$."

*Proof* Let $-B = \{-b: b \in B\}$. Then $-B$ is bounded from above so $\sup(-B)$ exists. Denoting this by $-n = \sup(-B)$, it follows that $n = \inf B$ ●

We can now prove the very reasonable:

**THEOREM 8** *The natural numbers are not bounded from above.*

*Proof* The proof will be by contradiction. Suppose there is a $u \in R$ satisfying

$$n \leq u \quad \text{for all } n \in N.$$

Then by Axiom 8, $\sup N = u_0$ exists, so it follows that $n \leq u_0$ for all $n \in N$. Since $n + 1 \in N$ for all $n \in N$, we also have $n + 1 \leq u_0$ for all $n \in N$ or $n \leq u_0 - 1$ for all $n \in N$. But according to Axiom 8, we must have $u_0 \leq u_0 - 1$, and this is clearly impossible ●

*"How often have I said to you that when you have eliminated the impossible, whatever remains, however improbable, must be the truth."*
                      (Sherlock Holmes to Dr. Watson)

**COROLLARY** *Given $\varepsilon > 0$ there is an $n \in N$ satisfying $\frac{1}{n} < \varepsilon$.*

**THEOREM 9** *Between any two distinct real numbers $x$ and $y$ there is a rational number.*

*Proof* We first suppose that $0 < x < y$. Applying the

*Proof* The real number $\frac{1}{\varepsilon}$ cannot be an upper bound for $N$ in view of the theorem. Hence, for some $n \in N$ we have $\frac{1}{\varepsilon} < n$. Rewriting this we have $\frac{1}{n} < \varepsilon$ ●

corollary of Theorem 8, choose $n_0 \in N$ such that $\frac{1}{n_0} < y - x$. Now consider the set

$$A = \left\{ n \in N: x < \frac{n}{n_0} \right\}.$$

By the corollary to Theorem 6 it follows that $A$ contains a least element $m_0$. Hence $x < \frac{m_0}{n_0}$ and it must be that $\frac{m_0 - 1}{n_0} \le x$ because in the contrary case, $x < \frac{m_0 - 1}{n_0}$, we would contradict the choice of $m_0$. Hence we have

$$x < \frac{m_0}{n_0} = \frac{m_0 - 1}{n_0} + \frac{1}{n_0} < x + (y - x) = y,$$

so $\frac{m_0}{n_0}$ is the desired rational number.

Now take arbitrary $x < y$, and choose some integer $n$ so that $0 < x + n < y + n$. We have just shown that there is a rational number $r$ satisfying $x + n < r < y + n$. Hence $x < r - n < y$ and $r - n$ is rational ●

→ The set $A = \left\{ n \in N: x < \frac{n}{n_0} \right\}$ is nonempty because if it were empty then we would have $\frac{n}{n_0} \le x$ for all $n \in N$. This means that $n \le n_0 x$ for all $n \in N$, from which it follows that $N$ is bounded from above. But this contradicts Theorem 8.

---

**Example 8**   a) Let $C = \{x \in R: 0 < x \le 1\}$. Then

$$\sup C = \max C = 1 \quad \text{and} \quad \inf C = 0.$$

The min $C$ does not exist.

b) If $D = \{x_1, x_2, \ldots, x_n\}$ is a finite set, then it is clearly bounded and both the max $D$ and min $D$ exist (Theorem 6).

---

## D. Exercises

**B1.** Given sets $A$, $B$, and $C$, prove
 a) $A \cup (B \cap C) = (A \cup B) \cap (A \cup C)$
 b) $(A \cap B) \cup C = A \cap (B \cup C)$ for all $B$ if and only if $C \subset A$

**B2.** Define the operation $\triangle$ by the formula

$$A \triangle B = A \cup B - A \cap B.$$

Prove that
 a) $A \triangle B = (A - B) \cup (B - A)$
 b) $(A \triangle B) \triangle C = A \triangle (B \triangle C)$

**B3.** Suppose $f: A \to B$ is a function, and $C$ and $D$ are subsets of $A$. Prove that
 a) $f(C \cup D) = f(C) \cup f(D)$
 b) $f(C \cap D) \subset f(C) \cap f(D)$
 c) $f(C \cap D) = f(C) \cap f(D)$ need not necessarily hold.

**B4.** Prove, using induction, the Binomial Theorem

$$(a + b)^n = \sum_{k=0}^{n} \binom{n}{k} a^k b^{n-k}$$

where

$$n \in N, a, b \in R, \quad \text{and} \quad \binom{n}{k} = \frac{n!}{k!(n-k)!}.$$

**B5.** Given $x \in R$ with $x > 1$, prove that the set

$$\{x, x^2, x^3, \ldots\}$$

is not bounded from above.

**M6.** Given sets $A$, $B$, and $C$, prove that

$$A \times (B \cap C) = (A \times B) \cap (A \times C).$$

**M7.** Prove or give a counterexample to each of the following.
 a) $A \cap (B \times C) = (A \cap B) \times (A \cap C)$
 b) $A \cap (B \triangle C) = (A \cap B) \triangle (A \cap C)$
 c) $A \triangle B = A \triangle C$ implies $B = C$

**M8.** Suppose $f: A \to B$ is a function and $A$ is a finite set. Prove that $f$ is injective if and only if $f$ is surjective.

**M9.** If we replace the set $R$ in Theorem 7 by the set of rational numbers, does the result still hold?

**M10.** Suppose $A$ is a finite set having $n$ elements.
    a) Prove that $A$ contains $2^n$ subsets. (Note that $\phi$ and $A$ are subsets of $A$.)
    b) Prove that the number of functions

$$f: A \to \{0, 1\}$$

    is equal to $2^n$.

**M11.** Prove that for any $n \in N$

    a) $(1 + 2 + 3 + \cdots + n)^2 = \dfrac{n^2(n + 1)^2}{4}$

    b) $1^3 + 2^3 + 3^3 + \cdots + n^3 = \dfrac{n^2(n + 1)^2}{4}$

**M12.** Prove that between any two positive rational numbers there is a square of a positive rational number.

**M13.** If $q$ is a positive rational number, prove that there is an integer $n$ such that $2^{-n} \le q$.

**M14.** Let $r$ be a real number and let $A$ be the set of all rational numbers less than $r$. Prove that $r$ is the supremum of $A$.

**M15.** We are given three spindles named $A$, $B$, and $C$, the first of which has on it a stack of $n$ discs of increasing diameter numbered $1, \ldots, n$.

(Here $n = 4$)

$A$      $B$      $C$

A game is played with them in which the only allowable move is to remove a disc from the top of a stack and place it on another spindle, but never on a disc of smaller diameter. The object of the game is to move the discs from spindle $A$ to spindle $B$ by means of a sequence of permissible moves. Prove by induction that this can be done for any number of discs.

**M16.** Two sets are said to have the same *cardinality* if there exists a bijection between them. Any set that is finite or has the same cardinality as $N$ is said to be *countable*.
    a) Show that $N$ and the set of even natural numbers have the same cardinality.
    b) Show that any infinite subset of $N$ is countable.
    c) Show that the set of all rational numbers is countable.

**M17.** Prove that the real numbers do not form a countable set by filling in the details in the following argument.
    Suppose $\phi: N \to [0, 1]$ is a bijection. For each $n$, using the decimal expansion for $\phi(n)$, we have

$$\phi(n) = .x_{n1}x_{n2}x_{n3}\cdots$$

    Now define $y = .y_1 y_2 y_3 \ldots \in [0, 1]$ with the property that for each $n \in N$, $|y_n - x_{nn}| = 2$. Show that $y$ cannot equal $\phi(n)$ for any $n \in N$.

**M18.** Given any set $A$, let $\mathcal{P}(A)$ denote the set of all subsets of $A$. Show that $A$ and $\mathcal{P}(A)$ do not have the same cardinality by completing the following argument.
    Assume $\phi: A \to \mathcal{P}(A)$ is a bijection and let

$$A_0 = \{a \in A: a \notin \phi(a)\}.$$

    Now show that there can be no $a_0 \in A$ satisfying $\phi(a_0) = A_0$.

---

## 9.2   *Topological Notions*

We begin this section with a precise definition of the limit of a sequence. A number of important concepts are then defined in terms of limits, and several important theorems are proven.

    →   *"The errors of definitions multiply themselves according as the reckoning proceeds; and lead men into absurdities, which at last they see but cannot avoid, without reckoning anew from the beginning."*

(Thomas Hobbes)

## A. Sequences

For $x \in R$, the *absolute value* $|x|$ is defined by the formula

$$|x| = \begin{cases} x & \text{if} & x \in R^+ \\ 0 & \text{if} & x = 0 \\ -x & \text{if} & -x \in R^+. \end{cases}$$

Exactly as in 1.1B it follows that for all $x, y \in R$

$$|x + y| \leq |x| + |y|,$$
$$||x| - |y|| \leq |x - y|,$$

and

$$|xy| = |x|\,|y|.$$

A function $a: N \to R$ is called a *sequence* of real numbers. For our purposes it will be convenient to describe it by listing the values of the function $a$ as

$$a(1), a(2), a(3), \dots.$$

This is also denoted by writing

$$\{a_1, a_2, a_3, \dots\},$$

or more simply by

$$\{a_n\}_{n=1}^{\infty} \quad \text{or even} \quad \{a_n\}.$$

A sequence $\{a_n\}$ is said to *converge to a*, or be *convergent*, if for any real number $\varepsilon > 0$ there is an $n_0 \in N$ such that

$$|a_n - a| < \varepsilon \quad \text{whenever } n > n_0.$$

The number $a$ is called the *limit* of the sequence. This is the precise version of the statement

"$a_n$ approaches $a$ as $n$ goes to infinity."

When the sequence $\{a_n\}$ converges to $a$ we denote this by writing $a_n \to a$ or $\lim_{n \to \infty} a_n = a$.

A sequence $\{a_n\}$ is said to be *bounded* if there exist $u, v \in R$ such that $u \leq a_n \leq v$ for all $n$. Equivalently we may say that a sequence is bounded if $\sup |a_n| < \infty$.

### THEOREM 1

a) *If $a_n \to a$ and $a_n \to b$, then $a = b$.*
b) *A convergent sequence is bounded.*

---

The absolute value is a special case of a *metric*, which is defined as follows:

Given a set $M$, a function

$$m: M \times M \to R$$

is called a *metric on M* if it satisfies

a) $m(x, y) \geq 0$  and equals zero iff  $x = y$,

b) $m(x, y) = m(y, x)$,

and

c) $m(x, z) \leq m(x, y) + m(y, z)$.

A set $M$ together with a metric $m$ is called a *metric space*, and many of the notions given below carry over to metric spaces. Notice that the function $m(x, y) = |x - y|$ is a metric on $R$.

When we denote a sequence by $\{a_n\}$ it is important to remember that this represents the function

$$a: N \to R$$

defined by the formula $a(n) = a_n$.

We could define convergence in any of the following equivalent ways: For any $\varepsilon > 0$ there is an $n_0 \in N$ such that

$$|a_n - a| \leq \varepsilon \quad \text{whenever} \quad n > n_0,$$
$$|a_n - a| \leq \varepsilon \quad \text{whenever} \quad n \geq n_0,$$

or

$$|a_n - a| < \varepsilon \quad \text{whenever} \quad n \geq n_0.$$

Also, this notion of convergence for sequences can be used to define the convergence of series in the following way:

The *series* $\sum_{k=1}^{\infty} x_k$ is said to be *convergent* if the sequence $\{S_n\}$, where $S_n = \sum_{k=1}^{n} x_k$, is a convergent sequence. (This makes precise the definition given in 7.1C.) More information about series will be given in 10.2, but we will also make a few additional comments about them in this chapter.

---

**Example 1**  Prove that

a) The sequence $\{1, \frac{1}{2}, \frac{1}{3}, \frac{1}{4}, \dots\}$ converges to 0.
b) The sequence $\{1, 0, 1, 0, 1, 0, \dots\}$ is bounded but not convergent.

*Proof*   a) Suppose $a \neq b$ and let $\varepsilon = \frac{|a-b|}{2}$. Since $a_n \to a$ there is an $n_1 \in N$ such that

$$|a_n - a| < \varepsilon \quad \text{for all } n > n_1.$$

Since $a_n \to b$ there is an $n_2 \in N$ such that

$$|a_n - b| < \varepsilon \quad \text{for all } n > n_2.$$

Then for $k > \max(n_1, n_2)$ we have

$$|a - b| = |a - a_k + a_k - b| \leq |a - a_k| + |a_k - b|$$
$$< \varepsilon + \varepsilon$$
$$= 2\varepsilon = |a - b|.$$

In other words $|a - b| < |a - b|$, and this is impossible. We therefore have that $a = b$.

b) If $a_n \to a$ then there is an $n_0 \in N$ such that $|a_n - a| < 1$ for $n > n_0$. Hence

$$|a_n| = |a_n - a + a| \leq |a_n - a| + |a| \leq 1 + |a|$$

for $n > n_0$. Letting $M = \max \{|a_1|, \ldots, |a_{n_0}|, 1 + |a|\}$, we have $|a_n| \leq M$ for all $n \in N$   ●

---

**THEOREM 2**   *Suppose $a_n \to a$ and $b_n \to b$. Then*
   a) $a_n + b_n \to a + b$   *and*
   b) $a_n b_n \to ab$.

*Proof*   a) Take any $\varepsilon > 0$. Then there is an $n_1$ such that

$$|a_n - a| < \frac{\varepsilon}{2} \quad \text{for all } n > n_1,$$

and there is an $n_2$ such that

$$|b_n - b| < \frac{\varepsilon}{2} \quad \text{for all } n > n_2.$$

Then for $n \geq \max \{n_1, n_2\}$ we have

$$|(a_n + b_n) - (a + b)| = |a_n - a + b_n - b|$$
$$\leq |a_n - a| + |b_n - b| < \frac{\varepsilon}{2} + \frac{\varepsilon}{2} = \varepsilon.$$

Hence $a_n + b_n \to a + b$.
   b) We have

$$|a_n b_n - ab| = |a_n b_n - a_n b + a_n b - ab|$$
$$\leq |a_n b_n - a_n b| + |a_n b - ab|$$
$$= |a_n| |b_n - b| + |a_n - a| |b|.$$

Since $\{a_n\}$ is a convergent sequence, it is a bounded

*Proof*   a) Take any $\varepsilon > 0$. By the corollary to Theorem 8 (9.1C) there is an $n_0 \in N$ such that $\frac{1}{n_0} < \varepsilon$.

Hence, if $n > n_0$ we have $\frac{1}{n} < \frac{1}{n_0} < \varepsilon$, so $\frac{1}{n} < \varepsilon$ whenever $n > n_0$.

   b) The sequence $\{1, 0, 1, 0, \ldots\}$ may be written as $\{a_1, a_2, \ldots\}$ where $a_n = 1$ if $n$ is odd, and $a_n = 0$ if $n$ is even. If $a_n \to a$, then, letting $\varepsilon = \frac{1}{2}$, it follows that there is an $n_0$ such that

$$|a_n - a| < \tfrac{1}{2} \quad \text{for all } n > n_0.$$

Hence, if $m, n > n_0$ we have

$$|a_n - a_m| = |a_n - a + a - a_m|$$
$$\leq |a_n - a| + |a - a_m| < \tfrac{1}{2} + \tfrac{1}{2} = 1.$$

Letting $n$ be even and $m$ be odd, we have that $1 = |0 - 1| < 1$, and this is a contradiction.

---

→   Notice that the sequence $\{a, a, a, \ldots\}$ obviously converges to $a$. Hence, from (b) we have that if $b_n \to b$, then $ab_n \to ab$ for any $a \in R$. In particular, $-b_n \to -b$, and we have $a_n - b_n \to a - b$.

→   It may appear to the reader that we are not using the definition of convergence properly. We could do the following: Let $\varepsilon_1 = \frac{\varepsilon}{2}$. Then $\varepsilon_1 > 0$ and since $a_n \to a$ there must be an $n_1$ such that $|a_n - a| < \varepsilon_1$ for all $n > n_1$. The point is that the "$\varepsilon > 0$" is taken as any positive real number, but the form in which it is written is not important.

---

**Problem 1**   a) Suppose $a_n \to a$ and no $a_n$ is equal to zero. Show that $\{a_n^{-1}\}$ is convergent if and only if $a \neq 0$.
b) Use Theorem 9 of 9.1C to show that if $x \in R$ is irrational (not rational) then there is a sequence of rational numbers $\{r_n\}$ satisfying $r_n \to x$.

sequence by Theorem 1b, so there is an $A > 0$ such that $|a_n| \leq A$ for all $n$. Choosing $B > 0$ with $|b| < B$, we then have

$$|a_n b_n - ab| \leq A|b_n - b| + |a_n - a|B \quad \text{for all } n.$$

Now let $\varepsilon > 0$ and choose $n_1$ and $n_2$ so that

$$|b_n - b| < \frac{\varepsilon}{2A} \quad \text{for } n > n_1$$

and

$$|a_n - a| < \frac{\varepsilon}{2B} \quad \text{for } n > n_2.$$

It follows that if $n > \max\{n_1, n_2\}$, then

$$|a_n b_n - ab| < A\frac{\varepsilon}{2A} + \frac{\varepsilon}{2B}B = \frac{\varepsilon}{2} + \frac{\varepsilon}{2} = \varepsilon.$$

Hence $a_n b_n \to ab$ ●

Suppose $\{a_1, a_2, a_3, \ldots\}$ is a sequence and

$$\{n_1 < n_2 < n_3 \ldots\}$$

is a subset of $N$. Then $\{a_{n_1}, a_{n_2}, a_{n_3}, \ldots\}$, or more briefly $\{a_{n_k}\}$, is called a *subsequence* of $\{a_n\}$.

---

*Example 2* Take the sequence $\{1, \frac{1}{2}, \frac{1}{3}, \frac{1}{4}, \frac{1}{5}, \ldots\}$. Then

a) $\{1, \frac{1}{3}, \frac{1}{5}, \frac{1}{7}, \ldots\}$ is a subsequence.
b) $\{(\frac{1}{2})^1, (\frac{1}{2})^2, (\frac{1}{2})^3, \ldots\}$ is a subsequence.
c) $\{\frac{1}{2}, 1, \frac{1}{3}, \frac{1}{4}, \frac{1}{5}, \ldots\}$ is *not* a subsequence.

d) $\{1, \frac{1}{2}, \frac{1}{3}, \frac{1}{3}, \frac{1}{3}, \frac{1}{3}, \ldots\}$ is *not* a subsequence.

a) $\{n_1, n_2, n_3, \ldots\} = \{1, 3, 5, 7, 9, \ldots\}$
b) $\{n_1, n_2, n_3, \ldots\} = \{2, 2^2, 2^3, 2^4, \ldots\}$
c) $\{n_1, n_2, n_3, \ldots\} = \{2, 1, 3, 4, 5, \ldots\}$, so does not satisfy $n_1 < n_2 < n_3 < \cdots$.
d) $\{n_1, n_2, n_3, \ldots\} = \{1, 2, 3, 3, 3, 3, \ldots\}$ and again does not satisfy $n_1 < n_2 < n_3 < n_4 < \cdots$.

---

Given the subsequence $\{a_{n_k}\}$ of $\{a_n\}$, let the sequence $\{b_1, b_2, b_3, \ldots\}$ be given by $b_k = a_{n_k}$. In this sense we may consider a subsequence as a sequence, and in particular, speak of the convergence of a subsequence. Indeed, the subsequence $\{a_{n_k}\}$ is said to converge to $a$ if for any $\varepsilon > 0$ there is a $k_0$ such that

$$|a_{n_k} - a| < \varepsilon \quad \text{for all } k > k_0.$$

Let us emphasize this. Suppose $N_0 = \{n_1 < n_2 < n_3 < \cdots\} \neq N$, and $\{a_{n_k}\}$ is a subsequence of $\{a_n\}$. Then the mapping

$$n_k \mapsto a_{n_k}: N_0 \to R$$

is not a sequence, but the mapping

$$k \mapsto a_{n_k}: N \to R \text{ is a sequence.}$$

A sequence can have convergent subsequences without itself being convergent. For example, the sequence $\{1, 0, 1, 0, 1, \ldots\}$ is not convergent but each of the subsequences $\{1, 1, 1, \ldots\}$ and $\{0, 0, 0, \ldots\}$ is obviously convergent.

**THEOREM 3** *If the sequence $\{a_n\}$ converges to $a$, then every subsequence converges to $a$.*

*Proof* Let $\{a_{n_k}\}$ be a subsequence of $\{a_n\}$ and let

$\varepsilon > 0$. There is an $n_0$ such that $|a_n - a| < \varepsilon$ whenever $n > n_0$. Hence if $k_0$ is chosen so that $n_{k_0} \geq n_0$ it follows that

$$|a_{n_k} - a| < \varepsilon \quad \text{for all } k > k_0.$$

Therefore $\{a_{n_k}\}$ converges to $a$ ●

## B. Nontrivial Theorems

**THEOREM 4**  *Suppose $\{a_n\}$ is a bounded sequence and either*

$$a_1 \leq a_2 \leq a_3 \leq \cdots$$

*or*

$$a_1 \geq a_2 \geq a_3 \geq \cdots.$$

*Then $\{a_n\}$ is convergent.*

*Proof*  Assume $a_1 \leq a_2 \leq a_3 \leq \cdots$. Since $\{a_n\}$ is bounded, the number $a = \sup a_n$ exists. We will show that $a_n \to a$. Let $\varepsilon > 0$. Since $a - \varepsilon < a = \sup a_n$, there must be an $a_k > a - \varepsilon$. Because $a_n \geq a_k$ for $n > k$ and $a_n \leq a$ for all $n$ we have

$$a - \varepsilon < a_n \leq a \quad \text{for all } n > k.$$

It follows that

$$|a_n - a| < \varepsilon \quad \text{whenever } n > k,$$

and $\{a_n\}$ is therefore a convergent sequence. The argument for the case when $a_1 \geq a_2 \geq a_3 \geq \cdots$ is similar ●

A sequence $\{a_n\}$ is said to be *Cauchy* if for any $\varepsilon > 0$ there is an $n_0 \in N$ such that

$$|a_n - a_m| < \varepsilon \quad \text{whenever } n, m \geq n_0.$$

An equivalent formulation of this is

$$|a_{n+k} - a_n| < \varepsilon \quad \text{for all } n \geq n_0 \quad \text{and all } k \geq 0.$$

The fundamental result about Cauchy sequences is given in the theorem below, but first we prove the following lemma.

**LEMMA**  *A Cauchy sequence is bounded.*

*Proof*  Let $\{a_n\}$ be a Cauchy sequence and let $\varepsilon = 1$. Then there is an $n_0$ such that

$$|a_n - a_m| < 1 \quad \text{whenever } n, m \geq n_0.$$

→ This result justifies the statement that an infinite decimal defines a real number. For example, take the infinite decimal

$$.d_1 d_2 d_3 \ldots$$

where $d_n$ is a digit (0, 1, 2, . . . , or 9). Letting $a_n = .d_1 d_2 \ldots d_n$ we clearly have that

$$a_1 \leq a_2 \leq a_3 \leq \cdots.$$

Moreover, the sequence $\{a_n\}$ is bounded ($0 \leq a_n \leq 1$) for all $n$. By Theorem 4, the sequence $\{a_n\}$ converges to some real number $a$, which is denoted by the infinite decimal

$$a = .d_1 d_2 d_3 \ldots.$$

→ In Theorem 1 of 7.2A we stated without proof that if $x_k \geq 0$ and the partial sums $S_n$ are bounded, then $\sum_{k=1}^{\infty} x_k$ is convergent. We may prove this now using Theorem 4. Indeed, we have

$$S_n = \sum_{k=1}^{n} x_k \leq \sum_{k=1}^{n+1} x_k = S_{n+1}$$

since $x_{n+1} \geq 0$. Hence, if $S_1 \leq S_2 \leq S_3 \leq \cdots$ is bounded, it is convergent.

→ Starting with the rational numbers, it is possible to construct the real numbers using Cauchy sequences of rationals. The basic idea is very simple. One first defines a real number to be a Cauchy sequence of rational numbers, and then defines two Cauchy sequences $\{a_n\}$ and $\{b_n\}$ to be equal real numbers if $a_n - b_n \to 0$. The identification of these sequences with numbers is then obtained via the limits of the sequences.

In particular,    $|a_n - a_{n_0}| < 1$   for all $n \geq n_0$,

so that      $|a_n| = |a_n - a_{n_0} + a_{n_0}|$

$\leq |a_n - a_{n_0}| + |a_{n_0}| < 1 + |a_{n_0}|$.

Hence $|a_n| < \max\{|a_1|, \ldots, |a_{n_0-1}|, 1 + |a_{n_0}|\}$ for all $n \in N$, so the sequence is bounded ●

**THEOREM 5**  *A sequence is Cauchy if and only if it is convergent.*

*Proof* (*Convergent implies Cauchy*)  This is the easy half of the theorem.

Assume $a_n \to a$ and $\varepsilon > 0$. Then there is an $n_0$ such that

$$|a_n - a| < \frac{\varepsilon}{2} \quad \text{whenever } n > n_0.$$

Hence for $m, n > n_0$ we have

$|a_n - a_m| = |a_n - a + a - a_m|$

$\leq |a_n - a| + |a - a_m| < \frac{\varepsilon}{2} + \frac{\varepsilon}{2} = \varepsilon,$

and this says that $\{a_n\}$ is Cauchy ●

**THEOREM 6**  *A bounded sequence contains a convergent subsequence.*

*Proof*  Let $\{a_n\}$ be the bounded sequence, and let

$$b_n = \sup\{a_n, a_{n+1}, a_{n+2}, \ldots\}.$$

Then, as in Theorem 5, it follows that

$$b_1 \geq b_2 \geq b_3 \geq \cdots.$$

Since $\{b_n\}$ is also bounded it follows from Theorem 4 that it is convergent. Denoting its limit by $b$ we will now find a subsequence $\{a_{n_k}\}$ that converges to $b$. To do so we inductively choose subsequences $\{a_{n_r}\}$ and $\{b_{m_r}\}$ satisfying

$$|a_{n_r} - b_{m_r}| < \frac{1}{r} \quad \text{for all } r \in N.$$

Let $m_1 = 1$. Since

$$b_{m_1} = \sup_{n \geq 1}\{a_n\},$$

*Proof* (*Cauchy implies convergent*)  Let $\{a_n\}$ be a Cauchy sequence. By the lemma the sequence is bounded, so that for some constant $K$,

$$|a_n| \leq K \quad \text{for all } n \in N.$$

Letting $b_k = \sup\{a_k, a_{k+1}, a_{k+2}, \ldots\}$, it then follows that

$$b_1 \geq b_2 \geq b_3 \geq \cdots \geq -K.$$

By Theorem 4 we have that $\{b_n\}$ is a convergent sequence, so $b_n \to b$ for some $b \in R$. We will show that $a_n \to b$. Let $\varepsilon > 0$. Choose $n_1$ so that $|a_n - a_m| < \frac{\varepsilon}{4}$ for $n, m \geq n_1$. We then have $|a_n - a_{n_1}| < \frac{\varepsilon}{4}$ for all $n \geq n_1$, or

$$a_{n_1} - \frac{\varepsilon}{4} < a_n < a_{n_1} + \frac{\varepsilon}{4} \quad \text{for all } n \geq n_1.$$

Since this is true for all $n \geq n_1$, we obtain

$$a_{n_1} - \frac{\varepsilon}{4} < b_n \leq a_{n_1} + \frac{\varepsilon}{4} \quad \text{for all } n \geq n_1.$$

Hence, both $a_n$ and $b_n$ are within $\frac{\varepsilon}{4}$ of $a_n$, so we have

$$|a_n - b_n| \leq \frac{\varepsilon}{2} \quad \text{for } n \geq n_1.$$

Now choose $n_2$ such that $|b_n - b| < \frac{\varepsilon}{2}$ for $n \geq n_2$. Then for $n \geq \max\{n_1, n_2\}$ we have

$$|a_n - b| \leq |a_n - b_n| + |b_n - b| < \frac{\varepsilon}{2} + \frac{\varepsilon}{2} = \varepsilon,$$

which proves that $a_n \to b$ ●

**COROLLARY** (*to Theorem 5*)  *An absolutely convergent series is convergent.* (See the corollary to Theorem 2 (7.2B).)

*Proof*  Suppose that $\sum_{k=1}^{\infty} |x_k| < \infty$,   let

it follows that $b_{m_1} - 1 < a_{n_1} \leq b_{m_1}$ for some $a_{n_1}$ with $n_1 \geq 1$. Hence

$$a_{n_1} - b_{m_1} < 1.$$

Suppose $a_{n_k}$ and $b_{m_k}$ have been chosen and pick $m_{k+1} > n_k$. Since

$$b_{m_{k+1}} = \sup_{n \geq m_{k+1}} \{a_n\},$$

it follows that for some $n_{k+1} \geq m_{k+1}$

$$b_{m_{k+1}} - \frac{1}{k+1} < a_{n_{k+1}} \leq b_{m_{k+1}}.$$

We therefore have

$$|a_{n_{k+1}} - b_{m_{k+1}}| < \frac{1}{k+1},$$

so by induction it follows that

$$|a_{n_r} - b_{m_r}| < \frac{1}{r} \quad \text{for all } r \in N.$$

Since $b_{m_r} \to b$ it follows that the $a_{n_r} \to b$. The easy proof of this last statement is left for the reader (Problem 2) ●

## C. Open Sets and Closed Sets

A subset $O \subset R$ is said to be *open* if for any $x \in O$ there is an $\varepsilon > 0$ such that

$$(x - \varepsilon, x + \varepsilon) \subset O.$$

A subset $C \subset R$ is said to be *closed* if its complement $C^c$ is an open set.

A characterization of closed sets is given in the following theorem:

**THEOREM 7**  *A set $C \subset R$ is closed if and only if whenever*

$$\{x_n\} \subset C$$

*is a convergent sequence, then its limit also belongs to $C$.*

*Proof*  Suppose first that $C$ is closed and $\{x_n\} \subset C$ converges to $x$. We will show that $x \in C$. If this is not the case, then the open set $C^c$ contains $x$. Hence there

$$T_n = |x_1| + |x_2| + \cdots + |x_n|,$$

and let

$$S_n = x_1 + x_2 + \cdots + x_n.$$

We will show that $\{S_n\}$ is Cauchy. Let $\varepsilon > 0$. Since

$$\sum_{k=1}^{\infty} |x_k|$$

is convergent, the sequence $\{T_n\}$ is Cauchy, so there is an $n_0$ such that

$$|T_{n+m} - T_n| < \varepsilon \quad \text{for all } n \geq n_0 \text{ and } m \geq 0.$$

Since

$$|S_{n+m} - S_n| = \left| \sum_{k=n+1}^{n+m} x_k \right| \leq \sum_{k=n+1}^{n+m} |x_k|$$

$$= |T_{n+m} - T_n| < \varepsilon,$$

it follows that $\{S_n\}$ is Cauchy. By Theorem 5, $\{S_n\}$ is therefore convergent, so $\sum_{k=1}^{\infty} x_n$ is convergent ●

---

**Problem 2**  Suppose $b_k \to b$, $\{a_k\}$ is a sequence, and

$$|a_k - b_k| \to 0.$$

Show that $a_k \to b$.

---

→  Recall that $(a, b)$ represents the interval defined by

$$(a, b) = \{x \in R : a < x < b\}.$$

→  We will use the notation $C^c$ in place of $R - C$. In other words, $C^c = \{x \in R : x \notin C\}$. It follows immediately that $(C^c)^c = C$, $C \cup C^c = R$, and $C \cap C^c = \phi$. A less trivial fact about complements is given in Example 3.

---

**Example 3**  We will show that for any subsets $A, B \subset R$

$$(A \cup B)^c = A^c \cap B^c.$$

*Proof*  If $x \in (A \cup B)^c$ then $x \notin A \cup B$, so $x$ belongs to neither $A$ nor $B$. Hence $x \in A^c$ and $x \in B^c$, so $x \in A^c \cap B^c$. This shows that $(A \cup B)^c \subset A^c \cap B^c$.

is an $\varepsilon > 0$ such that $(x - \varepsilon, \; x + \varepsilon) \subset C^c$. Since $\{x_n\} \subset C$, no $x_n$ belongs to $C^c$ and therefore no $x_n \in (x - \varepsilon, x + \varepsilon)$. This means that

$$|x_n - x| \geq \varepsilon \quad \text{for all } n$$

and contradicts the assumption that $x_n \to x$.

Conversely, suppose the condition in the theorem is satisfied and $C$ is not closed. Then $C^c$ is not an open set and therefore the definition given for an open set must be violated for $C^c$. This means that there must be an $x \in C^c$ with the property that for any $\varepsilon > 0$ the set $(x - \varepsilon, x + \varepsilon)$ is not contained in $C^c$. In particular, for each $n \in N$ the set $(x - \frac{1}{n}, x + \frac{1}{n})$ is not contained in $C^c$, so the interval must contain an element, call it $x_n$, belonging to $C$. Since $|x_n - x| < \frac{1}{n}$, we have that $x_n \to x$. But $\{x_n\} \subset C$ by construction and it converges to a point not in $C$. This contradicts the hypothesis, so $C$ must be closed ●

**THEOREM 8**  *Given $a, b \in R$ with $a < b$, the interval*

$$(a, b) \text{ is an open set,}$$

*and the interval*

$$[a, b] \text{ is a closed set.}$$

*Proof*  Suppose $x \in (a, b)$ and let $\varepsilon = \min \{x - a, b - x\}$. Then it follows that $(x - \varepsilon, x + \varepsilon) \subset (a, b)$, so $(a, b)$ is an open set.

To show that $[a, b]$ is closed we will show that its complement $B = (-\infty, a) \cup (b, \infty)$, is open. Suppose $x \in (-\infty, a)$ and let $\varepsilon = a - x$. Then $(x - \varepsilon, x + \varepsilon) \subset B$. If $x \in (b, \infty)$, then letting $\varepsilon = x - b$ gives $(x - \varepsilon, \; x + \varepsilon) \subset B$. Hence $B$ is open, so $B^c = [a, b]$ is a closed set ●

It is quite easy to prove that the intervals $(-\infty, a)$, $(b, \infty)$, and $(-\infty, \infty)$ are also open sets, and that the intervals $(-\infty, a]$, $[b, \infty)$, and $(-\infty, \infty)$ are closed sets. The argument is similar to that in Theorem 8 and will be omitted. Notice, however, that this conforms to our definition of open and closed intervals given in 1.1B.

A subset of $R$ is said to be *compact* if it is closed and bounded.

Now suppose

$$x \in A^c \cap B^c.$$

Then $x \notin A$ and $x \notin B$, so $x \notin A \cup B$, which means

$$x \in (A \cup B)^c.$$

This shows that $A^c \cap B^c \subset (A \cup B)^c$, and together with the reverse inclusion above, means that the sets are equal.

---

**Problem 3**  Prove that for any subsets $A, B \subset R$,

$$(A \cap B)^c = A^c \cup B^c.$$

---

→ Both parts of this proof employed the method of "proof by contradiction," which is based on the "law of the excluded middle." We have used this technique before and will use it several more times. It is based on the assumption that for a given statement "$S$" exactly one of the statements "$S$" and "not $S$" must be true. Hence, if one shows that "not $S$" is false, it follows that "$S$" must be true. There is, however, a style of mathematics called constructive mathematics which rejects the "law of the excluded middle" and, hence, the technique of "proof by contradiction." We will say more of this at the end of Chapter 10.

→ See 1.1B for the definition of interval. Notice that the interval $[0, 1)$ is neither an open set nor a closed set. It is not open because $0 \in [0, 1)$ and there is no $\varepsilon > 0$ with the property that $(0 - \varepsilon, 0 + \varepsilon) = (-\varepsilon, \varepsilon) \subset [0, 1)$. It is not closed because the sequence $\{\frac{1}{2}, \frac{2}{3}, \frac{3}{4}, \frac{4}{5}, \frac{5}{6}, \ldots\}$ belongs to $[0, 1)$ and converges to 1, but $1 \notin [0, 1)$ (see Theorem 7).

→ Notice that $R = (-\infty, \infty)$ is both an open and a closed set. It follows that its complement $R^c = \phi$ is also an open and a closed set. Indeed, the definitions of both open and closed sets are satisfied by the null set. For example, it is true that if $x \in \phi$, then there is an $\varepsilon > 0$ such that

$$(x - \varepsilon, x + \varepsilon) \subset \phi$$

because there are no $x$ in $\phi$. This is an example of a vacuously satisfied condition.

**THEOREM 9**  *A subset A of R is compact if and only if any sequence $\{a_n\} \subset A$ contains a convergent subsequence whose limit is in A.*

*Proof* (*Compactness implies the sequence property.*)

Suppose $A$ is compact. Then $A$ is bounded so $\{a_n\}$ must be a bounded sequence. Hence by Theorem 6, $\{a_n\}$ contains a convergent subsequence $\{a_{n_k}\}$. Since $A$ is closed it follows from Theorem 7 that the limit of $\{a_{n_k}\}$ belongs to $A$  ●

**THEOREM 10**  *If A is a compact set, there exist $c \in A$ and $d \in A$ such that*

$$c \leq x \leq d \quad \text{for all } x \in A.$$

*Proof* Since $A$ is compact it is bounded and therefore has a supremum $d$. For each $n \in N$, $d - \frac{1}{n}$ is not the supremum of $A$ so there is an $a_n \in A$ satisfying $d - \frac{1}{n} < a_n \leq d$. Then $|d - a_n| < \frac{1}{n}$, so $a_n \to d$. Since $A$ is a closed set, it follows that $d \in A$, so $d$ is the desired point. The point $c$ is obtained in a similar manner (see Problem 4)  ●

**THEOREM 11**  *Suppose $C_1 \supset C_2 \supset C_3 \supset \ldots$ are non-empty compact sets. Then there is a point x with the property that*

$$x \in C_k \quad \text{for all } k.$$

*Proof* For each $n$ choose $x_n \in C_n$. The sequence

$$\{x_1, x_2, \ldots\} \subset C_1$$

so by Theorem 9 it contains a convergent subsequence $\{x_{n_k}\}$. Let $x_{n_k} \to x$. Now for each $k$, $\{x_{n_k}, x_{n_{k+1}}, \ldots\} \subset C_k$ so its limit, which is $x$, also belongs to $C_k$ since $C_k$ is compact. Hence $x \in C_k$ for every $k$  ●

→ *Proof* (*The sequence condition implies compactness.*)

Suppose that every sequence in $A$ has a convergent subsequence with limit point in $A$. By Theorem 7, $A$ must be closed. Assume that $A$ is unbounded and, without loss of generality, that it has no upper bound. Then for each $k$ there must be an $a \in A$ with $a > k$. In particular, there is an $a_1 \in A$ with $a_1 > 1$. Assuming $a_k$ has been defined, we then inductively choose $a_{k+1}$ satisfying $a_{k+1} > a_k + 1$. Clearly the sequence $\{a_k\}$ is not Cauchy and has no Cauchy subsequence. Therefore the sequence $\{a_k\}$ contradicts the assumption that every sequence in $A$ has a convergent subsequence. It must be, then, that $A$ is bounded, and since it is closed, it is compact  ●

→ Notice that the theorem is false if the $C_k$ are not compact. For example, letting $O_n = (0, \frac{1}{n})$ we have that

$$O_1 \supset O_2 \supset O_3 \supset \ldots$$

but there is no $x$ that belongs to all $O_n$. Similarly, letting $A_n = [n, \infty)$, we have $A_1 \supset A_2 \supset \ldots$ but no $x \in R$ is contained in all $A_n$.

**Problem 4**  Given a compact set $A$, show that $\inf A \in A$.

## D. Exercises

**B1.** a) Show that if $\lim_{n \to \infty} x_n = x \neq 0$ and $\lim_{n \to \infty} x_n y_n$ exists, then $\lim_{n \to \infty} y_n$ exists.

b) Give an example showing that $\lim_{n \to \infty} x_n$ and $\lim_{n \to \infty} x_n y_n$ can exist without $\lim_{n \to \infty} y_n$ existing.

c) Give an example showing that $\lim_{n \to \infty} x_n y_n$ can exist even though neither $\lim_{n \to \infty} x_n$ nor $\lim_{n \to \infty} y_n$ exists.

**B2.** a) If $x_n \to x$ show that $\dfrac{x_1 + \cdots + x_n}{n} \to x$.

    b) Give an example to show that the converse to (a) is false.

**B3.** Prove the following:
    a) The union of any collection of open sets is open.
    b) The intersection of a finite number of open sets is open.
    c) The intersection of any collection of closed sets is closed.
    d) The union of a finite number of closed sets is closed.

**B4.** a) Give an example of a collection of closed sets whose union is an open set.
    b) Give an example of a collection of open sets whose intersection is a closed set.

**M5.** A *boundary point* of a set $A$ is a point with the property that every open set containing it contains points of $A$ and points of $A^c$. Prove that a set $C$ is closed if and only if it contains all its boundary points.

**M6.** Show that for any open set $U$ of $R$ there is a sequence of disjoint open intervals $\{I_n\}$ satisfying

$$U = \bigcup_{n=1}^{\infty} I_n.$$

**M7.** Given a set $A$ let $\partial A$ denote the collection of all its boundary points (see Exercise B5).
    a) Prove that $\partial(A \cup B) \subset \partial A \cup \partial B$
    b) Prove that $\partial(A \cap B) \subset \partial A \cap \partial B$
    c) Give examples to show that each of the inclusions in (a) and (b) can be strict.

**M8.** Suppose $A$ is a compact set and $\{U_n\}$ is a sequence of open sets with the property $A \subset \cup_{n=1}^{\infty} U_n$. Prove there is a finite set $\{n_1, n_2, \ldots, n_k\}$ with the property that

$$A \subset U_{n_1} \cup U_{n_2} \cup \cdots \cup U_{n_k}.$$

**M9.** a) Suppose $C$ is a closed set and $U$ is an open set. Show that the set

$$C + U = \{c + u \colon c \in C, u \in U\}$$

    is an open set.
    b) Show that $C + D$ need not be a closed set even though both $C$ and $D$ are closed sets.

**M10.** a) If $A$ and $B$ are compact sets show that $A + B$ is a compact set.
    b) If $A$ is a compact set and $B$ is a closed set, show that $A + B$ is a closed set.

---

## 9.3   *Continuous Functions*

In this section the concept of a continuous function is precisely defined and a number of basic results about such functions are proven.

**A. Definitions and Basic Properties**   Suppose $A$ is a subset of $R$ and $f \colon A \to R$ is a function. The following two definitions will be shown to be equivalent in Theorem 1.

*Sequential definition of continuity*   $f$ is said to be continuous at $x_0 \in A$ if

$$f(x_n) \to f(x_0) \quad \text{whenever} \quad x_n \in A \quad \text{and} \quad x_n \to x_0.$$

*$\varepsilon$-$\delta$ definition of continuity*   $f$ is said to be continuous  →

Recall that $f$ was said to be continuous at $x_0 \, \varepsilon \, A$ if

$$\lim_{x \to x_0} f(x) = f(x_0).$$

We leave as an exercise for the reader to prove that this definition is equivalent to the $\varepsilon$-$\delta$ definition of continuity if there are points in $A$ arbitrarily close to $x_0$.

at $x_0 \in A$ if for any $\varepsilon > 0$ there is a $\delta > 0$ such that

$$|f(x) - f(x_0)| < \varepsilon \quad \text{whenever} \quad x \in A$$

and

$$|x - x_0| < \delta.$$

(Notice that the above condition in terms of $\varepsilon$ and $\delta$ can be written as

$$f((x_0 - \delta, x_0 + \delta)) \subset (f(x_0) - \varepsilon, f(x_0) + \varepsilon).)$$

---

**Example 1** To illustrate the formal definition of limit, we will show that $\lim_{x \to 0} x^2 = 0$.

Choose $\varepsilon > 0$. We wish to find a $\delta$ such that $|x^2 - 0| < \varepsilon$ whenever $|x - 0| < \delta$. But $|x^2 - 0| < \varepsilon$ is equivalent to $x^2 < \varepsilon$ or $-\sqrt{\varepsilon} < x < \sqrt{\varepsilon}$; hence, letting $\delta = \sqrt{\varepsilon}$ will guarantee that $|f(x) - 0| < \varepsilon$. This completes the argument.

---

**THEOREM 1** *The sequential definition and $\varepsilon$-$\delta$ definition of continuity are equivalent.*

*Proof* ($\varepsilon$-$\delta$ *implies sequential.*) Suppose that $f: A \to R$ satisfies the $\varepsilon$-$\delta$ criterion and let $\{x_n\}$ be a sequence in $A$ that converges to $x_0$. Let $\varepsilon > 0$. By assumption there exists a $\delta > 0$ such that $|f(x_0) - f(x)| < \varepsilon$ whenever $0 < |x - x_0| < \delta$. Since $x_n \to x_0$, there is an $n_0$ such that $|x_n - x_0| < \delta$ whenever $n > n_0$. But this implies that $|f(x_n) - f(x_0)| < \varepsilon$ for $n > n_0$, so $f(x_n) \to f(x_0)$.

(*Sequential implies $\varepsilon$-$\delta$.*) Assume that $f: A \to R$ satisfies the sequential hypothesis but that the $\varepsilon$-$\delta$ criterion does not hold. Then there must be an $\varepsilon_0 > 0$ with the property that for each $\delta > 0$ there is an $x \in A$ satisfying $|x - x_0| < \delta$ and $|f(x) - f(x_0)| \geq \varepsilon_0$. In particular, for $\delta = \frac{1}{n}$ there is an $x_n \in A$ satisfying

$$|x_n - x| < \frac{1}{n} \quad \text{and} \quad |f(x_n) - f(x_0)| \geq \varepsilon_0.$$

Letting $n \to \infty$ we clearly have that $x_n \to x_0$, but the sequence $\{f(x_n)\}$ does not converge to $f(x_0)$. This contradicts the sequential definition. Hence the $\varepsilon$-$\delta$ definition must be satisfied ●

**Example 2** The function $f(x) = \begin{cases} x^2, x \neq 0 \\ 1, x = 0 \end{cases}$ is not continuous at $x = 0$ because, as we have shown,

$$\lim_{x \to 0} f(x) = 0 \neq 1 = f(0).$$

---

**Example 3** The function

$$f(x) = \begin{cases} 1 & \text{if } x \text{ is irrational} \\ 0 & \text{if } x \text{ is rational} \end{cases}$$ is not continuous at any point.

*Proof* If $x$ is irrational, then there is a sequence of rationals $\{r_n\}$ satisfying $r_n \to x$ (see Problem 1b of 9.2A). Since

$$0 = \lim f(r_n) \neq f(x) = 1,$$

$f$ is not continuous at $x$.

If $r$ is rational, then $x_n = r + \frac{\sqrt{2}}{n}$ is not rational and $x_n \to r$. Since $1 = \lim f(x_n) \neq f(r) = 0$, $f$ is not continuous at $r$.

---

**Example 4** Define $f: R \to R$ in the following manner. If $x \in R$ is rational, express it as the quotient of integers $\frac{p}{q}$ reduced to lowest terms and let $f(x) = \frac{1}{q}$. If $x \in R$ is irrational, define $f(x) = 0$. Then $f$ is continuous at every irrational $x \in R$.

*Proof* Let $x \in R$ be irrational and let $x_n \to x$ ($x_n$ rational). Express $x_n$ as $p_n/q_n$ with $p_n$, $q_n$ integers and the fraction in lowest terms. Since $f(x_n) = 1/q_n$, we need only show that $|q_n| \to \infty$, for then $f(x_n) \to 0 = f(x)$. If it is not true that $|q_n| \to \infty$, then there must be an $M > 0$ with the property that $q_n \in [-M, M]$ for infinitely many $n$. Since the $q_n$'s are integers, it must be that infinitely many of the $q_n$'s are equal. Therefore $q = q_{n_k}$ for $k \in N$. Since $x_{n_k} = p_{n_k}/q_{n_k} = p_{n_k}/q \to x$, it follows that $p_{n_k} \to xq$. This can occur only when $x$ is rational, so we have obtained a contradiction.

The following corollary will be used quite often.

**COROLLARY** *If $f: A \to R$ is continuous at $x_0$ and $f(x_0) > 0$, then there is a $\delta > 0$ such that $f(x) > 0$ whenever*

$$x_0 - \delta < x < x_0 + \delta \quad and \quad x \in A.$$

*Proof* Let $0 < \varepsilon < f(x_0)$. Since $f$ is continuous at $x_0$ there exists a $\delta$ such that $|f(x) - f(x_0)| < \varepsilon$ whenever $|x - x_0| < \delta$. But this just says that

$$-\varepsilon < f(x) - f(x_0) < \varepsilon$$

or

$$f(x_0) - \varepsilon < f(x) < f(x_0) + \varepsilon.$$

Since $f(x_0) - \varepsilon > 0$, we have $0 < f(x)$ whenever $|x - x_0| < \delta$, that is, whenever

$$x_0 - \delta < x < x_0 + \delta \quad \bullet$$

A function $f: A \to R$ will be said to be *continuous on A*, or simply *continuous,* if $f$ is continuous at every point $x \in A$. For example, the function $f$ in Example 2 is *not* continuous at $x = 0$, but $f$ restricted to $R - \{0\}$ is continuous.

**THEOREM 2** *Suppose $f: A \to R$ and $g: A \to R$ are continuous at $x_0 \in A$. Then $f + g$ and $fg$ are continuous at $x_0$.*

*Proof* Suppose $x_n \to x_0$. Then, since $f$ and $g$ are continuous at $x_0$, $f(x_n) \to f(x_0)$ and $g(x_n) \to g(x_0)$. By Theorem 2 (9.2A) it follows that

$$f(x_n) + g(x_n) \to f(x_0) + g(x_0)$$

and

$$f(x_n)g(x_n) \to f(x_0)g(x_0).$$

Hence

$$(f + g)(x_n) \to (f + g)(x_0)$$

and

$$(fg)(x_n) \to (fg)(x_0),$$

which proves that $f + g$ and $fg$ are continuous at $x_0$ $\bullet$

It follows immediately that if $f, g: A \to R$ are continuous on $A$ then $f + g$ and $fg$ are continuous on $A$.

**Example 5** The function in Example 4 is discontinuous at every rational $x \in R$.

*Proof* Let $x = \frac{p}{q}$ where $p$ and $q$ are integers $(q \neq 0)$ and the fraction is in lowest terms, and let $x_n = x + (\sqrt{2}/n)$. Since $x_n$ is not rational and $x_n \to x$, we have

$$0 = \lim f(x_n) \neq f(x) = \frac{1}{q}.$$

Therefore $f$ is not continuous at the rational $x$.

**Problem 1** Suppose $f: A \to R - \{0\}$ is continuous. Prove that $\frac{1}{f}: A \to R - \{0\}$ is continuous.

**COROLLARY 1** *If $f_1, f_2, \ldots, f_n: A \to R$ are continuous then $f_1 + f_2 + \cdots + f_n$ and $f_1 \cdot f_2 \cdot \cdots \cdot f_n$ are continuous.*

→ The proof is a simple exercise using induction.

**COROLLARY 2** *Any polynomial (2.3B) is continuous.* →

This follows from Corollary 1 and the fact that constant functions $f(x) = a$ and the identity function $g(x) = x$ are continuous.

**COROLLARY 3** *If $p$ and $q$ are polynomials, then $\dfrac{p}{q}$ is continuous at each point $x_0$ where $q(x_0) \neq 0$.* →

This follows from Theorem 2 and Problem 1.

**THEOREM 3** *Suppose $f: C \to R$ is continuous on $C$ and $C$ is a compact set. Then $f(C) = \{f(x): x \in C\}$ is a compact set.*

*Proof* We will show that any sequence $\{y_n\} \subset f(C)$ contains a subsequence converging to a point in $f(C)$. Choose $x_n \in C$ such that $f(x_n) = y_n$. Now $\{x_n\}$ is a sequence in $C$, so by Theorem 9 (9.2C) there is a subsequence $\{x_{n_k}\}$ such that

$$x_{n_k} \to x \in C.$$

Since $f$ is continuous, it follows that

$$f(x_{n_k}) \to f(x) \in f(C).$$

Since $f(x_{n_k}) = y_{n_k}$, we have that $y_{n_k} \to f(x)$. Hence by Theorem 9 (9.2C), $f(C)$ is compact ●

**THEOREM 4** *Suppose $A \subset R$ is compact and $f: A \to R$ is a function. If $f$ is continuous, then $f$ is uniformly continuous.*

*Proof* The proof is by contradiction. Suppose the result is false. Then for some $\varepsilon_0 > 0$ there is no $\delta > 0$ satisfying the stated condition. Hence for each $\frac{1}{n}$, $n \in N$, there must be points $x_n, y_n \in A$ such that

$$|x_n - y_n| < \frac{1}{n} \quad \text{and} \quad |f(x_n) - f(y_n)| \geq \varepsilon_0.$$

Since $\{x_n\}$ is a sequence in the compact set $A$ it must, by Theorem 9 (9.2C), have a convergent subsequence $\{x_{n_k}\}$ with

$$x_{n_k} \to x \in A.$$

Since $|x_{n_k} - y_{n_k}| < \dfrac{1}{n_k} \to 0$ it follows from Problem 2 (9.2B) that $y_{n_k} \to x$. Since $f$ is continuous, we have

**COROLLARY 1** (Extreme Value Theorem) *If $f: C \to R$ is continuous and $C$ is compact, then there are points $a, b, \in C$ such that*

$$f(a) \leq f(x) \leq f(b)$$

*for all $x \in C$.*

*Proof* By Theorem 3, $f(C)$ is compact, so by Theorem 10 (9.2C) $f(C)$ has a maximum $c$ and a minimum $d$. Choosing $a$ and $b$ such that $f(a) = c$ and $f(b) = d$, we have the result ●

Recall that we used this corollary in a fundamental way in the proof of the Mean Value Theorem (4.1B).

**COROLLARY 2** *If $a < b$ and $f: [a, b] \to R$ is continuous, then $f$ is a bounded function (that is, the image of $f$ is bounded).*

*Proof* $[a, b]$ is a compact set ●

→ Given a subset $S \subset R$, a function $f: S \to R$ is said to be *uniformly continuous* if for any $\varepsilon > 0$ there is a $\delta > 0$ such that $|f(x) - f(y)| < \varepsilon$ whenever $|x - y| < \delta$ ($x, y \in S$).

Clearly, uniform continuity implies continuity and Theorem 4 says that the converse is true if the function is defined on a compact set. In general these two notions are distinct. See Example 6.

**Example 6** The function $f(x) = x^2$ is continuous but not uniformly continuous on $R$.

*Proof* Let $\varepsilon = 1$ and take any $\delta > 0$. Let $0 < h < \delta$ and consider the points $x$ and $x + h$. We clearly have

$$|(x + h) - x| < \delta.$$

Then $f(x + h) - f(x) = (x + h)^2 - x^2 = 2xh + h^2$. Hence for $x > \dfrac{1 - h^2}{2h}$ we have $|f(x + h) - f(x)| =$

$f(x_{n_k}) \to f(x)$ and $f(y_{n_k}) \to f(x)$. In particular it follows that

$$|f(x_{n_k}) - f(y_{n_k})| \leq |f(x_{n_k}) - f(x)|$$
$$+ |f(x) - f(y_{n_k})| \to 0,$$

contradicting the assumption that

$$|f(x_n) - f(y_n)| \geq \varepsilon_0 > 0 \text{ for all } n \quad \bullet$$

**Example 7** The function $f(x) = x$ is uniformly continuous. To see this, choose $\varepsilon > 0$. Let $\delta = \varepsilon$. Then for $|x_1 - x_2| < \delta$ we have

$$|f(x_1) - f(x_2)| = |x_1 - x_2| < \delta = \varepsilon,$$

and $f$ satisfies the criterion for uniform continuity.

## B. The Intermediate Value Theorem

**LEMMA** *Suppose* $g: [a, b] \to R$ *is continuous and*

$$g(a) < 0 < g(b).$$

*Then there exists a point* $x_0 \in [a, b]$ *with* $g(x_0) = 0$.

*Proof* Let $A = \{x \in [a, b]: g(x) \leq 0\}$. Then $A$ is bounded, nonempty (because $a \in A$), and has a supremum. Let

$$x_0 = \sup A.$$

We will show that $g(x_0) = 0$ by showing that neither $g(x_0) < 0$ nor $g(x_0) > 0$ is possible.

Suppose $g(x_0) < 0$. Then by the corollary to Theorem 1 (9.3A) there is a $\delta > 0$ such that $g(x) < 0$ for all

$$x \in (x_0 - \delta, x_0 + \delta).$$

In particular $g(x_0 + \tfrac{1}{2}\delta) < 0$ and this contradicts the fact that $x_0 = \sup A$.

Now suppose $g(x_0) > 0$. By the corollary to Theorem 1 (9.3A) it follows that there is a $\delta > 0$ such that $g(x) > 0$ whenever $x \in (x_0 - \delta, x_0 + \delta)$. Since $x_0 = \sup A$ there must exist a $y \in A$ satisfying

$$x_0 - \delta < y \leq x_0.$$

But this means that

$$g(y) > 0,$$

$2xh + h^2 > 1$ no matter how small we pick $\delta$. This contradicts the definition of uniform continuity.

---

**Problem 2** Show that the function $f(x) = \frac{1}{x}$ is not uniformly continuous on the open interval $(0, 1)$.

---

**Example 8** We will show that the polynomial

$$p(x) = x^3 + ax^2 + bx + c$$

has a real root.

*Proof*

$$\frac{p(x)}{x^3} = \left(1 + \frac{a}{x} + \frac{b}{x^2} + \frac{c}{x^3}\right) \to 1$$

as $x \to \pm\infty$. Hence there exist numbers $L < 0$ and $M > 0$ such that

$$\frac{p(L)}{L^3} \quad \text{and} \quad \frac{p(M)}{M^3}$$

are positive. Since $L^3 < 0$ it follows that $p(L) < 0$, and since $M^3 > 0$ it follows that $p(M) > 0$. By the lemma we know that $p(x) = 0$ for some $x \in [L, M]$.

(This result can be generalized to any odd degree polynomial, but is not necessarily true when the degree is even. For example, $q(x) = 1 + x^2$ has no real zero.)

---

contradicting the fact that $g(y) \leq 0$ since $y \in A$. Hence the assumption that $g(x_0) > 0$ is impossible, so we conclude that $g(x_0) = 0$ ●

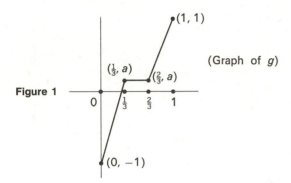

(Graph of g)

**Figure 1**

$(\frac{1}{3}, a)$    $(\frac{2}{3}, a)$

$(1, 1)$

$(0, -1)$

**THEOREM 5** (Intermediate Value Theorem)  *Suppose*

$$f: [a, b] \to R$$

*is continuous and $y_0$ lies between $f(a)$ and $f(b)$; that is,*

$$f(a) < y_0 < f(b) \quad or \quad f(b) < y_0 < f(a).$$

*Then there is an $x_0 \in [a, b]$ satisfying*

$$f(x_0) = y_0.$$

*Proof* Suppose first that $f(a) < f(b)$ and let $g(x) = f(x) - y_0$. Then $g: [a, b] \to R$ satisfies

$$g(a) = f(a) - y_0 < 0 < g(b) = f(b) - y_0.$$

By the lemma we see that there is an $x_0 \in [a, b]$ with $g(x_0) = 0$. But then $0 = g(x_0) = f(x_0) - y_0$, so $f(x_0) = y_0$.

If $f(a) > f(b)$, let $h(x) = -f(x)$. Then $h(a) < -y_0 < h(b)$, so by the first argument $h(x_0) = -y_0$ for some $x_0 \in [a, b]$. But this gives $f(x_0) = y_0$ ●

*COROLLARY 1*  *Suppose $f: [0, 1] \to [0, 1]$ is continuous. Then there is an $x_0 \in [0, 1]$ with $f(x_0) = x_0$.*

*Proof* If $f(0) = 0$ or $f(1) = 1$, there is nothing to prove; so assume $f(0) > 0$ and $f(1) < 1$. Let $g(x) = x - f(x)$. Then $g(0) < 0$ and $g(1) > 0$. By Theorem 4 there is an $x_0$ such that $g(x_0) = 0$. Hence $0 = g(x_0) = x_0 - f(x_0)$, or $f(x_0) = x_0$ ●

→ The following example is important from the point of view of constructive mathematics, and will be discussed more fully in the history section at the end of Chapter 10.

*Example 9*  For $a \in (-1, 1)$ define $g: [0, 1] \to R$ (Fig. 1) by

$$g(x) = \begin{cases} (3 + 3a)x - 1 & (0 \leq x \leq \frac{1}{3}) \\ a & (\frac{1}{3} \leq x \leq \frac{2}{3}) \\ (3 - 3a)x - 2 + 3a & (\frac{2}{3} \leq x \leq 1) \end{cases}$$

Now, $g$ satisfies the hypothesis of the lemma, and therefore has a zero in the interval $(0, 1)$. But consider the dependence of this zero on $a$. If $a > 0$ the zero lies in $(0, \frac{1}{3})$, and if $a < 0$ the zero lies in $(\frac{2}{3}, 1)$. In other words, a small change in $a$ (positive to negative) will produce a large change (at least $\frac{1}{3}$ unit) in the position of the zero of $g$. This discontinuous phenomenon will be used to show that, for suitably chosen $a$, no procedure exists for finding the zero of $g$.

→ This is the special case (for dimension one) of *Brouwer's Fixed Point Theorem*. In two dimensions the result says that if $A = \{(x, y): 0 \leq x, y \leq 1\}$ and $f: A \to A$ is continuous, then there exists an $x_0 \in A$ with $f(x_0) = x_0$. The theorem is very useful, but excepting the one-dimensional case, is difficult to prove.

**COROLLARY 2** *Suppose $I$ is an interval and $f$:* $I \to R$ *is continuous.* *Then $f(I)$ is an interval.*

*Proof* We will use the fact that a subset $A \subset R$ is an interval if and only if for any $c, d \in A$ with $c < d$, it follows that

$$[c, d] \subset A.$$

Suppose $c, d \in f(I)$, $f(a) = c$, and $f(b) = d$; let $y \in [c, d]$. By Theorem 5 there is an $x$ between $a$ and $b$ with $f(x) = y$. Hence $y \in f(I)$ for all $y \in [c, d]$, so $f(I)$ is an interval ●

→ The corollary does *not* say that $f(I)$ is the same type of interval as $I$. For example, let $f(x) = e^{-x^2}$. Then $f((-\infty, \infty)) = (0, 1]$, which is neither open nor closed, whereas $(-\infty, \infty)$ is both open and closed.

**COROLLARY 3** *For each $n \in N$ the function*

$$f\colon [0, \infty) \to [0, \infty)$$

*defined by $f(x) = x^n$ is bijective.*

*Proof* We will first show that for any $y \in [0, \infty)$ there exists an $x \in [0, \infty)$ with $f(x) = y$. For $y = 0$ we may take $x = 0$, so assume that $y > 0$. Choose $z > 0$ with $z^n > y$. It follows that $y$ lies between $f(0)$ and $f(z)$, so by Theorem 5 there exists an $x \in [0, z]$ satisfying $x^n = y$.

We will next show that this $x$ is unique. Suppose $x_1^n = y$. Then

$$0 = x^n - x_1^n$$
$$= (x - x_1)(x^{n-1} + x^{n-2}x_1 + \cdots + xx_1^{n-2} + x_1^{n-1}).$$

Since both $x$ and $x_1$ cannot equal zero, the second factor is positive. The first factor, $x - x_1$, must then be zero, so we have $x = x_1$ ●

→ In other words, the *n*th root of a nonnegative number exists and is unique. The corollary can also be stated in terms of inverse functions (9.1B). Since $f\colon [0, \infty) \to [0, \infty)$ is bijective, its inverse function

$$f^{-1}\colon [0, \infty) \to [0, \infty)$$

is defined. We will show, following Theorem 6, that the inverse function is also continuous.

**Problem 3** Using the above definition, show that if $x \geq 0$ and $m, n \in N$, then

$$(x^{1/n})^{1/m} = x^{1/mn}.$$

## C. On Inverse Functions

**LEMMA 1** *Suppose $f\colon [a, b] \to R$ is continuous and injective.*
  a) *If $f(a) < f(b)$, then $f(a) < f(x) < f(b)$ for all $x \in (a, b)$.*
  b) *If $f(a) > f(b)$, then $f(b) < f(x) < f(a)$ for all $x \in (a, b)$.*

*Proof* a) If $f(x) \leq f(a)$, then $f(a) \in [f(x), f(b)]$ and by Theorem 5 (9.3B) there is an $x_0 \in [x, b]$ satisfying $f(x_0) = f(a)$. This contradicts the assumption that $f$ is injective since $x_0 \neq a$.

If $f(x) \geq f(b)$, then $f(b) \in [f(a), f(x)]$ and by

→ **Example 10** The assumption of continuity is essential in this lemma. Consider the following function. Define $g\colon [0, 2] \to R$ by

$$g(x) = \begin{cases} -x & (0 \leq x \leq 1) \\ x - 1 & (1 < x \leq 2). \end{cases}$$

Then $g$ is injective and $0 = g(0) < g(2) = 1$, but $g$ does not satisfy

$$0 < g(x) < 1 \quad \text{for all } 0 < x < 2.$$

Theorem 5 (9.3B) there is a $y_0 \in [a, x]$ satisfying $f(y_0) = f(b)$. This contradicts the assumption that $f$ is injective since $y_0 \neq b$. The proof of (b) is similar and is left to the reader (Problem 4)   ●

**LEMMA 2**   *Suppose $f: [a, b] \to R$ is continuous and injective.*
    a) *If $f(a) < f(b)$, then $f$ is strictly increasing on $[a, b]$.*
    b) *If $f(a) > f(b)$, then $f$ is strictly decreasing on $[a, b]$.*

*Proof*   a) If $f$ is not strictly increasing, then there are points $x, y \in [a, b]$ with $x < y$ and $f(x) \geq f(y)$. By Lemma 1a $f(a) \leq f(y)$, so $f(y) \in [f(a), f(x)]$. By Theorem 5 (9.3B) there is a $z_0 \in [a, x]$ satisfying $f(z_0) = f(y)$. This contradicts the assumption that $f$ is injective because $z_0 \neq y$. The proof of (b) is similar and is left to the reader (Problem 5a)   ●

**LEMMA 3**   *Suppose $f: [a, b] \to R$ is continuous and injective.*
    a) *If $f(a) < f(b)$, then*

$$f([a, b]) = [f(a), f(b)] \quad and$$
$$f((a, b)) = (f(a), f(b)).$$

    b) *If $f(a) > f(b)$, then*

$$f([a, b]) = [f(b), f(a)] \quad and$$
$$f((a, b)) = (f(b), f(a)).$$

*Proof*   a) By Lemma 1a, $f([a, b]) \subset [f(a), f(b)]$, and by Theorem 5 (9.3B) it follows that

$$f([a, b]) \supset [f(a), f(b)].$$

Hence

$$f([a, b]) = [f(a), f(b)].$$

Now $[a, b] = \{a\} \cup (a, b) \cup \{b\}$, so

$$f([a, b]) = f(a) \cup f((a, b)) \cup f(b),$$

and because $f([a, b]) = [f(a), f(b)]$ we have

$$[f(a), f(b)] = \{f(a)\} \cup f((a, b)) \cup \{f(b)\}.$$

Because $f$ is injective no point in $(a, b)$ can map onto $f(a)$ or $f(b)$, so we must have $f((a, b)) = (f(a), f(b))$. The proof of (b) is similar and is left to the reader (Problem 5b)   ●

---

*Problem 4*   Prove the result stated in Lemma 1b. (Hint: Replace $f$ by $-f$.)

---

    ➡    The definitions of "strictly increasing" and "strictly decreasing" are given in 3.3A.

    ➡    A strictly monotone function is injective, but the converse is false, as Example 10 shows. Lemma 2 merely says that injectivity plus continuity does imply strict monotonicity.

---

*Problem 5*   a) Prove the result stated in Lemma 2b.
               b) Prove the result stated in Lemma 3b.

We now collect the above results to prove the following theorem about inverse functions.

**THEOREM 6** *Suppose $I$ is an open interval and $f\colon I \to R$ is continuous and injective. Then $f(I)$ is an open interval and the inverse function*

$$f^{-1}\colon f(I) \to I$$

*is continuous.*

*Proof* By Corollary 2 of Theorem 5 (9.3B), $f(I)$ is an interval. To show it is open, let $y \in f(I)$, and pick $x \in I$ satisfying

$$f(x) = y.$$

Because $I$ is open there is a $b > 0$ satisfying $(x - b, x + b) \subset I$, and according to Lemma 3 we have $f((x - b, x + b))$ is equal to

$$(f(x - b), f(x + b)) \quad \text{or} \quad (f(x + b), f(x - b)).$$

In either case, $f((x - b, x + b))$ is an open interval in $f(I)$ that contains $y$. Hence $f(I)$ is open.

To show that $f^{-1}\colon f(I) \to I$ is continuous, let

$$y_0 = f(x_0) \in f(I) \quad \text{and} \quad \varepsilon > 0.$$

Then $(x_0 - \varepsilon, x_0 + \varepsilon) \cap I = J$ is an open interval, and by Lemma 3, $f(J)$ is an open interval and contains $f(x_0)$. Choose $\delta > 0$ so that

$$(f(x_0) - \delta, f(x_0) + \delta) \subset f(J).$$

If $f(x) \in (f(x_0) - \delta, f(x_0) + \delta)$, it follows that

$$x \in (x_0 - \varepsilon, x_0 + \varepsilon).$$

In other words

$$|f(x) - f(x_0)| < \delta \quad \text{implies} \quad |x - x_0| < \varepsilon \quad \text{or}$$
$$|y - y_0| < \delta \quad \text{implies} \quad |f^{-1}(y) - f^{-1}(y_0)| < \varepsilon$$

where $y = f(x)$. This shows that $f^{-1}$ is continuous at $y_0$ ●

Now consider the function $f(x) = x^n$, $x \in (0, \infty)$, $n \in N$. It is injective (Corollary 3, Theorem 5 (9.3B)) and continuous, so by Theorem 6 its inverse function $g(x) = x^{1/n}$ is continuous on $(0, \infty)$. It is in fact continuous on $[0, \infty)$ and we leave this for the reader (Problem 6).

→ If $I$ is a closed interval it does not necessarily follow that $f(I)$ is a closed interval. For example, the function $f(x) = \mathrm{Tan}^{-1} x$ is continuous and injective on the closed interval $(-\infty, \infty)$, but $f((-\infty, \infty)) = (-\frac{\pi}{2}, \frac{\pi}{2})$ is not a closed interval. On the other hand, if $I$ is a compact interval, then from Lemma 3 it follows that $f(I)$ is a compact interval.

**Problem 6** Prove that the function $g(x) = x^{1/n}$, $x \in [0, \infty)$, $n \in N$, is continuous at $x = 0$.

## D. Exercises

**B1.** Suppose $f(0) = g(0) = 0$, $f(x) = \sin \frac{1}{x}$ $(x \neq 0)$, and $g(x) = x \sin \frac{1}{x}$ $(x \neq 0)$.
  a) Show that $f$ is not continuous at $0$.
  b) Show that $g$ is continuous at $0$.

**B2.** a) Suppose $f: (a, b) \to R$ is continuous and $\alpha \in R$. Show that the sets

$$\{x \in (a, b): f(x) > \alpha\}$$

and

$$\{x \in (a, b): f(x) < \alpha\}$$

are open.
  b) Suppose $f: [a, b] \to R$ is continuous and $\alpha \in R$. Show that the sets

$$\{x \in [a, b]: f(x) \geq \alpha\}$$

and

$$\{x \in (a, b): f(x) \leq \alpha\}$$

are closed.

**B3.** Prove that if $p$ is a polynomial of odd degree then

$$p(R) = R.$$

**B4.** Given $r \in (0, \infty)$, define $f_r: [0, \infty) \to R$ by $f_r(x) = x^r$.
  a) Show that $f_r$ is uniformly continuous on $[0, \infty)$ whenever $0 < r \leq 1$.
  b) Show that $f_r$ is not uniformly continuous on $[0, \infty)$ if $r > 1$.

**B5.** Suppose $f, g: R \to R$ are uniformly continuous. Prove that $f \circ g$ is uniformly continuous.

**M6.** A function $f$ is called *linear* if $f(cx + y) = cf(x) + f(y)$ for all real numbers $c$. Show that a linear function is continuous. (Hint: Show that there is a positive real number $d$ such that $|f(x) - f(y)| \leq d \cdot |x - y|$.)

**M7.** A function $f$ is called *additive* if $f(x + y) = f(x) + f(y)$. Show that an additive function that is continuous at $x = 0$ is continuous on $R$.

**M8.** (Universal Chord Problem)
  a) Suppose $f: [0, 1] \to R$ is continuous and satisfies $f(0) = f(1) = 0$. Prove that for any $n \in N$ there is an $x_n \in [0, 1]$ for which $f(x_n) = f(x_n + \frac{1}{n})$.

  b) For each $a \in (0, 1)$, $a \neq \frac{1}{n}$ $(n \in N)$, construct a continuous function $f_a: [0, 1] \to R$ satisfying

$$f_a(0) = f_a(1) = 0,$$

and $f(x + a) - f(x) \neq 0$ for any $x \in [0, 1 - a]$.

**M9.** Suppose $f: R \to R$ is continuous. Show that for every open interval $I$, the set

$$\{x \in R: f(x) \in I\}$$

is open. (See Exercises B2 and M11.)
Suppose $f: R \to R$ is a function, and for every open interval $I$, the set

$$\{x \in R: f(x) \in I\}$$

is open. Prove that $f$ is continuous on $R$.

**M10.** A continuous function $f: [a, b] \to R$ is said to be *piecewise linear* if there are points

$$a = x_0 < x_1 < \cdots < x_n = b$$

and numbers $a_k, b_k \in R$ $(k = 0, 1, \ldots, n - 1)$ such that

$$f(x) = a_k x + b_k \quad \text{for all } x \in [x_k, x_{k+1}].$$

Given a continuous function $g: [a, b] \to R$ and $\varepsilon > 0$, prove that there is a continuous piecewise linear function $f: [a, b] \to R$ satisfying $|f(x) - g(x)| \leq \varepsilon$.

**M11.** a) Construct a function $f: R \to R$ with the property that for any $y \in R$ there are exactly two points $x$ that satisfy $f(x) = y$.
  b) Show that no continuous function $f$ can satisfy the condition stated in (a).

**M12.** Suppose $f: R \to R$ is a strictly increasing function.
  a) Prove that for each $x \in R$, the following limits exist.

$$\lim_{\substack{h \to 0 \\ h > 0}} f(x + h) \quad \text{and} \quad \lim_{\substack{h \to 0 \\ h < 0}} f(x + h)$$

  b) Show that $f$ is continuous except at a countable number of points.

## 9.4 The Integral

**A. The Definition** Suppose $f: [a, b] \to R$ is continuous. For $n \in N$ define

$$A_n(f) = A_n = \left(\frac{b-a}{n}\right) \sum_{k=0}^{n-1} f\left(a + \frac{k}{n}(b-a)\right).$$

→ Recall that we defined the definite integral as the limit as $n \to \infty$ of the sum

$$A_n = \sum_{k=0}^{n-1} f(x_k)\, \Delta x$$

where $\Delta x = \frac{b-a}{n}$ and $x_k = a + k\,\Delta x$. Substituting for $\Delta x$ and $\frac{b-a}{n}$ gives the expression on the left.

**THEOREM 1** *The sequence $\{A_n\}$ is convergent. We denote its limit by*

$$\int_a^b f \quad or \quad \int_a^b f(x)\, dx$$

*and call it* the definite integral of $f$ from $a$ to $b$.

*Proof* Let

$$g(x) = (b-a)f(a + x(b-a)).$$

Then $g: [0, 1] \to R$ is continuous and we have

→ The device of letting $g(x) = (b-a)f(a + x(b-a))$ has the effect of making $x = 0$ correspond to $(b-a)f(a)$ and $x = 1$ correspond to $(b-a)f(b)$. This simplifies notation in the proof of the theorem.

$$A_n = \frac{1}{n} \sum_{k=0}^{n-1} g\left(\frac{k}{n}\right).$$

We will show that $\{A_n\}$ is a Cauchy sequence.

Let $\varepsilon > 0$. Since $g: [0, 1] \to R$ is continuous, it follows (Theorem 4, 9.3A) that there exists a $\delta > 0$ such that

$$|g(x_1) - g(x_2)| < \frac{\varepsilon}{2} \quad \text{whenever } |x_1 - x_2| < \delta.$$

For $n, m > \frac{1}{\delta}$ we will show that

$$|A_n - A_m| < \varepsilon.$$

We have

→ We will actually show that $|A_{mn} - A_n| < \frac{\varepsilon}{2}$ and that $|A_{mn} - A_m| < \frac{\varepsilon}{2}$. It then follows that

$$|A_m - A_n| = |A_m - A_{mn} + A_{mn} - A_n|$$

$$\leq |A_m - A_{mn}| + |A_{mn} - A_n| < \frac{\varepsilon}{2} + \frac{\varepsilon}{2} = \varepsilon.$$

$$A_n - A_{mn} = \frac{1}{n} \sum_{k=0}^{n-1} g\left(\frac{k}{n}\right) - \frac{1}{mn} \sum_{r=0}^{mn-1} g\left(\frac{r}{mn}\right)$$

$$= \frac{1}{n} \sum_{k=0}^{n-1} g\left(\frac{k}{n}\right) - \frac{1}{mn} \sum_{r=0}^{m-1} g\left(\frac{r}{mn}\right)$$

$$- \frac{1}{mn} \sum_{r=m}^{2m-1} g\left(\frac{r}{mn}\right) - \cdots$$

$$-\frac{1}{mn} \sum_{r=(n-1)m}^{mn-1} g\left(\frac{r}{mn}\right)$$

$$= \frac{1}{mn} \sum_{r=0}^{m-1} \left(g\left(\frac{0}{n}\right) - g\left(\frac{r}{mn}\right)\right)$$

$$+ \frac{1}{mn} \sum_{r=m}^{2m-1} \left(g\left(\frac{1}{n}\right) - g\left(\frac{r}{mn}\right)\right)$$

$$+ \frac{1}{mn} \sum_{r=2m}^{3m-1} \left(g\left(\frac{2}{n}\right) - g\left(\frac{r}{mn}\right)\right) + \cdots$$

$$+ \frac{1}{mn} \sum_{r=(n-1)m}^{mn-1} \left(g\left(\frac{n-1}{n}\right) - g\left(\frac{r}{mn}\right)\right).$$

Now consider the terms $g(\frac{k}{n}) - g(\frac{r}{mn})$ where

$$km + 1 \le r \le (k+1)m - 1.$$

We have that

$$0 \le \frac{r}{mn} - \frac{k}{n} \le \frac{(k+1)m - 1}{mn} - \frac{k}{n}$$

$$= \frac{1}{n} - \frac{1}{mn} < \frac{1}{n}.$$

By the choice of $\delta$ it follows that

$$\left|g\left(\frac{k}{n}\right) - g\left(\frac{r}{mn}\right)\right| < \frac{\varepsilon}{2},$$

and taking absolute values in the above identity we obtain

$$|A_n - A_{mn}| \le \frac{1}{mn} \sum_{r=0}^{m-1} \left|g\left(\frac{0}{n}\right) - g\left(\frac{r}{mn}\right)\right|$$

$$+ \frac{1}{mn} \sum_{r=m}^{2m-1} \left|g\left(\frac{1}{n}\right) - g\left(\frac{r}{mn}\right)\right|$$

$$+ \frac{1}{mn} \sum_{r=2m}^{3m-1} \left|g\left(\frac{2}{n}\right) - g\left(\frac{r}{mn}\right)\right| + \cdots$$

$$+ \frac{1}{mn} \sum_{r=(n-1)m}^{mn-1} \left|g\left(\frac{n-1}{n}\right) - g\left(\frac{r}{mn}\right)\right|$$

$$< \frac{1}{mn} \left(m\frac{\varepsilon}{2} + m\frac{\varepsilon}{2} + \cdots + m\frac{\varepsilon}{2}\right)$$

$$= \frac{1}{mn} \left(mn\frac{\varepsilon}{2}\right) = \frac{\varepsilon}{2}.$$

In a similar fashion we show that $|A_m - A_{mn}| < \frac{\varepsilon}{2}$. Hence $|A_m - A_n| < \varepsilon$ when $m, n > \frac{1}{\delta}$, so the sequence $\{A_n\}$ is Cauchy, and by Theorem 5 (9.2B) it is convergent ●

**THEOREM 2** *Suppose $f$: $[a, b] \to R$ is continuous. Then*

$$\min f \leq \frac{1}{b - a} \int_a^b f \leq \max f.$$

→

$$\min f = \inf \{f(x): x \in [a, b]\}$$

and

$$\max f = \sup \{f(x): x \in [a, b]\}.$$

Since $f$ is continuous and $[a, b]$ is compact, we know from Corollary 1, Theorem 3 (9.3A) that $\min f \in f([a, b])$ and $\max f \in f([a, b])$.

*Proof* For all $n \in N$ we have

$$(b - a) \min f = \left(\frac{b - a}{n}\right) \sum_{k=0}^{n-1} \min f$$

$$\leq \left(\frac{b - a}{n}\right) \sum_{k=0}^{n-1} f\left(a + \frac{k}{n}(b - a)\right) = A_n$$

$$\leq \left(\frac{b - a}{n}\right) \sum_{k=0}^{n-1} \max f = (b - a) \max f.$$

Since $A_n \to \int_a^b f$ the desired result follows ●

**THEOREM 3** *Suppose $f$: $[a, b] \to R$ is continuous; then*

$$\left|\int_a^b f\right| \leq \int_a^b |f|.$$

*Proof* For all $n \in N$ we have

$$\left|\left(\frac{b - a}{n}\right) \sum_{k=0}^{n-1} f\left(a + \frac{k}{n}(b - a)\right)\right|$$

$$\leq \left|\left(\frac{b - a}{n}\right) \sum_{k=0}^{n-1} |f|\left(a + \frac{k}{n}(b - a)\right)\right|.$$

Hence

$$|A_n(f)| \leq A_n(|f|),$$

and since

$$A_n(f) \to \int_a^b f \quad \text{and} \quad A_n(|f|) \to \int_a^b |f|,$$

we have

$$\left|\int_a^b f\right| \leq \int_a^b |f| \quad ●$$

Given $f$: $[a, b] \to R$, the function $|f|$: $[a, b] \to R$ is defined by the formula $|f|(x) = |f(x)|$. If $f$ is continuous it follows that $|f|$ is continuous. To see this let $x_n \to x_0$. Since $f$ is continuous, we have

→

$$f(x_n) \to f(x_0) \quad \text{or} \quad |f(x_n) - f(x_0)| \to 0.$$

Since $||f(x_n)| - |f(x_0)|| \leq |f(x_n) - f(x_0)|$, we have

$$||f(x_n)| - |f(x_0)|| \to 0, \quad \text{so} \quad |f(x_n)| \to |f(x_0)|.$$

On the other hand, $|f|$ can be continuous without $f$ being continuous. For example, let

$$f(x) = \begin{cases} 1 & \text{if } x \text{ is rational} \\ -1 & \text{if } x \text{ is irrational.} \end{cases}$$

Then $f$ is continuous nowhere, but $|f|(x) = 1$ for all $x \in R$, and is therefore obviously continuous on $R$.

**Problem 1** Given continuous functions $f$, $g$: $[a, b] \to R$, show that

$$A_n(f + g) = A_n(f) + A_n(g)$$

and then that

$$\int_a^b (f \pm g) = \int_a^b f \pm \int_a^b g.$$

**COROLLARY** *If $f: [a, b] \to R$ is continuous, then*

$$\left| \int_a^b f \right| \le (b - a) \sup |f|.$$

(This follows from Theorems 2 and 3.)

**B. Uniform Convergence** Suppose $f, f_n: A \to R$ ($n \in N$) are functions. The sequence of functions $\{f_n\}$ is said to converge *uniformly* on $A$ to $f$ if for any $\varepsilon > 0$ there is an $n_0 \in N$ such that

$$|f_n(x) - f(x)| < \varepsilon$$

for all $x \in A$ and $n > n_0$.

**THEOREM 4** *Suppose $f_n, f: [a, b] \to R$ are continuous and the sequence $\{f_n\}$ converges uniformly to $f$. Then*

$$\int_a^b f_n \to \int_a^b f.$$

**Proof** Let $\varepsilon > 0$. Then there is an $n_0$ such that

$$|f_n(x) - f(x)| < \frac{\varepsilon}{b - a} \quad \text{for all } n > n_0 \quad \text{and} \quad x \in A.$$

Hence for $n > n_0$,

$$\sup |f_n - f| < \frac{\varepsilon}{b - a}.$$

Then

$$\left| \int_a^b f_n - \int_a^b f \right| = \left| \int_a^b (f_n - f) \right| \quad \text{(by Problem 1)}$$

$$\le \int_a^b |f_n - f| \quad \text{(by Theorem 3)}$$

$$\le (b - a) \sup |f_n - f| \quad \text{(by the Corollary to Theorem 3)}$$

$$\le (b - a) \left( \frac{\varepsilon}{b - a} \right) = \varepsilon.$$

Hence $\int_a^b f_n \to \int_a^b f$ ●

**THEOREM 5** *Suppose $f_n: A \to R$ is continuous for each $n \in N$ and the sequence $\{f_n\}$ converges uniformly to $f: A \to R$. Then $f$ is continuous.*

**Proof** Let $x_0 \in A$ and $\varepsilon > 0$. Then

→ This contrasts with the related notion of pointwise convergence of functions:

The sequence of functions $\{f_n\}$ is said to converge pointwise to $f$ if for each $x \in A$, $f_n(x) \to f(x)$. The notions of pointwise convergence and uniform convergence are distinct. Clearly uniform convergence implies pointwise convergence, but the converse is false, as Example 1 shows.

**Example 1** Let $f_n: [0, 1] \to R$ be defined by $f_n(x) = x^n$, and let $f(x) = \begin{cases} 0, & x \in [0, 1) \\ 1, & x = 1 \end{cases}$. Then the sequence $\{f_n\}$ converges to $f$ pointwise, but not uniformly.

**Proof** For each $x \in [0, 1)$ we have that $x^n \to 0$ as $n \to \infty$. Hence $f_n(x) \to f(x)$ for each $x \in [0, 1)$. When $x = 1$, $f_n(x) = 1$ for all $n$ and $f(x) = 1$, so $f_n(1)$ obviously converges to $f(1)$. We therefore have pointwise convergence. Since $f$ is not continuous and each $f_n$ is continuous, it follows from Theorem 5 that $\{f_n\}$ cannot converge uniformly to $f$.

**Problem 2** Referring to Example 1, show directly (without the use of Theorem 5) that the sequence $\{f_n\}$ does not converge uniformly to $f$.

$$|f(x) - f(x_0)|$$
$$= |f(x) - f_n(x) + f_n(x) - f_n(x_0) + f_n(x_0) - f(x_0)|$$
$$\leq |f(x) - f_n(x)| + |f_n(x) - f_n(x_0)| + |f_n(x_0) - f(x_0)|.$$

Since $f_n \to f$ uniformly, there is an $n_0$ such that

$$|f(a) - f_n(a)| < \frac{\varepsilon}{3} \quad \text{for all } n > n_0 \quad \text{and all } a \in A.$$

In particular, the first and third terms above are each bounded by $\frac{\varepsilon}{3}$ for each $n \geq n_0$. Hence for all $x \in A$,

$$|f(x) - f(x_0)| < \frac{2\varepsilon}{3} + |f_n(x) - f_n(x_0)| \quad \text{if } n \geq n_0.$$

This is then true for $n = n_0$, so

$$|f(x) - f(x_0)| < \frac{2\varepsilon}{3} + |f_{n_0}(x) - f_{n_0}(x_0)|.$$

Since $f_{n_0}$ is continuous, there is a $\delta$ satisfying

$$|f_{n_0}(x) - f_{n_0}(x_0)| < \frac{\varepsilon}{3} \quad \text{.whenever } |x - x_0| < \delta.$$

Hence for $|x - x_0| < \delta$, we have

$$|f(x) - f(x_0)| < \frac{2\varepsilon}{3} + \frac{\varepsilon}{3} = \varepsilon \quad \bullet$$

**THEOREM 6**   *Suppose* $f: [a, b] \to R$ *is continuous and* $c \in (a, b)$. *Then*

$$\int_a^b f = \int_a^c f + \int_c^b f.$$

*Proof Step* 1   Suppose first that $c$ has the form

$$c = a + \frac{k}{m}(b - a)$$

where $k, m \in N$ and $0 < k < m$.
Let

$$A_n = \frac{b - a}{n} \sum_{i=0}^{n-1} f\left(a + \frac{i}{n}(b - a)\right),$$

$$B_n = \frac{c - a}{n} \sum_{i=0}^{n-1} f\left(a + \frac{i}{n}(c - a)\right),$$

and

**LEMMA**   *This result is true if* $f$ *is the function* $f(x) = \alpha x + \beta$ *where* $\alpha$ *and* $\beta$ *are constants.*

*Proof*   We will show that

$$\int_r^s (\alpha x + \beta)\, dx = \frac{\alpha}{2}(s^2 - r^2) + \beta(s - r).$$

We have

$$A_n(f) = \frac{s - r}{n} \sum_{k=0}^{n-1} f\left(r + \frac{k}{n}(s - r)\right)$$

$$= \frac{s - r}{n} \sum_{k=0}^{n-1} \left[\alpha\left(r + \frac{k}{n}(s - r)\right) + \beta\right]$$

$$= \frac{s - r}{n}\left[\alpha n r + n\beta + \frac{\alpha(s - r)}{n} \sum_{k=0}^{n-1} k\right]$$

$$= \frac{s - r}{n}\left[\alpha n r + n\beta + \frac{\alpha(s - r)}{n} \cdot \frac{n(n - 1)}{2}\right]$$

$$= \alpha r(s - r) + \beta(s - r) + \frac{\alpha(s - r)^2}{2}\left(\frac{n - 1}{n}\right).$$

Now as $n \to \infty$, $\dfrac{n - 1}{n} \to 1$, so

$$A_n \to \alpha r(s - r) + \beta(s - r) + \frac{\alpha(s - r)^2}{2}$$

$$C_n = \frac{b-c}{n} \sum_{i=0}^{n-1} f\left(c + \frac{i}{n}(c-a)\right).$$

For $n = qm$ $(q \in N)$, we have

$$A_{qm} = \frac{b-a}{qm} \sum_{i=0}^{qm-1} f\left(a + \frac{i}{qm}(b-a)\right)$$

$$= \frac{b-a}{qm} \sum_{i=0}^{qk-1} f\left(a + \frac{i}{qm}(b-a)\right)$$

$$+ \frac{b-a}{qm} \sum_{i=qk}^{qm-1} f\left(a + \frac{i}{qm}(b-a)\right)$$

$$= \frac{c-a}{qk} \sum_{i=0}^{qk-1} f\left(a + \frac{i}{qk}(c-a)\right)$$

$$+ \frac{b-c}{q(m-k)} \sum_{i=qk}^{qm-1} f\left(a + \frac{i}{qk}(c-a)\right)$$

$$= \frac{c-a}{qk} \sum_{i=0}^{qk-1} f\left(a + \frac{i}{qk}(c-a)\right)$$

$$+ \frac{b-c}{q(m-k)} \sum_{i=0}^{q(m-k)-1} f\left(a + \frac{(i+qk)(c-a)}{qk}\right)$$

$$= B_{qk} + \frac{b-c}{q(m-k)} \sum_{i=0}^{q(m-k)-1} f\left(c + \frac{i}{qk}(c-a)\right)$$

$$= B_{qk} + C_{q(m-k)}.$$

We know that $A_n \to \int_a^b f$, $B_n \to \int_a^c f$, $C_n \to \int_c^b f$, as $n \to \infty$, so all subsequences of $\{A_n\}$, $\{B_n\}$, and $\{C_n\}$ converge to the same limit (Theorem 3, 9.2A). Thus, letting $q \to 0$ we obtain

$$\int_a^b f = \int_a^c f + \int_c^b f.$$

**Step 2** Now take any $c \in (a, b)$ and choose $r_j$, $s_j \in (a, b)$ satisfying $r_j < c$, $s_j > c$, $r_j \to c$, $s_j \to c$, and having the form

$$a + \frac{k}{m}(b-a)$$

where $k, m \in N$. Define $f_j: [a, b] \to R$ as follows:

$$= \alpha(s-r)\left(r + \frac{(s-r)}{2}\right) + \beta(s-r)$$

$$= \alpha(s-r)\frac{(s+r)}{2} + \beta(s-r)$$

$$= \frac{\alpha}{2}(s^2 - r^2) + \beta(s - r).$$

Since $A_n \to \int_r^s (\alpha x + \beta)\, dx$, we have established the desired formula. The proof is completed by simply evaluating

$$\int_a^c f + \int_c^b f \quad \text{and showing that it equals} \quad \int_a^b f \quad \bullet$$

This is obtained from the preceding expression by replacing

$$a + \frac{i}{qm}(b-a) \quad \text{by} \quad a + \frac{i}{qk}(c-a),$$

the first $\frac{b-a}{qm}$ by $\frac{c-a}{qk}$, and the second $\frac{b-a}{qm}$ by $\frac{b-c}{q(m-k)}$.

This follows from the identity

$$\sum_{i=qk}^{qm-1} h(i) = \sum_{i=0}^{q(m-k)-1} h(i + qk),$$

which is valid for any function $h$.

This follows from the fact that between any two reals there is a rational. For each $j \in N$ we choose $r_j \in (c - \frac{1}{j}, c)$ and $s_j \in (c, c + \frac{1}{j})$.

$$f_j(x) = \begin{cases} f(x) & (x \in [a, r_j]) \\ \alpha_j x + \beta_j & (x \in [r_j, s_j]) \\ f(x) & (x \in [s_j, b]) \end{cases}$$

Then,

$$\int_a^b f_j$$

$$= \int_a^{r_j} f_j + \int_{r_j}^b f_j \qquad \text{(by Step 1)}$$

$$= \int_a^{r_j} f_j + \left( \int_{r_j}^{s_j} f_j + \int_{s_j}^b f_j \right) \qquad \text{(by Step 1)}$$

$$= \int_a^{r_j} f_j + \int_{r_j}^c f_j + \int_c^{s_j} f_j + \int_{s_j}^b f_j \quad \text{(by the lemma)}$$

$$= \int_a^c f_j + \int_c^b f_j \qquad \text{(by Step 1)}.$$

Finally, we have $f_j \to f$ uniformly on $[a, b]$, so

$$\int_a^b f_j \to \int_a^b f, \quad \int_a^c f_j \to \int_a^c f, \quad \text{and} \quad \int_c^b f_j \to \int_c^b f$$

as $j \to \infty$, and the result follows ●

**THEOREM 7** *Suppose $f: [a, b] \to R$ is continuous. Given any $\varepsilon > 0$, there is a $\delta > 0$ such that if $a = x_0 < x_1 < x_2 < \cdots < x_n = b$, $x_{i+1} - x_i < \delta$, and $\xi_i \in [x_{i-1}, x_i]$, then*

$$\left| \sum_{i=1}^n f(\xi_i)(x_i - x_{i-1}) - \int_a^b f \right| < \varepsilon.$$

*Proof* Since $f: [a, b] \to R$ is uniformly continuous (Theorem 4, 9.3A), given $\varepsilon > 0$ there is a $\delta > 0$ such that

$$|f(x) - f(y)| < \frac{\varepsilon}{b - a}$$

whenever $|x - y| < \delta$. Now suppose

$$a = x_0 < x_1 < \cdots < x_n = b,$$

$$x_i - x_{i-1} < \delta \quad \text{and} \quad \xi_i \in [x_{i-1}, x_i].$$

We have

$$\left| \int_a^b f(x)dx - \sum_{i=1}^n f(\xi_i)(x_i - x_{i-1}) \right|$$

→ Here $\alpha_j = \dfrac{f(s_j) - f(r_j)}{s_j - r_j}$ and $\beta_j = \dfrac{s_j f(r_j) - r_j f(s_j)}{r_j - s_j}$. The graph of $f_j$ on $[r_j, s_j]$ is simply the line segment joining $(r_j, f(r_j))$ to $(s_j, f(s_j))$.

→ We will show this. Given $\varepsilon > 0$, pick $\delta > 0$ (Theorem 4, 9.3A) so that

$$|f(x) - f(y)| < \frac{\varepsilon}{2} \quad \text{whenever} \quad |x - y| < \delta.$$

Choose $j_0 \in N$ so that

$$|r_j - s_j| < \delta \quad \text{whenever} \quad j \geq j_0.$$

Then for $x \in [r_j, s_j]$ and $j \geq j_0$ we have

$$|f(x) - f_j(x)| \leq |f(x) - f(s_j)| + |f(s_j) - f_j(s_j)| + |f_j(s_j) - f_j(x)|$$

$$< \frac{\varepsilon}{2} + 0 + \frac{\varepsilon}{2} = \varepsilon.$$

For $x \notin [r_j, s_j]$ we have $f(x) - f_j(x) = 0$. Hence, for all $j \geq j_0$ and all $x \in [a, b]$ we have

$$|f(x) - f_j(x)| < \varepsilon.$$

---

*Problem 3* Prove Theorem 6 using Theorem 7.

●

$$= \left| \sum_{i=1}^{n} \int_{x_{i-1}}^{x_i} f(x)\, dx - \sum_{i=1}^{n} f(\xi_i)(x_i - x_{i-1}) \right|$$

(by Theorem 6)

$$\leq \sum_{i=1}^{n} \left| \int_{x_{i-1}}^{x_i} f(x)\, dx - f(\xi_i)(x_i - x_{i-1}) \right|$$

→ The quantity $f(\xi_i)(x_i - x_{i-1})$ is rewritten in the following step in the form

$$\int_{x_{i-1}}^{x_i} f(\xi_i)\, dx.$$

$$= \sum_{i=1}^{n} \left| \int_{x_{i-1}}^{x_i} (f(x) - f(\xi_i))\, dx \right|$$

$$\leq \sum_{i=1}^{n} \int_{x_{i-1}}^{x_i} |f(x) - f(\xi_i)|\, dx.$$

Since $|f(x) - f(\xi_i)| < \frac{\varepsilon}{b-a}$ for all $x \in [x_{i-1}, x_i]$, we have

$$\int_{x_{i-1}}^{x_i} |f(x) - f(\xi_i)|\, dx < \frac{\varepsilon}{(b-a)} (x_i - x_{i-1}).$$

Hence the above sum is bounded by

If you have finished this chapter, do something other than mathematics before going on to the final one.

$$\sum_{i=1}^{n} \frac{\varepsilon}{(b-a)} (x_i - x_{i-1}) = \varepsilon$$

"Logic, like whiskey, loses its beneficial effect when taken in too large quantities."

and the proof is complete ●  →

(Lord Dunsany)

## C. Exercises

**B1.** If $f: [0, 1] \to R$ is continuous, show that

$$\int_0^1 f(x)\, dx = f(c) \quad \text{for some } c \in [0, 1].$$

**B2.** Suppose $f: [a, b] \to R$ is continuous, $f(x) \geq 0$ for all $x \in [a, b]$, and $\int_a^b f(x)\, dx = 0$. Prove that $f(x) = 0$ for all $x \in [a, b]$.

**B3.** (Cauchy-Schwarz Inequalities)
a) For $a_i, b_i \in R$ $(i = 1, 2, \ldots, n)$ prove that

$$(a_1 b_1 + \cdots + a_n b_n)^2 \leq (a_1^2 + \cdots + a_n^2)(b_1^2 + \cdots + b_n^2).$$

b) Suppose $f, g: [a, b] \to R$ are continuous. Prove that

$$\left( \int_a^b f(x)g(x)\, dx \right)^2 \leq \left( \int_a^b f(x)^2\, dx \right)\left( \int_a^b g(x)^2\, dx \right).$$

**B4.** Let $f_n(x) = 1 + x + x^2 + \cdots + x^n$ and

$$g(x) = \frac{1}{1-x} \quad (-1 < x < 1).$$

a) Prove that $f_n \to g$ uniformly on $[-r, r]$ for $0 \leq r < 1$.
b) Show that the sequence $\{f_n\}$ does not converge uniformly to $g$ on the interval $(-1, 1)$.

**M5.** Show that the convergence of the sequence of functions $f_n(x) = \sin^n x$ to its limit on $[0, \pi]$ is not uniform.

**M6.** Assume $f$ is continuous on $[a, b]$ and that $|f(x)| \leq c$ for all $x \in [a, b]$. Show that $|\int_a^b f| \leq c(b - a)$.

**M7.** If $f$ and $g$ are continuous on $[a, b]$ and $f(x) \leq g(x)$ for all $x \in [a, b]$, show that $\int_a^b f \leq \int_a^b g$.

**M8.** If $f$ and $g$ are continuous on $[a, b]$ and $f(x) \leq g(x)$ for all $x \in [a, b]$, but $f$ and $g$ are not identical there, show that $\int_a^b f < \int_a^b g$.

**M9.** Suppose $f: [a, b] \to R$ is continuous. Find a $c \in [a, b]$ satisfying

$$\int_a^c f(x)\, dx = \int_c^b f(x)\, dx.$$

**M10.** (Minkowski Inequality) Use the Cauchy-Schwarz Inequality to prove that

$$\left(\int_a^b (f+g)^2\right)^{1/2} \leq \left(\int_a^b f^2\right)^{1/2} + \left(\int_a^b g^2\right)^{1/2}.$$

**M11.** Show that if a sequence of functions converges uniformly on a set $A$, then any subsequence converges uniformly on $A$.

**M12.** Show that if $\{f_n\}$ and $\{g_n\}$ converge uniformly on $A$, then $\{f_n + g_n\}$ converges uniformly on $A$.

**M13.** Show that there cannot exist a continuous function $g: [-1, 1] \to R$ that satisfies

$$\int_{-1}^1 f(x)g(x)\, dx = f(0)$$

for all continuous functions $f: [-1, 1] \to R$. (See Exercise M7.)

**M14.** For $n \in N$ define $g_n: [-1, 1] \to R$ by the following:

$$g_n(x) = \begin{cases} 0 & \text{if } |x| > 1/n \\ n + n^2x & \text{if } -1/n \leq x \leq 0 \\ n - n^2x & \text{if } 0 < x \leq 1/n \end{cases}$$

a) Show that

$$\int_{-1}^1 g_n(x)\, dx = 1.$$

b) Show that for any continuous function $f: [-1, 1] \to R$,

$$\lim_{n\to\infty} \int_{-1}^1 f(x)g_n(x)\, dx = f(0).$$

**M15.** Suppose $f: [a, b] \to R$ is continuous and $f(x) \geq 0$ for all $x \in [a, b]$. Show that

$$\lim_{n\to\infty} \left(\int_a^b (f(x))^n\, dx\right)^{1/n} = \sup_{a \leq x \leq b} f(x).$$

**M16.** Construct a sequence of continuous functions $\{f_n\}$ which converges pointwise (but not uniformly) to a limit $f$ on $[a, b]$ and for which $\int_a^b f_n$ does not converge to $\int_a^b f$.

We begin this chapter by reformulating the definition of the derivative. We then prove a number of important theorems including the Mean Value Theorem, the Fundamental Theorem of Calculus, and Taylor's Theorem. Series of complex numbers are then studied, and some basic facts about power series are established. The exponential function is defined as a power series, and in terms of it the logarithmic and trigonometric functions are introduced. We then rederive some of the well known properties of these functions. In the last section two theorems are proven that are important in more advanced branches of mathematics. One is an approximation result, and the other is an existence theorem. The chapter closes with a history section indicating two radically different approaches to the material presented in this book. They are called constructive calculus and nonstandard calculus, and each is currently a subject of active research.

# Differentiability

## 10.1 *The Derivative*

**A. Definitions and Old Theorems** Suppose $O$ is an open subset of $R$ and $f: O \to R$ is a function. Then $f$ is said to be *differentiable at* $x_0 \in O$ if there is a number $f'(x_0)$ such that for any $\epsilon > 0$ there is a $\delta > 0$ satisfying

$$(1) \qquad |f(x_0 + h) - f(x_0) - f'(x_0)h| < \epsilon|h|$$

whenever $|h| < \delta$.

→ This is equivalent to the usual definition of the derivative. Assuming $h \neq 0$ and dividing by $|h|$ throughout Equation (1) gives

$$\left|\frac{f(x_0 + h) - f(x_0)}{h} - f'(x_0)\right| < \epsilon \quad \text{for} \quad |h| < \delta.$$

This means that

$$(2) \qquad \lim_{\substack{h \to 0 \\ h \neq 0}} \frac{f(x_0 + h) - f(x_0)}{h} = f'(x_0). \qquad \neq$$

Note that from Equation (2) it follows that $f'(x_0)$, if it exists, is unique. The definition in terms of Equation (1) has two advantages. One is that we do not divide by $h$ and therefore need not worry about $h$ being zero. The second is that this form of the definition of the derivative is the one that naturally extends to higher dimensions. (In this extension the number $f'(x_0)$ is replaced by what is called a linear mapping.)

**THEOREM 1** *If* $f: O \to R$ *is differentiable at* $x_0 \in O$, *then* $f$ *is continuous at* $x_0$.

*Proof* For $\epsilon > 0$ there is a $\delta > 0$ such that

$$|f(x_0 + h) - f(x_0) - f'(x_0)h| < \epsilon|h|,$$

so

$$|f(x_0 + h) - f(x_0)|$$
$$\leq |f(x_0 + h) - f(x_0) - f'(x_0)h| + |f'(x_0)h|$$
$$\leq \epsilon|h| + |f'(x_0)| \, |h| \quad \text{for all} \quad |h| < \delta.$$

Letting $|h| < \dfrac{\epsilon}{\epsilon + |f'(x_0)|}$, we then have

$$|f(x_0 + h) - f(x_0)| < (\epsilon + |f'(x_0)|)|h| < \epsilon.$$

Hence $f$ is continuous at $x_0$ ●

→ The function $f(x) = |x|$ shows that the converse of Theorem 1 is false.

A function $f: O \to R$ is said to be *differentiable in* $O$, or simply differentiable, if $f$ is differentiable at every point in $O$, and in this case the function $x \to f'(x): O \to R$ is called the *derivative of* $f$. The higher derivatives $f^{(n)}$, if they exist, are defined inductively by setting

$$f^{(n+1)} = (f^{(n)})'.$$

The function $f$ is said to be *n-times differentiable at* $x_0 \in O$ if $f^{(n)}(x_0)$ exists. If $f^{(n)}$ exist for all $n$, then $f$ is called a $C^\infty$*-function.*

Unfortunately there are still more definitions.

→ A function $f: O \to R$ is called a $C^n$*-function* if $f^{(n)}: O \to R$ exists and is continuous. Hence, $f$ is a $C^1$-function in $O$ if $f'$ exists and $f': O \to R$ is continuous. With the convention that $f^{(0)} = f$, we may speak of the $C^0$-functions as the continuous ones. (As usual, we often write $f'$ and $f''$ in place of $f^{(1)}$ and $f^{(2)}$ respectively.)

**THEOREM 2** *Suppose* $f: U \to R$ *and* $g: V \to R$ *are n-times differentiable and* $f(U) \subset V$. *Then the composition* $g \circ f: U \to R$ *is n-times differentiable.*

*Proof* We will use induction. For $n = 1$ the Chain Rule (see the alternative proof given in 3.1B) asserts that $g \circ f$ is differentiable and

$$(1) \qquad (g \circ f)' = (g' \circ f)f'.$$

This gives the result for $n = 1$, so assume it is true for $n = k$, and suppose that $f$ and $g$ are $(k + 1)$-times differentiable. Then $g'$ is $k$-times differentiable, so by the induction assumption it follows that $g' \circ f$ is $k$-times differentiable, and hence the product $(g' \circ f)f'$ (scc Problem 1) is $k$-times differentiable. Equation (1) now gives that $(g \circ f)'$ is $k$-times differentiable, so $g \circ f$ must be $(k + 1)$-times differentiable. By induction it follows that the result is true for all $n$ ●

**Problem 1** Suppose $f, g: O \to R$ are $n$-times differentiable. Show, using induction, that $f + g$ and $fg$ are $n$-times differentiable.

→ When $f$ and $g$ are $C^n$-functions, their composition is also a $C^n$-function. We leave the proof of this as an exercise.

**THEOREM 3** If $f(x) = x^n$ then $f'(x) = nx^{n-1}$ $(n \in N)$.

*Proof* By induction. The result is trivial for $n = 1$, so assume it is true for $k$. Then

$$(x^{k+1})' = (x^k x)' = (x^k)x' + (x^k)'x = x^k + kx^{k-1}x$$
$$= x^k + kx^k = (k + 1)x^k \qquad ●$$

→ Here we have used the Product Rule (2.5B), the proof of which will not be reproduced.

**THEOREM 4** Suppose $I$ is an open interval and $f: I \to R$ is differentiable with $f'$ never zero on $I$. Then the inverse function $f^{-1}: f(I) \to I$ exists, is differentiable, and satisfies

$$(f^{-1})' = \frac{1}{f' \circ f^{-1}}.$$

Moreover, if $f$ is $n$-times differentiable, then so is $f^{-1}$.

*Proof* Since $f'$ is never zero it follows from the Mean Value Theorem (see Theorem 6) that $f$ is injective. By Theorem 6 of 9.3C we have that $f(I)$ is an open interval and $f^{-1}: f(I) \to I$ is continuous. To show $f^{-1}$ is differentiable, let $y_0 = f(x_0) \in f(I)$ and $y = f(x)$. Then

$$\frac{f^{-1}(y) - f^{-1}(y_0)}{y - y_0} = \frac{x - x_0}{y - y_0} = \frac{1}{\dfrac{f(x) - f(x_0)}{x - x_0}}.$$

Indeed, if $f$ is not injective, then there are points $x, y \in I$ $(x < y)$ with $f(x) = f(y)$. Using the Mean Value Theorem, we then have

$$f'(c) = \frac{f(y) - f(x)}{y - x} = 0$$

for some $c \in (x, y) \subset I$, and this contradicts the hypothesis. Hence a nonzero derivative on an interval implies that the function is injective there. The converse is false. For example, $f(x) = x^3$ is injective on $R$, but $f'(0) = 0$.

**Problem 2** Suppose $g: I \to R$ is $n$-times differentiable and $g$ is never zero. Prove by induction that $\frac{1}{g}$ is $n$-times differentiable.

Letting $y \rightarrow y_0$ we then have

$$(f^{-1})'(y_0) = 1/f'(x_0) = 1/f'(f^{-1}(y_0)),$$

which is the desired formula.

By induction we now show that if $f$ is $n$-times differentiable then so is $f^{-1}$. For $n = 1$ this has just been shown, so assume the result for $n = k$ and suppose $f$ is $(k + 1)$-times differentiable. Then $f'$ and $f^{-1}$ are $k$-times differentiable and, by Theorem 2, $f' \circ f^{-1}$ is $k$-times differentiable. It follows (Problem 2) that $1/(f' \circ f^{-1}) = (f^{-1})'$ is $k$-times differentiable. But this means $f^{-1}$ is $(k + 1)$-times differentiable $\bullet$

$\longrightarrow$ Since $f^{-1}$ is continuous it follows that

$$x - x_0 = f^{-1}(y) - f^{-1}(y_0) \rightarrow 0 \quad \text{as} \quad y \rightarrow y_0.$$

Hence

$$\lim_{y \rightarrow y_0} \frac{1}{(f(x) - f(x_0))/(x - x_0)} = \lim_{x \rightarrow x_0} \frac{1}{(f(x) - f(x_0))/(x - x_0)}$$
$$= \frac{1}{f'(x_0)}$$

## B. The Mean Value Theorem

**THEOREM 5** (Rolle's Theorem) *Suppose that $h:[a, b] \rightarrow R$ is continuous, $h(a) = h(b) = 0$, and $h$ is differentiable on $(a, b)$. Then there is an $m \in (a, b)$ satisfying*

$$h'(m) = 0.$$

$\longrightarrow$ This is a special case of the Mean Value Theorem.

*Proof* If $h = 0$ the result is obvious, so assume $h(x) \neq 0$ for some $x$. Since $h$ is continuous on $[a, b]$, there are points $m_1, m_2 \in [a, b]$ such that

$$h(m_1) \leq h(x) \leq h(m_2)$$

for all $x \in [a, b]$. Since $h$ has nonzero image points, we must have either $h(m_1) \neq 0$ or $h(m_2) \neq 0$, say $h(m_1) \neq 0$. Then $m = m_1$ cannot equal $a$ or $b$ (because $h(a) = h(b) = 0$) and we have from Theorem 1 (4.1A) that $h'(m) = 0$ $\bullet$

$\longrightarrow$ We are using here the Extreme Value Theorem, Corollary 1 of Theorem 3 (9.3A).

**THEOREM 6** (Mean Value Theorem) *Suppose $f: [a, b] \rightarrow R$ is continuous, and $f$ is differentiable on $(a, b)$. Then there is an $m \in (a, b)$ satisfying*

$$\frac{f(b) - f(a)}{b - a} = f'(m).$$

*Proof* Let $h(x) = f(x) - f(a) - \frac{f(b)-f(a)}{b-a}(x - a)$. Then $h$ satisfies the hypothesis of Theorem 5, so there is an $m \in (a, b)$ with $h'(m) = 0$. Since

$$h'(x) = f'(x) - \frac{f(b) - f(a)}{b - a},$$

it follows that $f'(m) = \dfrac{f(b) - f(a)}{b - a}$ $\bullet$

$\longrightarrow$ **THEOREM 7** (Extended Mean Value Theorem) *Suppose*

$$f, g: [a, b] \rightarrow R$$

*are continuous, and $f$ and $g$ are differentiable on $(a, b)$. Then there is an $m \in (a, b)$ satisfying*

$$(f(b) - f(a))g'(m) = (g(b) - g(a))f'(m).$$

*Proof* Let $h(x)$ equal

$$f(x)(g(b) - g(a)) - g(x)(f(b) - g(a))$$
$$- f(a)g(b) - g(a)f(b)$$

and reason as in Theorem 6 $\bullet$

(Notice that when $g(x) = x$ Theorem 7 reduces to Theorem 6.)

**COROLLARY 1** *If f is differentiable on an open interval I and the derivative is never zero, then f is injective.*

*Proof* If $x \neq y$ and $f(x) = f(y)$, then there is an $m$ between $x$ and $y$ such that

$$f'(m) = \frac{f(y) - f(x)}{y - x} = 0 \quad \bullet$$

**COROLLARY 2** *If I is an open interval and f: I → R has zero derivative for all $x \in I$, then $f(x) = C$ for all $x \in I$ and some constant C.*

## C. Some Basic Results

**THEOREM 8** (The Fundamental Theorem of Calculus) *Suppose I is an open interval and f: I → R is a continuous function. Pick any $a \in I$ and let*

$$F(x) = \int_a^x f.$$

*Then F' = f.*

*Proof* Pick any $x_0 \in I$, let $x_0 + h \in I$, and set

$$A = \begin{cases} [x_0, x_0 + h] & \text{if } h \geq 0 \\ [x_0 + h, x_0] & \text{if } h < 0. \end{cases}$$

We then have

(1) $\quad |F(x_0 + h) - F(x_0) - hf(x_0)|$
$$\leq |h| \sup_{x \in A} |f(x) - f(x_0)|.$$ →

Now let $\varepsilon > 0$. Since $f$ is continuous at $x_0$ there is a $\delta > 0$ such that

$$|f(x) - f(x_0)| < \varepsilon \quad \text{whenever} \quad |x - x_0| < \delta.$$

Hence, for $|h| < \delta$, $\sup_{x \in A} |f(x) - f(x_0)| \leq \varepsilon$, and Equation (1) takes the form

$$|F(x_0 + h) - F(x_0) - hf(x_0)| \leq \varepsilon|h| \quad \text{for} \quad |h| < \delta.$$

By the definition of the derivative (10.1A) it follows that $F'(x_0) = f(x_0)$ $\quad \bullet$

**COROLLARY 1** *Suppose I is an open interval and g: I → R is a $C^1$-function. Then for*

$$a, x \in I, \quad g(x) - g(a) = \int_a^x g'(t) \, dt.$$

**COROLLARY** (L'Hôpital's Rule) *Suppose f and g are differentiable in an interval I containing zero, and*

$$f(0) = g(0) = 0, \quad \text{and} \quad \lim_{x \to 0} \frac{f'(x)}{g'(x)}$$

*exists. Then*

$$\lim_{x \to 0} \frac{f(x)}{g(x)} = \lim_{x \to 0} \frac{f'(x)}{g'(x)}.$$

*Proof* For any $x \neq 0$ we have

$$\frac{f(x)}{g(x)} = \frac{f(x) - f(0)}{g(x) - g(0)},$$

and by Theorem 7 this equals $f'(m)/g'(m)$ for some $m \in (0, x)$. If we let $x \to 0$ it follows that $m \to 0$, so

$$\lim_{x \to 0} \frac{f(x)}{g(x)} = \lim_{x \to 0} \frac{f'(x)}{g'(x)} \quad \bullet$$

This inequality is established as follows:

$$|F(x_0 + h) - F(x_0) - hf(x_0)|$$

$$= \left| \int_a^{x_0+h} f(x) \, dx - \int_a^{x_0} f(x) \, dx - hf(x_0) \right|$$

$$= \left| \int_{x_0}^{x_0+h} f(x) \, dx - hf(x_0) \right| \quad \text{(by Theorem 6 of 9.4B)}$$

$$= \left| \int_{x_0}^{x_0+h} (f(x) - f(x_0)) \, dx \right| \quad \left( \text{since } \int_{x_0}^{x_0+h} f(x_0) \, dx = hf(x_0) \right)$$

$$\leq |h| \sup \{|f(x) - f(x_0)| : x \text{ between } x_0 \text{ and } x_0 + h\}$$

$$= |h| \sup_{x \in A} |f(x) - f(x_0)| \quad \begin{array}{l} \text{(by the Corollary to} \\ \text{Theorem 3 of 9.4A).} \end{array}$$

The following variation on Corollary 1 will be used in the proof of Theorem 9.

→ **COROLLARY 2** *Suppose I is an open interval and g: I → R is a $C^1$-function. Then for a constant $a \in R$,*

$$f(a + x) = f(a) + \int_0^1 f'(a + tx)x \, dt.$$

*Proof* Letting $F(x) = \int_a^x g'(t)\, dt$, we have by Theorem 8 that $F' = g'$ in $I$. Hence, by Corollary 2 of Theorem 6 (10.1B), $F = g + C$ for some constant $C \in R$, so

$$g(x) + C = \int_a^x g'(t)\, dt.$$

Letting $x = a$ gives

$$g(a) + C = \int_a^a g' = 0,$$

so $C = -g(a)$ and the desired result follows $\bullet$

*Proof* Let $g(t) = f(a + tx)$. Then $g(0) = f(a)$, $g(1) = f(a + x)$ and $g'(t) = f'(a + tx)x$. By Corollary 1,

$$g(1) - g(0) = \int_0^1 g'(t)\, dt,$$

so $f(a + x) - f(a) = \int_0^1 f'(a + tx)x\, dt$

or $\qquad f(a + x) = f(a) + \int_0^1 f'(a + tx)x\, dt$ $\bullet$

**THEOREM 9** (Taylor's Theorem) *Suppose $I$ is an open interval and $f\colon I \to R$ is a $C^n$-function. Then for $a, a + x \in I$*

$$f(a + x) = \sum_{r=0}^{n-1} \frac{f^{(r)}(a)x^r}{r!} + R_n(a, x)$$

*where*

$$R_n(a, x) = \int_0^1 \frac{(1 - t)^{n-1}}{(n - 1)!} f^{(n)}(a + tx)x^n\, dt.$$

*Proof* The proof is by induction. For $n = 1$ the formula reduces to that of Corollary 2 of Theorem 8. Assume the result is true for $n = k$, and that $f\colon I \to R$ is a $C^{k+1}$-function. Then

$$f(a + x) = \sum_{r=0}^{k-1} f^{(r)}(a)\frac{x^r}{r!}$$

$$+ \int_0^1 \frac{(1 - t)^{k-1}}{(k - 1)!} f^{(k)}(a + tx)x^k\, dt.$$

Integrating this by parts yields

$$\sum_{r=0}^{k} f^{(r)}(a)\frac{x^r}{r!} + \int_0^1 \frac{(1 - t)^k}{k!} f^{(k+1)}(a + tx)x^{k+1}\, dt.$$

Hence the result is true for $n = k + 1$, so by induction the proof is complete $\bullet$

$\longrightarrow$   If $R_n(a, x) \to 0$ as $n \to \infty$, this formula takes the form

$$f(a + x) = \sum_{r=0}^{\infty} \frac{f^{(r)}(a)x^r}{r!},$$

and if $a = 0$ it becomes

$$f(x) = \sum_{r=0}^{\infty} \frac{f^{(r)}(0)x^r}{r!}.$$

The formula is basic for finding the power series expansions of functions. See Example 1 and Problems 3 and 4 below.

We let

$$u(t) = f^{(k)}(a + tx)x^k \quad \text{and} \quad v'(t) = \frac{(1 - t)^{k-1}}{(k - 1)!}.$$

$\longrightarrow$ Then

$$u'(t) = f^{(k+1)}(a + tx)x^{k+1}, \quad v(t) = \frac{-(1 - t)^k}{k!},$$

and

$$u(t)v(t)\Big|_0^1 = \frac{f^{(k)}(a)x^k}{k!}.$$

**COROLLARY** *If $|f^{(n)}(y)| \leq M$ for all $y \in I$ and $n \in N$, then*

$$f(a + x) = \sum_{r=0}^{\infty} \frac{f^{(r)}(a)x^r}{r!}.$$

*Proof* In this case we have

$$|R_n(a, x)| = \left| \int_0^1 \frac{(1 - t)^{n-1}}{(n - 1)!} f^{(n)}(a + tx)x^n \, dt \right|$$

$$\leq \frac{M|x^n|}{(n - 1)!} \int_0^1 (1 - t)^{n-1} \, dt = \frac{M|x|^n}{n!},$$

which converges to zero as $n \to \infty$ for each $x$  ●

---

**Example 1** (The exponential function) Suppose $f: R \to R$ is differentiable, $f(0) = 1$, and $f' = f$. Then

$$f(x) = \sum_{r=0}^{\infty} \frac{x^r}{r!}.$$

*Proof* Since $f' = f$ we have $f^{(n)} = f$, so $f^{(n)}(0) = f(0) = 1$ for all $n$. Because $f$ is differentiable on $R$ it is continuous on $R$ (Theorem 1, 10.1A) and therefore bounded on any compact interval. Since $f^{(n)} = f$ it follows that for any bounded interval $I$ containing 0

$$\sup_{y \in I} |f^{(n)}(y)| \leq M \quad \text{for all } n \in N.$$

Therefore the hypothesis of the corollary is satisfied and the desired result follows.

This corollary can be used to show that the power series expansions of the functions given in Chapter 7 actually converge to the functions. We illustrate this in Example 1 and Problems 3 and 4, with the exponential, sine, and cosine functions. These functions were introduced earlier, but will be defined again, in another way, in 10.3.

---

**Problem 3** (The sine function) Suppose $g: R \to R$ is twice differentiable, $g'' = -g$, and $g(0) = 0$. Prove that

$$g(x) = \sum_{r=0}^{\infty} (-1)^r \frac{x^{2r+1}}{(2r + 1)!}.$$

---

**Problem 4** (The cosine function) Suppose $h: R \to R$ is twice differentiable, $h'' = -h$, and $h(0) = 1$. Prove that

$$h(x) = \sum_{r=0}^{\infty} (-1)^r \frac{x^{2r}}{(2r)!}.$$

---

## D. Exercises

**B1.** Suppose $f$ is differentiable on the open interval $I$. Prove that $f'$ is continuous at $x_0 \in I$ if and only if for any $\varepsilon > 0$ there is a $\delta > 0$ such that

$$\left| \frac{f(x_0 + h) - f(x_0 - k)}{h + k} - f'(x_0) \right| < \varepsilon$$

whenever $0 < |h|, |k| < \delta$.
(Compare this with Exercises M6 and M7.)

**B2.** Define $g(x) = 0$ for $x \leq 0$ and $g(x) = e^{-1/x^2}$ for $x > 0$. Prove that $g$ is infinitely differentiable on $R$.

(Hint: Use L'Hôpital's Rule for computing the derivatives at $x = 0$.)

**B3.** Suppose $f, g, h: R \to R$ are continuous and $f$ and $g$ are differentiable. Show that

$$H(x) = \int_{f(x)}^{g(x)} h(t) \, dt \quad \text{is differentiable}$$

and

$$H'(x) = g'(x)h(g(x)) - f'(x)h(f(x)).$$

**B4.** Prove that the remainder in Taylor's Theorem

$$R_n(a, x) = \int_0^1 \frac{(1 - t)^{n-1}}{(n - 1)!} f^{(n)}(a + tx)x^n \, dt$$

can be written in the form

$$R_n(a, x) = \frac{f^{(n)}(c)x^n}{n!}$$

where $c$ is between $a$ and $a + x$.

**M5.** Verify Rolle's Theorem for the polynomial $f(x) = x^m(1 - x)^n$ on the interval $[0, 1]$.

**M6.** Suppose $I$ is an open interval and $f: I \to R$ is a function. Prove that $f$ is differentiable at $x_0 \in I$ if and only if the following condition is satisfied: There is a number $L$ with the property that for any $\varepsilon > 0$ there exists a $\delta > 0$ such that

$$\left| \frac{f(x_0 + h) - f(x_0 - k)}{h + k} - L \right| < \varepsilon$$

whenever $0 < h, k < \delta$.

**M7.** a) Suppose $I$ is an open interval and $f: I \to R$ is a function. Prove that if $f''(x_0)$ exists, then

$$f''(x_0) = \lim_{h \to 0} \frac{f(x_0 + h) - f(x_0 - h) - 2f(x_0)}{h^2}.$$

b) Use the function

$$f(x) = \begin{cases} x^2 & (x \geq 0) \\ -x^2 & (x < 0) \end{cases}$$

to show that the converse to (a) is false.

**M8.** Let $g$ be as in Exercise B2 and

$$h(x) = g(1 + x)g(1 - x).$$

Show that the Taylor expansion of $h(x)$,

$$\sum_{r=0}^{\infty} \frac{h^{(r)}(0)}{r!} x^r,$$

is equal to $h(x)$ only when $x = 0$.

**M9.** Use the Mean Value Theorem to show that $10.243 < \sqrt{105} < 10.250$. (Hint: Let $f(x) = \sqrt{x}$, $a = 100$, $b = 105$.)

**M10.** Given $a < b < c < d$, construct a $C^\infty$-function $f: R \to R$ that satisfies

$$f(x) = 1 \quad \text{for} \quad b \leq x \leq c,$$

and

$$f(x) = 0 \quad \text{for} \quad x \leq a \quad \text{or} \quad x \geq d.$$

(Hint: Start with the function

$$k(x) = \int_{-1}^{x} h(t) \, dt$$

where $h$ is as in Exercise M8.)
Use Taylor's Theorem to prove the Binomial Theorem,

$$(a + x)^n = \sum_{k=0}^{n} \binom{n}{k} a^{n-k} x^k.$$

**M11.** (A continuous, nowhere differentiable function) Define $f: R \to R$ by the formula:

$$f(x) = \begin{cases} 2(x - n) & \text{for} \quad n \leq x \leq n + \frac{1}{2} \\ 2 - 2(x - n) & \text{for} \quad n + \frac{1}{2} < x < n + 1 \end{cases}$$

where $n$ is an integer.

Let
$$f_k(x) = \frac{f(2^k x)}{2^k} \quad (k = 1, 2, \ldots)$$

and

$$g_m = f_1 + \cdots + f_k.$$

a) Prove that $\{g_m\}$ converges uniformly on $R$ to a continuous function $g$. (Hint: Note that $\{f_k\}$ converges uniformly to 0 on $R$.)

b) Prove that $g$ is not differentiable at any point $x_0 \in R$. (Hint: Show that the condition given in Exercise M6 is not satisfied.)

---

## 10.2 Theorems on Series

In this section we establish a number of basic results about series of complex numbers.  ➡  In particular, we will prove all the theorems left unproven in Chapter 7.

**A. Complex Series**  A sequence $\{z_1, z_2, \ldots\}$ of complex numbers is said to *converge* to $z \in C$ if for any $\varepsilon > 0$ there is an $n_0 \in N$ such that

$$|z_n - z| < \varepsilon \quad \text{whenever } n > n_0.$$

In this case we write

$$z_n \to z \quad \text{or} \quad \lim_{n \to \infty} z_n = z.$$

The sequence is said to be *Cauchy* if for any $\varepsilon > 0$ there is an $n_0$ such that

$$|z_n - z_m| < \varepsilon \quad \text{whenever } n, m > n_0.$$

**THEOREM 1**  *Suppose $z_n = a_n + ib_n$ ($a_n, b_n \in R$, $n \in N$). Then the sequence $\{z_n\}$ is convergent if and only if both of the sequences $\{a_n\}$ and $\{b_n\}$ are convergent. Moreover, we have*

$$\lim z_n = \lim a_n + i \lim b_n.$$

*(See also Problem 1.)*

*Proof*  Suppose $z_n \to z = a + ib$ and $\varepsilon > 0$. Then there is an $n_0$ such that $|z_n - z| < \varepsilon$ whenever $n > n_0$. Since

$$|a_n - a| \le |z_n - z| \quad \text{and} \quad |b_n - b| \le |z_n - z|$$

it follows that

$$|a_n - a|, \quad |b_n - b| < \varepsilon \quad \text{for} \quad n > n_0.$$

Hence

$$a_n \to a \quad \text{and} \quad b_n \to b.$$

Conversely, suppose $a_n \to a$ and $b_n \to b$. Let $z = a + ib$, and suppose $\varepsilon > 0$. Since $a_n \to a$ there is an $n_1$ such that

$$|a_n - a| < \varepsilon/\sqrt{2} \quad \text{for } n > n_1,$$

and since $b_n \to b$ there is an $n_2$ such that

$$|b_n - b| < \varepsilon/\sqrt{2} \quad \text{for } n > n_2.$$

Hence for $n > \max\{n_1, n_2\}$,

$$|z_n - z| = ((a_n - a)^2 + (b_n - b)^2)^{1/2}$$
$$< ((\varepsilon/\sqrt{2})^2 + (\varepsilon/\sqrt{2})^2)^{1/2} = \varepsilon,$$

so we have that $z_n \to z$  ●

→  The only facts we will use about the complex numbers $C$ are the following:

a) The set $C$ is equal to the set, $R \times R$ and *addition* and *multiplication* of complex numbers are defined by the following formulas:

$$(a, b) + (c, d) = (a + c, b + d)$$
$$(a, b)(c, d) = (ac - bd, ad + bc).$$

b) The *absolute value* of a complex number $(a, b)$ is defined as

$$|(a, b)| = (a^2 + b^2)^{1/2}.$$

A complex number $(a, b)$ will usually be denoted by a single letter $z$ and written in the form $z = a + ib$. Two properties of the absolute value that will often be used then take the form

$$|z_1 + z_2| \le |z_1| + |z_2| \quad \text{(Triangle Inequality)},$$

and

$$|z_1 z_2| = |z_1| \, |z_2|.$$

(See 7.5A and 7.5B for a more complete discussion.)

---

**Problem 1**  Prove that the sequence $\{z_n\}$ of Theorem 1 is Cauchy if and only if each of the sequences $\{a_n\}$ and $\{b_n\}$ is Cauchy.

---

The properties of real sequences given in Theorems 1, 2, and 3 of 9.2A carry over to complex sequences and no change in the earlier proofs is required. We will use these
→  facts whenever necessary.

**COROLLARY** *A sequence $\{z_n\}$ is convergent if and only if it is Cauchy.*

*Proof* Let $z_n = a_n + ib_n$ ($a_n, b_n \in R$). Then, by Theorem 1, $\{z_n\}$ is convergent if and only if $\{a_n\}$ and $\{b_n\}$ are convergent. By Theorem 5 of 9.2B, the sequences $\{a_n\}$ and $\{b_n\}$ are convergent if and only if they are Cauchy. From Problem 1 we know that the sequences $\{a_n\}$ and $\{b_n\}$ are Cauchy if and only if the sequence $\{z_n\}$ is Cauchy. The proof is now complete ●

The series of complex numbers (or "series" for short) $\sum_{k=1}^{\infty} z_k$ is said to *converge* to $z \in C$ if the sequence of partial sums $\{S_n = \Sigma_{k=1}^n z_k\}$ converges to $z$. We will often denote this by writing $\Sigma_{k=1}^{\infty} z_k = z$. The following theorem about series is the analog of Theorem 1(a) of 9.2A which deals with sequences.

**THEOREM 2** *If $\sum_{n=1}^{\infty} z_n$ and $\sum_{n=1}^{\infty} w_n$ converge, then*

$$\sum_{n=1}^{\infty} (z_n + w_n) = \sum_{n=1}^{\infty} z_n + \sum_{n=1}^{\infty} w_n.$$

*Proof* Suppose $\Sigma_{n=1}^{\infty} z_n = z$ and $\sum_{n=1}^{\infty} w_n = w$. Let $\varepsilon > 0$. Then there exists an $n_1$ such that

$$\left| \sum_{k=1}^n z_k - z \right| < \frac{\varepsilon}{2} \quad \text{for } n > n_1,$$

and there exists an $n_2$ such that

$$\left| \sum_{k=1}^n w_k - w \right| < \frac{\varepsilon}{2} \quad \text{for } n > n_2.$$

Then for $n > \max\{n_1, n_2\}$, we have

$$\left| \sum_{k=1}^n (z_k + w_k) - (z + w) \right| \le \left| \sum_{k=1}^n z_k - z \right|$$

$$+ \left| \sum_{k=1}^n w_k - w \right| < \frac{\varepsilon}{2} + \frac{\varepsilon}{2} = \varepsilon,$$

**Problem 2** Show that a bounded sequence $\{z_n\}$ of complex numbers contains a Cauchy subsequence. (Hint: Use Theorem 1, Problem 1, Theorem 6 of 9.2B, and the argument given in the Corollary to Theorem 1.)

Stated directly in terms of series, this says that the series $\Sigma_{k=1}^{\infty} z_k$ converges to $z$ if for any $\varepsilon > 0$ there is an $n_0 \in N$ with the property that

$$\left| \sum_{k=1}^n z_k - z \right| < \varepsilon \quad \text{whenever} \quad n > n_0.$$

**Problem 3** Suppose $z_n = a_n + ib_n$. Then $\Sigma_{n=1}^{\infty} z_n$ is convergent if and only if the series $\Sigma_{k=1}^{\infty} a_k$ and $\Sigma_{k=1}^{\infty} b_k$ are convergent. We then have

$$\sum_{k=1}^{\infty} z_k = \sum_{k=1}^{\infty} a_k + i \sum_{k=1}^{\infty} b_k.$$

so $\displaystyle\sum_{n=1}^{\infty}(z_n + w_n) = z + w = \sum_{n=1}^{\infty} z_n + \sum_{n=1}^{\infty} w_n$ ●

**THEOREM 3**  *If* $\displaystyle\sum_{k=1}^{\infty} z_k$ *converges, then* $z_k \to 0$.

*Proof*  Let $z = \Sigma_{k=1}^{\infty} z_k$ and $S_n = \Sigma_{k=1}^{n} z_k$. Given $\varepsilon > 0$, there is an $n_0 \in N$ such that $|S_n - z| < \frac{\varepsilon}{2}$ whenever $n > n_0$. Hence, for $n > n_0$,

$$|z_{n+1}| = |S_{n+1} - S_n| = |S_{n+1} - z + z - S_n|$$

$$\leq |S_{n+1} - z| + |z - S_n| < \frac{\varepsilon}{2} + \frac{\varepsilon}{2} = \varepsilon.$$

In other words $|z_r| < \varepsilon$ whenever $r > n_0 + 1$, so $z_k \to 0$ ●

➡ This proof is simply a rigorous version of the proof given for Theorem 1 of 7.1C.

It is important to remember that the converse of this theorem is false. The harmonic series $\Sigma_{k=1}^{\infty} \frac{1}{k}$ provides an example of a series that is divergent but whose terms converge to zero. A more obvious example is given by the series

➡ $$1 + \left(\frac{1}{2} + \frac{1}{2}\right) + \cdots + \underbrace{\left(\frac{1}{n} + \frac{1}{n} + \cdots + \frac{1}{n}\right)}_{n \text{ terms}} + \cdots.$$

**B. Absolute Convergence**  The series $\Sigma_{k=1}^{\infty} z_k$ is said to be *absolutely convergent* if the series $\Sigma_{k=1}^{\infty} |z_k|$ is convergent.

➡ By contrast, we have the following:
We will say the series is *absolutely divergent* if $\Sigma_{k=1}^{\infty} |z_k|$ is divergent.

**THEOREM 4**  *Absolute convergence implies convergence.*

*Proof*  Suppose $\Sigma_{k=1}^{\infty} z_k$ is absolutely convergent. Let

$$S_n = \sum_{k=1}^{n} z_k \quad \text{and} \quad T_n = \sum_{k=1}^{n} |z_k|.$$

We then have that $\{T_n\}$ is convergent so, by the Corollary to Theorem 1, $\{T_n\}$ is Cauchy. Using the Triangle Inequality, we have

$$|S_n - S_m| \leq |T_n - T_m|,$$

so it follows that $\{S_n\}$ is Cauchy. Then, by the Corollary to Theorem 1, $\{S_n\}$ is convergent. Hence, by definition, $\Sigma_{k=1}^{\infty} z_k$ converges ●

**THEOREM 5**  *Suppose* $\Sigma_{k=1}^{\infty} a_k$ *is a series of real numbers. Then it is absolutely convergent if and only if each of the series*

$$\sum_{k=1}^{\infty} a_k{}^{+} \quad \text{and} \quad \sum_{k=1}^{\infty} a_k{}^{-}$$

**Example 1**  The series $\Sigma_{k=1}^{\infty} z^k$ is absolutely convergent if $|z| < 1$, and absolutely divergent if $|z| \geq 1$.

*Proof*  We have $\Sigma_{k=1}^{\infty} |z^k| = \Sigma_{k=1}^{\infty} |z|^k$, and for $|z| < 1$ we know that $\Sigma_{k=1}^{\infty} |z|^k = |z|/(1 - |z|)$. Therefore $\Sigma_{k=1}^{\infty} |z^k|$ is convergent, so $\Sigma_{k=1}^{\infty} z^k$ is absolutely convergent. If $|z| \geq 1$, then $|z^k|$ does not converge to zero, so by Theorem 3, it follows that $\Sigma_{k=1}^{\infty} |z^k|$ is divergent.

Given $a \in R$, the numbers $a^+$ and $a^-$ are defined by the formulas

$$a^+ = \frac{|a| + a}{2} \quad \text{and} \quad a^- = \frac{|a| - a}{2}.$$

➡ Note that $a^+, a^- \geq 0$, $a^+ = a$ if $a \geq 0$, $a^- = -a$ if $a < 0$, and $|a| = a^+ + a^-$.

*is convergent. In this case we have*

$$\sum_{k=1}^{\infty} a_k = \sum_{k=1}^{\infty} a_k^{+} - \sum_{k=1}^{\infty} a_k^{-}$$

*and*

$$\sum_{k=1}^{\infty} |a_k| = \sum_{k=1}^{\infty} a_k^{+} + \sum_{k=1}^{\infty} a_k^{-}.$$

*Proof* If $\sum_{k=1}^{\infty} |a_k|$ is convergent, then by Theorem 4 $\sum_{k=1}^{\infty} a_k$ is convergent. Hence the series

$$\sum_{k=1}^{\infty} a_k^{+} = \sum_{k=1}^{\infty} \frac{|a_k| + a_k}{2}$$

and

$$\sum_{k=1}^{\infty} a_k^{-} = \sum_{k=1}^{\infty} \frac{|a_k| - a_k}{2}$$

are convergent and the above formulas are valid. Conversely, suppose

$$\sum_{k=1}^{\infty} a_k^{+} \quad \text{and} \quad \sum_{k=1}^{\infty} a_k^{-}$$

are convergent. Then, by Theorem 2,

$$\sum_{k=1}^{\infty} |a_k| = \sum_{k=1}^{\infty} a_k^{+} + \sum_{k=1}^{\infty} a_k^{-}$$

is convergent ●

Given the series $\sum_{k=1}^{\infty} z_k$ and a bijection $\pi: N \to N$, the series $\sum_{k=1}^{\infty} z_{\pi(k)}$ is called a *rearrangement* of the series $\sum_{k=1}^{\infty} z_k$.

**THEOREM 6** *The series $\sum_{k=1}^{\infty} z_k$ is absolutely convergent if and only if it is convergent and for any rearrangement $\sum_{k=1}^{\infty} z_{\pi(k)}$ we have*

$$\sum_{k=1}^{\infty} z_k = \sum_{k=1}^{\infty} z_{\pi(k)}.$$

*Proof* Assume $\sum_{k=1}^{\infty} z_k$ is absolutely convergent and suppose $\sum_{k=1}^{\infty} z_{\pi(k)}$ is a rearrangement of it. Let

$$S_n = \sum_{k=1}^{n} z_k, \quad S = \sum_{k=1}^{\infty} z_k, \quad \text{and} \quad T_n = \sum_{k=1}^{n} z_{\pi(k)}.$$

**Problem 4** Prove that $\sum_{k=1}^{\infty} z_k$ (with $z_k = a_k + ib_k$) is absolutely convergent if and only if each of the series $\sum_{k=1}^{\infty} a_k$ and $\sum_{k=1}^{\infty} b_k$ is absolutely convergent.

For example

$$z_2 + z_1 + z_4 + z_3 + z_6 + z_5 + \ldots,$$

and

$$z_2 + z_3 + z_1 + z_5 + z_6 + z_4 + z_8 + z_9 + z_7 + \ldots,$$

are rearrangements of $z_1 + z_2 + z_3 + \ldots$. However, the series $z_1 + z_3 + z_5 + z_7 + \ldots$ is *not* a rearrangement of $z_1 + z_2 + z_3 + \ldots$.

This result is in sharp contrast to the following:

**THEOREM 7** *Suppose $\sum_{k=1}^{\infty} a_k$ ($a_k \in R$) is convergent, but not absolutely convergent. Then for any $a \in R$ there is a rearrangement*

$$\sum_{k=1}^{\infty} a_{\pi(k)} \quad \text{with} \quad \sum_{k=1}^{\infty} a_{\pi(k)} = a.$$

*Proof* For convenience we will assume that no $a_n$ is zero and $a \geq 0$. Let $p_1, p_2, p_3, \ldots$ denote the positive terms and $n_1, n_2, n_3, \ldots$ be the negative terms listed in order of appearance. In view of Theorem 5, $\sum_{k=1}^{\infty} p_k$ and $\sum_{k=1}^{\infty} n_k$ are each divergent. But since $\sum_{k=1}^{\infty} a_k$ is

We will show that $T_n \to S$. Given $\varepsilon > 0$, there is an $n_0 \in N$ such that

$$|S_n - S| < \frac{\varepsilon}{2} \quad \text{whenever } n > n_0.$$

Since $\sum_{k=1}^{\infty} |z_k|$ is convergent, there is an $n_1 \in N$ such that

$$\sum_{k=m}^{\infty} |z_k| < \frac{\varepsilon}{2} \quad \text{whenever } m > n_1.$$

Now let $n_2 = \max \{n_0, n_1\}$ and choose $n_3$ so that

$$\{1, 2, \ldots, n_2\} \subset \{\pi(1), \pi(2), \ldots, \pi(n_3)\}.$$

Then for $n > n_3$,

$$T_n - S_n = \sum_{k=1}^{n} z_{\pi(k)} - \sum_{k=1}^{n} z_k$$

must equal a sum of $z_i$'s with $i > n_2$. Hence

$$|T_n - S_n| \le \sum_{i=n_2+1}^{\infty} |z_i| < \frac{\varepsilon}{2}.$$

Therefore, for $n > n_3$ we have

$$|T_n - S| \le |T_n - S_n| + |S_n - S| < \frac{\varepsilon}{2} + \frac{\varepsilon}{2} = \varepsilon,$$

so $T_n \to S$.

Conversely, if the condition in the theorem is satisfied for $\sum_{k=1}^{\infty} z_k$, it is satisfied for each of the series $\sum_{k=1}^{\infty} a_k$ and $\sum_{k=1}^{\infty} b_k$ where $z_k = a_k + ib_k$. It follows from the Corollary to Theorem 7 that each of these real series is absolutely convergent, so from Problem 3 of 10.2A we have that $\sum_{k=1}^{\infty} z_k$ is absolutely convergent ●

### C. On Power Series
The following notion arises in connection with power series.

For a sequence of real numbers $\{r_1, r_2, \ldots\}$, the *upper limit* (denoted by $\overline{\lim} \, r_n$) is defined as follows:

convergent, we have $a_k \to 0$, so it follows that $p_k \to 0$ and $n_k \to 0$.

Let $r_1$ be the smallest natural number satisfying

$$p_1 + \cdots + p_{r_1} > a,$$

and then let $s_1$ be the smallest natural number such that

$$p_1 + \cdots + p_{r_1} + n_1 + \cdots + n_{s_1} < a.$$

($r_1$ and $s_1$ exist since the series

$$\sum_{k=1}^{\infty} p_k \quad \text{and} \quad \sum_{k=1}^{\infty} n_k$$

are divergent.) Assuming $r_j$ and $s_j$ have been defined for $j \le k$, define $r_{k+1}$ to be the smallest natural number such that

$$\sum_{i=1}^{r_1} p_i + \sum_{i=1}^{s_1} n_i + \cdots + \sum_{i=r_{k-1}+1}^{r_k} p_i$$
$$+ \sum_{i=s_{k-1}+1}^{s_k} n_i + \sum_{i=r_k+1}^{r_{k+1}} p_i > a$$

and $s_{k+1}$ to be the smallest natural number satisfying

$$\sum_{i=1}^{r_1} p_i + \sum_{i=1}^{s_1} n_i + \cdots + \sum_{i=r_k+1}^{r_{k+1}} p_i + \sum_{i=s_k+1}^{s_{k+1}} n_i < a.$$

Because of the way the $r_j$'s and $s_j$'s were chosen, it follows that

$$\left| \sum_{i=1}^{r_1} p_i + \sum_{i=1}^{s_1} n_i + \cdots + \sum_{i=r_k+1}^{r_{k+1}} p_i - a \right| < p_{r_{k+1}}$$

and

$$\left| \sum_{i=1}^{r_1} p_i + \sum_{i=1}^{s_1} n_i + \cdots + \sum_{i=s_k+1}^{s_{k+1}} n_i - a \right| < n_{s_{k+1}}.$$

Since $p_{r_{k+1}} \to 0$ and $n_{s_{k+1}} \to 0$, it follows that

$$\sum_{i=1}^{r_1} p_i + \sum_{i=1}^{s_1} n_i + \cdots$$

converges to $a$ ●

$$\overline{\lim}\, r_n = \inf_k \sup \{r_k, r_{k+1}, \ldots\}.$$

→ If $A \subset R$ is not bounded from above, we write $\sup A = \infty$, and if $B \subset R$ is not bounded from below, we write $\inf B = -\infty$. It follows that $\overline{\lim}\, r_n$ could equal $\infty$ or $-\infty$.

Some properties of the upper limit that will be used are given below.

**THEOREM 8**   *Consider the series* $\sum\limits_{n=0}^{\infty} a_n z^n$ *and let*

$$\frac{1}{R} = \overline{\lim}\, |a_n|^{1/n}.$$

*Then for $|z| < R$ the series is absolutely convergent and for $|z| > R$ the series is absolutely divergent. (R is called the radius of convergence of the series.)*

*Proof*   If $0 < |z| < R$, then $\frac{|z|}{R} < 1$, so $|z|\overline{\lim}|a_n|^{1/n} < 1$. Choose $\theta$ satisfying

$$|z|\, \overline{\lim}\, |a_n|^{1/n} < \theta < 1.$$

Then $\overline{\lim}\, |a_n|^{1/n} < \frac{\theta}{|z|}$, and by Lemma 1 it follows that there is an $n_0 \in N$ with the property that

$$|a_n|^{1/n} < \frac{\theta}{|z|} \quad \text{whenever } n \geq n_0.$$

Therefore $|a_n z^n| < \theta^n \ (n \geq n_0)$, so

$$\sum_{n \geq n_0} |a_n z^n| < \sum_{n \geq n_0} \theta^n < \infty,$$

and it follows that $\sum_{n=0}^{\infty} |a_n z^n|$ is convergent.

If $|z| > R$, then $|z|\, \overline{\lim}\, |a_n|^{1/n} > 1$, so there is a $\phi$ satisfying

$$1 < \phi < |z|\, \overline{\lim}\, |a_n|^{1/n}.$$

Hence $\frac{\phi}{|z|} < \overline{\lim}\, |a_n|^{1/n}$, and by Lemma 2 there is a sequence $\{n_1, n_2, \ldots\}$ satisfying

$$\frac{\phi}{|z|} < |a_{n_k}|^{1/n_k} \quad \text{for all } k \in N.$$

It follows that $\phi^{n_k} < |a_{n_k} z^{n_k}|$ for $k \in N$. Since $\phi > 1$, $\sum_{k=1}^{\infty} \phi^{n_k}$ is divergent, and $\sum_{k=1}^{\infty} |a_{n_k} z^{n_k}|$ must also be divergent. We then have that $\sum_{n=1}^{\infty} |a_n z^n|$ is divergent, so $\sum_{n=1}^{\infty} a_n z^n$ is absolutely divergent ●

Using this theorem and Problem 6 we have:

**COROLLARY**   *The series*

---

**Example 2**   a) $\overline{\lim}\, \{1, 2, 3, \ldots\} = \infty$

b) $\overline{\lim}\, \{-1, -2, -3, \ldots\} = -\infty$

c) $\overline{\lim}\, \{1, 0, 1, 0, \ldots\} = 1$

d) $\overline{\lim}\, \{1, \frac{1}{2}, \frac{1}{3}, \ldots\} = 0$

---

→ In the proof we will make use of the following two lemmas.

**LEMMA 1**   *If $r > \overline{\lim}\, r_n$, then there is an $n_0 \in N$ with the property that $r > r_n$ whenever $n \geq n_0$.*

*Proof*   Since $r > \inf_k \sup \{r_k, r_{k+1}, \ldots\}$ we must have

$$r > \sup \{r_{n_0}, r_{n_0+1}, \ldots\} \quad \text{for some } n_0 \in N.$$

Hence $r > r_n$ for $n \geq n_0$ ●

**LEMMA 2**   *If $s < \overline{\lim}\, r_n$, then there is a sequence $\{n_1, n_2, \ldots\}$ with the property that $s < r_{n_k}$ for all $k \in N$.*

*Proof*   We will choose the $n_i$'s by induction. Since

$$s < \inf_k \sup \{r_k, r_{k+1}, \ldots\}$$

we have $s < \sup \{r_k, r_{k+1}, \ldots\}$ for all $k$. In particular,

$$s < \sup_{n \geq 1} \{r_n\},$$

so there is an $n_1 \in N$ with $s < r_{n_1}$. Suppose $n_1, \ldots, n_p$ have been chosen satisfying $s < r_{n_i}$ for $1 \leq i \leq p$. Since

$$s < \sup_{n > n_p} \{r_n\},$$

there is an $n_{p+1} \in N$ with $s < r_{n_{p+1}}$ ●

---

**Problem 5**   Prove that if $\lim r_n$ exists, then $\overline{\lim}\, r_n = \lim r_n$.

$$\sum_{k=0}^{\infty} a_k z^k \quad \text{and} \quad \sum_{k=1}^{\infty} k a_k z^{k-1}$$

*have the same radii of convergence.*

**THEOREM 9**  *Suppose $R$ is the radius of convergence of $\Sigma_{k=0}^{\infty} a_k z^k$, let*

$$f(z) = \sum_{k=0}^{\infty} a_k z^k,$$

*and let*

$$A_n(z) = \sum_{k=0}^{n} a_k z^k.$$

*Given $0 < r < R$ and $\varepsilon > 0$, there is an $n_0 \in N$ such that*

$$|f(z) - A_n(z)| < \varepsilon$$

*for all $n > n_0$ and $|z| \leq r$.*

**Proof**  Choose $\theta$ with $r < \theta < R$. Then

$$|f(z) - A_n(z)| = \left| \sum_{k>n} a_k z^k \right| \leq \sum_{k>n} |a_k z^k| \leq \sum_{k>n} |a_k| \theta^k.$$

Since $\Sigma_{k>n} |a_k| \theta^k < \infty$ there is an $n_0$ satisfying

$$\sum_{k>n} |a_k| \theta^k < \varepsilon \quad \text{whenever } n > n_0.$$

This gives $|f(z) - A_n(z)| < \varepsilon$ for $n > n_0$ ●

→ We may restate this theorem as "$A_n(z) \to f(z)$ uniformly on $\{z: |z| \leq r\}$ for any $r$ less than the radius of convergence."

**COROLLARY 1**  *Suppose $f(z) = \Sigma_{k=0}^{\infty} a_k z^k$ has radius of convergence $R$. Then the real-valued function $f(x) = \Sigma_{k=0}^{\infty} a_k x^k$ is continuous on $(-R, R)$.*

**Proof**  For each $0 < r < R$, the continuous functions

$$A_n(x) = \sum_{k=0}^{n} a_k x^k$$

converge uniformly to $f(x)$ on $[-r, r]$. Hence by Theorem 5 of 9.4B, $f$ is continuous on $[-r, r]$. Since this is true for each $0 < r < R$ and any $x \in (-R, R)$ satisfying $x \in [-r, r]$ for some such $r$, it follows that $f$ is continuous on $(-R, R)$ ●

*Problem 6*  a) Prove that $k^{1/k} \to 1$ as $k \to \infty$.
b) Prove that $\overline{\lim} |k a_k|^{1/(k-1)} = \overline{\lim} |a_k|^{1/k}$ for any sequence $\{a_1, a_2, \ldots\}$ in $C$.

→ In fact, the function $f(z) = \Sigma_{k=0}^{\infty} a_k z^k$ is continuous on the set $\{z: |z| < R\}$. The only reason we did not state this result in the more general form is that we have not defined the notion of continuity for functions of a complex variable. This definition and the proof of Corollary 1 can be carried out in a straightforward manner.

The following two corollaries simply say that convergent power series can be integrated and differentiated just as if they were polynomials.

**COROLLARY 2**  *If $f(x) = \Sigma_{k=0}^{\infty} a_k x^k$ has radius of convergence R, then for $x \in (-R, R)$*

$$\int_0^x f(t)\, dt = \sum_{k=0}^{\infty} \frac{a_k x^{k+1}}{k + 1}.$$

*Proof*  Since $A_n \to f$ uniformly on $[-|x|, |x|]$ it follows from Theorem 4 of 9.4B that

$$\int_0^x A_n(t)\, dt \to \int_0^x f(t)\, dt.$$

But

$$\int_0^x A_n(t)\, dt = a_1 x + \frac{a_2 x^2}{2} + \cdots + \frac{a_k x^{k+1}}{k + 1}.$$

Letting $n \to \infty$ we have the desired result  ●

**COROLLARY 3**  *If $f(x) = \Sigma_{k=0}^{\infty} a_k x^k$ has radius of convergence R, then f is differentiable on $(-R, R)$ and $f'(x) = \Sigma_{k=1}^{\infty} k a_k x^{k-1}$.*

*Proof*  Let $g(x) = \Sigma_{k=1}^{\infty} k a_k x^{k-1}$. By the Corollary to Theorem 8 the radius of convergence of this power series is also $R$. Using Corollary 2 of Theorem 9, we have

$$\int_0^x g(t)\, dt = \sum_{k=1}^{\infty} a_k x^k = f(x).$$

Differentiating this equation we obtain

$$g(x) = f'(x) \quad ●$$

## D. Exercises

**B1.** Suppose $a_1 \geq a_2 \geq a_3 \geq \cdots \geq 0$. Prove that $\Sigma_{k=1}^{\infty} a_k$ is convergent if and only if $\Sigma_{k=1}^{\infty} 2^k a_{2^k}$ is convergent.

**B2.** Show that if $\Sigma_{m=1}^{\infty} \Sigma_{n=1}^{\infty} |z_{mn}| < \infty$ then

$$\sum_{m=1}^{\infty} \sum_{n=1}^{\infty} z_{mn} = \sum_{n=1}^{\infty} \sum_{m=1}^{\infty} z_{mn}.$$

(See also Exercise M10.)

**B3.** Given the power series $\Sigma_{k=0}^{\infty} a_n z^n$, prove that if

$$L = \lim_{n \to \infty} \left| \frac{a_{n+1}}{a_n} \right|$$

exists, then $L^{-1}$ is equal to the radius of convergence.

**B4.** Find power series expansions $\Sigma_{k=0}^{\infty} a_k x^k$ for each of the following functions over the interval $(-1, 1)$.

a) $f(x) = \dfrac{1}{(x - 1)(x - 2)}$

b) $g(x) = \dfrac{1}{(1 - x)^2}$

c) $h(x) = \dfrac{1 - x}{1 - x + x^2}$

**M5.** a) Suppose

$$a_n, b_n \in C, \quad a_n \to 0, \quad \text{and} \quad \sum_{n=1}^{\infty} |b_n| < \infty.$$

Show that

$$\left| \sum_{n=1}^{\infty} a_n b_n \right| \leq \left( \sup_n |a_n| \right) \sum_{n=1}^{\infty} |b_n|.$$

b) Suppose $\{b_n\} \subset C$ and for every sequence $\{a_n\} \subset C$ with $a_n \to 0$ we have $\Sigma_{n=1}^{\infty} a_n b_n$ convergent. Prove that

$$\sum_{n=1}^{\infty} |b_n| < \infty.$$

**M6.** a) Suppose

$$b_n, c_n \in C, \quad \sum_{n=1}^{\infty} |b_n| < \infty, \quad \text{and} \quad \sup_n |c_n| < \infty.$$

Show that

$$\left| \sum_{n=1}^{\infty} b_n c_n \right| \leq \left( \sup_n |c_n| \right) \sum_{n=1}^{\infty} |b_n|.$$

b) Suppose $\{c_n\} \subset C$ and for every sequence $\{b_n\} \subset C$ with $\sum_{n=1}^{\infty} |b_n| < \infty$ we have $\sum_{n=1}^{\infty} b_n c_n$ convergent. Prove that $\sup_n |c_n| < \infty$.

**M7.** Given a sequence $\{r_n\} \subset R$, its *lower limit* is defined as $\varliminf_n r_n = \sup_k \inf \{r_k, r_{k+1}, \ldots\}$.

a) Prove that $\varliminf_n r_n \leq \varlimsup_n r_n$ and

b) $\varliminf_n r_n = \varlimsup_n r_n$ if and only if $\lim_n r_n$ exists.

**M8.** Given real sequences $\{r_n\}$ and $\{s_n\}$, show that

a) $\varlimsup_n (r_n + s_n) \leq \varlimsup_n r_n + \varlimsup_n s_n$ and

b) $\varliminf_n (r_n + s_n) \geq \varliminf_n r_n + \varliminf_n s_n$.

**M9.** (Abel's Convergence Test) Suppose

$$a_1 \geq a_2 \geq a_3 \geq \cdots \geq 0 \quad \text{and} \quad a_n \to 0.$$

Show that if $\sum_{k=1}^{\infty} b_k$ is a real sequence whose partial sums are bounded, then

$$\sum_{k=1}^{\infty} a_k b_k \quad \text{is convergent.}$$

**M10.** Construct a counterexample to the result stated in Exercise B2 when the assumption that

$$\sum_{m=1}^{\infty} \sum_{n=1}^{\infty} |z_{mn}| < \infty$$

is dropped.

## 10.3 *Special Functions*

We now give precise definitions of the exponential, logarithmic, and trigonometric functions.

**A. The Exponential Function** We define

$$\exp z = 1 + z + \frac{z^2}{2!} + \frac{z^3}{3!} + \cdots = \sum_{n=0}^{\infty} \frac{z^n}{n!}.$$

This converges for all $z \in C$ since its radius of convergence is

$$R = 1/\varlimsup_n \left(\frac{1}{n!}\right)^{1/n} = \infty.$$

**THEOREM 1** *For any* $z, w \in C$

$$(\exp z)(\exp w) = \exp(z + w).$$

*Proof* Since the series converge absolutely for all $z$ and $w$, we may freely rearrange the terms (Theorem 6, 10.2B). We have

$$(\exp z)(\exp w) = \left(\sum_{r=0}^{\infty} \frac{z^r}{r!}\right)\left(\sum_{s=0}^{\infty} \frac{w^s}{s!}\right)$$

$\longrightarrow$ This follows because $\lim_{n \to \infty} (n!)^{1/n} = \infty$, and this is proven in the following way. Suppose first that $n = 2k$ is even. Then

$$(n!)^{1/n} = ((2k)!)^{1/2k} = ((2k)(2k-1) \cdots (k+1)(k!))^{1/2k}$$
$$> (k^k(k!))^{1/2k} \geq (k^k)^{1/2k} = \sqrt{k} \to \infty.$$

The result will now follow if we show that the sequence is increasing. Since

$$n! < (n+1)^n,$$

we have

$$(n!)^{n+1} = (n!)^n n! < (n!)^n (n+1)^n = ((n+1)!)^n.$$

Hence

$$(n!)^{1/n} < ((n+1)!)^{1/(n+1)},$$

so the sequence is increasing.

$$= \sum_{r=0}^{\infty} \sum_{s=0}^{\infty} \frac{z^r}{r!} \frac{w^s}{s!} = \sum_{n=0}^{\infty} \left( \sum_{r+s=n} \frac{z^r}{r!} \frac{w^s}{s!} \right)$$

$$= \sum_{n=0}^{\infty} \left( \sum_{k=0}^{n} \frac{z^{n-k}}{(n-k)!} \frac{w^k}{k!} \right)$$

$$= \sum_{n=0}^{\infty} \frac{1}{n!} \left( \sum_{k=0}^{n} \frac{n!}{(n-k)!k!} z^{n-k} w^k \right)$$

$$= \sum_{n=0}^{\infty} \frac{1}{n!} \left( \sum_{k=0}^{n} \binom{n}{k} z^{n-k} w^k \right)$$

$$= \sum_{n=0}^{\infty} \frac{1}{n!} (z + w)^n = \exp (z + w) \quad \bullet$$

**Problem 1** Using Theorem 1 and mathematical induction, show that

$$\exp (z_1 + z_2 + \cdots + z_n)$$
$$= (\exp z_1)(\exp z_2) \cdots (\exp z_n)$$

for all $z_1, \ldots, z_n \in C$.

---

**COROLLARY** a) $\exp z$ *is never zero and*

$$\frac{1}{\exp z} = \exp (-z).$$

b) $\exp (nz) = (\exp z)^n$ *for all $z \in C$ and any integer n.*

*Proof* a) Given $z \in Z$, we have

$$1 = \exp 0 = \exp (z - z)$$
$$= \exp (z + (-z))$$
$$= (\exp z)(\exp (-z)).$$

Hence $\exp z \neq 0$ and the formula follows.
b) This follows from Problem 1 by letting

$$z_i = z \quad \text{for } i = 1, 2, \ldots, n \quad \bullet$$

The number

$$\exp (1) = 1 + \frac{1}{1!} + \frac{1}{2!} + \frac{1}{3!} + \cdots$$

is denoted by $e$, and from now on we will write

$$e^z \quad \text{in place of} \quad \exp z.$$

The function $z \mapsto e^z$ is called the *exponential function.*

**THEOREM 2** *Let $f(x) = e^x$ $(x \in R)$. Then*
a) *f is injective;*

In this notation these results take the more familiar form

$$e^{z+w} = e^z e^w \quad \text{(Theorem 1)}$$

$$e^{-z} = \frac{1}{e^z} \quad \text{(Corollary a)}$$

$$(e^z)^n = e^{nz} \quad \text{(Corollary b)}.$$

Note that for $a \neq e$, $a^z$ has not been defined. It will be defined in 10.3B.

b) *f is real-valued, differentiable, and satisfies $f' = f$;*  →
c) $f((-\infty, \infty)) = (0, \infty)$.

*Proof*  a) Suppose $x, y \in R$ ($x < y$). Then

$$e^y e^{-x} = e^{y-x} = 1 + \sum_{n=1}^{\infty} \frac{(y-x)^n}{n!} > 1.$$

Hence

$$e^y e^{-x} \neq 1, \quad \text{so} \quad e^y \neq e^x.$$

b) $f$ is obviously real valued, and by Corollary 2 of Theorem 8 of 10.2C, we have that $f$ is differentiable and

$$f'(x) = \left(1 + x + \frac{x^2}{2!} + \frac{x^3}{3!} + \cdots\right)'$$

$$= 0 + 1 + \frac{2x}{2!} + \frac{3x^2}{3!} + \cdots = f(x).$$

c) For any $x \in R$, $e^{x/2 + x/2} = (e^{x/2})^2 > 0$ since $e^{x/2} \neq 0$. Hence $f((-\infty, \infty)) \subset (0, \infty)$. For

$$x > 0, \quad e^x = \sum_{n=0}^{\infty} \frac{x^n}{n!} > 1 + x \to \infty \quad \text{as} \quad x \to \infty,$$

so $e^x$ assumes arbitrarily large values. Since $e^{-x} = (1/e^x) \to 0$ as $x \to \infty$, it follows that $f$ assumes values arbitrarily near zero. Hence the image of $f$ is $(0, \infty)$ ●

**B. The Logarithm Function**  By Theorem 2, the function $f(x) = e^x$ has domain $(-\infty, \infty)$, image $(0, \infty)$, and is injective. Its inverse function (9.1B) is therefore defined, and we will denote it by $\ln x$. It is called the *logarithm function*.

**THEOREM 3**  *Letting $g(x) = \ln x$ we have:*
  a) *The domain of $g$ is $(0, \infty)$, its image is $(-\infty, \infty)$, and $g$ is injective.*
  b) *$y = \ln x$  if and only if  $x = e^y$.*
  c) *$(\ln x)' = \frac{1}{x}$.*

*Proof*  (a) and (b) follow immediately from the properties of inverse functions.
c) By Theorem 4 of 10.1A we have

$$(\ln x)' = g'(x) = \frac{1}{(f' \circ g)(x)},$$

The property (b) characterizes the exponential function up to a constant multiple. Indeed, suppose $g: R \to R$ is differentiable and $g' = g$. Then

$$\left(\frac{g(x)}{e^x}\right)' = \frac{e^x g'(x) - g(x)e^x}{(e^x)^2} = \frac{1}{e^x}(g'(x) - g(x)) = 0,$$

and we have that $(g(x)/e^x)$ must equal a constant $C$. Hence $g(x) = Ce^x$ for all $x$. (See Example 1 of 10.2C for another proof of this fact.)

More exactly, we have $f(\frac{1}{n})$, $f(n) \in f((0, \infty))$, so by the Intermediate Value Theorem (Theorem 5 of 9.3B) it follows that $[f(\frac{1}{n}), f(n)] \subset f((0, \infty))$. Since $f(\frac{1}{n}) \to 0$ and $f(n) \to \infty$ as $n \to \infty$, we have $(0, \infty) = \cup_{n=1}^{\infty} [f(\frac{1}{n}), f(n)] \subset f((0, \infty))$.

---

**Problem 2**  For $x, y > 0$ and $n \in N$, show that
a) $\ln xy = \ln x + \ln y$,    b) $\ln x^n = n \ln x$.
(Hint: See 3.4C.)

---

**Problem 3**  Prove that
a) $a^{z_1 + z_2} = a^{z_1} a^{z_2}$,    b) $a^{z_1} = a^{z_2}$ iff $z_1 = z_2$,
c) $2^2 = 4$.

and because $f' = f$, this equals

$$\frac{1}{(f \circ g)(x)} = \frac{1}{f(g(x))} = \frac{1}{x} \quad \bullet$$

For $a > 0$ and $z \in C$, we now define

$$a^z = e^{z \ln a}.$$

**THEOREM 4**  For $z_1, z_2 \in C$,  $(a^{z_1})^{z_2} = a^{z_1 z_2}$.

*Proof*  $(a^{z_1})^{z_2} = (e^{z_1 \ln a})^{z_2} = e^{z_2 \ln (e^{z_1 \ln a})}$. Since $\ln (e^{z_1 \ln a}) = z_1 \ln a$, we have

$$(a^{z_1})^{z_2} = e^{z_2 z_1 \ln a} = a^{z_1 z_2} \quad \bullet$$

Given $a > 0$ and $x > 0$, define

$$\log_a x = \frac{\ln x}{\ln a}.$$

This gives a logarithm function for any base $a > 0$.

## C. The Trigonometric Functions

For $x \in R$ we define the cosine and sine functions as

$$\cos x = \frac{e^{ix} + e^{-ix}}{2} \quad \text{and} \quad \sin x = \frac{e^{ix} - e^{-ix}}{2i}.$$

**THEOREM 5**  a) $\sin x$ and $\cos x$ *are real-valued.*
  b) $\sin^2 x + \cos^2 x = 1$  *for all* $x \in R$.
  c) $(\sin x)' = \cos x$,  $(\cos x)' = -\sin x$.

*Proof*  a) Starting with $e^z = \sum_{n=0}^{\infty} \frac{z^n}{n!}$ we obtain

$$\cos x = 1 - \frac{x^2}{2!} + \frac{x^4}{4!} - \cdots$$

and

$$\sin x = x - \frac{x^3}{3!} + \frac{x^5}{5!} - \cdots,$$

which proves that $\sin x, \cos x \in R$ for $x \in R$.

b) $\sin^2 x + \cos^2 x = \left( \frac{e^{ix} - e^{-ix}}{2i} \right)^2 + \left( \frac{e^{ix} + e^{-ix}}{2} \right)^2$

*Problem 4*  Show that $\log_a x$ is the inverse function of $a^x$.

*Problem 5*  Use the current definitions of the sine and cosine functions to establish the following double angle formulas:

$$\sin (x + y) = \sin x \cos y + \sin y \cos x$$

and

$$\cos (x + y) = \cos x \cos y - \sin x \sin y.$$

$$= \frac{e^{2ix} - 2 + e^{-2ix}}{-4} + \frac{e^{2ix} + 2 + e^{-2ix}}{4}$$

$$= 1.$$

c)  $(\sin x)' = \left(x - \frac{x^3}{3!} + \frac{x^5}{5!} - \cdots\right)'$

$$= \left(1 - \frac{3x^2}{3!} + \frac{5x^4}{5!} - \cdots\right) = \cos x.$$

The proof that $(\cos x)' = -\sin x$ is similar  ●

We need the following lemma in order to define the number $\pi$. Some of the properties of $\pi$ are listed in Problem 6.

**LEMMA**  $\cos x$ *has a zero in* $[0, \infty)$.

*Proof*  Suppose this is false. Since $\cos 0 = 1 > 0$, it follows from the Intermediate Value Theorem that $\cos x > 0$ for all $x \geq 0$. By the Mean Value Theorem it then follows that $\sin x$ is strictly increasing on $(0, \infty)$. In particular $0 < \sin 1 < \sin c$ if $1 < c$. Using the Mean Value Theorem we have, for $1 < x$,

$$\frac{\cos x - \cos 1}{x - 1} = (\cos d)' = -\sin d < -\sin 1.$$

Hence

$$-\cos 1 < \cos x - \cos 1 < (x - 1)(-\sin 1),$$

so $\frac{\cos 1}{\sin 1} > x - 1$. But this obviously cannot be true for all $x$, so we have a contradiction  ●

**THEOREM 6**  *Let* $A = \{x > 0 : \cos x = 0\}$ *and define* $\pi = 2 \inf A$. *Then*

$$\pi > 0 \quad and \quad \sin \frac{\pi}{2} = 1.$$

*Proof*  By the lemma, $A$ is nonempty, and because $\cos 0 = 1$ we must have that $\pi = 2 \inf A > 0$. Since $\cos x$ is continuous it follows that $\cos \frac{1}{2}\pi = 0$, and because $\sin^2 x + \cos^2 x = 1$ we obtain $\sin(\frac{1}{2}\pi) = \pm 1$. Because the derivative of $\sin x$ is $\cos x$ and $\cos x > 0$ for $0 \leq x < \frac{1}{2}\pi$, $\sin x$ must increase on $(0, \frac{1}{2}\pi)$, so $\sin \frac{1}{2}\pi > 0$. Hence $\sin \frac{1}{2}\pi = 1$  ●

→ This proof is taken from George W. Mackey's *Lectures on The Theory of Functions of a Complex Variable*, D. van Nostrand Co. (1967).

**Problem 6**  Prove the following:
a) $e^{i\pi/2} = i$     b) $e^{i\pi} = -1$     c) $e^{2\pi i} = 1$
d) If $e^{ix} = 1$  $(x \in R)$, then $x = 2\pi n$ for some integer $n$.

## D. Exercises

**B1.** a) Prove that $z \in C$ has absolute value 1 if and only if it can be written in the form $z = e^{ix}$ for some $x \in R$.

b) Prove that any nonzero $w \in C$ can be written in a unique way in the form $w = re^{ix}$ where $r > 0$ and $0 \leq x < 2\pi$.

**B2.** Establish the following limits.

a) $\lim\limits_{x \to 1} \dfrac{\ln x}{x - 1} = 1$

b) $\lim\limits_{x \to 0} x \ln x = 0$

c) $\lim\limits_{\substack{x \to 0 \\ x > 0}} x^{1/x} = 1$

**B3.** Establish the following limits.

a) $\lim\limits_{h \to 0} \dfrac{x^h - 1}{h} = \ln x$

b) $\lim\limits_{h \to 0} (1 + hx)^{1/h} = e^x$

**B4.** Using the definition of $\pi$ given in 10.3C, show that:

a) The area of a circle of radius $a$ is equal to $\pi a^2$. (Hint: See Example 6 of 5.5B.)

b) The circumference of a circle of radius $a$ is equal to $2\pi a$. (Hint: See Example 3 of 6.3A.)

**M5.** a) Show that $\cos x > 0$ if $-\frac{\pi}{2} < x < \frac{\pi}{2}$.

b) Define $\tan x = \dfrac{\sin x}{\cos x}$ and show that $\tan x$ is strictly increasing on $(-\frac{1}{2}\pi, \frac{1}{2}\pi)$.

**M6.** Let $\mathrm{Tan}^{-1} x$ denote the inverse of the function described in Exercise M5b. Prove that

$$(\mathrm{Tan}^{-1} x)' = \frac{1}{1 + x^2}.$$

**M7.** Prove that for $n \in N$

$$\left(1 + \frac{1}{n}\right)^n < e < \left(1 + \frac{1}{n}\right)^{n+1}.$$

**M8.** Establish the identity

$$\tfrac{1}{2} + \cos x + \cos 2x + \cdots + \cos nx = \frac{\sin (n + \frac{1}{2})x}{2 \sin (\frac{1}{2}x)}.$$

$\left(\text{Hint: Write } \cos kx \text{ in the form } \dfrac{e^{ikx} + e^{-ikx}}{2}.\right)$

**M9.** Prove that

$$\mathrm{Tan}^{-1} x = \sum_{k=1}^{\infty} (-1)^{k-1} \frac{x^{2k-1}}{2k - 1}$$

for $-1 \leq x \leq 1$. Hence, establish the formula

$$\frac{\pi}{4} = 1 - \frac{1}{3} + \frac{1}{5} - \frac{1}{7} + \cdots.$$

---

## 10.4 Two Important Theorems

The two theorems proven in this section are very useful in more advanced areas of mathematics.

### A. The Weierstrass Approximation Theorem

*THEOREM 1* *Suppose $f: [a, b] \to R$ is a continuous function. Then for any $\varepsilon > 0$ there is a polynomial $p$ with the property that*

$$|f(x) - p(x)| < \varepsilon \quad \text{for all } x \in [a, b].$$

$\longrightarrow$ In other words this theorem says that a continuous function can be approximated arbitrarily closely by a polynomial. The assumption that the interval $[a, b]$ is closed and bounded is essential, as the following examples show.

*Proof* The proof given below is due to the Russian mathematician Serge N. Bernstein.

We first restrict ourselves to the case $[a, b] = [0, 1]$.

Given $f: [0, 1] \to R$, define the (Bernstein) polynomials

$$B_n(x) = \sum_{k=0}^{n} f\left(\frac{k}{n}\right)\binom{n}{k} x^k (1 - x)^{n-k}.$$

$\longrightarrow$ Recall that $\binom{n}{k} = \dfrac{n!}{k!(n-k)!}$.

Using the Binomial Theorem, we then write 1 in the following complicated fashion:

$$1 = 1^n = (x + (1 - x))^n$$

$$= \sum_{k=0}^{n} \binom{n}{k} x^k (1 - x)^{n-k}.$$

Hence

$$f(x) = f(x) \cdot 1 = \sum_{k=0}^{n} f(x)\binom{n}{k} x^k (1 - x)^{n-k},$$

so

$$f(x) - B_n(x) = \sum_{k=0}^{n} \left(f(x) - f\left(\frac{k}{n}\right)\right)\binom{n}{k} x^k (1 - x)^{n-k}$$

and

$$|f(x) - B_n(x)| \le \sum_{k=0}^{n} \left|f(x) - f\left(\frac{k}{n}\right)\right| \binom{n}{k} x^k (1 - x)^{n-k}.$$

Now let $\varepsilon > 0$. We will find an $n_0 \in N$ such that

$$|f(x) - B_n(x)| < \varepsilon \quad \text{for all } n > n_0 \quad \text{and} \quad x \in [0, 1].$$

Since $f$ is continuous on $[0, 1]$ it is bounded (Corollary 1 of Theorem 3 (9.3A)), so there is an $M > 0$ such that $|f(x)| \le M$ for all $x \in [0, 1]$. Since $f$ is uniformly continuous (Theorem 4 of 9.3A), there is a $\delta > 0$ such that $|f(x) - f(y)| < \varepsilon$ whenever $|x - y| < \delta$. Now choose $n_0 \in N$ satisfying

$$n_0 > \max\left\{\frac{1}{\delta^4}, \frac{M^2}{4\varepsilon^2}\right\}$$

and let $n > n_0$. Since $n > n_0 > \dfrac{1}{\delta^4}$, we have $n^{-1/4} < \delta$.

**Example 1**  Let $f(x) = \frac{1}{x}$ for $x \in (0, 1)$. Then for any polynomial $p$ there is an $x_0 \in (0, 1)$ with

$$|f(x_0) - p(x_0)| \ge 1.$$

*Proof*  Since a polynomial $p$ is continuous on $R$ it is bounded on any closed interval (Corollary 1 of Theorem 3 (9.3A)). Hence there is a constant $M$ such that $|p(x)| \le M$ for all $x \in [0, 1]$. Letting $x_0 = \frac{1}{M+1}$, we have $f(x_0) = M + 1$, so

$$|p(x_0) - f(x_0)| \ge 1.$$

Hence, if $|x - \frac{k}{n}| < n^{-1/4} < \delta$, we have $|f(x) - f(\frac{k}{n})| < \varepsilon$, so

$$\sum_{k=0}^{n} \left| f(x) - f\left(\frac{k}{n}\right) \right| \binom{n}{k} x^k (1-x)^{n-k}$$

$$\leq \sum_{k=0}^{n} \varepsilon \binom{n}{k} x^k (1-x)^{n-k} = \varepsilon.$$

If, on the other hand,

$$\left| x - \frac{k}{n} \right| \geq n^{-1/4}, \quad \text{then} \quad \frac{1}{(x - k/n)} \leq n^{1/4}.$$

It follows that

$$\sum_{k=0}^{n} \left| f(x) - f\left(\frac{k}{n}\right) \right| \binom{n}{k} x^k (1-x)^{n-k}$$

$$\leq \sum_{k=0}^{n} \left( |f(x)| + \left| f\left(\frac{k}{n}\right) \right| \right) \binom{n}{k} x^k (1-x)^{n-k}$$

$$\leq 2M \sum_{k=0}^{n} \binom{n}{k} x^k (1-x)^{n-k}$$

$$= 2M \sum_{k=0}^{n} \binom{n}{k} \frac{(x - k/n)^2}{(x - k/n)^2} x^k (1-x)^{n-k}$$

$$\leq 2M \sum_{k=0}^{n} \binom{n}{k} (x - k/n)^2 \sqrt{n} \, x^k (1-x)^{n-k}$$

$$= 2M\sqrt{n} \left( \frac{1}{n} x(1-x) \right) \leq \frac{M}{2\sqrt{n}} < \frac{M}{2\sqrt{n_0}} < \varepsilon. \quad \rightarrow$$

Now, given $f: [a, b] \to R$, define

$$g(t) = f(a + t(b - a)) = f(x).$$

Given $\varepsilon > 0$, we choose $B_n$ so that $|g(t) - B_n(t)| < \varepsilon$ for all $t \in [0, 1]$. Letting

$$p(x) = B_n\left(\frac{x - a}{b - a}\right),$$

it follows that

$$|f(x) - p(x)|$$

$$= \left| f(a + t(b - a)) - B_n\left(\frac{x - a}{b - a}\right) \right|$$

$$= |g(t) - B_n(t)| < \varepsilon \quad \text{for all } x \in [a, b] \quad \bullet$$

**Example 2** Let $f(x) = 0$ for $x \leq 0$, $f(x) = x$ for $0 \leq x \leq 2$, and $f(x) = 2$ for $x \geq 2$. Then for any polynomial $p$ there is an $x_0 \in R$ with

$$|f(x_0) - p(x_0)| \geq 1.$$

*Proof* If $p$ is of degree 0, and hence a constant function, the result is trivial. Hence assume

$$p(x) = a_n x^n + a_{n-1} x^{n-1} + \cdots + a_0.$$

where $a_n \neq 0$ and $n > 0$. Then

$$|p(x)| = |x^n| \left| \left( a_n + \frac{a_{n-1}}{x} + \cdots + \frac{a_0}{x^n} \right) \right|.$$

As $x \to \infty$, $|x^n| \to \infty$ and

$$a_n + \frac{a_{n-1}}{x} + \cdots + \frac{a_0}{x^n} \to a_n \neq 0.$$

Hence $|p(x)|$ can be made arbitrarily large. Choosing $x_0$ with $|p(x_0)| \geq 3$, we have $|f(x_0) - p(x_0)| \geq 1$. (We have simply shown that any nonconstant polynomial must be unbounded on $R$.)

**Problem 1** Using induction, verify the first identity given below.

To obtain this we used the identities

$$\sum_{k=0}^{n} \left( x - \frac{k}{n} \right)^2 \binom{n}{k} x^k (1-x)^{n-k} = \frac{1}{n} x(1-x)$$

and $x(1-x) \leq \frac{1}{4}$ for all $x$.

**Problem 2** Verify, using differentiation, that

$$x(1-x) \leq \frac{1}{4} \quad \text{for all } x.$$

**B. An Existence Theorem** We now show that in reasonable cases the differential equation

$$y' = f(x, y)$$

can always be solved.

*THEOREM 2   Let*

$$B = \{(x, y) \in R \times R : |x|, |y| < 1\}$$

*and suppose $f: B \to R$ is a continuous function and for some constant $L$ satisfies*

$$|f(x, y_1) - f(x, y_2)| \le L |y_1 - y_2|$$

*for all $(x, y_1), (x, y_2) \in B$. Then for any $(x_0, y_0) \in B$ there is a function $\phi$ defined on some open interval $I$ containing $x_0$ that satisfies*

$$\phi(x_0) = y_0$$

*and*

$$\phi'(x) = f(x, \phi(x)) \quad \text{for all } x \in I.$$

*Moreover, if $\psi(x_0) = y_0$ and $\psi'(x) = f(x, \psi(x))$ for all $x \in I$, then*

$$\phi(x) = \psi(x) \quad \text{for all } x \in I.$$

*Proof* We may choose an $a > 0$ such that if

$$K = \{(x, y): |x - x_0| \le a, |y - y_0| \le a\}$$

then $K \subset B$. Since $f$ is continuous on $K$ there is a constant $M$ (Problem 4) such that $|f(x, y)| \le M$ for all $(x, y) \in K$. Let $b = \min \{a, a/M\}$ and let $K' = \{(x, y) \in K : |x - x_0| \le b\}$.

We will now inductively define a sequence of continuous functions

$$\phi_n: [x_0 - b, x_0 + b] \to [y_0 - a, y_0 + a]$$

for $n = 0, 1, 2, \ldots$ that satisfy

$$|\phi_{n+1}(x) - \phi_n(x)| \le \frac{ML^n |x - x_0|^{n+1}}{(n + 1)!}$$

and

$$\phi_n(x_0) = y_0.$$

We begin by letting $\phi_0(x) = y_0$. Now, assuming $\phi_k$ has been defined, set

$$\phi_{k+1}(x) = y_0 + \int_{x_0}^{x} f(t, \phi_k(t)) \, dt.$$

→ A function $f: B \to R$ is called continuous if whenever $(x_n, y_n) \to (x, y)$ (that is, when $((x_n - x)^2 + (y_n - y)^2)^{1/2} \to 0$), then $f(x_n, y_n) \to f(x, y)$.

→ The uniqueness of the solution follows from the assumption

$$|f(x, y_1) - f(x, y_2)| \le L |y_1 - y_2|,$$

which is called a *Lipschitz* condition (see Problem 7) and is essential (see Example 3).

---

**Example 3** Let $f(x, y) = y^{1/3}$ and $x_0 = y_0 = 0$. Then the functions

$$\phi(x) = 0 \quad \text{and} \quad \psi(x) = \left(\frac{2x}{3}\right)^{3/2}$$

each satisfy the equation

$$\phi'(x) = f(x, \phi(x)).$$

(The $f$ in this case does not satisfy a Lipschitz condition.)

---

**Problem 3** Suppose $u: (-1, 1) \to R$ and $f: B \to R$ are continuous. Prove that $g(x) = f(x, u(x))$ is continuous on $(-1, 1)$.

Then

$$|\phi_{k+1}(x) - y_0| = \left| \int_{x_0}^{x} f(t, \phi_k(t))\, dt \right|$$

$$\leq \left| \int_{x_0}^{x} M\, dt \right| = M|x - x_0| \leq Mb \leq a.$$

Hence $\phi_{k+1}: [x_0 - b, x_0 + b] \to [y_0 - a, y_0 + a]$ is defined and is continuous (Problem 3), and clearly $\phi_{k+1}(x_0) = y_0$. Now

$$\phi_{k+1}(x) - \phi_k(x) = \int_{x_0}^{x} (f(t, \phi_k(t)) - f(t, \phi_{k-1}(t)))\, dt,$$

so

$$|\phi_{k+1}(x) - \phi_k(x)| \leq \left| \int_{x_0}^{x} |f(t, \phi_k(t)) - f(t, \phi_{k-1}(t))|\, dt \right|$$

$$\leq \left| \int_{x_0}^{x} L|\phi_k(t) - \phi_{k-1}(t)|\, dt \right|$$

$$\leq \left| \int_{x_0}^{x} L \frac{ML^{k-1}(t - x_0)^k}{k!}\, dt \right|$$

(by the inductive hypothesis)

$$\leq \frac{ML^k}{k!} \left| \int_{x_0}^{x} (t - x_0)^k\, dt \right|$$

$$= \frac{ML^k}{(k + 1)!} |x - x_0|^{k+1}.$$

By induction it then follows that

$$|\phi_{n+1}(x) - \phi_n(x)| \leq \frac{ML^n b^{n+1}}{(n + 1)!} \quad \text{for all } |x - x_0| \leq b.$$

From Problem 5 we have that for each $x$ satisfying $|x - x_0| \leq b$ the sequence $\{\phi_n(x)\}$ is Cauchy, and hence converges to a limit which we denote by $\phi(x)$. From Problem 6 and Theorem 5 (9.4B) it follows that $\phi$ is continuous.

Finally, given $\varepsilon > 0$, choose $n_1$ such that

$$|\phi_n(x) - \phi(x)| < \varepsilon/bL$$

for all $|x - x_0| \leq b$ and $n > n_1$. Then

$$\left| \int_{x_0}^{x} |f(t, \phi_n(t)) - f(t, \phi(t))|\, dt \right|$$

$$\leq \left| \int_{x_0}^{x} L|\phi_n(t) - \phi(t)|\, dt \right| \leq \frac{L\varepsilon}{bL} |x - x_0| \leq \varepsilon.$$

**Problem 4**  (Refer to the first paragraph of the proof of Theorem 2.) If $M$ does not exist, construct a sequence

$$\{(x_n, y_n) \in K\} \quad \text{with} \quad |f(x_n, y_n)| \to \infty.$$

Then construct subsequences $\{x_{n_k}\}$ and $\{y_{n_k}\}$ with $x_{n_k} \to x$ and $y_{n_k} \to y$. Consider $f(x, y)$ and obtain a contradiction.

**Problem 5**  Prove that

$$|\phi_{n+k}(x) - \phi_n(x)| \leq \sum_{r=n}^{n+k-1} \frac{ML^r b^{r+1}}{(r + 1)!}.$$

**Problem 6**  Prove that for any $\varepsilon > 0$ there is an $n_0 \in N$ such that $|\phi_n(x) - \phi(x)| < \varepsilon$ for all $|x - x_0| \leq b$ and $n > n_0$.

Hence

$$\int_{x_0}^{x} f(t, \phi_n(t))\, dt \rightarrow \int_{x_0}^{x} f(t, \phi(t))\, dt,$$

and since $\phi_n(x) \rightarrow \phi(x)$, letting $n \rightarrow \infty$ in the equation

$$\phi_{n+1}(x) = y_0 + \int_{x_0}^{x} f(t, \phi_n(t))\, dt$$

gives

$$\phi(x) = y_0 + \int_{x_0}^{x} f(t, \phi(t))\, dt.$$

Differentiating this equation, we obtain

$$\phi'(x) = f(x, \phi(x))$$

(since $\phi(x_0)$ must equal $y_0$), which gives us the existence of the solution. The uniqueness of the solution is established in Problem 7 ●

In place of exercises we close with a history section that will at least provide the reader with some things to think about.

*Problem 7*   If $\psi$ is another solution show that

$$\psi(x) = y_0 + \int_{x_0}^{x} f(t, \psi(t))\, dt,$$

so

$$|\psi(x) - \phi_{n+1}(x)| \leq L \int_{x_0}^{x} |\psi(t) - \phi_n(t)|\, dt.$$

Then by induction show that

$$|\psi(x) - \phi_n(x)| \leq \frac{aL^n|x_0 - x|^n}{n!}.$$

Hence $\phi_n(x) \rightarrow \psi(x)$ for all $x$, so $\phi(x) = \psi(x)$ for all $x$.

## *Two Other Versions of Calculus*

The presentation of calculus based on the axioms of Chapter 9 (see also the back cover) was developed in the nineteenth century, and most of modern analysis is rooted in this classical approach. But there are other ways to look at the foundations of analysis, and in the last ten years two very different theories have emerged. Ironically, each is based on ideas that long ago fell into disrepute.

### Constructive Analysis

When a constructivist says that he has defined something or has proven that some entity exists, he means that he has a definite procedure available for finding the object in question. This is in contrast to the formalist, who may claim an object exists without knowing a method for finding it, and may make this claim even when no such method exists. We will illustrate this difference with examples of quantities that exist in the formal sense, but not in the constructive sense.

For each positive integer $k$, define $x_k = 0$ if $2k$ is the sum of two prime numbers, and let $x_k = 1$ otherwise. The sequence $\{x_k\}$ is then constructively defined because for each $k$ there is a definite method available for computing $x_k$. We need only list all the primes less than $2k$, and then check to see if some pair adds up to $2k$. If so, then $x_k = 0$, and if not, then $x_k = 1$.

Now define

$$m = \begin{cases} 0 & \text{if } x_k = 0 \text{ for all } k \\ 1 & \text{if } x_k = 1 \text{ for some } k. \end{cases}$$

The number $m$ exists in the formal sense but there is no known method for computing it, because no one knows whether or not every even integer is the sum of two primes. The affirmative statement is known as Goldbach's conjecture. It has been an open problem for over 200 years and its truth or falsity depends on $m$ being 0 or 1 respectively. Since we have not provided $m$ with a method for computing it, $m$ is not constructively defined.

Do not be misled into believing that unsolved problems in mathematics are somehow intimately connected with constructive mathematics. An object is defined constructively when some method for finding it is provided, and that is all there is to it. The above bizarre example furnishes a number that is not constructively defined and, if anything, testifies to the oddness of formal, not constructive mathematics.

Many fundamental results of classical mathematics are not true in the constructive sense. For example, Axiom 8 of 9.1C states that a set of real numbers that is bounded from above has a supremum. The sequence $\{x_k\}$ shows that this is constructively false because $m = \sup \{x_k\}$ is not constructively defined. Another example is the classical fact that if $a$ is a real number, then either $a < 0$, $a = 0$, or $a > 0$. To see that this is false in the constructive sense, let

$$a = \sum_{k=1}^{\infty} \left(\frac{-x_k}{10}\right)^k.$$

This number is constructively defined because it can be computed to any desired degree of accuracy (to find $a$ to $n$ decimal places we need only evaluate $x_1, x_2, \ldots, x_n$). On the other hand, there is no way of showing that $a < 0$, $a = 0$, or $a > 0$. Indeed, if we had a method of showing that $a = 0$, then all $x_k$'s would have to be zero and this would furnish a proof of Goldbach's conjecture. And if we could prove either $a < 0$ *or* $a > 0$, this would disprove the conjecture. Other results that are false from the constructive viewpoint include the Extreme, Mean, and Intermediate Value Theorems (a counterexample to the Intermediate Value Theorem is given by the function described in Example 9 of 9.3B using the number $a$ described above).

In this century the leading exponent of constructive mathematics was the Dutch mathematician L. E. J. Brouwer (1882–1966). He did a great deal of fundamental work in the subject and attracted a small number of followers. Unfortunately their efforts were largely ignored and the formal viewpoint has dominated the mathematics of this century. Today, however, a renewed interest in constructivism is definitely noticeable and it stems from two quite different sources.

One reason for this interest is the appearance, in 1967, of Errett Bishop's book, *Foundations of Constructive Analysis* (McGraw-Hill, 1967). Major criticisms of constructivism have been that its approach was too restrictive, that it did not produce new results and, in fact, offered little hope of doing so. These objections have been laid to rest by Bishop, for he shows in his book how many important areas of analysis (calculus, in particular) can be rewritten from the constructive viewpoint. Moreover, he has obtained several new results, and even more significantly, has raised a large number of new and interesting questions.

The other reason is the increasing availability of computers. Constructive mathematics, with its emphasis on computability, is much more suited to the science of computers than is formal mathematics. To see this, let us return to the sequence $\{x_k\}$ and ask a digital computer to find $m$ for us. We program the computer to successively compute $x_k$ and halt if some $x_k$ is equal to one. If the computer does stop, then it will have succeeded in computing $m$, but as long as it keeps running it cannot decide. The computer may grind away for a billion years and never be able to prove that $m = 0$. It can determine that $m = 0$ only after it has computed every $x_k$ and found it to be zero, but this is something that it cannot do. It is also something that we cannot do, and the view of the constructivist is that we should not attempt to do it. In the words of Bishop:

We are not interested in properties of the positive integers that have no descriptive meaning for finite man. When a man proves an integer to exist, he should show how to find it. If God has mathematics of his own that needs to be done, let him do it himself.

### Non-Standard Analysis

As explained in "The Story of Little 0" (3.5D) Leibniz introduced infinitesimals to calculus but was never able to define them in an accurate way. Their dubious existence was tolerated until the rigorous era of the nineteenth century. At that time, because no one was able to precisely define Leibniz's infinitesimals, they passed out of the domain of respectable mathematics. Then, in 1960, a remarkable development occurred, for Abraham Robinson, using certain ideas from mathematical logic, rigorously defined them. He has rewritten calculus from the point of view originally envisioned by Leibniz, and has applied the same techniques to other branches of analysis with great success. Robinson's theory, called non-standard analysis, appears as a very exciting new development in mathematical analysis. The theory takes place in the setting of the non-standard real numbers, and our main concern will be the creation of this set. This requires the following axiom, which postulates the existence of a certain set called a *non-trivial ultrafilter*.

**Axiom**   *There exists a set $\mathfrak{M}$ whose elements are subsets of the natural numbers $N = \{1, 2, 3, \ldots\}$ satisfying the following properties:*

$M_1$) *$\mathfrak{M}$ contains all cofinite sets. (A set $A \subset N$ is called cofinite if $A^c = N - A$ is finite or void.)*

$M_2$) *If $A, B \in \mathfrak{M}$, then $A \cap B \in \mathfrak{M}$.*

$M_3$) *For any $A \subset N$, either $A \in \mathfrak{M}$ or $A^c \in \mathfrak{M}$, but not both.*

A very peculiar state of affairs exists concerning the set $\mathfrak{M}$, because no one has ever constructed it. On the other hand, no one has ever proven that it cannot exist, and the mathematician Kurt Gödel has even shown that the assumption of its existence does not lead to a contradiction in mathematics. We will not go more deeply into this interesting situation for it is both complicated and controversial. However, the above axiom is similar to the axiom of choice, and the reader is referred to a text on set theory for additional information.

**Definition**   A sequence $x = (x_1, x_2, x_3, \ldots)$ of real numbers, $x_n \in R$, is called a *non-standard real number*. The set of all such numbers is called the non-standard reals and is denoted by *$R$. Two non-standard real numbers $x = \{x_n\}$ and $y = \{y_n\}$ are said to be *equal* if $\{n : x_n = y_n\} \in \mathfrak{M}$. (It is this definition of equality that necessitated the postulation of the set $\mathfrak{M}$.)

The set of ordinary real numbers $R$ will be considered a subset of *$R$ by identifying $r \in R$ with $(r, r, r, \ldots) \in$ *$R$. In particular, we may consider the natural numbers as a subset of *$R$. As will be seen below, with one vital exception *$R$ looks very much like $R$.

Given $x = \{x_n\}$, $y = \{y_n\} \in {}^*R$, we define their *sum* and *product* by the formulas $x + y = \{x_n + y_n\}$ and $xy = \{x_n y_n\}$. These operations on ${}^*R$ behave exactly like the similar operations on $R$, and using $(M_2)$, it is not difficult to show that Axioms 1 to 5 of 9.1C hold for ${}^*R$. A non-standard real number $x = \{x_n\}$ is said to be *positive* if $\{n: x_n > 0\} \in \mathfrak{M}$. Denoting all such numbers by ${}^*R^+$ it can be shown, using $(M_2)$ and $(M_3)$, that Axiom 6 also holds. Moreover, we even find that axiom 7 is valid, but here the similarity ends because Axiom 8 is not true for ${}^*R$. We will show this by an indirect argument. Axiom 8 was used to prove Theorem 8 of 9.1C, which states that the natural numbers as a subset of $R$ are not bounded from above. This theorem is false in ${}^*R$ because the number $x = (1, 2, 3, \ldots) \in {}^*R$ satisfies $n < x$ for all $n \in N$. More exactly, it follows from $(M_1)$ that $(n, n, n, \ldots) < (1, 2, 3, \ldots)$ for any $n \in N$. The number $x$ is an example of an *infinitely large* number, and there are also numbers that are infinitely small. These infinite numbers should not be confused with the infinite cardinal numbers of Cantor, for the latter refer to magnitudes of sets. It is interesting to note that Cantor believed that no rigorous formulation of Leibniz's infinitesimal calculus was possible.

A number $x \in {}^*R$ is said to be an *infinitesimal* if $|x| < \varepsilon$ for any ordinary real number $\varepsilon > 0$. An example of a non-zero infinitesimal is $(1, \frac{1}{2}, \frac{1}{3}, \ldots)$, and more generally, if $x_n \to 0$ then $\{x_n\}$ is an infinitesimal. (This can be seen using $(M_1)$.) Two numbers $x, y \in {}^*R$ are said to be *infinitesimally close* if $x - y$ is an infinitesimal.

A function $f: R \to R$ has a natural extension to a function from ${}^*R$ to ${}^*R$, obtained by defining $f(x_1, x_2, x_3, \ldots) = (f(x_1), f(x_2), f(x_3), \ldots)$. In terms of the above notions we now state the following two theorems, whose proofs can be found in Robinson's book, *Non-Standard Analysis* (North-Holland, 1966).

**Theorem** *A function $f: R \to R$ is continuous at $x_0 \in R$ if and only if $f(y)$ is infinitesimally close to $f(x_0)$ whenever $y \in {}^*R$ is infinitesimally close to $x_0$.*

**Theorem** *A function $f: R \to R$ has the derivative $f'(x_0) \in R$ at $x_0 \in R$ if and only if for every infinitesimal $h \neq 0$, the infinitesimal $f(x_0 + h) - f(x_0)$ divided by the infinitesimal $h$ is infinitesimally close to $f'(x_0)$.*

To summarize Robinson's theory of non-standard analysis we quote from an authority:

the analysis of the present work penetrates as far as infinity itself. It compares the infinitely small differences of finite quantities; it discovers the relation between these differences; and in this way makes known the relations between finite quantities, which are, as it were, infinite compared with the infinitely small quantities.

The above quote is taken from the foreword of the first calculus text ever written. It was by L'Hôpital, a student of Leibniz.

Bonic

# Epilogue:
# The Evolution of
# a Textbook

As this last portion of the manuscript for the second edition of *Freshman Calculus* is being written, the spring of 1975 is finally putting in its much-delayed appearance. September 27, 1968, the date on which *Freshman Calculus* may be said to have been born, is nearly seven long years in the past.

For the freshman or sophomore college student who may be our reader, a few words about the development of this book might be in order. It is an unusual book. While your instructor may well be familiar with the ways in which it is unusual, especially if he or she taught from the previous edition, each new edition has a new story. This one is no exception.

The story of this book officially began, as we indicated, on September 27, 1968. On that date, Bob Bonic, who was then teaching at the University of California at Santa Cruz, stood up before the first meeting of his freshman calculus class and told the assembled students, "This year, *we're* going to write a calculus text." The reaction of the class was, as the chronicler of the first edition wrote, "shades of disbelief, curiosity, and incomprehension."

Nonetheless, it came to pass. The forty students, who initially comprised the student portion of the writing team, were weeded down to thirteen in the interest of getting things done. An engineer, Vahan Hajian, was added to the group early in the first quarter.

By the end of the first quarter, the project was well underway. Daily meetings of all the students were held for the purpose of criticizing the manuscript as it flowed from the pen of the major authors, Bonic and Hajian. The students had been divided into two committees, each meeting for two hours daily to perform his specialized task.

The second quarter of the year brought change and sobriety. After considerable negotiation, D. C. Heath and Company of Lexington, Massachusetts was selected as the publisher. Deadlines became serious. In the interest of organizational efficiency, two advanced students in mathematics, Steve Krantz and Estelle Cranford, were added to the team.

And so the work which was begun in September was virtually completed by June, the end of the academic year. By that time a complete draft of the first eight chapters, which are the heart of the book, was in existence. The summer group that finished the manuscript was smaller than the original thirteen, and by late August the students' role was done.

There is, of course, more to the story. While the students were finished in August 1969, Bonic and Hajian weren't through until early 1970 when the manuscript was finally deemed ready for publication. Publication occurred in early 1971 adding another entry to *Books in Print*.

The sales history of the first edition was nearly as interesting as the story of how it was written. Upon publication the book was controversial in a field where textbooks are noted for anonymity and blandness. Some thought the two column format was confusing. Some thought the idea of using students as "authors" was either a gimmick or outrageous (or both). But some were impressed. The book was adopted at a large number of major universities as well as many smaller schools. It also sold particularly well north of the border in Canadian universities. Where it was adopted it was usually retained for several years before another book was selected in its place; at some universities it has yet to be replaced. The sales were steady and sufficiently large that the publisher was pleased (and financially rewarded). The publisher said, "Let there be a new edition."

The fable of a camel resulting when a committee set out to design a horse typifies the way our committees functioned. There were no shortages of good ideas, but when the committee and not the individual assumed responsibility the ideas became diluted and the student's interest waned.

Vahan

The odds and ends proved far more formidable than we had anticipated. Vahan and I spent nine months of the following year completing the exercises and sample exams. Also, Vahan used copies of the typed manuscript in a calculus course he taught at Northeastern University during 1969–1970, and this testing resulted in our making many minor revisions and corrections.

Steve and Estelle

Preparing a new edition, however, was a task not easily begun. Bonic had moved to New York City; he was too busy to write. Vahan was working at M.I.T.'s Draper Laboratories; he was too busy to write. Estelle was married and in graduate school at Santa Cruz. Steve, who had just received his Ph.D. from Princeton, was just beginning his first year teaching at U.C.L.A. The students were scattered to the four winds.

Confronted with a rather difficult organizational problem, the publisher turned to Bonic for ideas on how to proceed. Bonic promptly suggested that two of his colleagues, Martin Lipschutz of New York University and Edgar DuCasse of Brooklyn College might be up to the task. A meeting in August 1974 between the publisher, Ed, and Marty resulted in sufficiently good vibrations that a contract was signed.

By that date the schedule for the new edition was tight. The areas where improvement was needed were known. The major changes to be implemented included the introduction of limits and continuity in Chapter 2, an improved discussion of the definite integral, and the expansion of the problem sets.

And so it was done. A last minute burst of writing in Newton, Massachusetts in March of 1975 polished the textbook. Ahead lay "only" the reading of copyedited manuscript, galley proofs, page proofs, and the completion of the student book that accompanies the text. After that, it was up to the publisher's representatives to worry about sales.

Marty Lipschutz

Ed DuCasse

# Answers

## 1.1

**B3.** a) $<$ or $\leq$    d) $>$ or $\geq$    g) $<$ or $\leq$
    b) $>$ or $\geq$    e) $<$ or $\leq$    h) $=$ or $\leq$ or $\geq$
    c) $>$ or $\geq$    f) $<$ or $\leq$    i) $<$ or $\leq$

**B4.** a) C    b) D    c) A    d) E    e) B

**B5.** a) $(1, 2]$    c) $[5, 8]$    e) $[0, 2]$
    b) $(2, 4]$    d) $(-3, 2]$

**M12.** a) $\frac{3}{2}$    b) $-1$    c) $\frac{7}{4}$    d) $\frac{8}{3}$    e) $-\frac{9}{4}$

**M13.** a) $(2, 5)$    b) $(-3, 2)$    c) $(-3, -1)$

**M14.** a) $(-\infty, 0)$
    b) $(-\infty, 1] \cup [2, 3]$

**M15.** Canceling $x - 1$ amounts to dividing by zero, which is an illegal operation.

## 1.2

**B2.** $(4, 4)$, $(1, 7)$

**B3.** a) $5$    b) $\sqrt{109}$    c) $5\sqrt{5}$    d) $5$

**B6.** $(x + 1)^2 + (y - 1)^2 = 4$

**M14.** $26$

**M15.** $3\sqrt{3}$

**M16.** $(2\sqrt{3}, 2\sqrt{3})$

**M17.** $(0, 9)$,   $(8, 3)$

**M18.** Vertex: $(a + b, c)$
    Lengths of diagonals:
$$\sqrt{(a + b)^2 + c^2}, \ \sqrt{(a - b)^2 + c^2}$$

**M19.** a) $(a, -b)$    c) $(-a, -b)$
    b) $(-a, b)$    d) $(b, a)$

**M20.** They do.

**M21.** $(0, 7)$,   $(6, 5)$,   $(2, 1)$

**M23.** $(-5, 0)$,   $(4, 3)$

## 1.3

**B1.** a) $0$    b) $\frac{1}{2}$    c) $-2$    d) Undefined

**B3.** $x - 3y = 23$,    $3x + y = -1$

**B4.** $2x - y = 2$,   $(1, 0)$,   $(0, -2)$

**B6.** $(2, 1)$,   $(1, 2)$,   $(3, 4)$

**M22.** $\dfrac{6}{\sqrt{17}}, \ \dfrac{2}{\sqrt{5}}, \ \dfrac{3}{\sqrt{2}}$

**M25.** $x + 2y = 8$   or   $2x - y = 1$

## 1.4

**B1.** a) All values except $x = 1$ and $x = -2$
    b) $x \leq -2$   or   $x \geq 2$    c) $x > -4$

**B2.** $f$ is, $g$ is not

**B4.** a) $3 - 4x$    c) $9 - 12x + 4x^2$
    b) $1 + 2x$    d) $4x - 3$

**B5.** $f(x) = g(x)$ for all values except at $x = -1$,
    where $f$ is not defined.

**B6.** $(f + g)(x) = x^2 + x + 2$,    $(gh)(x) = x^2 - x$,
    $(g - h)(x) = 1$

**B7.** $f(a + b) = a^2 + 2ab + b^2 + a + b - 1$,
    $f(a^2 - 2) = a^4 - 3a^2 + 1$,

$$f\left(\frac{a}{2b}\right) = \frac{a^2 + 2ab - 4b^2}{4b^2}$$

**M14.** a) 5    b) 19    c) 8

**M15.** $a = c$

**M16.** a) $acx + ad + b$
    b) $acx + cb + d$
    c) When $ad + b = cb + d$
    d) $c = d = \frac{1}{2}$

**M17.** a) When $a \geq 0$
    b) When $a = 0$ and $b \geq 0$.

**M21.** $g(x) = (x - b)/a$

**M22.** a) 1    b) 2    c) 6
    d) 720    $f(n) = n!$

## 1.5

**B1.** a) $\frac{1}{2}\pi$    b) $\pi$    c) $\pi$    d) $2\pi$    e) $\frac{1}{4}\pi$    f) $\frac{1}{6}\pi$

**B4.** a) Parabola    b) Hyperbola    c) Ellipse

**B5.** $(\sqrt{2}, 2)$,    $(-\sqrt{2}, 2)$

**M13.** a) $(n\pi, 0)$ and $(2n\pi \pm \frac{1}{3}\pi, \pm\frac{1}{2}\sqrt{3})$, where $n$ is any
    integer.
    b) $(\frac{3\pi}{2} + 2n\pi, -1)$, $(\frac{\pi}{6} + 2n\pi, \frac{1}{2})$, and
    $(\frac{5\pi}{6} + 2n\pi, \frac{1}{2})$, where $n$ is any integer.

**M14.** $x = 0$ and $y = mx$ for $m \geq \frac{2}{3}$ or
    $m \leq -\frac{2}{3}$.

**M15.** $3x + 2y = 11$

**M16.** $2x + y + 2 = 0$,    $2x - 3y + 18 = 0$

**M17.** $y = x^2 - 2x + 4$

**M18.** $a = c = 0$, $b = -4$

**M23.** a) $2yb = 4x + b^2$, where $P = (a, b)$

## 1.6

**B2.** a) $r = 4\cos\theta$    b) $r = 2\sin\theta\tan\theta$
    c) $r = \sin 2\theta$

**B3.** a) $y = 6$    b) $x^2 + (y - 3)^2 = 9$
    c) $xy = 4$

**B4.** a) $x^2 + y^2 = 4$    b) $r = 2$
    c) $x = 2\sin t$, $y = 2\cos t$   but other answers are
    possible

**B5.** a) $x^2 + y^2 = 9$    b) $y = x^2$

**B6.** $(6, \frac{1}{3}\pi)$, $(2, \frac{2}{3}\pi)$, $(2, \frac{4}{3}\pi)$, $(6, \frac{5}{3}\pi)$

**M19.** $(r_1{}^2 + r_2{}^2 - 2r_1r_2 \cos(\theta_1 - \theta_2))^{1/2}$

**M20.** The first pair of equations define a parabola
    $(y = x^2)$, while the second pair represent only a
    portion of the parabola (for $-1 \leq x \leq 1$).

## 1.7

**Exam 1**    **1.** $2x + y = 7$
    **3.** $(2, 6)$

**Exam 2**    **1.** a) $[-4, 6]$    b) $(-4, 6)$
    **2.** $(\frac{1}{2}, 0)$
    **3.** a) 0    b) $4x^2 - 2x$    c) $2x^2$
    **4.** $\frac{1}{2}\sqrt{5}$
    **5.** a) $x = n\pi$ ($n$ any integer)
    b) $-\sqrt{3} < x < \sqrt{3}$

**Exam 3**    **2.** a) $(-1, 3)$    b) $(-1, 5)$
    **3.** a) Hyperbola    b) Parabola    c) Circle
    **5.** 18
    **6.** a) $\frac{9}{2}$    b) $-\frac{2}{3}$    c) $\frac{3}{2}$
    **7.** a) $0, -2$

**Exam 4**    **1.** $h(x) = \begin{cases} 3 & \text{for } x \leq -1 \\ 4 - x & \text{for } x > -1 \end{cases}$
    **2.** $(\frac{5}{7}, \frac{31}{7})$
    **3.** $(x - 1)^2 + (y - 2)^2 = 8$
    **4.** The graph is the circle $x^2 + y^2 = 1$ minus
    the point $(-1, 0)$.

## 2.1

**B1.** a) $f'(x) = 4x - 1$    b) $g'(x) = -x^{-2}$

**B2.** a) $y = 2x - 4$    b) $y = -2$

**B3.** a) $x = -1$    b) $x = -1$

**B4.** $(0, -1)$

**B5.** $x + 12y = 110$

**M11.** $x + 4y = 2$

**M12.** $6x - y = 3$ is the common tangent.

**M13.** $12x - 3y = 1$

**M20.** At $x = -1, 0$, and 1

## 2.2

**B1.** a) $12, 12$     b) $4, -4$     c) $3, 27$

**B2.** a) $y = 0$                c) $6x + y = -5$
   b) $6x - y = 5$          d) $192x - y = 320$

**B4.** $(2, 8)$

**B5.** a) $f'(x) = 3nx^{3n-1}$
   b) $f'(x) = (m + n)x^{m+n-1}$

**M8.** The area equals 1.

**M9.** $n = 7$

**M10.** $y = 1 - 2n - 2nx$

**M11.** $x = 0, x = n$

**M13.** $A_n = 1 - \frac{1}{2}n$

**M14.** $f'(x) = \dfrac{n|x|^n}{x}$

**M16.** a) $4x^3$     b) $4x^{12}$     c) $256x^{12}$     d) $256x^9$

**M17.** $a = 2, \ b = -1$

## 2.3

**B1.** a) $25$          b) $-56$          c) $9$

**B2.** a) $20$          b) $6$          c) $225$

**B3.** a) $9x + y + 15 = 0$     b) $y = 6x + 1$

**B4.** $t = 1, 3$          **B5.** $-\frac{28}{9}$

**B6.** $a = -\frac{1}{2}, \ b = 3, \ c = 4$

**B7.** a) $f'(x) = 3mx^{3m-1} + 3(2m + n)x^{2m+n-1}$
   $\quad\quad + 3(m + 2n)x^{m+2n-1} + 3nx^{3n-1}$
   b) $f'(x) = 2mx^{2m-1} - 2(m + n)x^{m+n-1}$
   $\quad\quad + 2m(n + 1)x^{m(n+1)-1}$
   $\quad\quad - 2n(m + 1)x^{n(m+1)-1}$
   $\quad\quad + 2mnx^{2mn-1} + 2nx^{2n-1}$

**M12.** $5x - y + 2 = 0, \ 5x - y - 2 = 0$

**M13.** $(-\frac{1}{2}, 8)$

**M15.** $y = -\dfrac{x^2}{2} + 3x - \dfrac{1}{2}$

**M16.** $b = 2, \ c = 1$

**M17.** b) Whenever $b^2 < 3ac$

**M18.** $a = \frac{1}{2}, \ b = -1, \ c = -\frac{1}{2}$

## 2.4

**B1.** $f'(0) = 0, \ f'(\frac{1}{3}\pi) = -\frac{\sqrt{3}}{2}, \ f'(\frac{1}{2}\pi) = -1$

**B2.** $f'(0) = 1, \ f'(\frac{1}{2}\pi) = 0, \ f'(3\pi) = -1,$
   $f'(-\frac{3}{4}\pi) = -\frac{1}{2}\sqrt{2}$

**B3.** $y - \frac{1}{2} = \frac{\sqrt{3}}{2}(x - \frac{\pi}{6})$

**B4.** $y - \frac{1}{2} = \frac{2}{\sqrt{3}}(x - \frac{\pi}{3})$

**B5.** $\dfrac{\pi}{2} + n\pi$   where $n$ is any integer

**M9.** $\pi + 2n\pi$   where $n$ is any integer

**M10.** $\dfrac{\pi}{4} + n\pi$   where $n$ is any integer

**M11.** $y + \pi^2 = \dfrac{1}{2\pi + 3}(x - \pi)$

**M12.** a) $-\cos x$     b) $-\sin x$     c) $0$
   d) $\cos x$     e) $\cos x$

## 2.5

**B1.** a) $y' = \dfrac{3x^2 \cos x + x^3 \sin x}{\cos^2 x}$
   b) $y' = 3 + x^3 \sec^2 x + 3x^2 \tan x$

**B2.** a) $-9$          b) $1$

**B3.** a) Nowhere     b) At $x = 0$ and $x = -3$

**B4.** $8x + 8\pi y = 2\pi^2 + \pi^3$

**B6.** a) $6(2x + 3)^2$          b) $-3 \sin x \cos^2 x$
   c) $\dfrac{-6x - 6}{(x^2 + 2x + 3)^4}$

**B7.** a) $\dfrac{dy}{dx} = 2x \sin x \cos x + \sin^2 x$
   b) $\dfrac{dy}{dx} = -3x^2 \cos^2 x \sin x + 2x \cos^3 x$

**M11.** $n\pi$   or   $2n\pi \pm \frac{\pi}{3}$   where $n$ is any integer

**M12.** $\frac{\pi}{4} + n\pi$   where $n$ is any integer

**M14.** The intervals $(n\pi, (n + \frac{1}{2})\pi)$ where $n$ is any integer

**M15.** a) $2 \cos^2 x - 2 \sin^2 x$     b) $-4 \sin x \cos x$
   c) $\dfrac{2}{(\cos^2 x - \sin^2 x)^2}$

**M16.** $x \sec x \tan x - \sec x \tan x + \sec x$

**M19.** a) $\sqrt{x}(2x - 3) + \dfrac{x^2 - 3x + 9}{2\sqrt{x}}$
   b) $\dfrac{-x - 3}{2\sqrt{x}(x - 3)^2}$

**M22.** $(f/(g/h))' = \dfrac{fgh' + f'gh - fg'h}{g^2}$
   $((f/g)/h)' = \dfrac{f'gh - fgh' - fg'h}{g^2 h^2}$

## 2.6

**B1.** a) $f$ is a polynomial and thus is continuous for all real numbers.

b) $f$ is discontinuous at $x = 0$, $f$ is not defined at $x = 0$, nor is there a limit as $x \to 0$. $f$ is continuous for all other real numbers $x$ however.

c) $f$ is discontinuous at $x = 2$ since $\lim_{x \to 2} f(x)$ does not exist. $f$ is continuous for all other real numbers $x$ however.

d) $f$ is discontinuous at $x = 2$ since $f(2)$ is not defined. Note, however, there is a limit equal to 4 as $x \to 2$. $f$ is continuous for all other real numbers $x$.

**B2.** a) 19     b) $-6$     c) $3x^2$     d) $-\dfrac{1}{x_0{}^2}$

**B3.** a) 2     b) $\frac{3}{5}$     c) 0

**M7.** a) $\frac{4}{3}$     b) $-\frac{1}{2}$

## 2.7

**Exam 1**   1. a) $6x + \dfrac{6}{x^3} + \sqrt{2}$

  b) $\cos\theta - \sin\theta + \sec^2\theta$

  2. 42

  4. $x = -4, 3$     5. $4x - y = 2$

**Exam 2**   1. $x^2 \sec x \tan x + 2x \sec x$
  2. 24
  3. $x - y = 2$
  4. $(2, 10), (-2, -10)$
  5. $24x^2 + 120x + 150$

**Exam 3**   1. $a = -\frac{5}{2}, \ b = 3, \ c = 5$
  2. At $y = -\frac{9}{10}$   3. $x = -3, -1$
  4. $(-1, 1), (0, 0)$
  5. a) $x \sec^2 x + \tan x$
   b) $3t^2 - 2\csc^2 t \cot t$
  6. a) 6     b) 32

**Exam 4**   1. $x = -1, 1$     2. 0
  3. $y = 0, \ 8x - y = 16$
  4. $a = 3, \ b = -3$
  5. At $x = -2$ and $x = 0$

**Exam 5**   1. $y = 2x, \ y = -6x$
  2. c) At $x = 0$
  3. a) 3     b) 1     c) $\frac{1}{4}$

**Exam 6**   1. $(\frac{3}{2}, \frac{5}{2})$
  2. $3x - 4y - 3 = 0$
  3. a) $5x - 10x^{-4}$
   b) $-\csc^3 x - \csc x \cot^2 x$
   c) $4t^3 - 10t$
  4. a) At $x = -5$ and $x = -1$
   b) For $x < -5$ or $x > -1$
  6. $(1, 1), (1, 3), (5, 1), (5, 3)$
  9. a) 0     b) $-5$
  10. $a = -3, \ b = 3, \ c = 2$

## 3.1

**B1.** a) $4x^2, 2x^2$     b) $\cos(x^3), \cos^3 x$

  c) $1 + \cos\left(\dfrac{x}{x-1}\right), \dfrac{1 + \cos x}{\cos x}$

**B2.** a) 3     b) 0     c) 13     d) 31

**B3.** a) $-16x(1 + x^2)^{-9}$
  b) $(-2x \sin x^2)(\cos(\cos x^2))$
  c) $(x^{-2} - x^2)^{-3}(4x^{-3} + 4x)$
  d) $b \sec(a + bx) \tan(a + bx)$

**B4.** a) $5 \sin^4 x \cos x$     c) $\dfrac{-4x - 4x^3}{1 + (1 + x^2)^2}$

  b) $5t^4 \cos t^5$     d) $\dfrac{4x + 1}{(3x + 1)^{2/3}}$

**B5.** $\dfrac{dh}{dx} = 54x - 24$

**B6.** a) 3     b) 441     c) 6912
  d) 3     e) 9     f) 3

**M11.** a) $f(x) = x^2, \ u(x) = \sin x$
  b) $f(x) = \sin x, \ u(x) = x^2$
  c) $f(x) = \sqrt{x}, \ u(x) = 3 - 2x + x^4$
  d) $f(x) = \frac{1}{x}, \ u(x) = 1 + x^2$

**M12.** a) $x^2$     b) $x^{\sqrt{2}}$     c) $\sin x$     d) $1 - 2x$     e) $x$

**M13.** a) $(1 + x^2)^3(2 - x^3)^4(16x - 15x^2 - 23x^4)$
  b) $(1 + x^2)^5 4 \sin^3 x \cos x + 10x(1 + x^2)^4 \sin^4 x$
  c) $(\cos x - x \sin x)(3 \sin^2 (x \cos x))(\cos(x \cos x))$

**M14.** $u'(0) = \frac{4}{3}$

**M15.** a) $\dfrac{dV}{dt} = 4\pi r^2 \cdot \dfrac{dr}{dt}$     b) $\dfrac{dr}{dt} = \dfrac{25}{32\pi}$

**M16.** a) $\dfrac{dI}{dt} = A\omega \cos(\omega t + \phi)$

  b) $\dfrac{dW}{dt} = -A\omega^2 \sin(\omega t + \phi)$

## 3.2

**B1.** $y = 2 + \sqrt{4 - x^2}, \quad y = 2 - \sqrt{4 - x^2}$

**B3.** a) $y' = \dfrac{3x + 4y}{y - 4x}$   b) $y' = \dfrac{y \sin x - \cos y}{\cos x - x \sin y}$

**B4.** a) $\dfrac{2}{9}$   b) $\dfrac{\pi}{4 - 2\pi}$

**B5.** $3x + 4y = 25, \quad 3x - 4y = 25$
**B7.** a) $\frac{7}{3}$   b) $\frac{1}{2}$

**B8.** a) $y' = \dfrac{2x^2 + 1}{(x^2 + 1)^{1/2}}$

   b) $y' = \dfrac{-4}{3(x + 1)^{1/3}(x - 1)^{5/3}}$

**M14.** a) $-2 < a < 2$   b) $|a| > 2$
   c) $a = \pm 2$
**M15.** a) $x - 3y + 2 = 0$   b) $9x - 8y = 52$
**M16.** a) $\frac{10}{3}$   b) 1
**M17.** $y = -1$   **M18.** 1   **M19.** $-\frac{5}{9}$

**M20.** a) $\dfrac{3x + 4y}{y - 4x}$   b) $\dfrac{1 - y \cos xy}{x \cos xy - 1}$

## 3.3

**B1.** a) $[2, \infty)$   b) $(-\infty, 2]$
**B2.** a) $f^{-1}(x) = 4x + 12$   b) $g^{-1}(x) = x^3$

**B5.** a) $\dfrac{2}{1 + 4x^2}$

   b) $\dfrac{-2x^2}{x^2 + 4} + 2x \, \mathrm{Tan}^{-1} \dfrac{2}{x}$

   c) $\dfrac{\cos^2 x - \sin^2 x}{(1 - \sin^2 x \cos^2 x)^{1/2}}$

   d) $\dfrac{2 \sec^2 x}{1 + 4 \tan^2 x}$

**B6.** a) $y' = 2x \sec^2 x^2$   b) $y' = -\cos x$
**M11.** $k = -1$
**M12.** $a + d = 0$ or $a = d$ and $b = c = 0$
**M16.** $(1 - x^2)^{-3/2}$

## 3.4

**B1.** a) 4   b) 2   c) 3   d) 1024

**B3.** a) $-\dfrac{1}{e^3}$   b) $-\dfrac{\sqrt{3}}{2} e^{-\pi/3}$

**B4.** a) $e$   b) 0   c) 1

**B5.** a) $e^x \, \mathrm{sech}^2 \, e^x$   b) $10 \sinh 5x \cosh 5x$
   c) $3 \cos 3x e^{\sin 3x}$
**B6.** $(1, e)$
**M10.** $(e, 1)$

**M11.** a) $\dfrac{dy}{dx} = \dfrac{2x + 1}{2(x(x + 1))^{1/2}}$

   b) $\dfrac{dy}{dx} = \dfrac{2(x^2 + 2x - 1)}{3(x(x - 1)(x^2 + 1)^5)^{1/3}}$

**M12.** a) $2^x \ln 2$   b) $3^{\sin x} \cos x \ln 3$

   c) $\dfrac{2x \cos x + 2 \sin x}{x \sin x \ln 10}$   d) $\dfrac{\log_2 e}{(1 + x^2) \, \mathrm{Tan}^{-1} x}$

**M13.** a) $2x$   b) 0   c) $\log_a b$
**M14.** a) $x^{(x^2)}(x + 2x \ln x)$   b) $x^{2x}(2 + 2 \ln x)$

**M17.** a) $\dfrac{dy}{dx} = \sinh \dfrac{x}{a}$

**M20.** $y = x + 1$

## 3.5

**B1.** a) $f'(x) = \dfrac{-2}{(1 + x)^2}, \quad f''(x) = \dfrac{4}{(1 + x)^3}$

   b) $g'(x) = e^x \cos e^x,$
   $g''(x) = e^x \cos e^x - e^{2x} \sin e^x$
   c) $h'(x) = 3x + 6x \ln x, \quad h''(x) = 9 + 6 \ln x$
**B2.** $f(2) = \frac{2}{5}, \quad f'(2) = \frac{-3}{25}, \quad f''(2) = \frac{4}{125}$
**B3.** a) $dy = 2dx$

   b) $dy = (3x^2 - 4x)dx$

   c) $dy = \dfrac{x + 3}{(1 - x)^3} dx$

**B5.** a) $\dfrac{-b^4}{a^2 y^3}$   b) $\dfrac{2y^3 - 8}{9y^5}$

**B6.** a) $-15$   b) 64   c) $-40320$
**M10.** $dy = 4(2t + 1) \, dt$
**M11.** 2.4 in.³
**M12.** 10.003333
**M15.** b) $p(x) = \frac{1}{6}(x^4 - 9x^3 + 36x^2 - 43x)$
**M16.** $p(x) = 4 - 3x - 3x^2$
**M17.** a) For no $x$   b) $x = 1$   c) $x = 2$
**M19.** a) $6x - 4$   b) $-18x$

   c) $-\dfrac{9}{a^2} \sin \dfrac{3x}{a}$

**M24.** b) $P_1(x) = x, \quad P_2(x) = 6x^2 - 4x,$
   $P_3(x) = 120x^3 - 120x^2 + 16$

## 3.6

**Exam 1**  1. a) $f'(x) = 8(2 - 3x^2 + 2x^{-3})^7$
$\times (-6x - 6x^{-4})$
   b) $g'(x) = -\frac{1}{3}\sin(\frac{x}{3} + 4)$
   c) $h'(x) = \cos x \sec^2(\sin x)$
2. a) $-\frac{8}{5}$   b) $\frac{3}{5}$
3. a) 2   b) 0
4. a) $4xe^{2x^2}\,dx$   b) $\frac{1}{x \ln x}\,dx$

5. a) 4   b) $\frac{1}{4}$   c) 4

**Exam 2**  1. a) $-9x^2(1 + x^3)^{-4}$
   b) $-9(1 + x)^{-10}$   c) 1
2. a) $y = -x + x\sqrt{1 + x}$,
$y = -x - x\sqrt{1 + x}$
   c) $(\frac{3\pi}{2} + 2n\pi, 0)$   where $n$ is any integer
3. a) $3\sin^2 t \cos t$
   b) $-3\cos^2 t \sin t$
   c) $-\tan t$
   d) $6\sin t \cos^2 t - 3\sin^3 t$
   e) $6\sin^2 t \cos t - 3\cos^3 t$

4. a) $f^{-1}(x) = \frac{x - 5}{4}$

   b) $f^{-1}(x) = \left(\frac{x - 7}{2}\right)^{1/3}$

5. a) 4   b) 1   c) 3   d) 0   e) 2

**Exam 3**  1. a) $(78 + 156x)(1 + x + x^2)^{77}$
   b) $e^x \cos e^x$
   c) $-3\sin 3x e^{\cos 3x}$

   d) $\frac{8x}{1 + 16x^4}$   e) $\frac{1}{2\sqrt{x - x^2}}$

   f) $\frac{e^x - e^{-x}}{e^x + e^{-x}}$

2. $a = 3$,  $b = 2$,  $c = 2$
3. a) $-\frac{2}{3}$   b) $y' = -1$
4. a) $(8, -7), (4, 25)$
   b) $(4, -2), (8, 20)$
5. a) $x = \pm\frac{1}{2}$
   b) $\frac{\pi}{4} + \frac{n\pi}{2}$   where $n$ is any integer
   c) $\frac{n\pi}{3}$   where $n$ is any integer

**Exam 4**  1. a) $3x^2 + 3^x \ln 3$

   b) $\frac{1}{\sqrt{1 - x^2}} - \frac{\cos x^{-1}}{x^2} - \frac{\cos x}{\sin^2 x}$

   c) $\frac{1}{x}\log_{10} e - \frac{1}{x \ln x}\log_x 10$

2. a) $\sqrt{\pi}$   b) 1   c) 1
3. a) $y = 9 + \sqrt{84 - 4x^2 + 16x}$,
$y = 9 - \sqrt{84 - 4x^2 + 16x}$
4. a) $(2, 3)$ and $(3, 4)$   are two possibilities.
   b) $(\frac{1}{4}\pi, \frac{1}{2}\pi)$ and $(\frac{1}{2}\pi, \frac{3}{4}\pi)$   are two possibilities.

**Exam 5**  1. a) $\left(\frac{n\pi}{2}, \frac{(n + 1)\pi}{2}\right)$ for any integer $n$

   b) For  $y = \mathrm{Sec}^{-1} x$   $(0 < y < \frac{1}{2}\pi)$,

$\frac{dy}{dx} = \frac{1}{x\sqrt{x^2 - 1}}.$

2. a) $x = 0$   b) $x = -3, 0$   c) $x = 0$
3. a) $(1 + g'(x))g'(x + g(x))$
   b) $(xg'(x) + g(x))g'(xg(x))$
   c) $3(x + 2)^2 g'((x + 3)^3)$
4. $P = (1, 2)$
5. b) $a = 1, 4$

**Exam 6**  1. $40x - y = 79$
2. a) 0   b) $-\frac{1}{2}$   c) $-\frac{1}{2}$   d) 16
3. $(1, 7)$
5. a) $(6x - 6x^2)e^{3x^2 - 2x^3}$
   b) $-4x \csc^2 x^2 \cot x^2$

   c) $\frac{x}{(2 + x^2)(1 + x^2)^{1/2}}$

6. a) $x = 1$   b) $x = e$
7. $-\pi$
8. a) $\frac{\pi}{2}$   b) $\frac{1}{9}$   c) 18

9. a) $\frac{-2x^4 - 2y^3 x}{y^5}$

   b) $\frac{-2x}{(1 + x^2)^2}$   or   $-2\sin y \cos^3 y$

10. a) All $x \neq \frac{2}{3}$   c) All $x \neq \frac{2}{3}$
   b) $f^{-1}(x) = f(x)$   d) 5

## 4.1

**B1.** a) $x = 2$   b) $x = -3$   c) $\frac{9}{4}$
**B2.** a) $m = \frac{\sqrt{3}}{3}$   b) $m = e - 1$
**B3.** $x = n\pi$   for any integer   $n$
**B4.** a) $x = \frac{1}{2}$ (minimum)   b) $x = -3$ (minimum)
   c) $x = 1$ (maximum)

**M11.** a) $f$ is not continuous at $x = 1$
c) $f$ is not differentiable at $x = 0$
**M22.** a) $3\sqrt{3}$   b) $(1, 3)$
**M23.** $g(x) = x$

## 4.2

**B1.** a) $(-\infty, -2]$  and  $[3, \infty)$
b) $[\pi(n + \frac{1}{2}), \pi(n + 1)]$  for any integer $n$
**B2.** a) $x = -1$ (local maximum),
$x = 1$ (local minimum)
b) $x = \frac{\pi}{4} + 2n\pi$ (local maximum),
$x = \frac{\pi}{4} + (2n + 1)\pi$ (local minimum)
**B3.** a) Vertical at  $x = -3$,  horizontal at  $y = 1$
b) Vertical at  $x = -5$  and  $x = -2$,
horizontal at  $y = 5$
**B4.** a) Convex upward on $(-\infty, -\frac{1}{2}]$  and  $[1, \infty)$,
inflection points at $(-\frac{1}{2}, \frac{27}{8})$ and $(1, 6)$
b) Convex upward everywhere, no inflection points
**M15.** It takes the ball $\frac{15}{8}$ sec. to reach maximum height. This maximum is $\frac{225}{4}$ feet.
**M16.** $I = \sqrt{10}$ amperes yields maximum efficiency. For this value of $I$, $E = 1 - \frac{\sqrt{10}}{50}$.

## 4.3

**B1.** $y + 2x = 4$        **B2.** $a^2$
**B4.** $\frac{5}{2}$              **B5.** 1
**M8.** 10 by 10 by 5
**M9.** $10\sqrt{2}$ by $10\sqrt{2}$ by $\frac{5}{2}$
**M10.** $(1, 2)$
**M11.** $2 + \sqrt{6}$ by $4 + 2\sqrt{6}$
**M12.** 6 items should be produced.
**M13.** 24 days
**M14.** Length 12 in., width 6 in., height 2 in.
**M15.** $\dfrac{2P}{\pi + 4}$
**M16.** 107     **M17.** 20     **M18.** 20
**M23.** After $\frac{80}{13}$ seconds     **M25.** Ten dollars

## 4.4

**B1.** a) When $t = 3$  b) At $x = -9$  c) At $x = -5$
**B2.** a) When $t = 1$        b) 176
c) When $t = 1 + \sqrt{11}$
**B3.** a) $(16, 12)$            b) 5
c) When $t = \frac{1}{2}$

**B4.** $\frac{3}{4}$ ft.
**B5.** a) $\frac{225}{4}$                b) When $t = \frac{15}{4}$
**B6.** a) $t > 2$  b) $-1 < t < 1$
**M11.** a) 37    b) 0    c) 6    d) 0    e) $-12$
f) none
**M12.** a) When $t \leq 2$ or $t \geq 4$
b) When $t \geq 3$    c) 20      d) 44
**M13.** a) 16    b) 2    c) $\frac{15}{4}$    d) $\frac{47}{8}$
**M18.** $\frac{6050}{9}$ ft.
**M19.** $32\sqrt{30}$ ft./sec.
**M20.** $32\sqrt{7}$ ft./sec.

## 4.5

**B1.** a) $4\pi$ in./min.        b) 24 in.$^2$/min.
**B2.** a) $\frac{5}{9\pi}$ ft./min        b) $18\pi$ min.
**B3.** $\frac{50\pi}{81}$ in./sec.
**B4.** $-2$ in./min.        **B5.** 8 ft./sec.
**B6.** $15\sqrt{3}$ in.$^2$/sec.
**B7.** a) 144                b) $\frac{11}{16}$
**M13.** $\frac{12}{5}$ ft./sec.        **M14.** $\frac{3}{2}$ in./sec.
**M15.** 320 ft./sec. while both are falling
**M16.** 10 ft./sec.
**M17.** $\dfrac{124}{\sqrt{61}}$
**M18.** a) $\dfrac{\sqrt{2}}{2}$ ft./min.        b) $\dfrac{\sqrt{2}}{4}$ ft./min.
c) When $t = \dfrac{6}{\sqrt{33}}$
**M19.** a) $20\pi$ ft./sec.        b) $18\pi$ ft.$^2$/sec.
**M20.** $\dfrac{dV}{dt} = -4$ in.$^3$/sec.
**M23.** $\dfrac{dR}{dt} = R_0\left(a + \dfrac{b^3}{2\sqrt{T}}\right)$

## 4.6

*Exam 1*   **1.** a) $x = -4, 1$  b) $x = -1, 0, 1$
**3.** $y = 2x + 3$
**4.** a) When $t = 2$      b) 64 ft.
c) $-64$ ft./sec.
**5.** $\frac{6}{5}$ in./min.
*Exam 2*   **1.** $x = 0$ (local maximum),
$x = \pm 1$ (local minimum)

**3.** $\dfrac{50}{6 - \sqrt{3}}$ by $\dfrac{150 - 50\sqrt{3}}{6 - \sqrt{3}}$

**4.** $\left[ -\dfrac{\sqrt{3}}{3}, \dfrac{\sqrt{3}}{3} \right]$

**5.** $(648\pi)^{-1/3}$

**Exam 3**  **2.** a) $x = 0$ (local maximum),
$x = 4$ (local maximum),
$(2, -16)$ (inflection point)
b) $x = -3$ (local maximum),
$x = 2$ (local minimum),
$(-\frac{1}{2}, \frac{47}{2})$ (inflection point)
**4.** a) 0  b) 12  c) $t = 1, 3$  d) $x = 7$
e) $-6$ and 6
**5.** 64 ft./sec.
**6.** $12\sqrt{3}$(base), 18 (height)

**Exam 4**  **1.** $\frac{16}{3}$ by $\frac{8}{3}$
**3.** 9 inches for the square and 8 for the rectangle.
**5.** $\frac{9}{20}$ ft./hr.

**Exam 5**  **1.** a) $x^2 - 3x + 2$  b) $x^2 - 5x + 7$
**2.** a) $2 \tan x \sec^2 x + 2x \sec^2 x^2$
b) $t^3 - 3t + 9$  c) $(3 + \cos y)^{-1}$

**3.** a) At $\dfrac{2\pi}{3}$, $\pi$, and $\dfrac{4\pi}{3}$

c) $\dfrac{2\pi}{3}$ and $\dfrac{4\pi}{3}$ (local minima),

$\pi$ (local maximum)

**4.** a) $y = \dfrac{-8}{3}t^2 + 80t + 576$

b) 1176 ft.  c) When $t = 36$.
**5.** 3 in./min.  **6.** $x = 2$
**7.** $(-\sqrt{10}, 3)$, $(\sqrt{10}, 3)$
**8.** $y = 4x$, $y = -4x$

**9.** a) $(x + 6)e^x$  b) $\dfrac{-7680}{(1 + 2x)^6}$

c) $-64 \sin 2x - 729 \cos 3x$

**Exam 6**  **1.** a) $x \neq \pm 2$  b) $x \leq -\frac{1}{4}$ or $x > 0$
**2.** a) $\frac{1}{4}$  b) $\frac{5}{4}$  c) 1  d) $\frac{1}{12}$
**3.** a) $x + 12y = 24$  b) $5y + 16x = 148$
**4.** $a = 1$,  $b = 8$,  $c = 6$,  $d = 11$

**7.** a) $\dfrac{\sin x - \cos x}{\sin x}$

b) $\cos x \cos (\sin x) + 2 \sin x \cos x$

c) $\dfrac{1}{6t^2}$

**8.** $2\sqrt{2}$ by $4\sqrt{2}$
**9.** 2  **10.** $\frac{1}{135}$ ft./min.
**11.** a) When $t = 2$
b) 4 units (to the left)
c) At $x = 12$  d) 2

## 5.1

**B1.** a) $\dfrac{x^3}{3} - \dfrac{1}{x} + x^2 + C$

b) $2x^4 + 12x^3 + 27x^2 + 27x + C$

c) $\dfrac{x^3}{3} + \dfrac{x^2}{2} + x + C$  d) $\dfrac{x^3}{3} + \dfrac{3x^2}{2} + C$

**B2.** a) $\dfrac{(x^4 + 1)^9}{36} + C$  b) $-\frac{1}{5}(x^{-2} + 1)^{5/2} + C$

**B3.** a) $\sin (\ln x) + C$  b) $-\frac{1}{2} \cos (\ln x^2) + C$
**B4.** a) $\frac{1}{2}e^{\tan 2x} + C$  b) $\frac{1}{2}e^{\text{Tan}^{-1} 2x} + C$
**B5.** a) $\frac{1}{6} \ln (2x^3 - 4) + C$  b) $\frac{1}{2}(\ln x)^2 + C$
**B6.** a) $\frac{1}{2} \text{Tan}^{-1} 2e^x + C$  b) $\text{Tan}^{-1} e^x + C$
**B7.** $\frac{1}{24}(1 + \sin^2 x^3)^4 + C$

**M13.** a) $y = \dfrac{x^3}{3} + x + C$

b) $y = \dfrac{(2x + 1)^4}{8} + C$

c) $y = \frac{4}{3}x^3 + 4x - \frac{1}{x} + C$
**M14.** $\sin (1 + e^{\cos (\ln x)}) + C$
**M15.** a) $e^{\sin^2 x} + C$
b) $\text{Tan}^{-1} (\sin x) + C$

c) $\dfrac{-1}{1 + \sin^2 x} + C$

## 5.2

**B1.** a) 129  b) 35  c) $-4$  d) $\frac{\pi}{2}$
**B2.** a) $\frac{7}{3}$  b) $\frac{\pi}{12}$  c) $\frac{196}{3}$  d) 2
**B3.** $\frac{24}{25}$  **B4.** $\frac{32}{3}$
**M8.** a) 9  b) $\frac{25}{3}$  c) $\frac{97}{3}$

**M9.** a) $\dfrac{3 - \sqrt{2}}{2}$  b) $\dfrac{3 - \sqrt{2}}{2}$

**M10.** a) $f'(x) = 2x \cos^4 x^2$

b) $g'(x) = 3e^{-9x^2} - 2e^{-4x^2}$

c) $h'(x) = \dfrac{x^2}{2} + 1 - \cos x$

**M11.** $\frac{1}{2}(1 - e^{-A^2}) \to \frac{1}{2}$ as $A \to \infty$

**M12.** 32

**M13.** $a = 2, \quad b = 3$

**M15.** $f(t) = 2t + 3$

**M16.** $a = \dfrac{\pi}{2} + n\pi$ where $n$ is any integer

## 5.3

**B1.** $\frac{3x}{4}e^{4x} - \frac{3}{16}e^{4x} + C$

**B2.** $\frac{x}{3}\sin 3x + \frac{1}{9}\cos 3x + C$

**B3.** $-x + x\ln 4x + C$

**B4.** $x\,\mathrm{Tan}^{-1}x - \ln\sqrt{1+x^2} + C$

**B5.** $e^x(x^2 - 2x + 2) + C$

**B6.** $-x^2\cos x + 2x\sin x + 2\cos x + C$

**B7.** $\frac{x}{2}\cos(\ln x) + \frac{x}{2}\sin(\ln x) + C$

**B8.** $\frac{1}{3}(1+x^2)^{3/2} - (1+x^2)^{1/2} + C$

**M11.** a) $\frac{x}{2}\sin(\ln x) - \frac{x}{2}\cos(\ln x) + C$

b) $\dfrac{e^{2x}}{13}(2\cos 3x + 3\sin 3x) + C$

**M12.** $\frac{1}{3}\cos^3 x - \cos x + C$

**M13.** $\frac{4}{15}(1 + \sqrt{2})$

**M14.** $\frac{\pi}{16}$       **M15.** b) $\frac{5\pi}{32}$

**M16.** a) $f(x) = x(\ln x)^n, \; C = -n$    b) $6 - 2e$

**M17.** a) $-\ln x + (\ln x)\ln(\ln x) + C$

b) $\frac{2}{9}e^{3/2} + \frac{4}{9}$

c) $\dfrac{x^2}{4}(1 + 2(\ln x)^2 - 2\ln x) + C$

**M19.** a) $\dfrac{\pi\sqrt{2} + 4\sqrt{2} - 8}{8}$    b) $\dfrac{\pi}{4} - \dfrac{1}{2}\ln 2$

## 5.4

**B1.** a) $\ln|x-3| + C$    b) $\frac{1}{2}\ln|2x-3| + C$

c) $\dfrac{1}{6 - 4x} + C$

**B2.** a) $\ln|x+2| + C$    b) $x - \ln(x+2)^2 + C$

c) $\dfrac{x^2}{2} - 2x + 4\ln|x+2| + C$

**B3.** a) $\frac{3}{4}\mathrm{Tan}^{-1}\frac{x}{4} + C$    b) $\frac{3}{2}\ln(x^2 + 16) + C$

c) $3x - 12\,\mathrm{Tan}^{-1}\dfrac{x}{4} + C$

**B4.** a) $\frac{1}{3}\ln\left|\dfrac{x-3}{x+3}\right| + C$    b) $\ln|x^2 - 9| + C$

c) $2x + 3\ln\left|\dfrac{x-3}{x+3}\right| + C$

**B5.** a) $\frac{1}{4}\mathrm{Tan}^{-1}\left(\dfrac{x-2}{4}\right) + C$

b) $\frac{1}{8}\ln\left|\dfrac{x-6}{x+2}\right| + C$

**B6.** a) $\frac{1}{7}\ln\left|\dfrac{x-3}{x+4}\right| + C$    b) $\frac{2}{7}\ln\left|\dfrac{x-3}{x+4}\right| + C$

**B7.** $\ln|(x-1)(x-2)(x+3)| + C$

**B8.** $\dfrac{x^2}{2} + 5x - \dfrac{31}{2}\mathrm{Tan}^{-1}\left(\dfrac{x-1}{2}\right) + C$

**M11.** $\dfrac{x^2}{2} - 2x + 3\ln|x+3| + C$

**M12.** $\dfrac{x^2}{2} + 2x + \dfrac{1}{1-x} + 3\ln|x-1| + C$

**M13.** $\frac{1}{3}\ln\frac{5}{4}$       **M14.** $\frac{3}{2} + \ln\frac{3}{2}$

**M15.** a) $\ln\frac{27}{16}$    b) $\frac{3}{2} + 2\ln\frac{5}{4}$

**M16.** $\ln\frac{5}{12}$

**M17.** $\ln|(3+\sin x)(4+\sin x)| + C$

**M18.** $\frac{1}{6}\mathrm{Tan}^{-1}\left(\frac{2}{3}\sec x\right) + C$

**M19.** $\frac{1}{2}\ln(x^2 - 4x + 8) + \frac{3}{2}\mathrm{Tan}^{-1}\left(\dfrac{x-2}{2}\right) + C$

**M20.** $\frac{1}{9}\ln(9x^2 - 12x + 8) + \frac{13}{18}\mathrm{Tan}^{-1}\left(\dfrac{3x-2}{2}\right) + C$

**M21.** $-\frac{1}{8}\ln|4x^2 + 4x - 3| + \frac{5}{16}\ln\left|\dfrac{2x-1}{2x+3}\right| + C$

**M23.** $\frac{1}{4}\ln\left|\dfrac{x}{4-x}\right| + C$

**M24.** a) $\frac{1}{10}\ln|x-3| - \frac{1}{20}\ln(1+x^2)$
$\qquad - \frac{3}{20}\mathrm{Tan}^{-1}x + C$

b) $\dfrac{5}{x^2 + 5} + \ln|x-1| + C$

c) $\ln\left|\dfrac{(x+1)^4}{2x^2 + 5x + 3}\right| + C$

**M25.** $\mathrm{Tan}^{-1}x + \frac{1}{2}\ln(x^2 + 2) + C$

**M26.** $\ln(x^2 + 4) + \dfrac{1}{2}\mathrm{Tan}^{-1}\left(\dfrac{x}{2}\right) + \dfrac{4}{x^2 + 4} + C$

## 5.5

**B1.** a) $2 \operatorname{Sin}^{-1} \frac{x}{2} + \frac{x}{2} \sqrt{4 - x^2} + C$

b) $-\frac{1}{3}(4 - x^2)^{3/2} + C$

**B2.** a) $\operatorname{Sin}^{-1} \frac{x}{2} + C$     b) $-\sqrt{4 - x^2} + C$

**B3.** a) $\frac{2}{3}x^{3/2} + 6x^{1/2} + C$   b) $\frac{2\sqrt{3}}{3} \operatorname{Tan}^{-1} \sqrt{\frac{x}{3}} + C$

**B4.** a) $\frac{2}{5}x^{5/2} - \frac{8}{3}x^{3/2} + C$

b) $\frac{2}{5}(x - 4)^{5/2} + \frac{8}{3}(x - 4)^{3/2} + C$

**B5.** a) $-2x \cos x + 2 \sin x + C$

b) $-2\sqrt{x} \cos \sqrt{x} + 2 \sin \sqrt{x} + C$

**B6.** 1192        **B7.** $2e^3 - 2e^2$

**B8.** $\frac{\pi}{12}$

**M12.** a) $\ln \left| \dfrac{\tan \frac{x}{2}}{1 + \tan \frac{x}{2}} \right| + C$

b) $\frac{2}{3} \operatorname{Tan}^{-1} \left( \frac{1}{3} \tan \frac{x}{2} \right) + C$

**M13.** $\frac{2}{3}(x + 1)^{3/2} - 2(x + 1)^{1/2} + C$

**M14.** $(36 - x^2)^{5/2} - 60(36 - x^2)^{3/2} + C$

**M15.** $\dfrac{2x}{5} + \dfrac{1}{5} \ln \left| \dfrac{2 + \tan x}{\sec x} \right| + C$

**M16.** $\ln \left| \dfrac{1 - \sqrt{1 + e^x}}{1 + \sqrt{1 + e^x}} \right| + C$

**M17.** $(1 + x^2)^{-1/2}(1 + x \operatorname{Tan}^{-1} x) + C$

**M18.** $x^x + C$        **M19.** $2 \ln \frac{4}{3}$

**M20.** $\dfrac{1}{2} \ln \left( \dfrac{3 + 2\sqrt{2}}{3} \right)$

**M21.** $-\frac{1}{2} + \ln 2$       **M22.** $\frac{162}{5}$

**M23.** $\operatorname{Sin}^{-1} \left( \dfrac{x - 4}{6} \right) + C$

**M24.** $-\sqrt{5 - 4x - x^2} + \operatorname{Sin}^{-1} \left( \dfrac{x + 2}{3} \right) + C$

**M25.** b) $\sqrt{x^2 + 9} + 2 \ln (x + \sqrt{x^2 + 9}) + C$

**M26.** b) $\sqrt{x^2 + 2x - 3} + \ln |x + 1 + \sqrt{x^2 + 2x - 3}| + C$

## 5.6

**B1.** a) 4            b) 4

**B3.** a) $\frac{\pi}{4}$           b) 1

**B5.** The integrand blows up at $x = 1$, and the integral is divergent.

**M8.** $\frac{\pi}{2}$    **M9.** $\frac{\pi}{8}$    **M10.** $\frac{\pi}{2}$    **M11.** $\frac{1}{2}$

**M12.** a) $a > -1$        b) $a < -1$

**M14.** 2              **M15.** $\pi$

**M16.** $\dfrac{1}{\ln 2}$        **M17.** $\dfrac{\pi}{2}$

**M18.** $\frac{2}{e}$         **M19.** $-6$

**M20.** a) Divergent     b) Divergent

**M21.** 1

**M22.** a) $-1$    b) $-\frac{1}{4}$    c) $-\frac{1}{9}$

**M23.** $\displaystyle\int_0^\infty x^n e^{-x} \, dx = n!$

## 5.7

**Exam 1**    **1.** a) $\frac{1}{5}(4 + 2x - x^3)^5 + C$

b) $e^{\sin x} + C$

c) $\ln (4x^2 + 5) + C$

**2.** $\frac{16}{15}$

**3.** a) $\ln \frac{5}{3}$      b) $4 \ln 3$

**4.** $2 \ln |x - 1| + 3 \ln |x + 2| + C$

**Exam 2**    **1.** a) $\frac{1}{18}(3x^2 - 5)^9 + C$

b) $\operatorname{Sin}^{-1} \dfrac{x}{3} + C$

c) $\frac{2}{5}x^5 + \frac{8}{3}x^3 + 8x + C$

**2.** $\frac{1}{5}(1 - x^2)^{5/2} - \frac{1}{3}(1 - x^2)^{3/2} + C$

**3.** $x(\ln x)^2 - 2x \ln x + 2x + C$

**4.** $\frac{\pi}{12}$          **5.** $\frac{\pi}{4}$

**Exam 3**    **1.** a) $\frac{\pi^2}{72}$         b) $\frac{\pi^2}{32}$

**2.** $\frac{1}{a} \operatorname{Tan}^{-1} (a \tan x) + C$

**3.** $\dfrac{1}{3} \operatorname{Tan}^{-1} \left( \dfrac{x + 1}{2} \right) + C$

**4.** $e^{\cos x}(1 - \cos x) + C$

**5.** $g''(0) = 16$

**Exam 4**    **1.** a) $-\ln |\cos x| + C$   b) $-x + \tan x + C$

c) $\dfrac{\tan^2 x}{2} + \ln |\cos x| + C$

**2.** $\dfrac{1}{\sqrt{5}} \ln \left| \dfrac{1 - \sqrt{5} \tan \frac{x}{2}}{1 + \sqrt{5} \tan \frac{x}{2}} \right| + C$

**3.** $\frac{1}{2} + \frac{1}{2}e^\pi$      **4.** $f(u) = u + 1$

**Exam 5**    **1.** a) $-16(9 + x^{-4})^3 x^{-5}$

b) $\dfrac{2e^{2x}}{1 + e^{4x}}$    c) $\dfrac{3}{x} \cos (\ln x^3)$

**2.** $y' = -\dfrac{x}{y^3}(1 + y^2)$

**3.** $y'' = \dfrac{-5}{8y^3}$

**4.** $0 < x < e$     **5.** $4\sqrt{3}$ in.²/hr.

**6.** $x = 2,\ y = \frac{4}{3}$

**7.** a) $\frac{1}{4}(\text{Tan}^{-1} 2x)^2 + C$

    b) $\cos x - \cos x \ln (\cos x) + C$

**8.** $\dfrac{x^2}{4} \ln x - \dfrac{x^2}{8} + C$

**9.** $\ln \left| \dfrac{(x + 4)^2}{x - 1} \right| + C$

**10.** a) $\dfrac{2}{e}$     b) $\dfrac{\pi^2}{8}$

*Exam 6*   **1.** a) $1$     b) $2x \cos x^2$

    c) $\frac{x}{2} + \frac{1}{4}\sin 2x + C$

    d) $-\sin (\cos^2 x) + C$

**2.** a) $2x$     b) $2x$

    c) $\frac{2}{x} x^{\ln x} \ln x$    d) $\frac{x}{2} e^{2x} - \frac{1}{4} e^{2x} + C$

**3.** a) $2x(1 + x^6)^{10}$   b) $\frac{2}{3} x^7$

    c) $\frac{1}{33}(1 + x^3)^{11} + C$

**4.** a) and b) $-2$ (local minimum),

      $0$ (local maximum), and

      $1$ (local minimum)

**5.** $18$ in.³      **6.** $13\sqrt{13}$

**7.** $\frac{27}{4}\sqrt{3}$      **8.** $\frac{8}{\pi}$

**9.** a) $\ln |\text{Tan}^{-1} x| + C$   b) $\ln |\text{Sin}^{-1} x| + C$

**10.** a) $\dfrac{1}{3} \ln \left| \dfrac{x - 1}{x + 2} \right| + C$

    b) $\dfrac{x^3}{3} + \dfrac{1}{3} \ln \left| \dfrac{x - 1}{x + 2} \right| + C$

## 6.1

**B1.** a) $\frac{56}{3}$      b) $\frac{32}{3}$

**B2.** $\frac{4}{3}$    **B3.** $\frac{64}{3}$    **B4.** $18$

**B5.** $\sqrt{2} - 1$    **B6.** $\frac{9\pi}{2}$

**M9.** a) $\frac{1}{6}$     b) $\frac{1}{6}$     c) $\frac{1}{3}$

**M10.** $\frac{4}{3}$     **M11.** $\frac{4}{3}$    **M12.** $6\pi - \frac{9}{2}\sqrt{3}$

**M13.** $2 + \frac{1}{4}\pi$    **M14.** $2$    **M15.** b) $1$

**M16.** a) $1$    b) The regions involved are congruent.

**M17.** $1$

**M18.** Approximately $3.36$

**M19.** $1$

**M20.** $k = \frac{3}{2}$

## 6.2

**B1.** $\frac{16\pi}{15}$    **B2.** $\frac{3\pi}{10}$    **B3.** $\frac{3\pi}{10}$

**B4.** $2\pi$    **B5.** $36\sqrt{3}$    **B6.** $\frac{384}{5}\pi$

**M9.** $\frac{16\pi}{15}$    **M10.** $\frac{27\pi}{2}$

**M11.** $4\pi a b^2 /3$

**M12.** $\frac{2\pi}{3}$

**M13.** $16 r^3 /3$

**M14.** $\pi^2 - 2\pi$

**M15.** $\frac{22\pi}{15}$

**M16.** $\frac{\pi}{2}$

**M17.** $\frac{\pi}{6}$

**M18.** $\frac{\pi}{6}$

**M19.** $\frac{7\pi}{15}$

**M20.** $\frac{8\pi}{15}$

**M21.** $2000/3$ cubic inches

## 6.3

**B1.** $\frac{335}{27}$    **B2.** $6$    **B3.** $\ln (2 + \sqrt{3})$

**B4.** $14$    **B5.** $\frac{109}{3}\pi$    **B6.** $\frac{12}{5}\pi$

**M9.** $\ln \left( \dfrac{2 + \sqrt{3}}{\sqrt{3}} \right)$      **M10.** $\ln (2 + \sqrt{3})$

**M11.** $\frac{1}{2}\sqrt{2} + \frac{1}{2}\ln (1 + \sqrt{2})$   **M12.** $\sqrt{2}$

**M13.** $\frac{1}{3}\sqrt{2} - \frac{1}{6}$

**M14.** $\frac{\pi}{4}$

**M15.** $2(2 - \sqrt{2})$

**M16.** $\frac{109\pi}{3}$

**M17.** $\frac{\pi}{2}(e^2 - e^{-2} + 4)$

**M18.** $\frac{26\pi}{9}$

**M19.** a) $\dfrac{49\pi}{4}$        b) $\left( \dfrac{820 - 81 \ln 3}{128} \right)\pi$

**M20.** $\frac{64\pi}{3}$

**M21.** $2\pi a r$

**M22.** $2\pi r^2 (1 - \cos \alpha)$

## 6.4

**B1.** a) $\frac{1}{2}$    b) $-\frac{1}{2}$    c) $0$

**B2.** $256\rho g$        **B3.** $\frac{9}{4}$ ft. lbs.

**B4.** $\frac{105}{2}$ in. lbs.      **B5.** $\frac{8}{3}\rho g \times 10^7$

**M8.** a) $2$    b) $\frac{\pi}{4}$    c) $\frac{\pi}{2}$    d) Indefinite

**M9.** $\frac{80}{3} g\rho$

**M10.** $\frac{2}{3} g\rho$

**M11.** $\frac{5}{3}\pi\rho g \sqrt{26}$

**M13.** $\frac{32}{15}\rho g(2 + 2\sqrt{2})$

**M15.** $\frac{k}{2}$, where $k$ is the constant of proportionality

**M16.** $\frac{16}{15}\pi g\rho$

## 6.5

**B1.** $A = \frac{1}{6}$, $\quad M_y = \frac{1}{12}\rho$, $\quad I_y = \frac{1}{20}\rho$, $\quad \bar{x} = \frac{1}{2}$

**B2.** $M_x = \frac{1}{15}\rho$, $\quad I_x = \frac{1}{28}\rho$, $\quad \bar{y} = \frac{2}{5}$, $\quad (\frac{1}{2}, \frac{2}{5})$ is center of mass

**B3.** $\frac{\pi}{2} + \frac{\pi}{4}\sqrt{2} - 1 - \sqrt{2}$  **B4.** $(\frac{2}{5}, \frac{2}{5})$

**B5.** $\left(\dfrac{9}{2\pi}, \dfrac{9\sqrt{3}}{2\pi}\right)$

**M9.** $(2a/5, 2a/5)$

**M10.** $M = 4 + 3\ln 3$, $\quad M_y = 14$

**M11.** 5

**M12.** 392

**M13.** $M_x = \rho \displaystyle\int_{\theta_1}^{\theta_2} f(\theta) \sin\theta \sqrt{(f(\theta))^2 + (f'(\theta))^2}\, d\theta$

**M14.** $M_x = \frac{1}{4}\rho\pi$, $\quad M_y = \frac{1}{2}\rho$

**M15.** b) $(\frac{4}{5}, \frac{1}{3})$

**M17.** On the axis, $\frac{3}{5}h$ from vertex.

## 6.6

**B1.** Exact value is $\frac{64}{3}$, $\quad T_4 = 22$, $\quad P_4 = \frac{64}{3}$.

**B2.** Exact value is 2, $\quad T_4 = (1 + \sqrt{2})\frac{\pi}{4}$ (about 1.896) $P_4 = (1 + 2\sqrt{2})\frac{\pi}{6}$ (about 1.999).

**B3.** Exact value is $\ln 15$ (about 2.71), $T_4 = 2.8$, $P_4 = \frac{49}{18}$ (about 2.72).

**B4.** $P_4 = \frac{4220}{3}$

**B5.** $A_{10} = 285$, $\quad T_{10} = 335$, $\quad P_2 = \frac{1000}{3}$

**M8.** 24.654

**M9.** 1.2821

**M10.** $A_4 = \frac{9}{5} = 1.8$, $\quad T_4 = \frac{113}{85}$ (about 1.33), $P_4 = \frac{328}{255}$ (about 1.29)

**M15.** $P_{26} = \displaystyle\int_0^5 (1 - 2x + 7x^2)\, dx = \frac{815}{3}$

**M16.** Find the expression for $A_n$ for the definite integral

$$\int_0^1 8x^3\, dx.$$

## 6.7

**Exam 1**  1. $\frac{32}{3}$  2. $\frac{4\pi}{3}$  3. $\frac{38}{3}$  4. $\frac{7\pi}{9}$
5. 64

**Exam 2**  1. $\frac{9}{2}$

2. a) $8\pi$    b) $\frac{128}{5}\pi$

3. $\frac{16}{3}$

4. a) $\displaystyle\int_0^2 \sqrt{1 + 4x^2}\, dx$

b) $\displaystyle\int_0^2 2\pi x^2 \sqrt{1 + 4x^2}\, dx$

c) $\displaystyle\int_0^4 \pi\sqrt{1 + 4y}\, dy$

5. $(0, -\frac{8}{5})$

**Exam 3**  1. $\frac{\pi}{4}$    2. $\sqrt{2}\,(e^5 - e^3)$
3. $\frac{\pi}{27}(10^{3/2} - 1)$
4. a) $\frac{1}{5}$    b) $\frac{1}{6}$
5. $64\pi g$

**Exam 4**  1. $y = 6x$ and $y = -6x$
2. 4    3. $\frac{4}{15}\rho g\pi$
4. $\frac{\pi}{30}(2\sqrt{2} - 1)$
5. $M_x = \frac{1}{8}\rho\pi$, $\quad M_y = \frac{1}{2}\rho\pi - \rho$

**Exam 5**  2. a) $\frac{3}{4}$   b) $\frac{5}{3}$   c) $\frac{5}{3}$   d) $\frac{17}{8}$
3. a) $\frac{35}{6}(\ln 6)^{1/2}$   b) 0
4. $\frac{75}{64}$    5. $\frac{12}{5}$
6. $\frac{\pi}{32}(15 + 16\ln 2)$   7. $\frac{3}{2}\pi\ln 2 - \frac{1}{2}\pi$
8. $\frac{25}{24}$    9. $\dfrac{\pi}{16}(15 + 16\ln 2)$

10. $\frac{1}{6}\rho(8\ln 3 - 4)$

**Exam 6**  1. $-3$    2. $52 + 24t$
4. a) 1   b) 4   c) 12

5. $\dfrac{1}{e}$

6. a) $\frac{1}{4}(x + x\sqrt{x})^{-1/2}$

b) $\dfrac{2x}{(1 + x^2)\ln(1 + x^2)}$

c) $e^x + e^{x^2} + C$

7. $\frac{189}{2}\rho g\pi$

8. $\ln\left(\dfrac{x^2 - 1}{x + 2}\right) + C$

9. $\ln(1 + \sqrt{2})$   10. $\frac{19}{3}$
11. $\frac{729}{10}\pi$   12. 320

## 7.1

**B1.** a) $\frac{1}{2}$   b) $\frac{1}{5}$   c) $\frac{25}{6}$   d) $\frac{8}{3}$

**B2.** a) $1, 3, 6, 10, 15, 21, 28$

**B3.** a) $S_{2000} = 1,\quad S_{2001} = 0$

b) $S_n = \begin{cases} 1 & \text{if } n \text{ is even} \\ 0 & \text{if } n \text{ is odd} \end{cases}$

**B5.** a) $\frac{20}{33}$                           b) $\frac{1}{33}$

**B6.** a) $\dfrac{1}{3} = \dfrac{3}{10} + \dfrac{3}{10^2} + \dfrac{3}{10^3} + \dfrac{3}{10^4} + \cdots$

b) $\dfrac{4}{15} = \dfrac{2}{10} + \dfrac{6}{10^2} + \dfrac{6}{10^3} + \dfrac{6}{10^4} + \cdots$

c) $\dfrac{23}{99} = \dfrac{2}{10} + \dfrac{3}{10^2} + \dfrac{2}{10^3} + \dfrac{3}{10^4} + \cdots$

d) $\dfrac{19}{33} = \dfrac{5}{10} + \dfrac{7}{10^2} + \dfrac{5}{10^3} + \dfrac{7}{10^4} + \cdots$

**M10.** $\dfrac{nx}{n+2}$

**M11.** $\dfrac{x^{n/2}}{2 \cdot 4 \cdots (2n)}$

**M12.** a) $\frac{38}{99}$                        b) $\frac{123}{999}$

**M13.** a) $\frac{123}{999} = \frac{41}{333}$   b) $\frac{56}{99}$   c) $\frac{78}{99} = \frac{26}{33}$

**M22.** 70 ft.

## 7.2

**B1.** a) Convergent        c) Convergent
b) Divergent        d) Divergent

**B2.** a) Convergent        c) Divergent
b) Convergent        d) Convergent

**B3.** a) Convergent        c) Divergent
b) Convergent        d) Convergent

**B4.** a) Convergent        c) Divergent
b) Convergent        d) Divergent

**B5.** a) Convergent, but not absolutely convergent
b) Absolutely convergent
c) Convergent, but not absolutely convergent
d) Convergent, but not absolutely convergent

**M10.** a) Divergent            b) Convergent

**M11.** a) Convergent, but not absolutely convergent
b) Absolutely convergent

**M14.** a) Convergent, but not absolutely convergent
b) Convergent, but not absolutely convergent
c) Divergent            e) Convergent
d) Divergent            f) Divergent

**M15.** a) Convergent        d) Divergent
b) Divergent          e) Convergent
c) Convergent        f) Divergent

## 7.3

**B4.** a) 2   b) 4   c) 1   d) 5

**M8.** a) 1                          b) 1

**M9.** a) 1                          b) 1

**M10.** a) 2                          b) $\frac{1}{3}$

**M11.** a) 1                          b) 1

**M12.** a) 1                          b) 1

**M13.** a) 1                          b) $\infty$

**M14.** a) 1                          b) $\infty$

**M15.** a) $\frac{1}{\pi}$                       b) $e$

**M16.** a) $\left(-\frac{2}{3}, \frac{2}{3}\right)$                b) $[0, 0]$

**M17.** a) $\frac{1}{e}$                        b) $e^{-a}$

## 7.4

**B1.** a) $\displaystyle\sum_{n=0}^{\infty} (-1)^n \frac{x^n}{2^{n+1}}$   b) $-\dfrac{1}{2} \displaystyle\sum_{n=0}^{\infty} \left(\frac{3x}{4}\right)^n$

c) $\displaystyle\sum_{n=0}^{\infty} x^{2n}$

**B2.** a) $2x - \dfrac{(2x)^3}{3!} + \dfrac{(2x)^5}{5!} - \cdots$

$+ (-1)^n \dfrac{(2x)^{2n+1}}{(2n+1)!} + \cdots$

b) $1 - \dfrac{x^4}{2!} + \dfrac{x^8}{4!} - \cdots + (-1)^n \dfrac{x^{4n}}{(2n)!} + \cdots$

**B4.** a) $x^2 - \dfrac{x^6}{3} + \dfrac{x^{10}}{5} - \cdots + (-1)^n \dfrac{x^{4n+2}}{2n+1} + \cdots$

b) $\dfrac{10!}{5}, 0, 0$

**M9.** a) $\displaystyle\sum_{n=0}^{\infty} \frac{x^{n+1}}{n!}$   b) $\displaystyle\sum_{n=0}^{\infty} (-1)^n \frac{x^{2n+3}}{(2n+1)!}$

c) $\displaystyle\sum_{n=0}^{\infty} (-1)^n \frac{x^{2n+3}}{(2n)!}$

**M10.** a) $x + x^2 + \dfrac{x^3}{3} - \dfrac{x^5}{30} - \dfrac{x^6}{90} + \cdots$

b) $1 + x - \dfrac{x^3}{3} - \dfrac{x^4}{6} - \dfrac{x^5}{30} + \cdots$

**M11.** a) $1 + \frac{1}{2}x^2 + \frac{5}{24}x^4 + \frac{61}{720}x^6 + \cdots, \ -\frac{\pi}{2} < x < \frac{\pi}{2}$

b) $x + \frac{1}{3}x^3 + \frac{2}{15}x^5 + \frac{17}{315}x^7 + \cdots, \ -\frac{\pi}{2} < x < \frac{\pi}{2}$

**M12.** a) $1 + \frac{x}{2} - \frac{x^2}{2^2 \cdot 2!} - \frac{x^3}{2^3 \cdot 3!} + \cdots$

b) $e - \frac{e}{2}x^2 + \frac{e}{6}x^4 - \frac{31e}{720}x^6 + \cdots$

**M15.** b) $\frac{26}{35}$

**M16.** a) 0    b) 0    c) 0

## 7.5

**B1.** a) $-1 + 12i$    c) $9 + 8i$    e) $\dfrac{9 - 8i}{29}$

b) $9 - 8i$    d) $-1 - 12i$    f) $\sqrt{145}$

**B2.** a) $4e^{i\pi/2}$    c) $2\sqrt{2}e^{i7\pi/4}$

b) $2e^{i4\pi/3}$    d) $2e^{i\pi/6}$

**B3.** a) $i$    b) $\frac{3}{2} + i\frac{3}{2}\sqrt{3}$    c) $-2\sqrt{3} - 2i$

d) $-1 + i\sqrt{3}$

**B4.** $3, \ -\frac{3}{2} + i\frac{3}{2}\sqrt{3}, \ -\frac{3}{2} - i\frac{3}{2}\sqrt{3}$

**B5.** a) $(1 - 3t^2) + i(3t - t^3)$

b) $(t^2 - 1)\cos t - 2t\sin t$
$$+ i((t^2 - 1)\sin t + 2t\cos t)$$

**B6.** a) $4i$    c) $64$

b) $16$    d) $128\sqrt{3} + 128i$

**M11.** a) $x = \frac{2}{5}, \ y = -\frac{1}{5}$

b) $x = -1, \ y = 0$

**M14.** a) $1, \ -\frac{1}{2} + i\frac{\sqrt{3}}{2}, \ -\frac{1}{2} - i\frac{\sqrt{3}}{2}$

b) $-1, \ \frac{1}{2} + i\frac{\sqrt{3}}{2}, \ \frac{1}{2} - i\frac{\sqrt{3}}{2}$

c) $-i, \ \frac{\sqrt{3}}{2} + \frac{i}{2}, \ -\frac{\sqrt{3}}{2} + \frac{i}{2}$

d) $i, \ \frac{\sqrt{3}}{2} - \frac{i}{2}, \ -\frac{\sqrt{3}}{2} - \frac{i}{2}$

**M15.** The twelve numbers listed as answers to M14.

**M16.** $-3, \ \pm i\sqrt{3}$

**M17.** $1 \pm i\sqrt{3}, \ -1 \pm i\sqrt{3}$

## 7.6

*Exam 1*    **1.** 15 feet

**2.** a) Divergent    b) Divergent

c) Convergent

**3.** a) 1    b) $\infty$

**4.** a) $\dfrac{1}{1 + x^2}$    b) $\dfrac{x^3}{1 - x^3}$    c) $\dfrac{1}{1 + 2x}$

**5.** $-\frac{7}{13} - \frac{22}{13}i$

*Exam 2*    **1.** a) 2    b) $\frac{64}{25}$    c) 1

**2.** a) $1 + x^2 + \dfrac{x^4}{2} + \dfrac{x^6}{6} + \dfrac{x^8}{24} + \dfrac{x^{10}}{120}$

b) 120

**3.** a) Converges    b) Converges

c) Diverges

**4.** a) $-1 \le x < 1$    b) all $x$

c) $-1 \le x < 1$

**5.** a) 5    b) $\sqrt{34}$    c) $10 + 5i$

d) $\frac{2}{5} + \frac{11i}{5}$

*Exam 3*    **1.** a) $\frac{9}{5}$    b) $\frac{81}{65}$    c) $\frac{36}{65}$    d) $\frac{3}{5}$

**2.** a) Divergent    b) Convergent

**3.** a) 1    b) 0

**5.** a) $\displaystyle\sum_{n=0}^{\infty} (-1)^n \frac{2^{2n+1}x^{4n+2}}{(2n + 1)!}$

b) 0    c) 0

*Exam 5*    **1.** a) $\frac{1}{9}$    b) $\frac{23}{99}$

**2.** a) Divergent    b) Convergent

c) Divergent

**3.** $f(x) = 1 - \dfrac{x^2}{2!} + \cdots$

**4.** a) $e^{-2x}$    b) $e^{x-1}$

**5.** a) Convergent    b) Convergent

**6.** a) 1    b) 3

**7.** a) $\displaystyle\sum_{n=0}^{\infty} x^n$    b) $-\displaystyle\sum_{n=0}^{\infty} \frac{x^n}{n}$

c) $\displaystyle\sum_{n=0}^{\infty} (n + 1)x^n$

**9.** a) $f'(x) = e^{2ix} - e^{-2ix}$

b) $g'(x) = \dfrac{-3}{x}(\ln ix)^2$

c) $h'(x) = 4i(ix)^3 + (2 + 3i)e^{(2+3i)x}$

*Exam 6*    **2.** $\frac{3}{4}$

**4.** a) Absolutely convergent

b) Absolutely convergent

c) Conditionally convergent

**5.** a) 1    b) 1    c) 1

**6.** a) $\displaystyle\sum_{n=0}^{\infty} (-3)^n \frac{x^{2n+1}}{(2n+1)!}$

b) $\displaystyle\sum_{n=0}^{\infty} \frac{2^n x^{2n+1}}{(2n+1)!}$

**7.** a) $2e^{i\pi/6}$    b) $10e^{\pi i}$    c) $15e^{\pi i/2}$
d) $2\sqrt{3}\,e^{i4\pi/3}$

**8.** a) $9 + 2i$    d) $6\sqrt{10} - i\sqrt{10}$
b) $9 - 2i$    e) $\sqrt{85}$
c) $17 - 9i$    f) $\sqrt{10} + \sqrt{37}$

**9.** $\dfrac{\sqrt{2}}{2} + i\dfrac{\sqrt{2}}{2}, \quad -\dfrac{\sqrt{2}}{2} + i\dfrac{\sqrt{2}}{2},$

$\dfrac{\sqrt{2}}{2} - i\dfrac{\sqrt{2}}{2}, \quad -\dfrac{\sqrt{2}}{2} - i\dfrac{\sqrt{2}}{2}$

**10.** a) $\displaystyle\sum_{n=0}^{\infty} \frac{3\cdot 5\cdots(2n-1)}{2\cdot 4\cdots(2n)} x^n$

b) 1

## 8.1

**B1.** a) $y = A\sqrt{1 + x^2}$
b) $y^2 = x^2 + \ln x^2 + C$

**B2.** $y = Ax^2 - A + 1$

**B3.** a) $\dfrac{du}{dx} + u = x$    b) $u = x - 1 + Ce^{-x}$

**B4.** $y^2 = Ax^3 - x^2$

**B5.** $\dfrac{3\ln 10}{\ln 2}$ weeks

**M8.** $3y - y^3 = 3x + x^3 + C$

**M9.** $y = -\frac{1}{3} + Ce^{3x^2/2}$

**M10.** $y = \dfrac{-1}{2x}\cos 2x + \dfrac{C}{x}$

**M11.** $y = \frac{-1}{3} + Cx^{-12}$

**M12.** $y = x\sin x + C\sin x$

**M13.** $y + xe^{y/x} = A$

**M14.** $(x + y)^3 = C(2x + y)^4$

**M15.** $4y - 2x + \sin(2x + 4y) = C$

**M16.** $Ce^{-Rt/L} + E(\omega L \sin \omega t + R \cos \omega t)/(R^2 + \omega^2 L^2)$

**M17.** \$1648.66

**M18.** a) $150(\frac{4}{3})^{5/2}$ million

b) $20\left(\dfrac{\ln 20 - \ln 3}{\ln 4 - \ln 3}\right)$ years after 1950.

**M19.** About 6.9 hours

**M21.** a) $y = \frac{x}{2\pi}\sqrt{1 - 4\pi^2} + C$
b) $y = 1 - Ae^{-x}$

**M22.** a) limiting velocity is $\dfrac{mg}{k}$

b) limiting velocity is $\sqrt{mg/k}$

## 8.2

**B1.** a) $y = 3x^2 + Ax^3 + B$
b) $y = x^3 + Ax^2 + B$

**B2.** a) $y = \frac{1}{3}\ln|3x + A| + B$

b) $y = \dfrac{2}{3A}\text{Tan}^{-1}\dfrac{x}{A} + B$

**B3.** a) $y^4 = Ax + B$    b) $y^2 = \dfrac{1}{Ax + B}$

**B4.** a) $y = Ae^x + Be^{2x}$    b) $y = 2e^x + 2e^{2x}$

**B5.** a) $y = e^{3x}(Ae^{2ix} + Be^{-2ix})$
b) $y = e^{3x}(C \cos 2x + D \sin 2x)$

**M9.** $y = \ln|\sec(2x + A)| + B$

**M10.** $y = Ae^{Bx} + \frac{1}{B}$

**M11.** $y = Ax^2 - x + B$

**M12.** a) $y = Ae^{2x} + Be^{-2x}$
b) $y = A\sin 2x + B\cos 2x$

**M13.** a) $y = A + Be^{-4x}$
b) $y = A + Be^{4x}$

**M14.** a) $y = e^{-2x}(Ae^{2x\sqrt{2}} + Be^{-2x\sqrt{2}})$
b) $y = (A + Bx)e^{-2x}$

**M15.** $(x + C_1)^2 + y^2 = C_2{}^2$

**M16.** $C_1 \text{Sin}^{-1}\left(\dfrac{y - C_1}{C_1}\right) - \sqrt{2C_1 y - y^2} = \pm x + C_2$

**M17.** $(x - A)^2 + (y - B)^2 = 1$

**M18.** a) $y = Ax + Bx\ln x$
b) $y = A\sin(\ln x) + B\cos(\ln x)$

**M20.** $y = \cos\dfrac{t\sqrt{2}}{4}$

**M21.** $y = (5/12)\cos(\sqrt{6}/2)t + (5\sqrt{6}/3)\sin(\sqrt{6}/2)t$

**M22.** $m = \frac{5}{18}$

**M23.** 4.5 lbs.

**M24.** $A\cos(1/\sqrt{2})t + B\sin(1/\sqrt{2})t$

## 8.3

**B1.** $y = Ax + Bxe^x$

**B2.** Particular solution is $y = \frac{1}{4}e^{2x}$.
General solution is $y = Ae^{-2x} + Be^{-x} + \frac{1}{4}e^{2x}$

**B3.** $y = A\cos x + B\sin x$
$\qquad - (\cos x)(\ln|\sec x + \tan x|)$

**B4.** $y = -x^2 + A(x + 1) + Be^x$

**B5.** a) $y = A + Be^{-2x} + \frac{1}{2}x^3 - \frac{3}{4}x^2 + \frac{3}{4}x$
$\quad$ b) $y = A + Be^{-2x} + \frac{2}{5}\sin x - \frac{1}{5}\cos x$

**M8.** $v = e^x$

**M9.** $v = x\ln\left(\dfrac{x-1}{x}\right)$

**M10.** a) $y = Ae^x + Be^{2x} + x + 1$
$\quad$ b) $y = Ae^x + Be^{2x} + 6xe^{2x}$

**M11.** a) $y = Ae^{4x} + Be^{-x} - \frac{x}{4} + \frac{3}{16}$
$\quad$ b) $y = Ae^{4x} + Be^{-x} - \frac{1}{5}xe^{-x}$

**M12.** $y = (A + Bx)e^{-x} + \frac{3}{25}e^x\cos x + \frac{4}{25}e^x\sin x$

**M13.** a) $y = A\cos x + B\sin x + e^x$
$\quad$ b) $y = A\cos x + B\sin x + \cos 3x$
$\quad$ c) $y = A\cos x + B\sin x + e^x + \cos 3x$

**M14.** $y = Ae^x + Be^{2x} + x + 1 + 6xe^{2x}$

**M15.** $y = C_1 + C_2e^{2x} - \frac{1}{2}e^x\sin x$

**M16.** $y = C_1e^{2x} + C_2xe^{2x} + \dfrac{x^5e^{2x}}{20} + \dfrac{x^3e^{2x}}{6}$

**M17.** $y = C_1\cos x + C_2\sin x + \sin x\ln\sin x - x\cos x$

**M18.** $y = C_1e^x + C_2e^{3x} + \dfrac{e^{2x}}{2} + \dfrac{e^x - e^{3x}}{2}\cdot\ln(1 + e^{-x})$

**M19.** $y = C_1e^x + C_2e^{-x} - e^x\sin e^{-x}$

**M20.** $y = C_1e^{-x} + C_2xe^{-x} + \dfrac{8\sin 2x}{25}$
$$- \frac{6\cos 2x}{25} + 3x - 4 + \frac{3e^x}{4}$$

**M21.** a) $y = \frac{1}{8}e^x$ $\quad$ b) $y = \frac{1}{4}xe^x$ $\quad$ c) $y = \frac{1}{2}x^2e^x$

**M22.** $y = \sin x\tan x - \dfrac{1}{4\cos^3 x}$

**M23.** $y = \frac{1}{4}x^2e^x - \frac{1}{4}xe^x + \frac{1}{8}e^x$

**M24.** $y = e^x$

**M25.** $y = \dfrac{\sqrt{x}}{4x}(3\cos 2x\sin x + 6x\cos x - 3\sin 2x\cos x)$

**M26.** a) $A\sin(1/\sqrt{LC})t + B\cos(1/\sqrt{LC})t$
$$+ C\sin\alpha t/(1 - LC\alpha^2)$$
$\quad$ b) $A\sin(1/\sqrt{LC})t + B\cos(1/\sqrt{LC})t$
$$- (\sqrt{C}/2L)t\cos(1/\sqrt{LC})t$$

## 8.4

**B3.** $y = a_0\left(1 - \dfrac{x^2}{2} + \dfrac{x^4}{2\cdot 4} - \dfrac{x^6}{2\cdot 4\cdot 6} + \cdots\right)$
$$+ a_1\left(x - \frac{x^3}{3} + \frac{x^5}{3\cdot 5} - \frac{x^7}{3\cdot 5\cdot 7} + \cdots\right)$$

**B4.** $a_{2n} = \dfrac{1}{(2n+1)!}, a_{2n+1} = \dfrac{2}{(2n+2)!}$

**M6.** a) $a_{2n+1} = 0 \quad (n = 1, 2, 3, \ldots)$
$$a_{2n} = -\frac{a_0}{(2n)!}(2\cdot 6\cdot 10\cdots(4n - 6))$$
$\quad$ b) $y = a_0 + a_1x + \sum\limits_{n=1}^{\infty}a_{2n}x^{2n}$
$\quad$ where $a_{2n}$ is as in (a).

**M7.** $y = 1 + \sum\limits_{n=1}^{\infty}a_{2n}x^{2n}$,
$\quad$ where $a_{2n} = 1/(1\cdot 3\cdot 5\cdot 7\cdots(2n - 1))$

**M8.** $y = x^2 - x + 1$

**M9.** $y = \cosh x$

## 8.6

**Exam 1**
$\quad$ **1.** $(1 - x^2)(1 - y^2) = A$
$\quad$ **2.** a) $y = Ae^{7x} + Be^{-5x}$
$\qquad$ b) $y = e^x(A\cos 2x\sqrt{3} + B\sin 2x\sqrt{3})$
$\quad$ **3.** Particular solution $= \frac{1}{2}\cos 2x$
$\qquad$ General solution is
$\qquad A\cos x\sqrt{6} + B\sin x\sqrt{6} + \frac{1}{2}\cos 2x$.
$\quad$ **4.** $y = Ax^5 + B$

**Exam 2**
$\quad$ **1.** $y = A(3x^2 + 1) + 1$
$\quad$ **2.** $Ax + B = e^y(y - 1)$
$\quad$ **3.** a) $y = A\cos 2x + B\sin 2x$
$\qquad$ b) $y = (A + Bx)e^{-2x}$
$\quad$ **4.** a) $y = 2x + 4$
$\qquad$ b) $y = e^{-x/2}\left(A\cos\dfrac{x\sqrt{3}}{2} + B\sin\dfrac{x\sqrt{3}}{2}\right)$
$$+ 2x + 4$$

**Exam 3**
$\quad$ **1.** a) $y = \frac{1}{7}e^x + e^{-2x}(Ae^{x\sqrt{2}} + Be^{-x\sqrt{2}})$
$\qquad$ b) $y = \frac{1}{3}xe^x + Ae^x + Be^{-2x}$
$\quad$ **2.** $y = 4x + B + Ae^x - Ae^{-x}$

**Exam 4**
$\quad$ **1.** $y = -\frac{1}{3}\sin x\cos x$
$\quad$ **2.** $y = A + Bx^2$ $\quad$ **3.** $y = x^2$

# Tables

**Table 1** *Trigonometric Functions*

| Angle | | | | | Angle | | | | |
|---|---|---|---|---|---|---|---|---|---|
| Deg. | Rad. | Sin | Cos | Tan | Deg. | Rad. | Sin | Cos | Tan |
| 0° | 0.000 | 0.000 | 1.000 | 0.000 | | | | | |
| 1° | 0.017 | 0.017 | 1.000 | 0.017 | 46° | 0.803 | 0.719 | 0.695 | 1.030 |
| 2° | 0.035 | 0.035 | 0.999 | 0.035 | 47° | 0.820 | 0.731 | 0.682 | 1.072 |
| 3° | 0.052 | 0.052 | 0.999 | 0.052 | 48° | 0.838 | 0.743 | 0.669 | 1.111 |
| 4° | 0.070 | 0.070 | 0.998 | 0.070 | 49° | 0.855 | 0.755 | 0.656 | 1.150 |
| 5° | 0.087 | 0.087 | 0.996 | 0.087 | 50° | 0.873 | 0.766 | 0.643 | 1.192 |
| 6° | 0.105 | 0.105 | 0.995 | 0.105 | 51° | 0.890 | 0.777 | 0.629 | 1.245 |
| 7° | 0.122 | 0.122 | 0.993 | 0.123 | 52° | 0.908 | 0.788 | 0.616 | 1.280 |
| 8° | 0.140 | 0.139 | 0.990 | 0.141 | 53° | 0.925 | 0.799 | 0.602 | 1.327 |
| 9° | 0.157 | 0.156 | 0.988 | 0.158 | 54° | 0.942 | 0.809 | 0.588 | 1.370 |
| 10° | 0.175 | 0.174 | 0.985 | 0.176 | 55° | 0.960 | 0.819 | 0.574 | 1.428 |
| 11° | 0.192 | 0.191 | 0.982 | 0.194 | 56° | 0.977 | 0.829 | 0.559 | 1.483 |
| 12° | 0.209 | 0.208 | 0.978 | 0.213 | 57° | 0.995 | 0.830 | 0.545 | 1.540 |
| 13° | 0.227 | 0.225 | 0.974 | 0.231 | 58° | 1.012 | 0.848 | 0.530 | 1.600 |
| 14° | 0.244 | 0.242 | 0.970 | 0.249 | 59° | 1.030 | 0.857 | 0.515 | 1.664 |
| 15° | 0.262 | 0.259 | 0.966 | 0.268 | 60° | 1.047 | 0.866 | 0.500 | 1.732 |
| 16° | 0.279 | 0.276 | 0.961 | 0.287 | 61° | 1.065 | 0.875 | 0.485 | 1.804 |
| 17° | 0.297 | 0.292 | 0.956 | 0.306 | 62° | 1.082 | 0.883 | 0.469 | 1.881 |
| 18° | 0.314 | 0.309 | 0.951 | 0.325 | 63° | 1.100 | 0.891 | 0.454 | 1.963 |
| 19° | 0.332 | 0.326 | 0.946 | 0.344 | 64° | 1.117 | 0.899 | 0.438 | 2.050 |
| 20° | 0.349 | 0.342 | 0.940 | 0.364 | 65° | 1.134 | 0.906 | 0.423 | 2.145 |
| 21° | 0.367 | 0.358 | 0.934 | 0.384 | 66° | 1.152 | 0.914 | 0.407 | 2.240 |
| 22° | 0.384 | 0.375 | 0.927 | 0.404 | 67° | 1.169 | 0.921 | 0.391 | 2.356 |
| 23° | 0.401 | 0.391 | 0.921 | 0.424 | 68° | 1.187 | 0.927 | 0.375 | 2.475 |
| 24° | 0.419 | 0.407 | 0.914 | 0.445 | 69° | 1.204 | 0.934 | 0.358 | 2.603 |
| 25° | 0.436 | 0.423 | 0.906 | 0.466 | 70° | 1.222 | 0.940 | 0.342 | 2.748 |
| 26° | 0.454 | 0.438 | 0.899 | 0.488 | 71° | 1.239 | 0.946 | 0.326 | 2.904 |
| 27° | 0.471 | 0.454 | 0.891 | 0.510 | 72° | 1.257 | 0.951 | 0.309 | 3.078 |
| 28° | 0.489 | 0.469 | 0.883 | 0.532 | 73° | 1.274 | 0.956 | 0.292 | 3.271 |
| 29° | 0.506 | 0.485 | 0.875 | 0.554 | 74° | 1.292 | 0.961 | 0.276 | 3.487 |
| 30° | 0.524 | 0.500 | 0.866 | 0.577 | 75° | 1.309 | 0.966 | 0.259 | 3.732 |
| 31° | 0.541 | 0.515 | 0.857 | 0.601 | 76° | 1.326 | 0.970 | 0.242 | 4.011 |
| 32° | 0.559 | 0.530 | 0.848 | 0.625 | 77° | 1.344 | 0.974 | 0.225 | 4.332 |
| 33° | 0.576 | 0.545 | 0.839 | 0.649 | 78° | 1.361 | 0.978 | 0.208 | 4.705 |
| 34° | 0.593 | 0.559 | 0.829 | 0.675 | 79° | 1.379 | 0.982 | 0.191 | 5.145 |
| 35° | 0.611 | 0.574 | 0.819 | 0.700 | 80° | 1.396 | 0.985 | 0.174 | 5.671 |
| 36° | 0.628 | 0.588 | 0.809 | 0.727 | 81° | 1.414 | 0.988 | 0.156 | 6.314 |
| 37° | 0.646 | 0.602 | 0.799 | 0.754 | 82° | 1.431 | 0.990 | 0.139 | 7.115 |
| 38° | 0.663 | 0.616 | 0.788 | 0.781 | 83° | 1.449 | 0.993 | 0.122 | 8.141 |
| 39° | 0.681 | 0.629 | 0.777 | 0.810 | 84° | 1.466 | 0.995 | 0.105 | 9.511 |
| 40° | 0.698 | 0.643 | 0.766 | 0.839 | 85° | 1.484 | 0.996 | 0.087 | 11.43 |
| 41° | 0.716 | 0.656 | 0.755 | 0.869 | 86° | 1.501 | 0.998 | 0.070 | 14.30 |
| 42° | 0.733 | 0.669 | 0.743 | 0.900 | 87° | 1.518 | 0.999 | 0.052 | 19.08 |
| 43° | 0.750 | 0.682 | 0.731 | 0.933 | 88° | 1.536 | 0.999 | 0.035 | 28.64 |
| 44° | 0.768 | 0.695 | 0.719 | 0.966 | 89° | 1.553 | 1.000 | 0.017 | 57.29 |
| 45° | 0.785 | 0.707 | 0.707 | 1.000 | 90° | 1.571 | 1.000 | 0.000 | |

**Table 2**  *Exponential Functions*

| x | $e^x$ | $e^{-x}$ | x | $e^x$ | $e^{-x}$ |
|------|--------|--------|-----|--------|---------|
| 0.00 | 1.0000 | 1.0000 | 2.5 | 12.182 | 0.0821 |
| 0.05 | 1.0513 | 0.9512 | 2.6 | 13.464 | 0.0743 |
| 0.10 | 1.1052 | 0.9048 | 2.7 | 14.880 | 0.0672 |
| 0.15 | 1.1618 | 0.8607 | 2.8 | 16.445 | 0.0608 |
| 0.20 | 1.2214 | 0.8187 | 2.9 | 18.174 | 0.0550 |
| 0.25 | 1.2840 | 0.7788 | 3.0 | 20.086 | 0.0498 |
| 0.30 | 1.3499 | 0.7408 | 3.1 | 22.198 | 0.0450 |
| 0.35 | 1.4191 | 0.7047 | 3.2 | 24.533 | 0.0408 |
| 0.40 | 1.4918 | 0.6703 | 3.3 | 27.113 | 0.0369 |
| 0.45 | 1.5683 | 0.6376 | 3.4 | 29.964 | 0.0334 |
| 0.50 | 1.6487 | 0.6065 | 3.5 | 33.115 | 0.0302 |
| 0.55 | 1.7333 | 0.5769 | 3.6 | 36.598 | 0.0273 |
| 0.60 | 1.8221 | 0.5488 | 3.7 | 40.447 | 0.0247 |
| 0.65 | 1.9155 | 0.5220 | 3.8 | 44.701 | 0.0224 |
| 0.70 | 2.0138 | 0.4966 | 3.9 | 49.402 | 0.0202 |
| 0.75 | 2.1170 | 0.4724 | 4.0 | 54.598 | 0.0183 |
| 0.80 | 2.2255 | 0.4493 | 4.1 | 60.340 | 0.0166 |
| 0.85 | 2.3396 | 0.4274 | 4.2 | 66.686 | 0.0150 |
| 0.90 | 2.4596 | 0.4066 | 4.3 | 73.700 | 0.0136 |
| 0.95 | 2.5857 | 0.3867 | 4.4 | 81.451 | 0.0123 |
| 1.0 | 2.7183 | 0.3679 | 4.5 | 90.017 | 0.0111 |
| 1.1 | 3.0042 | 0.3329 | 4.6 | 99.484 | 0.0101 |
| 1.2 | 3.3201 | 0.3012 | 4.7 | 109.95 | 0.0091 |
| 1.3 | 3.6693 | 0.2725 | 4.8 | 121.51 | 0.0082 |
| 1.4 | 4.0552 | 0.2466 | 4.9 | 134.29 | 0.0074 |
| 1.5 | 4.4817 | 0.2231 | 5 | 148.41 | 0.0067 |
| 1.6 | 4.9530 | 0.2019 | 6 | 403.43 | 0.0025 |
| 1.7 | 5.4739 | 0.1827 | 7 | 1096.6 | 0.0009 |
| 1.8 | 6.0496 | 0.1653 | 8 | 2981.0 | 0.0003 |
| 1.9 | 6.6859 | 0.1496 | 9 | 8103.1 | 0.0001 |
| 2.0 | 7.3891 | 0.1353 | 10 | 22026 | 0.00005 |
| 2.1 | 8.1662 | 0.1225 | | | |
| 2.2 | 9.0250 | 0.1108 | | | |
| 2.3 | 9.9742 | 0.1003 | | | |
| 2.4 | 11.023 | 0.0907 | | | |

**Table 3**  *Natural Logarithms*

| x | ln x | x | ln x | x | ln x |
|-----|--------|-----|--------|-----|--------|
| 0.0 | | 4.5 | 1.5041 | 9.0 | 2.1972 |
| 0.1 | 7.6974⌐ | 4.6 | 1.5261 | 9.1 | 2.2083 |
| 0.2 | 8.3906 | 4.7 | 1.5476 | 9.2 | 2.2192 |
| 0.3 | 8.7960 | 4.8 | 1.5686 | 9.3 | 2.2300 |
| 0.4 | 9.0837 | 4.9 | 1.5892 | 9.4 | 2.2407 |
| 0.5 | 9.3069 | 5.0 | 1.6094 | 9.5 | 2.2513 |
| 0.6 | 9.4892 | 5.1 | 1.6292 | 9.6 | 2.2618 |
| 0.7 | 9.6433 | 5.2 | 1.6487 | 9.7 | 2.2721 |
| 0.8 | 9.7769 | 5.3 | 1.6677 | 9.8 | 2.2824 |
| 0.9 | 9.8946⌐ | 5.4 | 1.6864 | 9.9 | 2.2925 |
| 1.0 | 0.0000 | 5.5 | 1.7047 | 10 | 2.3026 |
| 1.1 | 0.0953 | 5.6 | 1.7228 | 11 | 2.3979 |
| 1.2 | 0.1823 | 5.7 | 1.7405 | 12 | 2.4849 |
| 1.3 | 0.2624 | 5.8 | 1.7579 | 13 | 2.5649 |
| 1.4 | 0.3365 | 5.9 | 1.7750 | 14 | 2.6391 |
| 1.5 | 0.4055 | 6.0 | 1.7918 | 15 | 2.7081 |
| 1.6 | 0.4700 | 6.1 | 1.8083 | 16 | 2.7726 |
| 1.7 | 0.5306 | 6.2 | 1.8245 | 17 | 2.8332 |
| 1.8 | 0.5878 | 6.3 | 1.8405 | 18 | 2.8904 |
| 1.9 | 0.6419 | 6.4 | 1.8563 | 19 | 2.9444 |
| 2.0 | 0.6931 | 6.5 | 1.8718 | 20 | 2.9957 |
| 2.1 | 0.7419 | 6.6 | 1.8871 | 25 | 3.2189 |
| 2.2 | 0.7885 | 6.7 | 1.9021 | 30 | 3.4012 |
| 2.3 | 0.8329 | 6.8 | 1.9169 | 35 | 3.5553 |
| 2.4 | 0.8755 | 6.9 | 1.9315 | 40 | 3.6889 |
| 2.5 | 0.9163 | 7.0 | 1.9459 | 45 | 3.8067 |
| 2.6 | 0.9555 | 7.1 | 1.9601 | 50 | 3.9120 |
| 2.7 | 0.9933 | 7.2 | 1.9741 | 55 | 4.0073 |
| 2.8 | 1.0296 | 7.3 | 1.9879 | 60 | 4.0943 |
| 2.9 | 1.0647 | 7.4 | 2.0015 | 65 | 4.1744 |
| 3.0 | 1.0986 | 7.5 | 2.0149 | 70 | 4.2485 |
| 3.1 | 1.1314 | 7.6 | 2.0281 | 75 | 4.3175 |
| 3.2 | 1.1632 | 7.7 | 2.0412 | 80 | 4.3820 |
| 3.3 | 1.1939 | 7.8 | 2.0541 | 85 | 4.4427 |
| 3.4 | 1.2238 | 7.9 | 2.0669 | 90 | 4.4998 |
| 3.5 | 1.2528 | 8.0 | 2.0794 | 95 | 4.5539 |
| 3.6 | 1.2809 | 8.1 | 2.0919 | 100 | 4.6052 |
| 3.7 | 1.3083 | 8.2 | 2.1041 | | |
| 3.8 | 1.3350 | 8.3 | 2.1163 | | |
| 3.9 | 1.3610 | 8.4 | 2.1282 | | |
| 4.0 | 1.3863 | 8.5 | 2.1401 | | |
| 4.1 | 1.4110 | 8.6 | 2.1518 | | |
| 4.2 | 1.4351 | 8.7 | 2.1633 | | |
| 4.3 | 1.4586 | 8.8 | 2.1748 | | |
| 4.4 | 1.4816 | 8.9 | 2.1861 | | |

# Index

## The Derivative and the Integral

$$f'(x) = \lim_{h \to 0} \frac{f(x+h) - f(x)}{h}$$ is the *derivative of f at x* and is equal to the slope of the tangent at the point $(x, f(x))$.

$$\int_a^b f(x)\, dx = \lim_{n \to \infty} \left(\frac{b-a}{n}\right) \sum_{k=0}^{n-1} f\left(a + \frac{k}{n}(b-a)\right)$$ is the *definite integral of f from a to b*.

$$\int f(x)\, dx = F(x) + C \quad \text{where} \quad F'(x) = f$$

## Theorems and Formulas

*Fundamental Theorem of Calculus*

$$\int_a^b f(x)\, dx = F(b) - F(a) \quad \text{where} \quad F' = f$$

*Mean Value Theorem* $\dfrac{f(b) - f(a)}{b - a} = f'(c)$ for some $c \in (a, b)$ when $f$ is continuous on $[a, b]$ and differentiable on $(a, b)$

*Integration by Parts* $\displaystyle \int u \frac{dv}{dx}\, dx = uv - \int v \frac{du}{dx}\, dx$

*Maximum-Minimum Test* If $f'(x) = 0$, then
$f''(x) < 0$ implies local maximum
$f''(x) > 0$ implies local minimum

$M_2I_2ACIDS$  Maximum, Minimum, Intercepts, Inflections, Asymptotes, Convexity, Increasing, Decreasing, Symmetry

*Volumes* $\displaystyle \int_a^b \pi y^2\, dx$ (disc method about x-axis)

$\displaystyle \int_a^b 2\pi xy\, dx$ (shell method about y-axis)

*Arc Length* $\displaystyle \int_{t_0}^{t_1} \sqrt{\left(\frac{dx}{dt}\right)^2 + \left(\frac{dy}{dt}\right)^2}\, dt \quad (x = x(t), y = y(t))$

$\displaystyle \int_a^b \sqrt{1 + (y')^2}\, dx \quad (y = y(x)), \int_c^d \sqrt{1 + (x')^2}\, dy \quad (x = x(y))$

*Surface Area* $\displaystyle \int_a^b 2\pi y \sqrt{1 + (y')^2}\, dx$ (about x-axis)

*Moments* $\displaystyle (\bar{x}, \bar{y}) = \left(\frac{M_y}{M}, \frac{M_x}{M}\right)$

$\displaystyle M_y = \int_c^d \rho y k(y)\, dy, \quad M_x = \int_a^b \rho x h(x)\, dx$

## Differentiation Formulas

$$(fg)' = fg' + gf'$$

$$\left(\frac{1}{g}\right)' = \frac{-g'}{g^2}$$

$$\left(\frac{f}{g}\right)' = \frac{gf' - fg'}{g^2}$$

$$(x^r)' = rx^{r-1}$$

$$(\sin x)' = \cos x$$

$$(\cos x)' = -\sin x$$

$$(\tan x)' = \sec^2 x$$

$$(\cot x)' = -\csc^2 x$$

$$(\sec x)' = \sec x \tan x$$

$$(\csc x)' = -\csc x \cot x$$

$$(f \circ g)'(x) = f'(g(x))g'(x)$$

$$(e^x)' = e^x$$

$$(a^x)' = a^x \ln a$$

$$(\ln x)' = \frac{1}{x}$$

$$(\log_a x)' = \frac{1}{x} \log_a e$$

$$(\mathrm{Sin}^{-1} x)' = \frac{1}{\sqrt{1 - x^2}}$$

$$(\mathrm{Tan}^{-1} x)' = \frac{1}{1 + x^2}$$

## Integration Formulas (constant omitted)

$$\int x^r\, dx = \frac{x^{r+1}}{r+1} \quad (r \neq -1)$$

$$\int x^{-1}\, dx = \ln |x|$$

$$\int e^{ax}\, dx = \frac{1}{a} e^{ax}$$

$$\int \frac{dx}{x^2 - a^2} = \frac{1}{2a} \ln \left|\frac{x - a}{x + a}\right|$$

$$\int \frac{dx}{x^2 + a^2} = \frac{1}{a} \mathrm{Tan}^{-1} \frac{x}{a}$$

$$\int \frac{dx}{\sqrt{a^2 - x^2}} = \mathrm{Sin}^{-1} \frac{x}{a}$$

$$\int \frac{dx}{\sqrt{x^2 \pm a^2}} = \ln(x + \sqrt{x^2 \pm a^2})$$

$$\int \sin ax\, dx = -\frac{1}{a} \cos ax$$

$$\int \cos ax\, dx = \frac{1}{a} \sin ax$$

$$\int \tan ax\, dx = -\frac{1}{a} \ln |\cos ax|$$

$$\int \cot ax\, dx = \frac{1}{a} \ln |\sin ax|$$

$$\int \sec ax\, dx = \frac{1}{a} \ln |\sec ax + \tan ax|$$

$$\int \csc ax\, dx = -\frac{1}{a} \ln |\csc ax + \cot ax|$$

$$\int x \sin ax\, dx = \frac{-x \cos ax}{a} + \frac{\sin ax}{a^2}$$

$$\int x \cos ax\, dx = \frac{x \sin ax}{a} + \frac{\cos ax}{a^2}$$

$$\int x e^{ax}\, dx = \frac{x e^{ax}}{a} - \frac{e^{ax}}{a^2}$$

## Series

$$e^x = \sum_{n=0}^{\infty} \frac{x^n}{n!} \qquad \frac{1}{1-x} = \sum_{n=0}^{\infty} x^n \quad (-1 < x < 1)$$

$$\cos x = \sum_{n=0}^{\infty} (-1)^n \frac{x^{2n}}{(2n)!} \qquad \sin x = \sum_{n=0}^{\infty} (-1)^n \frac{x^{2n+1}}{(2n+1)!}$$

$$\ln(1-x) = -\sum_{n=1}^{\infty} \frac{x^n}{n} \quad (-1 < x < 1)$$